Edited by
Daniel B. Werz and Sébastien Vidal

Modern Synthetic Methods in Carbohydrate Chemistry

Related Titles

Boysen, M.M.K. (ed.)

Carbohydrates - Tools for Stereoselective Synthesis

2013

ISBN: 978-3-527-32379-1

Wrolstad, R.E.

Food Carbohydrate Chemistry

Series: Institute of Food Technologists Series

2012

ISBN: 978-0-8138-2665-3

Wang, B., Boons, G.-J. (eds.)

Carbohydrate Recognition

Biological Problems, Methods, and Applications

2011

ISBN: 978-1-118-01756-2

Hanessian, S., Giroux, S., Merner, B.L

Design and Strategy in Organic Synthesis

From the Chiron Approach to Catalysis

2013

ISBN: 978-3-527-33391-2

Edited by Daniel B. Werz and Sébastien Vidal

Modern Synthetic Methods in Carbohydrate Chemistry

From Monosaccharides to Complex Glycoconjugates

WILEY-VCH
Verlag GmbH & Co. KGaA

The Editors

Prof. Dr. Daniel B. Werz
Technische Universität Braunschweig
Institut für Organische Chemie
Hagenring 30
38106 Braunschweig
Germany

Dr. Sébastien Vidal
Université Claude Bernard Lyon 1
ICMBS-UMR-CNRS 5246
43 Boulevard du 11 Nov. 1918
69622 Villeurbanne
France

Library of Congress Card No.: applied for

British Library Cataloguing-in-Publication Data
A catalogue record for this book is available from the British Library.

Bibliographic information published by the Deutsche Nationalbibliothek
The Deutsche Nationalbibliothek lists this publication in the Deutsche Nationalbibliografie; detailed bibliographic data are available on the Internet at <http://dnb.d-nb.de>.

Print ISBN: 978-3-527-33284-7
ePDF ISBN: 978-3-527-65897-8
ePub ISBN: 978-3-527-65896-1
Mobi ISBN: 978-3-527-65895-4
oBook ISBN: 978-3-527-65894-7

Cover Design Bluesea Design, McLeese Lake, Canada
Typesetting Laserwords Private Ltd., Chennai, India
Printing and Binding Markono Print Media Pte Ltd, Singapore

Printed on acid-free paper

Contents

Foreword

The ever-expanding field of glycoscience is driven by the vast diversity of biological processes mediated by carbohydrates, their oligomers, and conjugates, thereby providing enormous opportunities for the use of carbohydrates and their derivatives as research tools and as therapeutic agents. The difficulties in isolating homogeneous glycoforms from nature in any significant quantity and the limitations of such methods to naturally occurring glycoforms underline the need for efficient and effective chemical synthesis of such molecules. The superficially simple mechanism of the glycosidation reaction masks what is, in reality, a very difficult problem and one that drives the need for innovative chemical solutions, which in turn is attracting an increasing number of bright young (and not so young) researchers to the area. Against this background, Sébastien Vidal and Daniel B. Werz have assembled in this book an outstanding collection of work of talented authors to give their individual perspectives on "Modern Synthetic Methods in Carbohydrate Chemistry," be they at the level of "simple" monosaccharides or complex glycosides. As the title suggests, oligosaccharide synthesis and carbohydrate chemistry, in general, are but one facet of modern organic chemistry. This is richly brought out in the content of the book from which it is clear that many recent advances in glycochemistry and indeed glycoscience depend very heavily on the power and ingenuity of contemporary synthetic organic methodology.

The first chapter, an authoritative contribution by George O'Doherty from Northeastern University in Boston, sets the tone by drawing the reader's attention to the fact that monosaccharides and complex glycans are not necessarily best prepared from actual sugars and that modern synthetic chemistry in the form of *de novo* synthetic approaches has an important role to play. The second chapter continues the theme as Stéphane Vincent from the University of Namur reminds us of the importance of the aldoheptoses and the need for their synthesis in terms of both biological probes and lipopolysaccharide (LPS) fragments. Mark Nitz from the University of Toronto then expounds on the protecting-group-free synthesis of glycoconjugates and in particular on the use of hydrazides and oxyamine derivatives for the synthesis of *N*-glycosides, before Alphert Christina, Gijs van der Marel, and Jeroen Codée from the Leiden University laboratory describe the recent evolution of the stereocontrolled synthesis of the 1,2-*cis*-glycosidic linkages, thereby nicely underlining the power of modern synthetic chemistry. Nikolay Nifantiev and his coauthors from the Zelinsky Institute of Organic Chemistry in Moscow continue with the theme of the stereocontrolled synthesis of the 1,2-*cis*-glycosides with particular emphasis on the role of participation by remote esters, a process for which the jury has yet to return a clear verdict. A chapter by Micha Fridman from Tel Aviv University then takes us into the realm of the aminoglycoside antibiotics and presents both useful chemistry and a perspective on novel applications of these compounds beyond their current use as antibacterials. Pierre-Alexandre Driguez at Sanofi in France takes us into the realm of complex oligosaccharide synthesis and tackles the vexing problems of the synthesis of heparin fragments, both natural and nonnatural, before Rajarshi Roychoudhury and Nicola Pohl from Indiana University set out the many advantages of light fluorous tag-assisted synthesis of oligosaccharides and the various strategies devised to date in order to take advantage of them. Samuel Guieu and Matthieu Sollogoub at the University of Aveiro and the Université Pierre and Marie Curie in Paris recount recent advances in cyclodextrin chemistry with an emphasis on selective functionalization methods, after which Anna Bernardi from the University of Milan takes up the theme of the design and synthesis of GM1 glycomimetics as cholera toxin ligands. This book continues with a chapter by Hiromune Ando and coworkers from Gifu University on recent advances in the synthesis of glycosphingolipids, from which much is to be learnt about the stereocontrolled synthesis of the once-difficult sialic acid glycosides. Finally, the volume closes with an important chapter from Daniel Varón Silva and his coauthors at the Max Planck Institute of Colloids and Interfaces in Berlin describing the need for, the many challenges in, and their numerous elegant solutions to, the synthesis of glycosylphosphatidylinositol (GPI) anchors and of GPI-anchored molecules.

The discerning reader will gain much from this volume not the least of which, hopefully, will be the recognition of the dynamic nature of the field of glycochemistry and of the many challenges still remaining. When viewed in the broader context, these many challenges are nothing more than problems in modern synthetic organic chemistry whose solution only awaits the arrival and ingenuity of fresh minds to the area.

Wayne State University — *David Crich*
November 2013

Preface

More than 100 years ago, pioneering achievements in carbohydrate chemistry were awarded the second Nobel Prize in Chemistry to Emil Fischer. Since that time this branch of organic chemistry has lost none of its fascination. The success story began with Fischer's brilliant logical deduction of the glucose structure (based on simple organic reactions performed with sugars) and was followed by basic glycosylation methods (such as the Fischer-Helferich or the Koenigs-Knorr procedures). Later, in 1937, another Nobel Prize was awarded for the investigation of disaccharides and vitamin C to Walter N. Haworth. After a long period of hibernation, carbohydrate chemistry has developed into an arsenal of highly sophisticated chemical methods, which have enabled the carbohydrate chemist to selectively build almost all of the possible glycosidic linkages. Since Nature has created an unbelievable variety of oligosaccharides and we can at least partially understand the biological significance of many of these structures today, it is very necessary to have a specialized chemical toolkit either to create mono- as well as oligosaccharides or to further modify any of these structures.

The present monograph comprises a fine selection of hot topics in carbohydrate chemistry that have found applications in biological studies and the preparation of complex natural glycans. The synthetic methodologies used for the preparation of these substrates have gathered modern skills from general organic chemistry to the more specific field of glycochemistry. Nevertheless, both organic chemistry and glycochemistry benefited from each other's experiences to push further the limits of the molecular architectures attainable.

In contrast to several other books dealing with carbohydrate chemistry, a collection of very recent synthetic developments in the field of carbohydrate chemistry are presented. Recent synthetic achievements including *de novo* synthesis of carbohydrates, highlights of cyclodextrin chemistry, synthesis of highly complex glycoconjugates such as glycosphingolipids and GPI anchors are treated, always with a strong focus on the synthetic aspects.

The idea for this book project came up at the European Young Investigators Workshop co-organized by us in April 2011 in Lyon (France). After this conference gathering young researchers in glycochemistry, we felt that putting together a book would help the scientific community to identify some key aspects of glycosciences in order to address the next scientific challenges in the post-genomics and post-proteomics era. Glycomics are now the next leap for scientists and this book is intended to bring together techniques from synthetic organic and carbohydrate chemistry so that each domain would benefit from each other.

This book was the collective work of a number of glycochemists. Most importantly, we would like to thank all contributors whose time, efforts, and expertise have made this book a useful scientific resource for beginners and advanced researchers both in organic chemistry and glycochemistry. We are grateful to Drs. Elke Maase and Lesley Belfit at Wiley-VCH for their help and useful advices in preparing this book.

Braunschweig, Germany
Lyon, France
November 2013

Daniel B. Werz
Sébastien Vidal

List of Contributors

Hiromune Ando
Gifu University
Department of Applied
Bioorganic Chemistry
Faculty of Applied Biological
Sciences
1-1 Yanagido
Gifu 501-1193
Japan

and

Kyoto University
Institute for Integrated
Cell-Material Sciences (iCeMS)
Yoshida Ushinomiya-cho
Sakyo-ku
Kyoto 606-8501
Japan

Yifat Berkov-Zrihen
Tel Aviv University
School of Chemistry
Raymond and Beverly Sackler
Faculty of Exact Sciences
Ramat Aviv
Tel Aviv 69978
Israel

Anna Bernardi
Università degli Studi di Milano
Dipartimento di Chimica
via Golgi 19
20133 Milano
Italy

Alphert E. Christina
Leiden University
Leiden Institute of Chemistry
Bio-organic Synthesis Group
Einsteinweg 55, P.O. Box 9502
2300 RA Leiden
The Netherlands

Melissa Cochran
University of Toronto
Department of Chemistry
Toronto
ON M5S 3H6
Canada

Jeroen D. C. Codée
Leiden University
Leiden Institute of Chemistry
Bio-organic Synthesis Group
Einsteinweg 55, P.O. Box 9502
2300 RA Leiden
The Netherlands

Michael F. Cuccarese
Northeastern University
Department of Chemistry and Chemical Biology
360 Huntington Ave.
Boston, MA 02115
USA

Pierre-Alexandre Driguez
Sanofi R&D
Early to Candidate Unit
195 Route d'Espagne
BP 13669
31036 Toulouse Cedex
France

and

1 Avenue Pierre Brossolette
91385 Chilly-Mazarin Cedex
France

Micha Fridman
Tel Aviv University
School of Chemistry
Raymond and Beverly Sackler Faculty of Exact Sciences
Ramat Aviv
Tel Aviv 69978
Israel

Sebastian Götze
Max Planck Institute of Colloids and Interfaces
Department of Biomolecular Systems
Am Mühlenberg 1
14476 Potsdam
Germany

and

Freie Universität Berlin
Institut für Chemie und Biochemie
Arnimallee 22
14195 Berlin
Germany

Samuel Guieu
University of Aveiro
CICECO and QOPNA
Department of Chemistry
Aveiro 3810-193
Portugal

Hideharu Ishida
Gifu University
Department of Applied Bioorganic Chemistry
Faculty of Applied Biological Sciences
1-1 Yanagido
Gifu 501-1193
Japan

Makoto Kiso
Gifu University
Department of Applied Bioorganic Chemistry
Faculty of Applied Biological Sciences
1-1 Yanagido
Gifu 501-1193
Japan

and

Kyoto University
Institute for Integrated Cell-Material Sciences (iCeMS)
Yoshida Ushinomiya-cho
Sakyo-ku
Kyoto 606-8501
Japan

Bozhena S. Komarova
N.D. Zelinsky Institute of
Organic Chemistry
Russian Academy of Sciences
Laboratory of Glycoconjugate
Chemistry
Leninsky prospect 47
Moscow 119991
Russia

Yoshiyuki A. Kwase
University of Toronto
Department of Chemistry
Toronto, ON M5S 3H6
Canada

Jiazhen J. Li
Northeastern University
Department of Chemistry and
Chemical Biology
360 Huntington Ave.
Boston, MA 02115
USA

Gijsbert A. van der Marel
Leiden University
Leiden Institute of Chemistry
Bio-organic Synthesis Group
Einsteinweg 55, P.O. Box 9502
2300 RA Leiden
The Netherlands

Nikolay E. Nifantiev
N.D. Zelinsky Institute of
Organic Chemistry
Russian Academy of Sciences
Laboratory of Glycoconjugate
Chemistry
Leninsky prospect 47
Moscow 119991
Russia

Mark Nitz
University of Toronto
Department of Chemistry
Toronto, ON M5S 3H6
Canada

George A. O'Doherty
Northeastern University
Department of Chemistry and
Chemical Biology
360 Huntington Ave.
Boston, MA 02115
USA

Rita Pal
Gifu University
Department of Applied
Bioorganic Chemistry
Faculty of Applied Biological
Sciences
1-1 Yanagido
Gifu 501-1193
Japan

and

Kyoto University
Institute for Integrated
Cell-Material Sciences (iCeMS)
Yoshida Ushinomiya-cho
Sakyo-ku
Kyoto 606-8501
Japan

Nicola L. B. Pohl
Indiana University
Department of Chemistry
212 S. Hawthorne Drive
Bloomington
IN 47405
USA

José J. Reina
Università degli Studi di Milano
Dipartimento di Chimica
via Golgi 19
20133 Milano
Italy

Rajarshi Roychoudhury
Indiana University
Department of Chemistry
212 S. Hawthorne Drive
Bloomington
IN 47405
USA

Peter H. Seeberger
Max Planck Institute of Colloids
and Interfaces
Department of Biomolecular
Systems
Am Mühlenberg 1
14476 Potsdam
Germany

and

Freie Universität Berlin
Institut für Chemie und
Biochemie
Arnimallee 22
14195 Berlin
Germany

Daniel Varón Silva
Max Planck Institute of Colloids
and Interfaces
Department of Biomolecular
Systems
Am Mühlenberg 1
14476 Potsdam
Germany

and

Freie Universität Berlin
Institut für Chemie und
Biochemie
Arnimallee 22
14195 Berlin
Germany

Matthieu Sollogoub
UPMC Univ Paris 06
Sorbonne Université
Institut Universitaire de France
Institut Parisien de Chimie
Moléculaire
UMR-CNRS 7201
C. 181, 4 place Jussieu
75005 Paris
France

Abdellatif Tikad
University of Namur (UNamur)
Chemistry Department
rue de Bruxelles 61
5000 Namur
Belgium

Yury E. Tsvetkov
N.D. Zelinsky Institute of
Organic Chemistry
Russian Academy of Sciences
Laboratory of Glycoconjugate
Chemistry
Leninsky prospect 47
Moscow 119991
Russia

Nadezhda E. Ustyuzhanina
N.D. Zelinsky Institute of
Organic Chemistry
Russian Academy of Sciences
Laboratory of Glycoconjugate
Chemistry
Leninsky prospect 47
Moscow 119991
Russia

Ivan Vilotijevic
Max Planck Institute of Colloids
and Interfaces
Department of Biomolecular
Systems
Am Mühlenberg 1
14476 Potsdam
Germany

and

Freie Universität Berlin
Institut für Chemie und
Biochemie
Arnimallee 22
14195 Berlin
Germany

Stéphane P. Vincent
University of Namur (UNamur)
Chemistry Department
rue de Bruxelles 61
5000 Namur
Belgium

1
De Novo Approaches to Monosaccharides and Complex Glycans

Michael F. Cuccarese, Jiazhen J. Li, and George A. O'Doherty

1.1
Introduction

Over the years, considerable effort has been made toward the development of new synthetic routes to monosaccharides [1]. This interest came primarily from the medicinal chemistry community, as these new routes often provided access to unnatural sugars, which could be of use in structure–activity relationship (SAR) studies. In addition, the synthesis of monosaccharides, and in particular hexoses, has served as a challenge and a measuring stick to the synthetic organic community. Of particular interest are the routes to hexoses that start from achiral starting materials, where asymmetric catalysis is used to install the stereochemistry. In the synthetic organic community, these routes are described as "*de novo*" or "*de novo* asymmetric" routes to carbohydrates, whereas in the carbohydrate community, the term *de novo* takes up other meanings. For the purposes of this review, the term *de novo* asymmetric synthesis refers to the use of catalysis for the asymmetric synthesis of carbohydrates from achiral compounds [2]. This then precludes the inclusion of *de novo* process that produced sugars from molecules with preexisting chiral centers (e.g., Seeberger and Reißig) [3, 4].

The challenge of a *de novo* synthetic approach to carbohydrates has been met by many groups (Scheme 1.1). These approaches begin most notably with the seminal work by Masamune and Sharpless [5] (**2** to **5**), which utilized iterative asymmetric epoxidation of allylic alcohols to prepare all eight possible hexoses. More recently, Danishefsky [6] demonstrated the power of asymmetric hetero-Diels–Alder reaction for the synthesis of several glycals (**3** and **4** to **5**), which inspired further studies toward oligosaccharide synthesis. Johnson and Hudlicky [7] turned to the use of enzyme catalysis for the oxidation/desymmetrization of substituted benzene rings to achieve hexopyranoses (**1** to **5**). Alternatively, Wong and Sharpless [8] used a combination of transition metal catalysis (asymmetric dihydroxylation) and an enzyme-catalyzed aldol reaction for the synthesis of several 2-keto-hexoses. More recently, this challenge has been engaged by MacMillan who utilized an iterative aldol reaction approach (a proline-catalyzed aldol followed by a subsequent diastereoselective aldol reaction) to produce various hexoses

Modern Synthetic Methods in Carbohydrate Chemistry: From Monosaccharides to Complex Glycoconjugates,
First Edition. Edited by Daniel B. Werz and Sébastien Vidal.

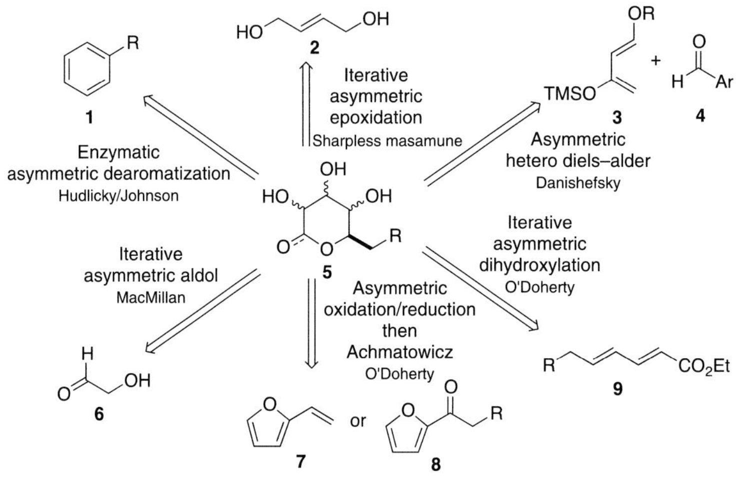

Scheme 1.1 *De novo* approaches to hexoses.

(**6** to **5**) [3]. Of these approaches, only the iterative epoxidation strategy of Masamune and Sharpless provides access to all eight hexoses, but it is also noteworthy that their route required the most steps and protecting groups. It is this latter point, the reduction of steps and avoidance of protecting groups, which guided the development of these synthetic endeavors [9].

We have also developed two practical methods for the *de novo* synthesis of hexoses. These efforts have resulted in the discovery of two orthogonal approaches to hexopyranoses with variable C-6 substitutions. These approaches entail an iterative dihydroxylation strategy to hexose sugar lactones (**9** to **5**) [10] and an Achmatowicz strategy that is amenable to all eight hexose diastereomers (**7** or **8** to **5**) [11] Of the two approaches, the iterative asymmetric dihydroxylation of dienoates (Scheme 1.1) is the most efficient in terms of steps (one step for racemic to three steps for asymmetric) and the minimal use of protecting groups. On the contrary, the Achmatowicz approach is superior in terms of synthetic scope. The potential of this approach can be seen in the highly efficient *de novo* route to various mono-, di-, tri- tetra-, and heptasaccharide motifs, allowing their use for biological and medicinal structure–activity studies.

Of the many ways to compare these *de novo* routes (e.g., number of steps, availability of starting materials, and atom economy), clearly the best metric is the scope of its use in synthetic and biological applications. The Achmatowicz approach is distinguished from the other approaches when it comes to practical application to rare sugars, medicinal chemistry, and more specially oligosaccharides. These features result from its compatibility with the Pd-π-allyl-catalyzed glycosylation for the stereospecific formation of the glycosidic bond [12, 13]. As outlined in Scheme 1.2, the Pd(0)-catalyzed glycosylation reaction is both general and stereospecific [14, 15]. The reaction occurs rapidly and in high yields for both the α-**10** to α-**11** and β-**10** to β-**11** systems and works best when $Pd_2(dba)_3{\cdot}CHCl_3$ is used as the Pd(0) source

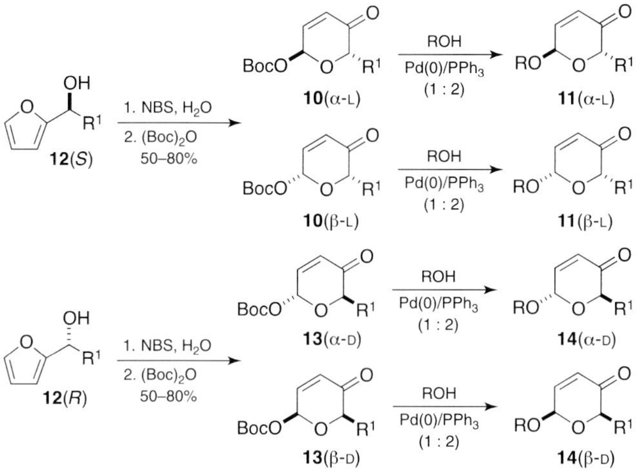

Scheme 1.2 Stereospecific Pd-catalyzed glycosylation.

with triphenylphospine as the ligand in a 1 : 2 Pd/PPh_3 ratio. While carboxylate-leaving groups also work, the *t*-butoxycarbonate group ($BocO^-$) is critical for the successful implementation of this reaction with alcohol nucleophiles. For example, when the *t*-butoxycarboxy group was replaced with a benzoyl or pivaloyl group, the palladium-catalyzed glycosylation reaction was significantly slower.

The *de novo* Achmatowicz approach to hexoses has great potential for preparing various D- and L-sugars because the starting 6-*t*-butoxycarboxy-2*H*-pyran-3(6*H*)-ones (**10** and **13**) can easily be prepared from optically pure furfuryl alcohols **12** (either (*R*) or (*S*) enantiomer) [16] by a one or two step procedure (Scheme 1.3). Depending on the temperature of the second step, the *t*-butylcarbonate acylation

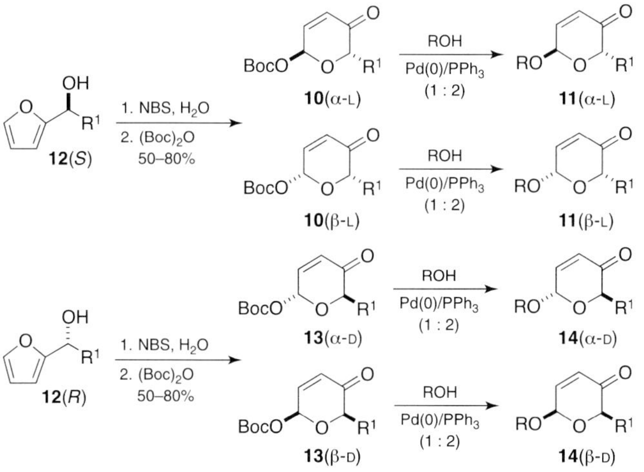

Scheme 1.3 Achmatowicz approach to the simplest hexoses.

reaction can selectively give the α-Boc pyranones **10α** and **13α** at −78 °C, whereas at room temperature, a 1 : 1 ratio of the α- and β-Boc protected enones were produced. Thus, this procedure can be used to prepare multigram quantities of both α- and β-pyranones in either enantiomeric form. When the Pd-glycosylation reaction was coupled with the Achmatowicz oxidation of a furan alcohol and diastereoselective *t*-butylcarbonate acylation reaction, a net three-step stereo-divergent pyranone-forming reaction resulted. Key to the success of the *de novo* asymmetric Achmatowicz approach is the practical access to all four possible pyranone diastereomers from either furan alcohol enantiomers **12**(*R*) or **12**(*S*).

An important aspect of this approach is the ease with which furan alcohols can be prepared in enantiomerically pure form from achiral furans (e.g., **7** and **8**). There are many asymmetric approaches to prepare furan alcohols. The two most prevalent approaches are (i) the Noyori reduction of acylfurans (**8** to **12**) and (ii) the Sharpless dihydroxylation of vinylfurans (**7** to **12**) (Scheme 1.4) [17]. Both routes are readily adapted to 100 g scale synthesis and use readily available reagents. While the Sharpless route is most amenable to the synthesis of hexoses with a C-6 hydroxy group, the Noyori route distinguishes itself in its flexibility to virtually any substitution at the C-6 position. Herein, we review the development of the Achmatowicz approach to the *de novo* synthesis of carbohydrates, with application to oligosaccharide assembly and medicinal chemistry studies.

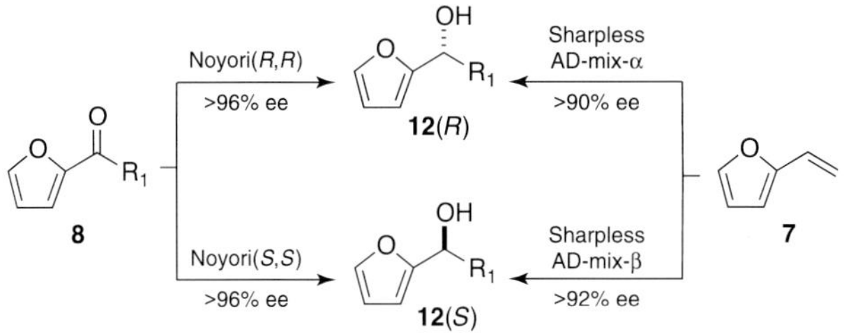

Scheme 1.4 *De novo* asymmetric approaches to chiral furan alcohols.

1.2
De Novo Synthesis of Monosaccharides

Putting this together, a very practical *de novo* approach to *manno*-hexoses can be carried out in six steps from achiral acylfuran **15**. The route began with a three-step synthesis of Boc-pyranone **18** in an ~50% overall yield. Thus, in only three highly diastereoselective steps, pyranone **18** was converted into *manno*-pyranose **21** in 45% overall yield. The three-step sequence consists of a Pd-catalyzed glycosylation (**18** to **19**) and two postglycosylation reactions, a Luche reduction ($NaBH_4/CeCl_3$, **19** to **20**) and Upjohn dihydroxylation (OsO_4(cat)/NMO, **20** to **21**) (NMO, *N*-methylmorpholine-*N*-oxide)(Scheme 1.5). This six-step sequence of achiral **15** to partially protected mannose **21** demonstrates the power of asymmetric synthesis.

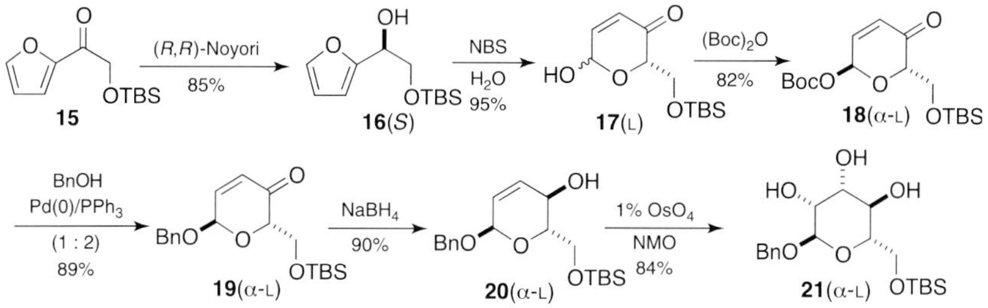

Scheme 1.5 *De novo* asymmetric Achmatowicz approach to L-*manno*-hexoses.

From the point of view of carbohydrate synthesis, this route consists of a three-step asymmetric synthesis of glycosyl donor **18**, which in three additional steps is converted into *manno*-sugar **21**. Viewing this approach from a synthetic perspective becomes more relevant in the subsequent schemes. Other hexose congeners were also prepared using other postglycosylation transformations [18]. Critical to the success of this approach is the chemoselective use of functionality of C–C and C–O π-bonds as atom-less protecting groups (i.e., enone as a protected triol) as well as an anomeric-directing group (via a Pd-π-allyl).

1.3 Iterative Pd-Catalyzed Glycosylation and Bidirectional Postglycosylation

The synthetic efficiency of the approach reveals itself when the Pd(0)-catalyzed glycosylation was applied in an iterative manner for oligosaccharide synthesis (Scheme 1.6 and Scheme 1.7) [19]. The step savings occurred because of the

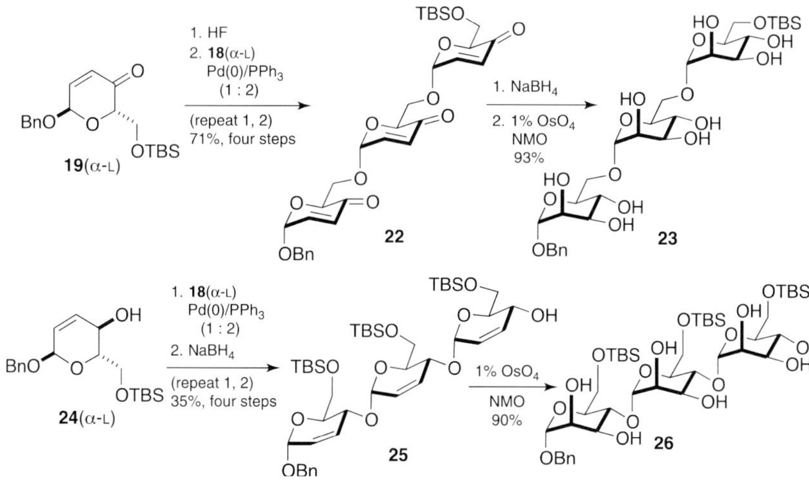

Scheme 1.6 *De novo* asymmetric approach to 1,6- and 1,4-oligo-hexoses.

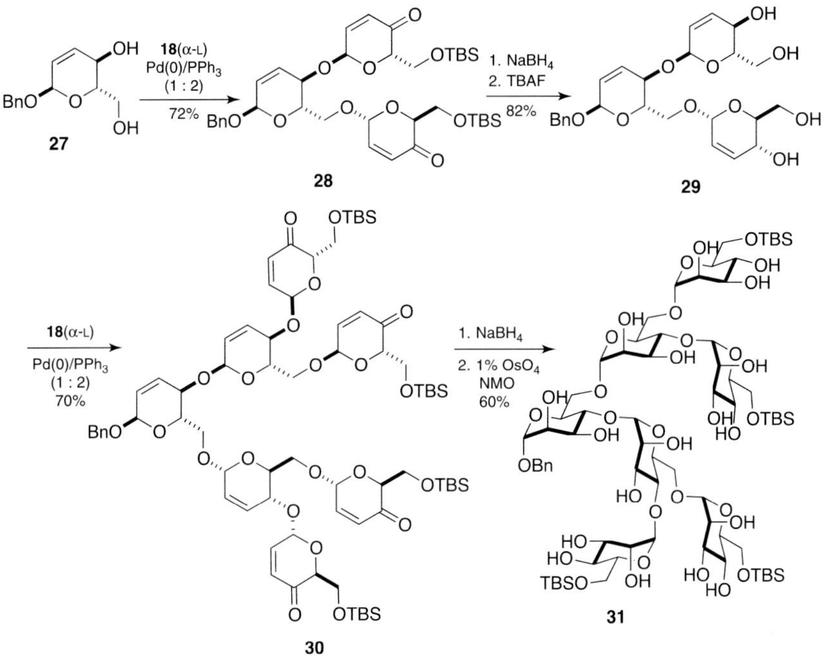

Scheme 1.7 *De novo* asymmetric approach to branched 1,4- and 1,6-oligo-L-hexoses.

bidirectional use of postglycosylation Luche and Upjohn reactions. While not always shorter, these routes compare favorably with more traditional carbohydrate approaches and offer exclusive access to enantiomers as well as D-/L-sugar diastereomers. In addition to reducing steps, this highly atom-economical approach avoids the extensive use of protection/deprotection steps. For example, the 1,6-*manno*-trisaccharide **23** was prepared from enone **18** in six steps (nine from achiral furan **15**). The synthesis was accomplished by an iterative use of a *t*-butyldimethylsilyl (TBS)-deprotection/glycosylation strategy to prepare trisaccharide **22**, followed by a tris-reduction and tris-dihydroxylation to install the *manno*-stereochemistry. By simply switching the order of the reduction and glycosylation steps, this route can also be used to prepare 1,4-manno-trisaccharide **23**. Key to the success of this sequence was the highly stereoselective reduction and dihydroxylation reaction, which installed six stereocenters in one transformation (**25** to **26**). This approach was successfully used in the medicinal chemistry SAR study of digitoxin, an anticancer agent [20].

1.3.1
Bidirectional Iterative Pd-Catalyzed Glycosylation and Postglycosylation

The synthetic efficiency was magnified when the glycosylation reaction was also applied in a bidirectional manner [21]. For example, when the TBS group of pyran

20 was removed, the resulting diol **27** can be bis-glycosylated to form tris-pyran **28**. This bidirectional application of ketone reduction and TBS deprotection gave tetraol **29**. Once again, the tetraol of **29** can be per-glycosylated with excess pyranone **18** to give heptasaccharide **30**. Finally, a ketone per-reduction and double bond per-dihydroxylation gave heptasaccharide **31**. It is worth noting, while there is similar local symmetry around each alkene, they exist in different stereochemical environments.

1.3.2
Synthesis of Monosaccharide Aminosugar Library

While these bidirectional approaches do have some significant synthetic efficiency over traditional approaches, they do suffer from the fact that the routes tend to be linear in nature and do not readily adapt to convergent synthesis. On the contrary, these routes most readily adapt to divergent synthesis. In particular to divergent synthesis, as it is applied to the synthesis of unnatural sugars. This becomes particularly advantageous when it is being used to address problems associated with medicinal chemistry. An example of this application can be seen in our application of this *de novo* Achmatowicz approach for the synthesis of a library of glycosylated methymycin analogs for eventual medicinal chemistry SAR studies [22].

Methymycin is one of the several 12-membered ring-glycosylated macrolide antibiotics isolated from *Streptomyces venezuelae* ATCC 15439 (ATCC, American Type Culture Collection; Scheme 1.8). Similar to other macrolide antibiotics, the rare deoxy-aminosugar portion of methymycin (desosamine) is important for their bioactivity. Thus, its modifications hold promise as a valuable approach

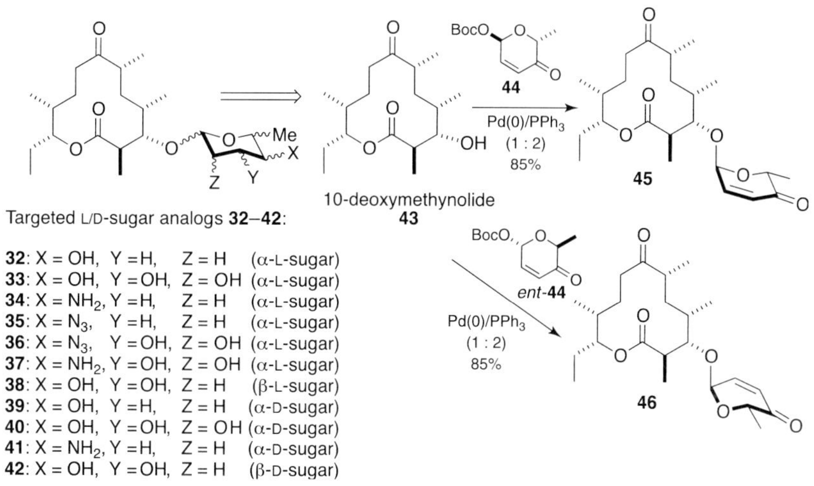

Scheme 1.8 *De novo* synthesis of a methymycin monosaccharide library.

toward preparing new macrolide antibiotics with improved and/or altered biological properties. Our *de novo* approach to a library of methymycin analogs is retrosynthetically outlined in Scheme 1.8, where the macrolide aglycon 10-deoxymethynolide was glycosylated in a stereo-divergent manner (with D- or L-Boc pyranones **44** or *ent*-**44**, respectively) to give either α-D-glycoside **45** or its diastereomer α-L-glycoside **46**. Subsequent postglycosylation transformations were used to provide various sugar congeners and stereoisomers, in particular, unnatural deoxy-aminosugar isomers.

The installation of amino-functional groups, in practice, was most easily accomplished at the C-4 position (Scheme 1.9). For instance, the α-D-pyranone ring on methymycin analog **45** could be converted into a 4-aminosugar **41** with α-D-*rhodino*-stereochemistry in four steps, via a reduction, activation of the resulting alcohol, azide inversion, and reduction strategy. Alternatively, this approach is

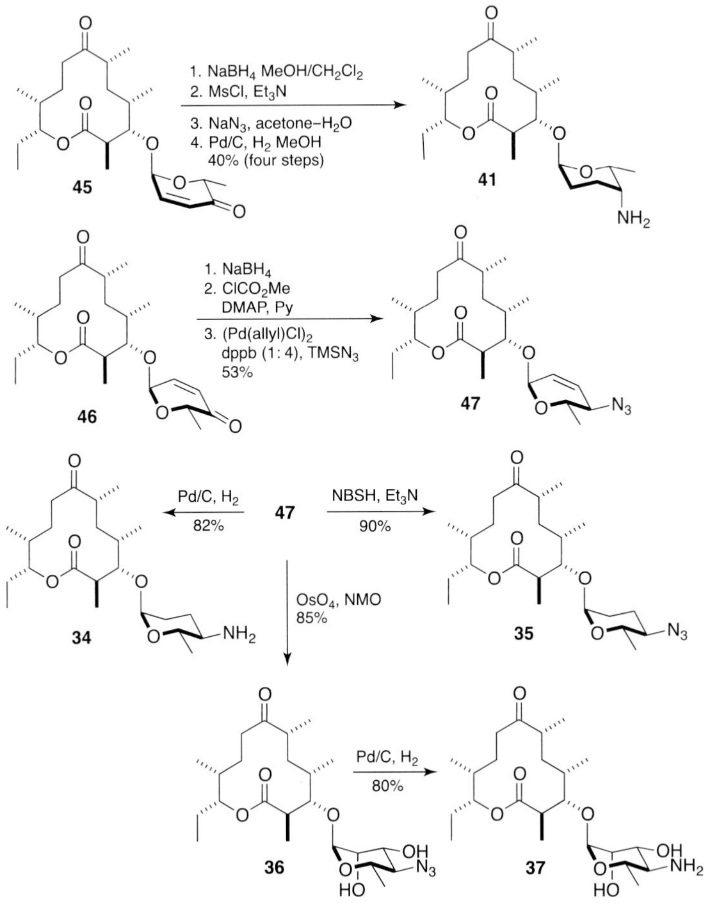

Scheme 1.9 *De novo* synthesis of a methymycin monosaccharide aminosugar library.

also compatible with the installation of equatorial amino groups at the C-4 position. This is accomplished by means of a net retention of stereochemistry in the substitution reaction at the C-4 position. The reaction occurred via a Pd-catalyzed π-allyl reaction with trimethylsilyl azide ($TMSN_3$) as the nucleophile to give allylic azide **47** from α-D methymycin analog **46**. In turn, azide **47** could be converted into azido-/azasugar methymycin analogs with α-L-*rhamno*- (**36** and **37**) and α-L-*amiceto*-stereochemistries (**34** and **35**).

1.4 Synthesis of Monosaccharide Azasugar

Among the polyhydroxylated indolizidine alkaloids, the most well-known member is D-swainsonine **60** (Scheme 1.10) [23–25]. Swainsonine is known as a *potent inhibitor* of both lysosomal α-mannosidase [26] and mannosidase II [27] and has shown promise as an anticancer agent [28]. Fleet *et al.* [29] have shown that the enantiomer (L-swainsonine) selectively inhibited narginase (L-rhamnosidase, $K_i = 0.45\ \mu M$), whereas the D-swainsonine showed no inhibitory activity toward this enzyme. Owing to the biological importance of both D- and L-swainsonines, several epimers and analogs have become attractive targets for syntheses [30, 31].

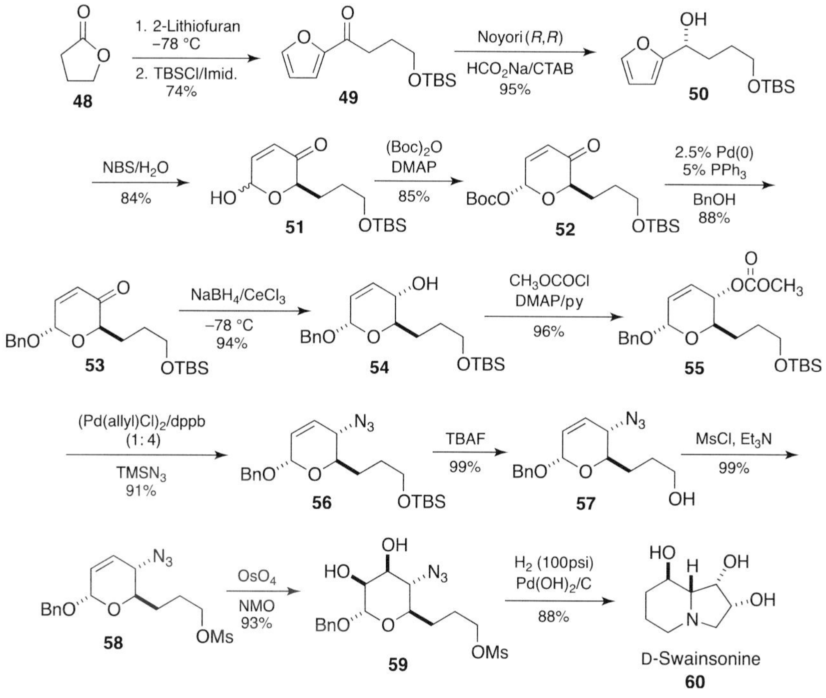

Scheme 1.10 *De novo* synthesis of D-swainsonine.

Thus, we became interested in the *de novo* asymmetric synthesis of both enantiomers of swainsonine and various epimers for further biological studies [32]. We envisioned a similar postglycosylation transformation, in which the installation of the 4-amino-*manno*-stereochemistry in methymycin analog **47** could be instrumental in the development of a *de novo* Achmatowicz approach to the indolizidine natural product swainsonine **60** (Scheme 1.10). In practice, this required access to acylfuran **49**, which was prepared in two steps from 2-lithiofuran and butyrolactone **48**. After Noyori reduction, Achmatowicz reaction, and diastereoselective acylation, **49** was converted into pyranone **52**. Glycosylation of benzyl alcohol with **52** installed the required C-1-protecting group. As before, Luche reduction, carbonate formation, and Pd-π-allyl allylic azide displacement installed the C-4 azido group in **56**. Before the C-2/C-3 double bond was dihydroxylated, the TBS ether is converted into a mesylate-leaving group as in **58**. Finally, a diastereoselective dihydroxylation (**58** to **59**) and exhaustive hydrogenolysis (**59** to **60**) cleanly provided D-swainsonine in an optically pure form. Thus, in only 13 steps, either D- or L-swainsonine can be prepared from achiral starting material. This is of particular note because both enantiomers have been valued as known inhibitors of glycosidase enzymes (D-swainsonine for α-D-mannosidases and α-L-swainsonine for α-L-rhamnosidases). This route has also been used to prepare various diastereoisomers of swainsonine, which are also known to be effective glycosidase inhibitors [32].

1.5
Oligosaccharide Synthesis for Medicinal Chemistry

As part of the continuing search for new antibiotics against bacterial resistance [33], the cyclic hexapeptide mannopeptimycin-ε **61** was isolated from the fermentation broths of *Streptomyces hygroscopicus* LL-AC98 and related mutant strains [34]. The key structural features of the mannopeptimycins are a cyclic hexapeptide core with alternating D- and L-amino acids, three of which are rare. Two of the amino acids (β-D-hydroxyenuricididine and D-tyrosine) are glycosylated with mannose sugars. The glycosylated amino acids are an N-glycosylated β-hydroxyenuricididine with an α-mannose, and an O-glycosylated tyrosine with an α-(1,4-linked)-bis-mannopyranose disaccharide portion. The unique structure and unprecedented biological activity have inspired both biological [35] and synthetic studies from us and others.

Our *de novo* asymmetric Achmatowicz approach was also applied to the synthesis of the glycosylated tyrosine portion of the antibiotic mannopeptimycin-ε [36]. Specifically, we targeted a protected tyrosine with bis-manno-1,4-disaccharide with an isovalerate at the C-4′ position. This approach was used to prepare the amino acid portion of the natural product **62** as well as the disaccharide portion in the unnatural L/L-configuration. The linear route involved the application of the iterative bis-glycosylation, acylation, and bis-dihydroxylation of protected tyrosine in only six steps (Scheme 1.11).

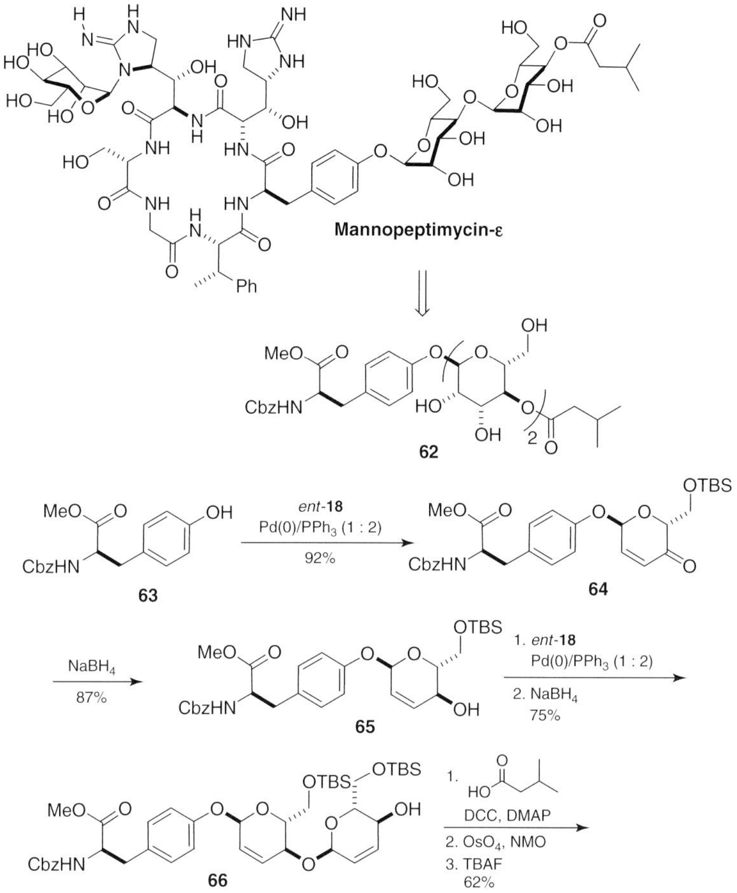

Scheme 1.11 Assembly of the mannopeptimycin-ε disaccharide.

As part of further SAR studies of the antibiotic mannopeptimycin, access to the C-4 amide analogs **42** was also desired [37]. As before, this was also accomplished using the Pd-π-allyl-catalyzed allylic azide alkylation in conjunction with azide reduction and acylation. Specifically, allylic alcohol **66** was converted into methyl carbonate **67**, and the allylic carbonate was converted into allylic azide **68**. In turn, the azide was selectively reduced and the corresponding amine **69** was acylated. The remaining double bonds of **70** were dihydroxylated, and the TBS groups were removed to provide the desired target tyrosine disaccharide **71** (Scheme 1.12). Key to the success of this approach was the somewhat surprisingly selective ionization of the equatorial allylic carbonate in **67** without any sign of ionizing the anomeric phenol in the axial configuration.

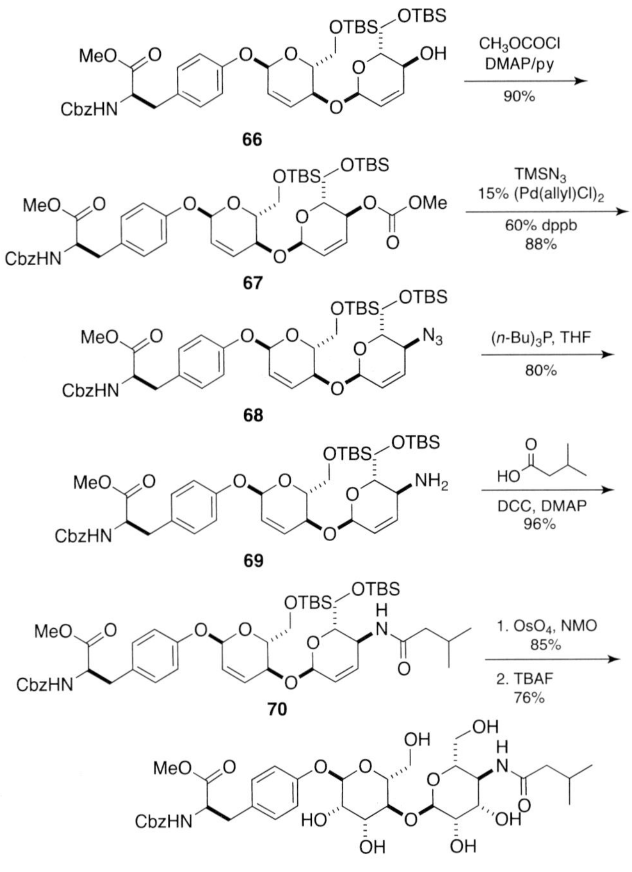

Scheme 1.12 Elaboration to C-4″ aza-mannopeptidomycin.

1.5.1
Tri- and Tetrasaccharide Library Syntheses of Natural Product

As part of a high-throughput-based search for new natural products with unique structures and interesting biological activity, two partially acetylated trisaccharide (cleistrioside-5/-6, **72** and **73**) and six partially acetylated tetrasaccharide (cleistetroside-2/-3/-4/-5/-6/-7, **74–81**) natural products were discovered (Figure 1.1) from the leaves and bark of trees with a folk medicinal tradition [38–40]. These dodecanyl tri- and tetrarhamnoside structures with various degrees of acylation were isolated from *Cleistopholis patens* and *Cleistopholis glauca* [36, 37]. The structures of the cleistriosides and cleistetrosides were assigned by detailed NMR analysis and later confirmed by total synthesis by us [41] and others [42, 43]. In addition to clarifying the structural issues, our synthetic interest in these

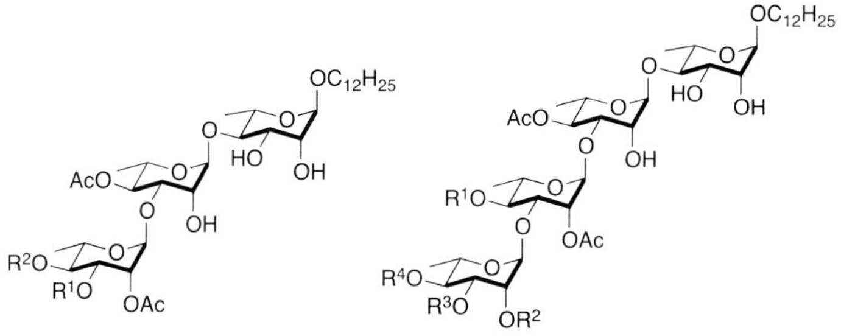

72: Cleistrioside-5: R^1 = H, R^2 = Ac
73: Cleistrioside-6: R^1 = Ac, R^2 = H

74: Cleistetroside-2: $R^1 = R^4$ = Ac, $R^2 = R^3$ = H
75: Cleistetroside-3: $R^1 = R^2$ = Ac, $R^3 = R^4$ = H
76: Cleistetroside-4: R^1 = Ac, $R^2 = R^3 = R^4$ = H
77: Cleistetroside-5: $R^1 = R^2 = R^3 = R^4$ = H
78: Cleistetroside-6: $R^1 = R^2 = R^4$ = Ac, R^3 = H
79: Cleistetroside-7: $R^1 = R^2 = R^3 = R^4$ = Ac
80: Cleistetroside-9 (new): $R^1 = R^2 = R^3$ = H, R^4 = Ac
81: Cleistetroside-10 (new): R^1 = H, $R^2 = R^3 = R^4$ = Ac

Figure 1.1 The cleistriosides, cleistetrosides, and analogs.

oligosaccharides was aimed at supplying sufficient material for SAR-type studies and as a test of our synthetic methodology [44].

The route began with pyranone *ent*-**44**, which was easily prepared in three steps from commercially available acetylfuran (Scheme 1.13). In four steps (glycosylation, reduction, dihydroxylation, and acetonide protection), glycosyl donor *ent*-**44** was

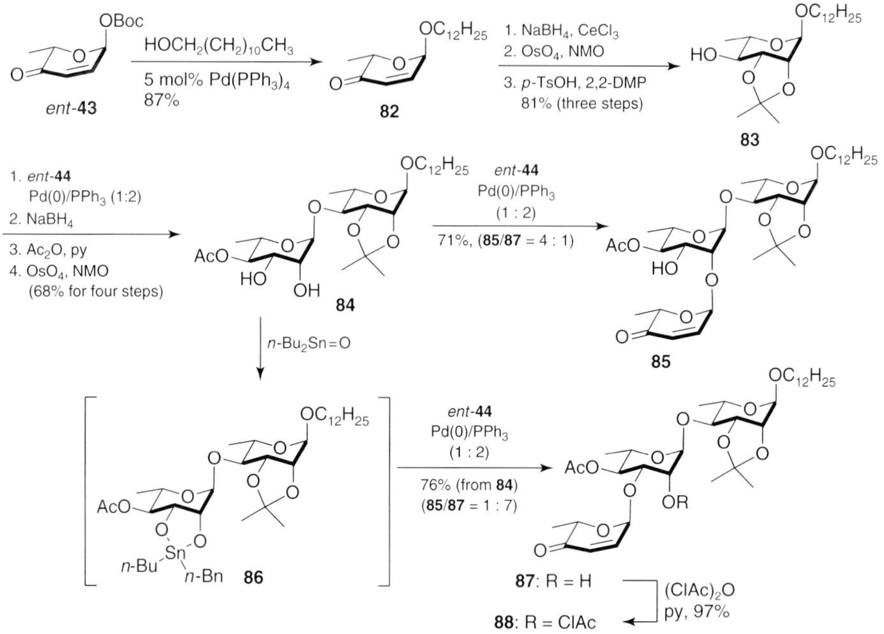

Scheme 1.13 Synthesis of trisaccharide **88**.

converted into protected rhamnose **83**. A subsequent glycosylation, reduction, acylation, and dihydroxylation gave diol disaccharide **84**. Unfortunately, when diol **84** was exposed to our typical Pd-catalyzed glycosylation conditions, the trisaccharide **85** with the wrong regiochemistry was formed. To our delight, this substrate regioselectivity (4 : 1) could be reversed via the formation of tin acetal **86**. Thus, a tin-directed regioselective (7 : 1) glycosylation gave trisaccharide **87** with the required carbohydrate at the C-3 position. A subsequent chloro-acylation gave trisaccharide **88**, which was ready for further elaboration into both the cleistrioside and cleistetroside natural products (Scheme 1.14, Scheme 1.16, and Scheme 1.17).

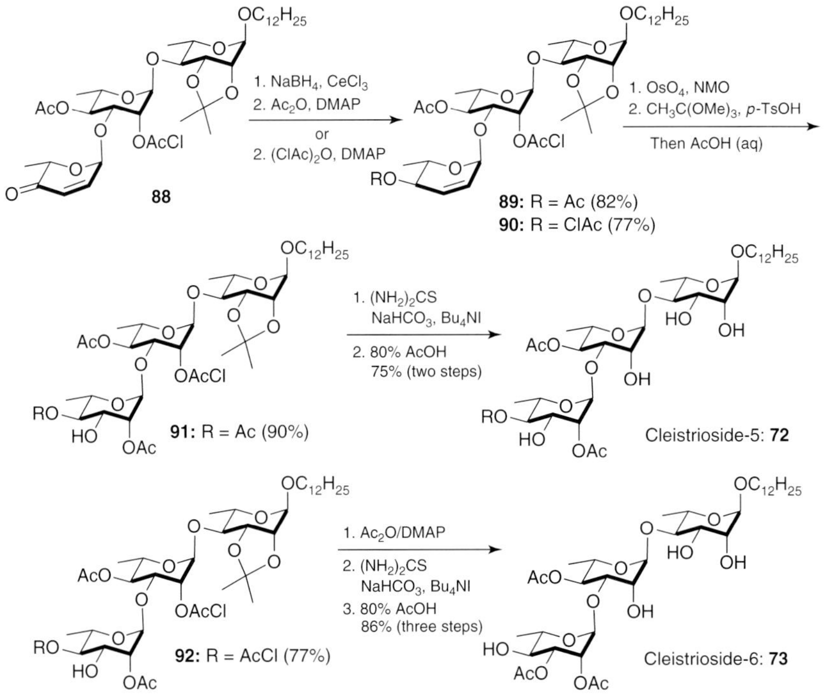

Scheme 1.14 Synthesis of cleistrioside-5 and -6.

The pyranone ring in trisaccharide **88** is perfectly situated for further elaboration into a rhamnose ring with the desired acylation pattern for cleistrioside-5 and -6. This was accomplished by Luche reduction and acylation or chloro-acylation to give **89** and **90**. A subsequent dihydroxylation and ortho-ester-mediated C-2 acylation gave trisaccharides **91** and **92**, which could be converted into cleistrioside-5 and -6 (**72** and **73**) by a selective removal of the chloro-acetates over the other acetates (removed with thiourea) and acetonide-protecting groups (hydrolyzed with AcOH/H_2O).

In addition, the trisaccharides **91** and **92** can be further converted into eight members of the cleistetroside family of natural products (Scheme 1.16 and Scheme 1.17). This divergent route began with the selective Pd-catalyzed glycosylation of the free C-3 alcohol in **91** and **92** and a subsequent Luche reduction to afford tetrasaccharides **93** and **94** (Scheme 1.15).

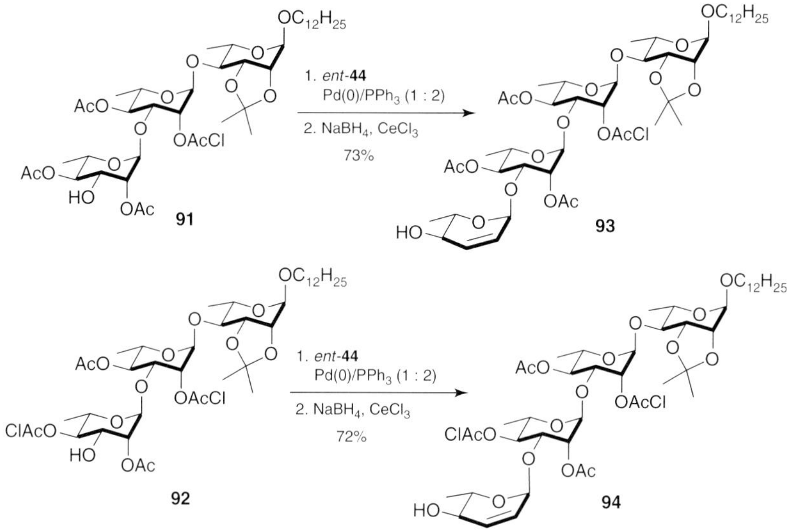

Scheme 1.15 Synthesis of tetrasaccharides **93** and **94**.

As outlined in Scheme 1.16, tetrasaccharide **93** was divergently converted into five of the six desired cleistetroside-2, -3, -4, -6, and -7. This was accomplished in a range of three to five steps by subtle variations of the postglycosylation reactions (dihydroxylation, acylation, and chloro-acylation), followed by global deprotection (thiourea then $AcOH/H_2O$).

By applying the same postglycosylation/deprotection reaction sequence on tetrasaccharide **94**, the remaining cleistetroside-5 was prepared in three steps. In addition, the revised route was also used to prepare two previously unknown analogs, cleistetroside-9 and -10. This was accomplished in a range of one to three steps by modular application of the postglycosylation reactions (dihydroxylation, acylation, and chloro-acylation) (Scheme 1.17).

It is worth noting that the route to any of the cleistetrosides is quite comparable to the two previously reported routes to cleistetroside-2 in terms of total number of steps. What distinguished it from these more traditional routes is its flexibility to diverge to any of the possible natural product isomers. Thus, we have found that this divergent approach is particularly amenable to medicinal chemistry studies. In this regard, our synthetic access to these eight natural products (two

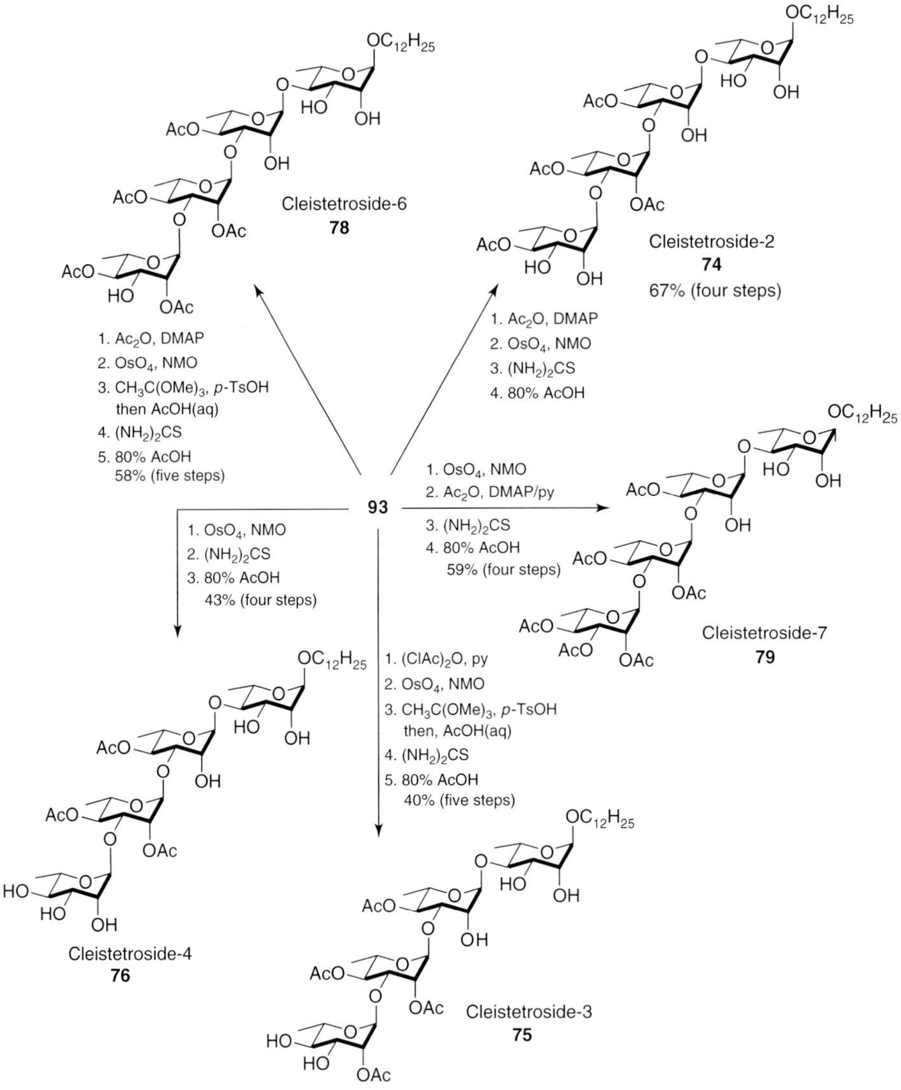

Scheme 1.16 Synthesis of cleistetroside-2, -3, -4, -6, and -7.

trisaccharides and six tetrasaccharides) and additional two natural product analogs enabled detailed medicinal chemistry SAR studies. The divergent nature of the approach is graphically displayed in Scheme 1.18, where in only 13 steps and in 20% overall yield, the key trisaccharide **88** could be prepared from achiral furan **95**. Trisaccharide **88** serves as the linchpin molecule that can, in 6–11 steps, be converted into any of the desired natural products, in sufficient quantities for further studies.

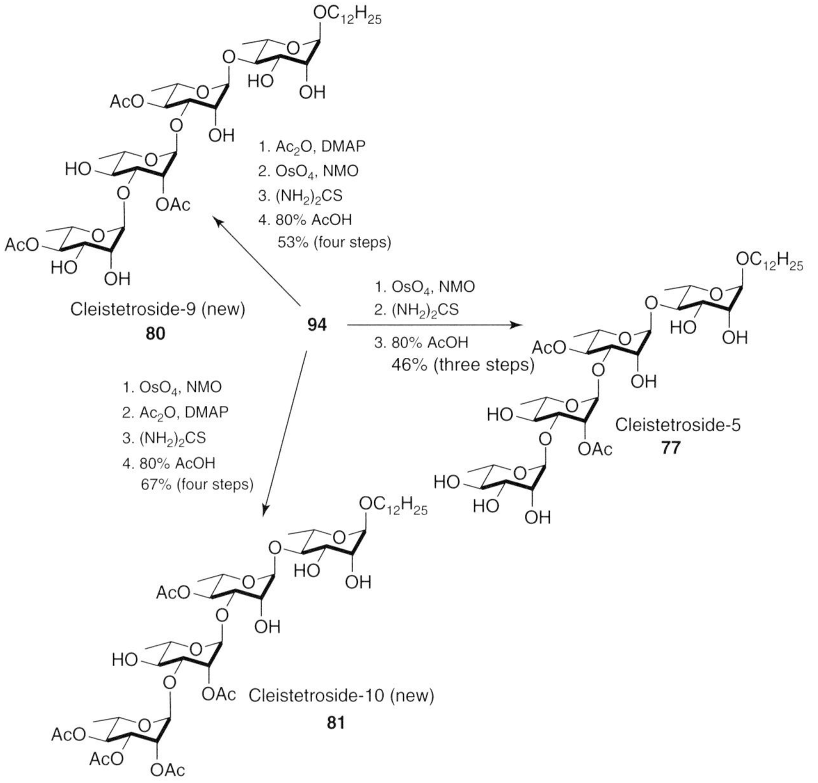

Scheme 1.17 Synthesis of cleistetroside-5, -8, and -10.

1.5.2 Anthrax Tetrasaccharide Synthesis

Possibly the most elaborate application of this *de novo* Achmatowicz approach to oligosaccharides was the synthesis of the anthrax tetrasaccharide **100**. The approach to this tetrasaccharide natural product merged well with our efforts to C-4 aminosugars with the synthesis of rhamnose-containing oligosaccharides (Scheme 1.19) [45, 46]. Anthrax is a zoonotic disease caused by the spore-forming bacterium *Bacillus anthracis* [47]. In an effort to find a unique structural motif associated with the bacterium, the anthrax tetrasaccharide **100** was discovered. The tetrasaccharide **100** consists of three L-rhamnose sugars and a rare sugar, D-anthrose [48]. The uniqueness of the D-anthrose sugar and the resistance of carbohydrate structures to evolutionary change make the anthrax tetrasaccharide an interesting target for synthesis [49]. Several carbohydrate approaches to the anthrax tetrasaccharide and one to a related trisaccharide have been reported [49a, 50], which derive their stereochemistry from the known but less common sugar L-rhamnose and the rare D-fucose. Our *de novo* approach to the tetrasaccharide **100** was envisioned as occurring through a traditional glycosylation between trisaccharide **96** with

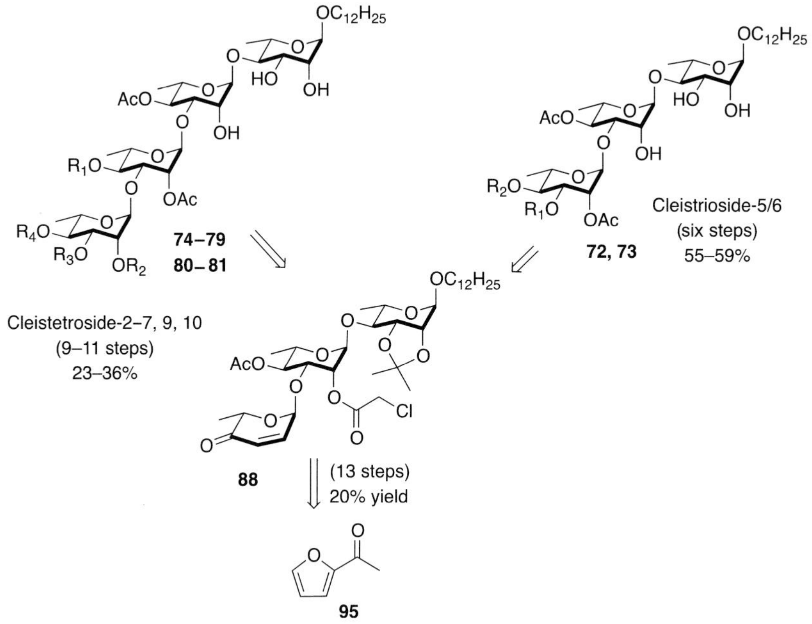

Scheme 1.18 Divergent synthesis retrosynthetic summary.

trichloroacetimidate **97** (Scheme 1.19). In turn, our *de novo* approach was planned to prepare both of these fragments (**96** and **97**) from the achiral acetylfuran **95**, which it is worth noting are significantly less expensive than either L-rhamnose or D-fucose.

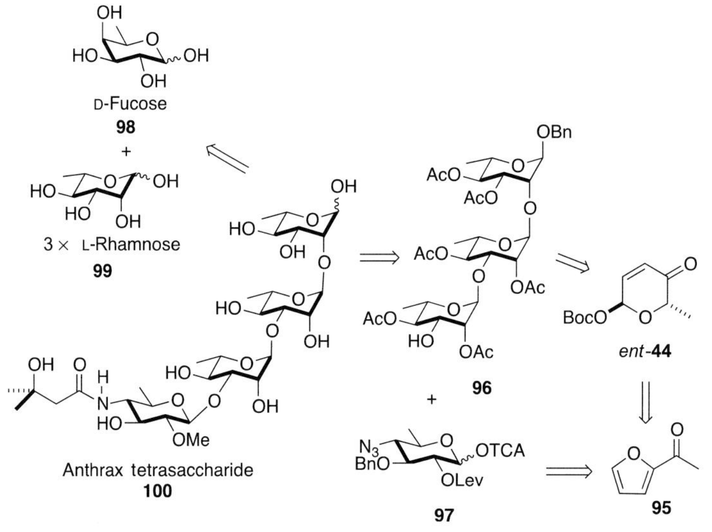

Scheme 1.19 Retrosynthesis of anthrax tetrasaccharide **100**.

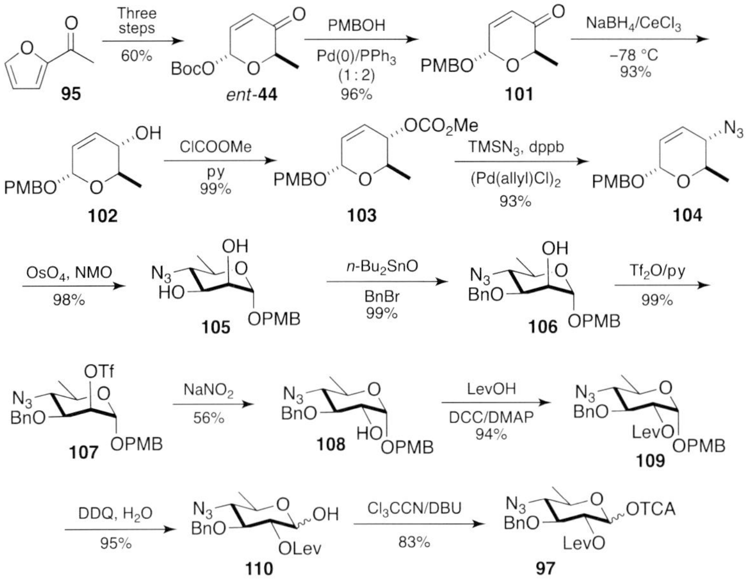

Scheme 1.20 *De novo* synthesis of a D-anthrose sugar trichloroacetimidate.

Our synthesis of the anthrose monosaccharide **97** is described in Scheme 1.20 and involved two Pd-π-allylation reactions. The route began with the synthesis of (*p*-methoxybenzyl)PMB-protected pyranone **101** from **95** via *ent*-**44**. Using the Pd-catalyzed C-4 allylic azide chemistry previously described, pyranone *ent*-**44** was converted into allylic azide **104** and dihydroxylated to give rhamno-sugar **105**. The 6-deoxy-*gluco*-stereochemistry is then installed by a protection/C-2 inversion strategy to give anthrose sugar **108**. Finally, a Lev-protection, PMB-deprotection strategy, and trichloroacetimidate formation were used to convert **108** into the glycosyl donor sugar **97** (14 steps from achiral acetylfuran **95**).

The *de novo* Achmatowicz approach to the tris-rhamno portion of the anthrax tetrasaccharide began with the synthesis of disaccharide **115** from pyranone *ent*-**44** and benzyl alcohol (Scheme 1.21). After glycosylation and postglycosylation transformations to install the *rhamno*-stereochemistry (*ent*-**44** to **111**), the 1,2-*trans*-diol of **111** was then protected with the Ley-spiroketal to provide monosaccharide **112** with a free C-2 hydroxyl group. After a similar three-step glycosylation (**112** and **113**) and postglycosylation sequence, **113** was converted into disaccharide **114**, which in a one-pot ortho-ester protocol was protected to give disaccharide **115** with a free C-3 alcohol.

Simply repeating the same three-step glycosylation and postglycosylation sequence converted disaccharide **115** into trisaccharide **116** (Scheme 1.22). Once again the one-pot ortho-ester formation/acylation/hydrolysis sequence gave trisaccharide **117** with the free C-3 alcohol, ready for glycosylation with an anthrose sugar fragment. Unfortunately, any attempt at glycosylation of trisaccharide **117** failed because of the instability of the Ley-spiroketal to the Lewis acidic nature of

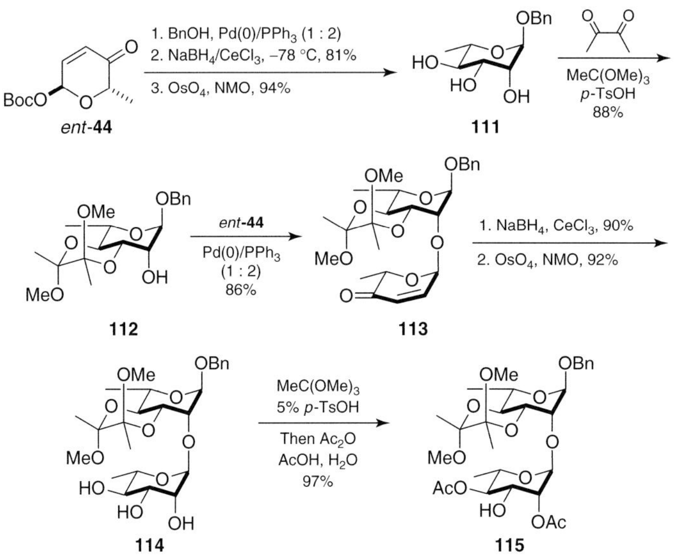

Scheme 1.21 Synthesis of disaccharide **115**.

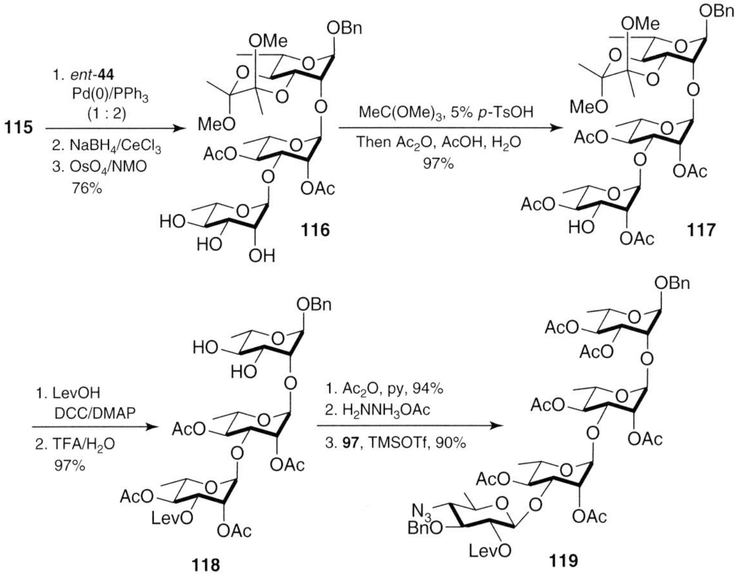

Scheme 1.22 Synthesis of tetrasaccharide **119**.

the traditional glycosylation conditions. Undaunted, we turned to an alternative protecting group strategy. Thus, the C-3 hydroxyl group of **117** was protected as a levulinate ester, and the Ley-spiroketal-protecting group was removed to form **118**. Then, the anthrose sugar was installed by an acylation, selective levulinate deprotection (using hydrazine), and glycosylation with anthrose monosaccharide **97** delivering the corresponding tetrasaccharide **119**.

Finally, we turned to the deprotection of tetrasaccharide **119** into anthrax tetrasaccharide **100**. Deprotection of levulinate-protecting groups followed by an etherification (MeI/Ag_2O) delivered the methyl ether **120**. A one-pot condition was employed to reduce and acylate azide **120** along with global deprotection of the acetate groups to generate the free alcohol (PEt_3/LiOH/H_2O), which upon selective peptide coupling of primary amine and 3-hydroxy-3-methylbutanoic acid (HBTU/Et_3N) (HBTU, *O*-(benzotriazol-1-yl)-*N*,*N*,*N*′,*N*′-tetramethyluronium hexafluorophosphate) afforded amide **121**. Removal of the benzyl groups in **121** under hydrogenolysis conditions provided the natural product anthrax tetrasaccharide **100** (Scheme 1.23).

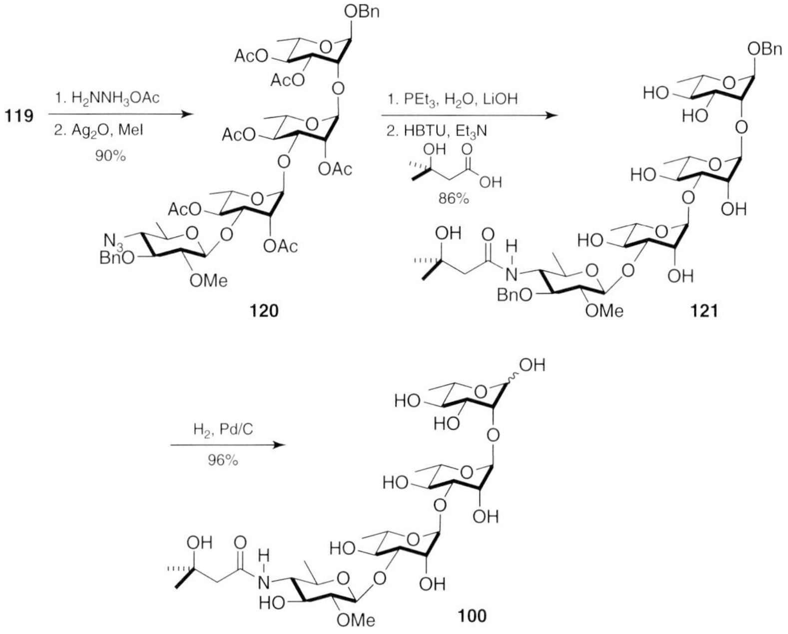

Scheme 1.23 Synthesis of anthrax tetrasaccharide **100**.

1.6
Conclusion and Outlook

For the last 25 years, various groups have been investigating the use of asymmetric catalysis for the synthesis of hexoses. When beginning with achiral starting materials and when asymmetric catalysis is used for the installation of asymmetry, these syntheses are called "*de novo asymmetric*" or "*de novo*" for short. While these *de novo* approaches have been quite impressive in terms of the scope of products prepared and the brevity of steps, they were lacking in terms of application to oligosaccharide synthesis. As part of these efforts, we have developed two orthogonal *de novo* asymmetric approaches to hexoses: an iterative dihydroxylation strategy and an Achmatowicz strategy. Owing to its compatibility with a Pd-catalyzed glycosylation reaction, this later Achmatowicz *de novo* approach to hexoses has seen significant application for the assembly of oligosaccharides.

Of particular note is the flexibility of the *de novo* Achmatowicz route to a myriad of carbohydrate motifs. The approach couples asymmetric catalysis with the Achmatowicz rearrangement for the synthesis of D- and L-pyranones and utilizes highly diastereoselective glycosylation and postglycosylation reactions for the assembly of oligosaccharides. Heavily featured in the glycosylation and postglycosylation reactions was the Pd-π-allyl-catalyzed allylic substitution reaction for substitution at both the anomeric C-1 and C-4 positions of the hexose. What made these allylic substitution reactions so powerful is the one-step double inversion mechanism of the Pd-catalyzed reaction, which in a highly stereospecific manner reliably provides products with net retention of stereochemistry.

Whether these approaches were used in linear and/or bidirectional manner, highly efficient syntheses of natural and unnatural mono-, di-, and oligosaccharides resulted. The overall efficiency of these approaches was the result of the strategic use of the enone functionality in the pyranone as atom-less protecting groups. While the use of atom-less protecting groups has been under-explored in carbohydrate chemistry, the success of the above-mentioned approaches suggest that this strategy should be given more attention from the synthetic organic chemistry community. The overall practicality of these syntheses can be seen in their ability to provide material for medicinal chemistry studies, often in ways that are not readily available by traditional carbohydrate syntheses [51].

1.7 Experimental Section

(*R*)-(+)-1-(2-furyl)ethanol: To a 250 ml flask, 2-acetylfuran (22.0 g, 0.20 mol), CH_2Cl_2 (100 ml), formic acid/triethylamine (1 : 1, 108 ml), and Noyori asymmetric transfer hydrogenation catalyst (*R*)-Ru(η^6-mesitylene)-(*R*,*R*)-TsDPEN (585 mg, 0.95 mmol) were added. The resulting solution was stirred at room temperature for 24 h. The reaction mixture was diluted with water (150 ml) and extracted with Et_2O (3 × 150 ml). The combined organic layers were washed with saturated aqueous $NaHCO_3$, dried over Na_2SO_4, and concentrated under reduced pressure. The crude product was purified by silica gel flash chromatography and eluted with 20% Et_2O/hexanes to give *(R)-(+)-1-(2-furyl)ethanol* (21.5 g, 0.19 mmol, 96%) as a clear liquid.

6-Hydroxy-2-methyl-(2*R*)-2*H*-pyran-3(6*H*)-one: *(R)-(+)-1-(2-furyl)ethanol* (8.9 g, 79.5 mmol): Tetrahydrofuran (THF) (100 ml) and H_2O (25 ml) were added to a round bottom flask and cooled to 0 °C. Solid $NaHCO_3$ (13.4 g, 159 mmol), $NaOAc{\cdot}3H_2O$ (10.8 g, 79.5 mmol) and *N*-bromosuccinimide

(NBS) (14.9 g, 83.5 mmol) were added to the solution and the mixture was stirred at 0 °C for 1 h. The reaction was quenched with saturated aqueous $NaHCO_3$ (200 ml), extracted with Et_2O (3 × 200 ml), dried over Na_2SO_4, and concentrated under reduced pressure. The crude product was purified by silica gel flash chromatography and eluted with 35% EtOAc/hexanes to give *6-hydroxy-2-methyl-(2R)-2H-pyran-3(6H)-one* (9.67 g, 75.5 mmol, 95%, α:β = 2.6 : 1) as a clear liquid.

Carbonic acid, (2*R*,6*R*)-5,6-dihydro-6-methyl-5-oxo-2*H*-pyran-2-yl-1,1-dimethylethyl ester (*ent*-44): 6-Hydroxy-2-methyl-(2*R*)-2*H*-pyran-3(6*H*)-one (3.70 g, 28.9 mmol) was dissolved in CH_2Cl_2 (15 ml) and the solution was cooled to −78 °C. A CH_2Cl_2 (15 ml) solution of $(Boc)_2O$ (9.46 g, 43.4 mmol), and a catalytic amount of dimethylaminopyridine (DMAP) (350 mg, 2.89 mmol) was added to the reaction mixture. The reaction was stirred at −78 °C for 5 h. The reaction was quenched with saturated aqueous $NaHCO_3$ (50 ml), extracted with Et_2O (3 × 50 ml), dried over Na_2SO_4, and concentrated under reduced pressure. The crude product was purified by silica gel flash chromatography and eluted with 7% EtOAc/hexanes to give two diastereomers of *tert*-butyl((6*R*)-6-methyl-5-oxo-5,6-dihydro-2*H*-pyran-2-yl) carbonate (5.72 g, 25.1 mmol, 87%) in 3 : 1 (α:β) ratio *ent*-**44**.

(2*R*,6*S*)-2*H*-6-[(4-Methoxyphenyl)methoxy]-2-methyl-pyran-3(6*H*)-one (101): To a solution of Boc-protected pyranone, *ent*-**44** (13 g, 57.0 mmol), and *para*-methoxy benzyl alcohol (157.3 g, 114.0 mmol) in dry CH_2Cl_2 (57 ml), $Pd_2(dba)_3{\cdot}CHCl_3$ (294 mg, 1 mol% Pd) and PPh_3 (297 mg, 2.0 mol%) at 0 °C were added under argon atmosphere. After stirring for 2 h, the solution was warmed to room temperature, the reaction mixture was quenched with 300 ml of saturated $NaHCO_3$, extracted (3 × 300 ml) with Et_2O, dried (Na_2SO_4), and concentrated under reduced pressure. The crude product was purified using silica gel flash chromatography and eluted with 5% EtOAc/hexane to give PMB ether **101** (13.6 g, 54.7 mmol, 96%) as a colorless oil.

(2*R*,3*S*,6*S*)-2*H*-3,6-Dihydro-6-[(4-methoxyphenyl)methoxy]-2-methyl-pyran-3-ol (102): A solution of pyranone **101** (13.5 g, 54.4 mmol) in dry CH_2Cl_2 (54.4 ml) and 0.4 M $CeCl_3$/MeOH (54.4 ml) was cooled to −78 °C. $NaBH_4$ (2.08 g, 55.5 mmol) was added, and the reaction mixture was stirred for 4 h at −78 °C. The resulting solution was diluted with Et_2O (400 ml) and was quenched with 200 ml of saturated $NaHCO_3$, extracted (3 × 400 ml) with Et_2O, dried (Na_2SO_4), and concentrated under reduced pressure. The crude product was purified using silica gel chromatography and eluted with 40% EtOAc/hexane to give 12.6 g (50.6 mmol, 93%) of allylic alcohol **102** as a white solid.

Carbonic acid, (2*R*,3*S*,6*S*)-3,6-Dihydro-6-[(4-methoxyphenyl)methoxy]-2-methyl-2*H*-pyran-3-yl methyl ester (103): To a stirred solution of allylic alcohol **102** (20 g, 80 mmol), pyridine (38.8 ml, 480 mmol), and DMAP (1.96 g) in dry CH_2Cl_2 (400 ml), methyl chloroformate was added dropwise (33.9 ml, 480 mmol) at 0 °C. After reacting for 1 h at 0 °C, water (300 ml) was added and the reacted mixture was extracted with CH_2Cl_2 (3 × 400 ml), dried (Na_2SO_4), and concentrated under reduced pressure. The crude product was purified using silica gel flash chromatography and eluted with 10% EtOAc/hexane to give 24.4 g (79.2 mmol, 99%) of carbonate **103** as colorless oil.

(2*R*,3*S*,6*S*)-2*H*-3-Azido-3,6-dihydro-6-[(4-methoxyphenyl)methoxy]-2-methyl-pyran (104): To a stirred solution of carbonate **103** (30 g, 97.4 mmol), allylpalladium chloride dimer (378 mg, 1.0 mmol%), and 1,4-bis(diphenylphosphino)butane (1.68 g, 4.0 mmol%) in dry THF (97.2 ml) $TMSN_3$ (15.5 ml, 116.9 mmol) was added under argon atmosphere. The reaction mixture was stirred at room temperature for 0.5 h, the solvent was evaporated under reduced pressure and purified using silica gel flash chromatography, and eluted with 7% EtOAc/hexane to obtain 24.9 g (90.6 mmol, 93%) allylic azide **104** as colorless oil.

(4-Methoxyphenyl)methyl-4-azido-4,6-dideoxy-α-D-mannopyranoside (105): To a mixture of *t*-butanol, acetone (145.4 ml, 1 : 1 (v/v), 1 M) and solution of allylic azide **104** (20 g, 72.7 mmol) at 0 °C, a solution of *N*-methyl morpholine *N*-oxide/water (50% w/v, 50 ml) was added. Crystalline OsO_4 (185 mg, 1 mol%) was added and the reaction mixture was allowed to stir for 24 h. The reaction mixture was quenched with 200 ml saturated $Na_2S_2O_3$ solution, extracted with EtOAc (3 × 500 ml), dried with (Na_2SO_4), concentrated under reduced pressure, and then purified using silica gel flash chromatography, eluting with 90% EtOAc/hexane to give diol **105** (22.0 g, 71.2 mmol, 98%).

List of Abbreviations

Ac	Acetyl
AD-mix-α	3 equiv $K_3Fe(CN)_6/K_2CO_3$, 1 equiv $MeSO_2NH_2$, 5% OsO_4, and 6% $(DHQ)_2PHAL$
AD-mix-β	3 equiv $K_3Fe(CN)_6/K_2CO_3$, 1 equiv $MeSO_2NH_2$, 5% OsO_4, and 6% $(DHQD)_2PHAL$
ATCC	American Type Culture Collection
Bn	Benzyl
Boc	*t*-Butoxycarbonyl
Bu	Butyl
Cbz	Carbobenzoxy
ClAc	Chloroacetyl
CTAB	Cetyltrimethylammonium bromide
DCC	Dicyclohexylcarbodiimide
DDQ	2,3-Dichloro-5,6-dicyano-1,4-benzoquinone
DMAP	Dimethylaminopyridine
DMP	Dess–Martin periodinane
dppb	1,4-Bis(diphenylphosphino)butane
Et	Ethyl

HBTU	*O*-(Benzotriazol-1-yl)-*N*,*N*,*N'*,*N'*-tetramethyluronium hexafluorophosphate
Lev	Levulinoyl
NBS	*N*-Bromosuccinimide
NMO	*N*-Methylmorpholine-*N*-oxide
Noyori	(*S*,*S*) (*R*)-Ru(η^6-mesitylene)-(*S*,*S*)-TsDPEN
p-TsOH	*p*-Toluenesulfonic acid
Ph	Phenyl
PMB	*p*-Methoxybenzyl
py	Pyridine
SAR	Structure Activity Relationship
TBAF	Tetrabutylammonium fluoride
TBS	*t*-Butyldimethylsilyl
Tf	Trifluoromethanesulfonyl
THF	Tetrahydrofuran
TMS	Trimethylsilyl
TsDPEN	PhCH(TsN)CH(NH)Ph

Acknowledgments

We are grateful to NIH (GM090259) and NSF (CHE-0749451) for their support of our research programs. MFC also acknowledges the NSF for his fellowship from the NSF-IGERT Nanomedicine Program (DGE-0965843) at Northeastern University.

References

1. (a) Bertozzi, C.R. and Kiessling, L.L. (2001) *Science*, **291**, 2357–2364. (b) Danishefsky, S.J., McClure, K.F., Randolph, J.T., and Ruggeri, R.B. (1993) *Science*, **260**, 1307–1309. (c) Plante, O.J., Palmacci, E.R., and Seeberger, P.H. (2001) *Science*, **291**, 1523–1527. (d) Sears, P. and Wong, C.-H. (2001) *Science*, **291**, 2344–2350. (e) Yamada, H., Harada, T., Miyazaki, H., and Takahashi, T. (1994) *Tetrahedron Lett.*, **35**, 3979–3982. (f) Wang, L.X. (2006) *Curr. Opin. Drug Discovery Dev.*, **9**, 194–202. (g) Ni, J., Song, H., Wang, Y., Stamatos, N., and Wang, L.X. (2006) *Bioconjugate Chem.*, **17**, 493–500. (h) Zeng, Y., Wang, J., Li, B., Hauser, S., Li, H., and Wang, L.X. (2006) *Chem. Eur. J.*, **12**, 3355–3364. (i) Wang, J., Le, N., Heredia, A., Song, H., Redfield, R., and Wang, L.X. (2005) *Org. Biomol. Chem.*, **3**, 1781–1786.
2. For reviews of other approaches to hexoses, see: (a) Gijsen, H.J.M., Qiao, L., Fitz, W., and Wong, C.-H. (1996) *Chem. Rev.*, **96**, 443–473. (b) Hudlicky, T., Entwistle, D.A., Pitzer, K.K., and Thorpe, A.J. (1996) *Chem. Rev.*, **96**, 1195–1220. (c) Yu, X. and O'Doherty, G.A. (2008) *ACS Symp. Ser.*, **990**, 3–28.
3. (a) Timmer, M.S.M., Adibekian, A., and Seeberger, P.H. (2005) *Angew. Chem. Int. Ed.*, **44**, 7605–7607. (b) Adibekian, A., Bindschädler, P., Timmer, M.S.M., Noti, C., Schützenmeister, N., and Seeberger, P.H. (2007) *Chem. Eur. J.*, **13**, 4510–4527. (c) Stallforth, P., Adibekian, A., and Seeberger, P.H. (2008) *Org. Lett.*, **10**, 1573–1576.
4. (a) Bouché, L. and Reißig, H.-U. (2012) *Pure Appl. Chem.*, **84**, 23–36.

(b) Pfrengle, F. and Reißig, H.-U. (2010) *Chem. Soc. Rev.*, **39**, 549–557.

5. (a) Ko, S.Y., Lee, A.W.M., Masamune, S., Reed, L.A. III,, Sharpless, K.B., and Walker, F.J. (1983) *Science*, **220**, 949–951. (b) Ko, S.Y., Lee, A.W.M., Masamune, S., Reed, L.A. III,, Sharpless, K.B., and Walker, F.J. (1990) *Tetrahedron*, **46**, 245–264.
6. (a) Danishefsky, S.J. (1989) *Chemtracts*, 273–297. For improved catalysis see: (b) Schaus, S.E., Branalt, J., and Jacobsen, E.N. (1998) *J. Org. Chem.*, **63**, 403–405.
7. (a) Johnson, C.R., Golebiowski, A., Steensma, D.H., and Scialdone, M.A. (1993) *J. Org. Chem.*, **58**, 7185–7194. (b) Hudlicky, T., Pitzer, K.K., Stabile, M.R., Thorpe, A.J., and Whited, G.M. (1996) *J. Org. Chem.*, **61**, 4151–4153.
8. Henderson, I., Sharpless, K.B., and Wong, C.H. (1994) *J. Am. Chem. Soc.*, **116**, 558–561.
9. For examples of protecting group free synthesis, see: (a) Crabtree, R.H. (2007) *Science*, **318**, 756. (b) Chen, M.S. and White, M.C. (2007) *Science*, **318**, 783.
10. (a) Ahmed, M.M. and O'Doherty, G.A. (2005) *J. Org. Chem.*, **67**, 10576–10578. (b) Ahmed, M.M. and O'Doherty, G.A. (2005) *Tetrahedron Lett.*, **46**, 4151–4155. (c) Gao, D. and O'Doherty, G.A. (2005) *Org. Lett.*, **7**, 1069–1072. (d) Zhang, Y. and O'Doherty, G.A. (2005) *Tetrahedron*, **61**, 6337–6351. (e) Ahmed, M.M. and O'Doherty, G.A. (2005) *Tetrahedron Lett.*, **46**, 3015–3019. (f) Ahmed, M.M., Berry, B.P., Hunter, T.J., Tomcik, D.J., and O'Doherty, G.A. (2005) *Org. Lett.*, **7**, 745–748.
11. (a) Harris, J.M., Keranen, M.D., Nguyen, H., Young, V.G., and O'Doherty, G.A. (2000) *Carbohydr. Res.*, **328**, 17–36. (b) Harris, J.M., Keranen, M.D., and O'Doherty, G.A. (1999) *J. Org. Chem.*, **64**, 2982–2983.
12. Babu, R.S. and O'Doherty, G.A. (2003) *J. Am. Chem. Soc.*, **125**, 12406–12407.
13. Concurrent with these studies was the similar discovery by Feringa, Lee: (a) Comely, A.C., Eelkema, R., Minnaard, A.J., and Feringa, B.L. (2003) *J. Am. Chem. Soc.*, **125**, 8714–8715. (b) Kim, H., Men, H., and Lee, C. (2004) *J. Am. Chem. Soc.*, **126**, 1336–1337.
14. The poor reactivity in Pd-catalyzed allylation reaction of alcohols as well as a nice solution to this problem was reported, see: Kim, H. and Lee, C. (2002) *Org. Lett.*, **4**, 4369–4372.
15. For a related Rh system, see: Evans, P.A. and Kennedy, L.J. (2000) *Org. Lett.*, **2**, 2213–2215.
16. We have had great success at the preparation of optically pure furan alcohols from the Noyori reduction of the corresponding acylfuran, see: Li, M., Scott, J.G., and O'Doherty, G.A. (2004) *Tetrahedron Lett.*, **45**, 6407–6411.
17. (a) Li, M., Scott, J.G., and O'Doherty, G.A. (2004) *Tetrahedron Lett.*, **45**, 1005–1009. (b) Bushey, M.L., Haukaas, M.H., and O'Doherty, G.A. (1999) *J. Org. Chem.*, **64**, 2984–2985.
18. (a) Wang, H.-Y.L. and O'Doherty, G.A. (2011) *Chem. Commun.*, **47**, 10251–10253. (b) Shan, M., Xing, Y., and O'Doherty, G.A. (2009) *J. Org. Chem.*, **74**, 5961–5966. (c) Zhou, M. and O'Doherty, G.A. (2007) *J. Org. Chem.*, **72**, 2485–2493. (d) Haukaas, M.H. and O'Doherty, G.A. (2002) *Org. Lett.*, **4**, 1771–1774. (e) Haukaas, M.H. and O'Doherty, G.A. (2001) *Org. Lett.*, **3**, 3899–3992.
19. (a) Babu, R.S., Zhou, M., and O'Doherty, G.A. (2004) *J. Am. Chem. Soc.*, **126**, 3428–3429. (b) Babu, R.S. and O'Doherty, G.A. (2005) *J. Carbohydr. Chem.*, **24**, 169–177.
20. (a) Wang, H.-Y.L., Rojanasakul, Y., and O'Doherty, G.A. (2011) *ACS Med. Chem. Lett.*, **2**, 264–269. (b) Wang, H.Y.L., Xin, W., Zhou, M., Stueckle, T.A., Rojanasakul, Y., and O'Doherty, G.A. (2011) *ACS Med. Chem. Lett.*, **2**, 73–78.
21. Babu, R.S., Chen, Q., Kang, S.W., Zhou, M., and O'Doherty, G.A. (2012) *J. Am. Chem. Soc.*, **134**, 11952–11955.
22. Borisova, S.A., Guppi, S.R., Kim, H.J., Wu, B., Liu, H.W., and O'Doherty, G.A. (2010) *Org. Lett.*, **12**, 5150–5153.
23. (a) Hino, M., Nakayama, O., Tsurumi, Y., Adachi, K., Shibata, T., Terano, H., Kohsaka, M., Aoki, H., and Imanaka, H. (1985) *J. Antibiot.*, **38**, 926–935. (b) Patrick, M., Adlard, M.W., and Keshavarz, T. (1993) *Biotechnol. Lett.*, **15**, 997–1000.

24. (a) Colegate, S.M., Dorling, P.R., and Huxtable, C.R. (1979) *Aust. J. Chem.*, **32**, 2257–2264. (b) Colegate, S.M., Dorling, P.R., and Huxtable, C.R. (1985) *Plant Toxicol.*, 249–254.
25. (a) Molyneux, R.J. and James, L.F. (1982) *Science*, **216**, 190–191. (b) Davis, D., Schwarz, P., Hernandez, T., Mitchell, M., Warnock, B., and Elbein, A.D. (1984) *Plant Physiol.*, **76**, 972–975.
26. Liao, Y.F., Lal, A., and Moremen, K.W. (1996) *J. Biol. Chem.*, **271**, 28348–28358.
27. (a) Elbein, A.D., Solf, R., Dorling, P.R., and Vosbeck, K. (1981) *Proc. Natl. Acad. Sci. U.S.A.*, **78**, 7393–7397. (b) Kaushal, G.P., Szumilo, T., Pastuszak, I., and Elbein, A.D. (1990) *Biochemistry*, **29**, 2168–2176. (c) Pastuszak, I., Kaushal, G.P., Wall, K.A., Pan, Y.T., Sturm, A., and Elbein, A.D. (1990) *Glycobiology*, **1**, 71–82.
28. (a) Goss, P.E., Baker, M.A., Carver, J.P., and Dennis, J.W. (1995) *Clin. Cancer Res.*, **1**, 935–944. (b) Das, P.C., Robert, J.D., White, S.L., and Olden, K. (1995) *Oncol. Res.*, **7**, 425–433. (c) Goss, P.E., Reid, C.L., Bailey, D., and Dennis, J.W. (1997) *Clin. Cancer Res.*, **3**, 1077–1086.
29. Davis, B., Bell, A.A., Nash, R.J., Watson, A.A., Griffiths, R.C., Jones, M.G., Smith, C., and Fleet, G.W.J. (1996) *Tetrahedron Lett.*, **37**, 8565–8568.
30. For the first syntheses, see: (a) Mezher, H.A., Hough, L., and Richardson, A.C. (1984) *J. Chem. Soc., Chem. Commun.*, 447–448. (b) Fleet, G.W.J., Gough, M.J., and Smith, P.W. (1984) *Tetrahedron Lett.*, **25**, 1853–1856.
31. For a review of swainsonine syntheses, see: (a) Nemr, A.E. (2000) *Tetrahedron Lett.*, **56**, 8579–8629. For more recent syntheses, see: (b) Martin, R., Murruzzu, C., Pericas, M.A., and Riera, A. (2005) *J. Org. Chem.*, **70**, 2325–2328. (c) Heimgaertner, G., Raatz, D., and Reiser, O. (2005) *Tetrahedron Lett.*, **61**, 643–655. (d) Song, L., Duesler, E.N., and Mariano, P.S. (2004) *J. Org. Chem.*, **69**, 7284–7293. (e) Lindsay, K.B. and Pyne, S.G. (2004) *Aust. J. Chem.*, **57**, 669–672. (f) Pearson, W.H., Ren, Y., and Powers, J.D. (2002) *Heterocycles*, **58**, 421–430. (g) Lindsay, K.B. and Pyne, S.G. (2002) *J. Org. Chem.*, **67**, 7774–7780. (h) Buschmann, N., Rueckert, A., and Blechert, S. (2002) *J. Org. Chem.*, **67**, 4325–4329. (i) Zhao, H., Hans, S., Cheng, X., and Mootoo, D.R. (2001) *J. Org. Chem.*, **66**, 1761–1767. (j) Ceccon, J., Greene, A.E., and Poisson, J.F. (2006) *Org. Lett.*, **8**, 4739–4712. (k) Au, C.W.G. and Pyne, S.G. (2006) *J. Org. Chem.*, **71**, 7097–7099. (l) Dechamps, I., Pardo, D., and Cossy, J. (2007) *ARKIVOC*, **5**, 38–45.
32. (a) Guo, H. and O'Doherty, G.A. (2006) *Org. Lett.*, **8**, 1609–1612. (b) Coral, J.A., Guo, H., Shan, M., and O'Doherty, G.A. (2009) *Heterocycles*, **79**, 521–529. (c) Abrams, J.N., Babu, R.S., Guo, H., Le, D., Le, J., Osbourn, J.M., and O'Doherty, G.A. (2008) *J. Org. Chem.*, **73**, 1935–1940.
33. Walsh, C.T. (2000) *Nature*, **406**, 775–781.
34. He, H., Williamson, R.T., Shen, B., Grazaini, E.I., Yang, H.Y., Sakya, S.M., Petersen, P.J., and Carter, G.T. (2002) *J. Am. Chem. Soc.*, **124**, 9729–9736.
35. (a) Petersen, P.J., Wang, T.Z., Dushin, R.G., and Bradford, P.A. (2004) *Antimicrob. Agents Chemother.*, **48**, 739–746. (b) Sum, P.E., How, D., Torres, N., Petersen, P.J., Lenoy, E.B., Weiss, W.J., and Mansour, T.S. (2003) *Bioorg. Med. Chem. Lett.*, **13**, 1151–1155. (c) He, H., Shen, B., Petersen, P.J., Weiss, W.J., Yang, H.Y., Wang, T.-Z., Dushin, R.G., Koehn, F.E., and Carter, G.T. (2004) *Bioorg. Med. Chem. Lett.*, **14**, 279–282.
36. Babu, R.S., Guppi, S.R., and O'Doherty, G.A. (2006) *Org. Lett.*, **8**, 1605–1608.
37. Guppi, S.R. and O'Doherty, G.A. (2007) *J. Org. Chem.*, **72**, 4966–4969.
38. Tané, P., Ayafor, J.P., Sondengam, B.L., Lavaud, C., Massiot, G., Connolly, J.D., Rycroft, D.S., and Woods, N. (1988) *Tetrahedron Lett.*, **29**, 1837–1840.
39. Seidel, V., Baileul, F., and Waterman, P.G. (1999) *Phytochemistry*, **52**, 465–472.
40. Hu, J.-F., Garo, E., Hough, G.W., Goering, M.G., O'Neil-Johnson, M., and Eldridge, G.R. (2006) *J. Nat. Prod.*, **69**, 585–590.
41. Wu, B., Li, M., and O'Doherty, G.A. (2010) *Org. Lett.*, **12**, 5466–5469.

42. Zhang, Z., Wang, P., Ding, N., Song, G., and Li, Y. (2007) *Carbohydr. Res.*, **342**, 1159–1168.
43. Cheng, L., Chen, Q., and Du, Y. (2007) *Carbohydr. Res.*, **342**, 1496–1501.
44. Shi, P., Silva, M., Wu, B., Wang, H.Y.L., Akhmedov, N.G., Li, M., Beuning, P., and O'Doherty, G.A. (2012) *ACS Med. Chem. Lett.*, **3**, 1086–1090.
45. Guo, H. and O'Doherty, G.A. (2007) *Angew. Chem. Int. Ed.*, **46**, 5206–5208.
46. Guo, H. and O'Doherty, G.A. (2008) *J. Org. Chem.*, **73**, 5211–5220.
47. (a) Mock, M. and Fouet, A. (2001) *Annu. Rev. Microbiol.*, **55**, 647–671. (b) Sylvestre, P., Couture-Tosi, E., and Mock, M. (2002) *Mol. Microbiol.*, **45**, 169–178.
48. Daubenspeck, J.M., Zeng, H., Chen, P., Dong, S., Steichen, C.T., Krishna, N.R., Pritchard, D.G., and Turnbough, C.L. Jr., (2004) *J. Biol. Chem.*, **279**, 30945–30953.
49. Several approaches to related tri- and pentasaccharides, see: (a) Saksena, R., Adamo, R., and Kovac, P. (2006) *Bioorg. Med. Chem. Lett.*, **16**, 615–617. (b) Saksena, R., Adamo, R., and Kovac, P. (2007) *Bioorg. Med. Chem.*, **15**, 4283–4310. (c) Wang, Y., Liang, X., and Wang, P. (2011) *Tetrahedron Lett.*, **52**, 3912–3915. (d) Tamborrini, M., Bauer, M., Bolz, M., Maho, A., Oberli, M.A., Werz, D.B., Schelling, E., Jakob, Z., Seeberger, P.H., Frey, J., and Pluschke, G. (2011) *J. Bacteriol.*, **193**, 3506–3511. (e) Milhomme, O., Dhenin, S.G.Y., Djedaini-Pilard, F., Moreau, V., and Grandjean, C. (2012) *Carbohydr. Res.*, **356**, 115–131.
50. (a) Werz, D.B. and Seeberger, P.H. (2005) *Angew. Chem. Int. Ed.*, **44**, 6315–6318. (b) Adamo, R., Saksena, R., and Kovac, P. (2005) *Carbohydr. Res.*, **340**, 2579–2582. (c) Adamo, R., Saksena, R., and Kovac, P. (2006) *Helv. Chim. Acta*, **89**, 1075–1089. (d) Mehta, A.S., Saile, E., Zhong, W., Buskas, T., Carlson, R., Kannenberg, E., Reed, Y., Quinn, C.P., and Boons, G.J. (2006) *Chem. Eur. J.*, **12**, 9136–9149. (e) Crich, D. and Vinogradova, O. (2007) *J. Org. Chem.*, **72**, 6513–6520. (f) Werz, D.B., Adibekian, A., and Seeberger, P.H. (2007) *Eur. J. Org. Chem.*, 1976–1982.
51. (a) Hinds, J.W., McKenna, S.B., Sharif, E.U., Wang, H.Y.L., Akhmedov, N.G., and O'Doherty, G.A. (2013) *Chem. Med. Chem.*, **8**, 63–69. (b) Tibrewal, N., Downey, T.E., van Lanen, S.G., Sharif, E.U., O'Doherty, G.A., and Rohr, J. (2012) *J. Am. Chem. Soc.*, **134**, 12402–12405. (c) Sharif, E.U. and O'Doherty, G.A. (2012) *Eur. J. Org. Chem.*, **11**, 2095–2108. (d) Elbaz, H., Stueckle, T.A., Wang, H.-Y.L., O'Doherty, G.A., Lowry, D.T., Sargent, L.M., Wang, L., Dinu, C.Z., and Rojanasakul, Y. (2012) *Toxicol. Appl. Pharmacol.*, **258**, 51–60. (e) Wang, H.-Y.L., Wu, B., Zhang, Q., Rojanasakul, Y., and O'Doherty, G.A. (2011) *ACS Med. Chem. Lett.*, **2**, 259–263.

2
Synthetic Methodologies toward Aldoheptoses and Their Applications to the Synthesis of Biochemical Probes and LPS Fragments

Abdellatif Tikad and Stéphane P. Vincent

2.1
Introduction

The preparation of "higher-carbon sugars" such as heptoses has been investigated for more than a century. At that time carbohydrates were already seen as starting natural molecules to build complex nonnatural chiral compounds. Thus, the construction of heptose skeletons has witnessed all the major (r)evolutions of organic chemistry that occurred during the twentieth century. Beyond the fact that carbohydrates constitute a major class of molecules to implement novel synthetic methodologies, the discovery of natural heptoses stimulated the synthetic chemists to develop robust technologies to synthesize them and prove their structures. For instance, the spicamycin [1], miharamycins [2], and desferrisalmycin [3] families possess a heptose subunit (Figure 2.1). Total syntheses of these natural products have been recently disclosed.

Furthermore, the discovery that the glycero-D-*manno*-heptoses are common constituents of lipopolysaccharides (LPSs) of many pathogenic bacteria, boosted the field by providing novel applications: synthetic heptosides or heptosylated oligosaccharides could be envisioned as antimicrobial agents or synthetic vaccines, respectively.

This chapter intends to provide an overview of both the synthetic strategies to build heptoses, as monosaccharides or as part of oligosaccharides, and their use as biochemical probes. Some of the synthetic methods described here have been reviewed by Oscarson [4] a decade ago, and by Kosma [5] in 2008.

2.2
Methods to Construct the Heptose Skeleton

The specific nomenclature of heptoses is illustrated in Scheme 2.1 by the structures of two biologically relevant bacterial heptoses. 6-*O*-Methyl-D-*glycero*-L-*gluco*-heptoses are found in the repeating unit of the capsular polysaccharide required for virulence of all campylobacters, such as *Campylobacter jenuni* [6]. As mentioned above,

Modern Synthetic Methods in Carbohydrate Chemistry: From Monosaccharides to Complex Glycoconjugates, First Edition. Edited by Daniel B. Werz and Sébastien Vidal.

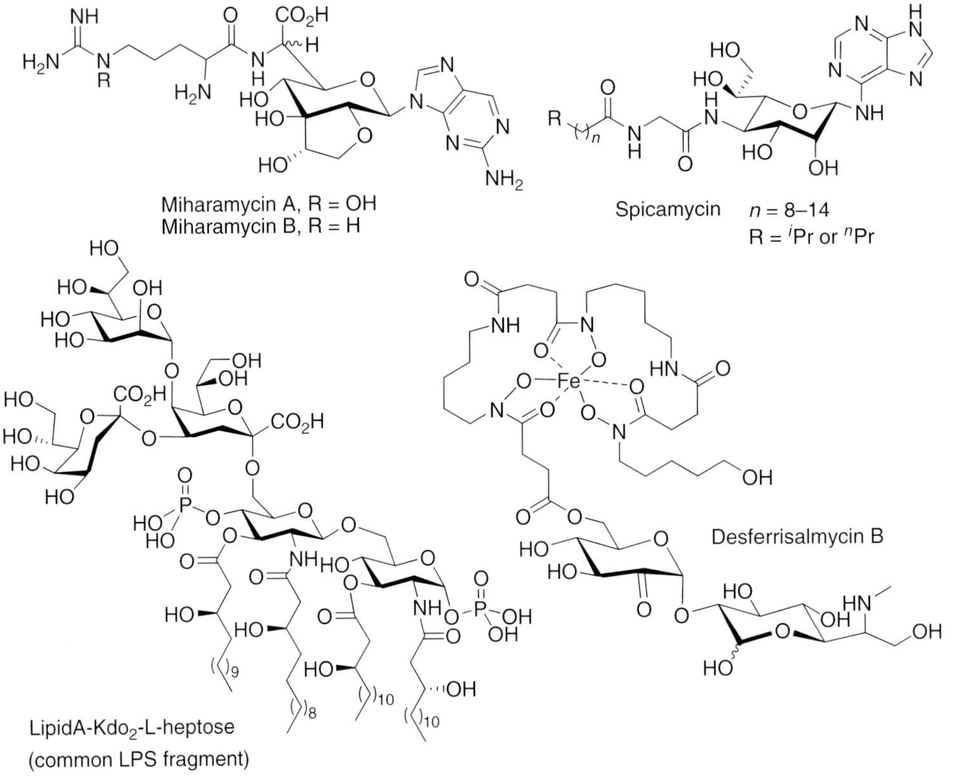

Figure 2.1 Natural molecules displaying heptosidic substructures.

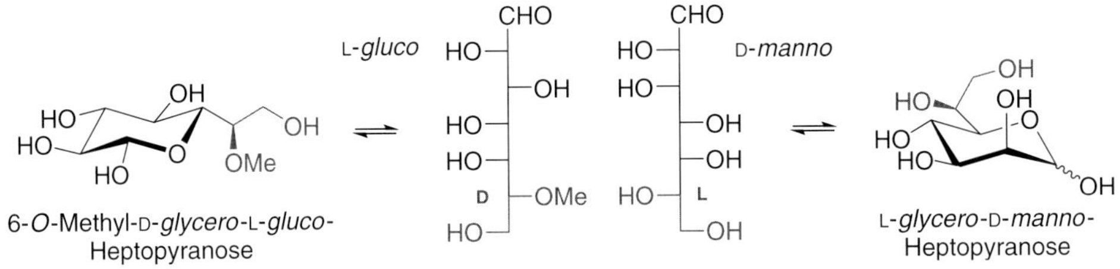

Scheme 2.1 Nomenclature of two typical heptoses.

L-*glycero*-D-*manno*-heptose represented in Scheme 2.1 is a major bacterial carbohydrate: this carbohydrate and its D-*glycero* isomer are very often coined L,D-heptose (L,D-hep) and D,D-heptose (D,D-hep), respectively.

The main synthetic strategies to construct heptose skeletons are illustrated in Scheme 2.2. C_1–C_3 elongations of pentoses or hexoses at the anomeric or at the "last" position are the most classical.

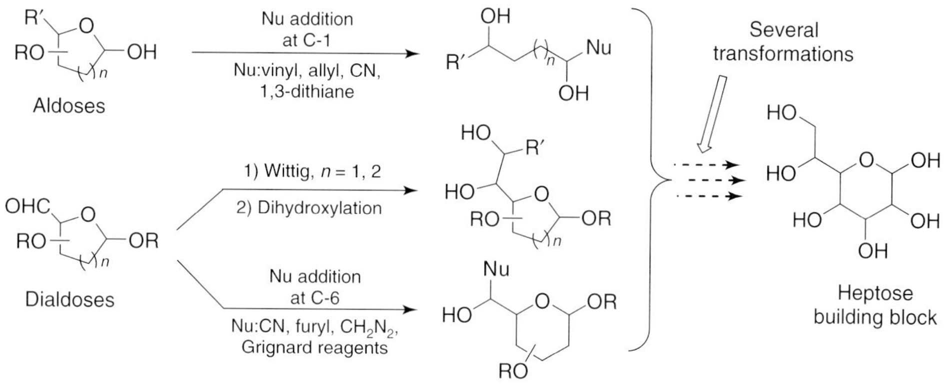

Scheme 2.2 Main synthetic strategies to construct heptoses.

The following paragraphs will attempt to analyze the main features of these synthetic strategies in terms of efficiency and stereoselectivity.

2.2.1 Olefination of Dialdoses Followed by Dihydroxylation

2.2.1.1 Olefination at C-5 Position of Pentodialdoses

Two-carbon homologation of pentoses can be achieved by combining a Wittig reaction with a diastereocontrolled dihydroxylation. For instance, the synthesis of both L- and D-*glycero*-D-*manno*-heptoses was reported by Brimacombe and Kabir [7], thanks to a Wittig olefination of α-D-*lyxo*-1,5-dialdo-furanoside **1** (Scheme 2.3).

In order to confirm that the stereoselectivity followed Kishi's empirical rule for dihydroxylations, the (*Z*)- and (*E*)-allylic alcohols were prepared using (methoxycarbonylmethylene)triphenylphosphorane **2** and formylmethylenetriphenylphosphorane **3**, respectively, followed by reduction with diisobutylaluminum hydride. The catalytic dihydroxylation in the presence of osmium tetroxide and *N*-methylmorpholine *N*-oxide (NMO) afforded the D-*glycero*-isomer **6** from olefin **4** whereas the L-*glycero*-isomer **9** was obtained, as a major product, from the (*E*)-isomer **7**. Finally, the separation by column chromatography or by protection/deprotection sequences provided the pure D-*glycero*-D-*manno*-heptose **10** and L-*glycero*-D-*manno*-heptose **11** (Scheme 2.3).

The same group has shown that the synthesis of D-*glycero*-D-*manno*-heptose **10** can also be elaborated from the (*E*)-allylic alcohol **7** via Sharpless epoxidation followed by hydroxide-mediated epoxide opening (Scheme 2.4). Using diisopropyl L-(+)-tartrate, the titanium-catalyzed epoxidation of **7** afforded as a single crystalline epoxide β-L-*glycero*-D-*manno*-heptofuranoside **12** in 88% yield. This epoxide was transformed into triol **14**, via a preferential alkaline hydrolysis of the terminal epoxide **13**, resulting from an *in situ* Payne rearrangement of **12** (Scheme 2.4) [7a].

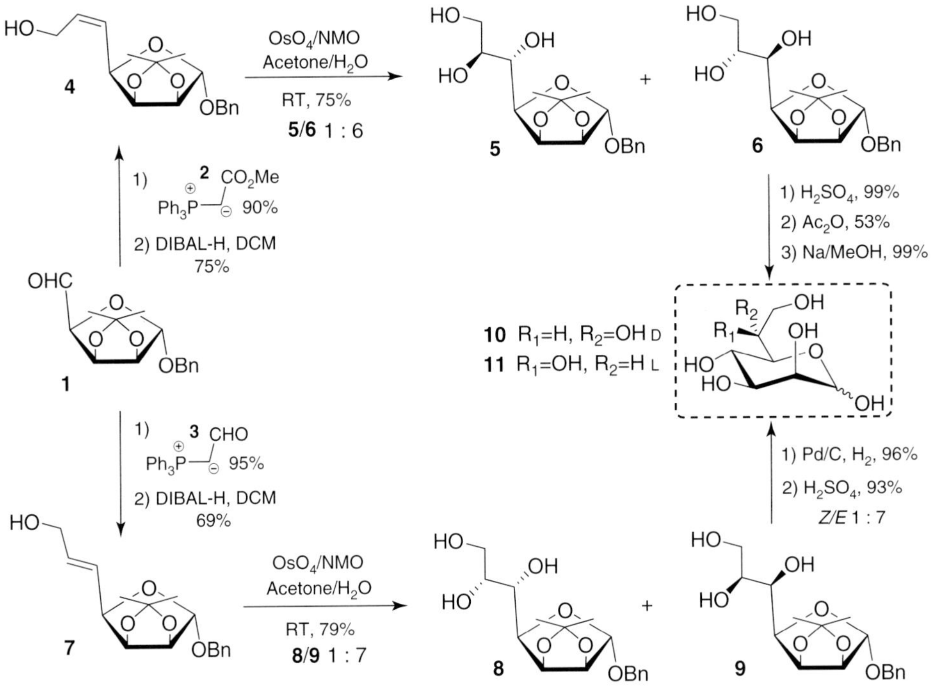

Scheme 2.3 Synthesis of D- and L-heptoses **10** and **11** via Wittig homologation followed by dihydroxylation.

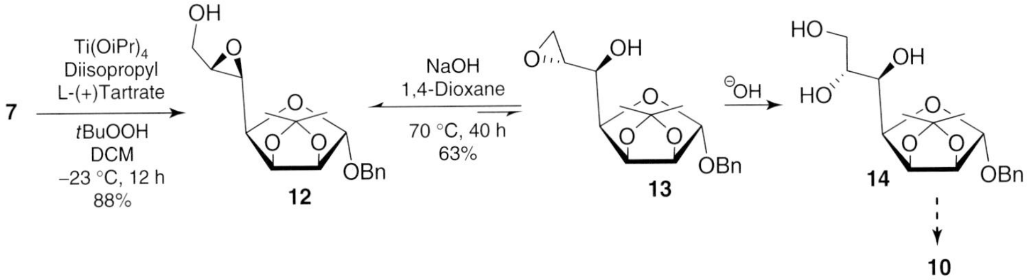

Scheme 2.4 Synthesis of **10** via Sharpless epoxidation followed by epoxide hydrolysis.

Later on, Dohi *et al.* [8] developed an L-selective epoxidation/hydrolysis sequence on a glycoheptopyranose intermediate.

In 2005, Kosma *et al.* [9] employed the same alcohols **4** and **7** as intermediates to prepare D-*glycero*-D-*manno*-heptose-7-phosphate **18** and other heptose derivatives [10]. The phosphorylation of alcohol **4** using the phosphoramidite procedure afforded the 7-*O*-phosphotriester **15** in 65% yield. The catalytic osmylation of the alkene provided a 1 : 4 mixture of **16** and **17** in 71% yield. After separation by chromatography, the catalytic hydrogenation of **17** in the presence of 10% Pd/C resulted in concomitant cleavage of the isopropylidene group, and produced in 81% yield the heptosyl phosphate **18**, the substrate of the bacterial kinases HIDE and HddA (Scheme 2.5).

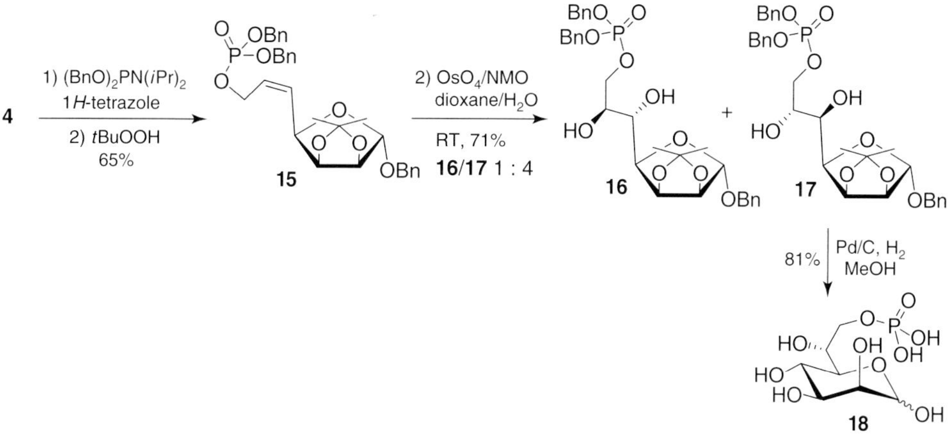

Scheme 2.5 Synthesis of D-*glycero*-D-*manno*-heptose-7-phosphate (**18**).

Despite the reliability of the reaction sequences described above on furanoside **1**, which requires four steps from D-mannose [7], the homologation of hexopyranosides seems to be the most straightforward pathway to give different heptopyranoside derivatives.

2.2.1.2 Olefination at C-1 Position of Hexoses

For the total synthesis of halichondrin B, Kishi and Duan [11] have developed the preparation of heptoside fragment **21** from L-mannonic γ-lactone **19**. In this strategy, the Wittig reaction was performed on an anomeric lactol with methoxymethylenetriphenylphosphorane. Osmylation and acetylation yielded the protected L-*glycero*-L-*manno*-heptose derivative in 58% overall yield after four steps (Scheme 2.6).

2.2.1.3 Olefination at C-6 Position of Hexodialdoses

The olefination at C-6 of a hexodialdose followed by osmylation was reported by Brimacombe and Kabir [7b]. This approach has been extensively exploited in the literature [3, 8, 12]. In order to perform the synthesis of trisaccharides displaying two heptose units, as fragments of the inner core LPS region of *Vibrio parahaemolyticus*, van Boom *et al.* [12a] described the preparation of several heptose donors and acceptors bearing different protecting groups using this method. Thus, the aldehyde generated at C-6 of hexoses was elongated by a Wittig olefination followed by dihydroxylation. For example, this synthetic sequence was applied to the terminal alcohol of **22** to give diol(D) and its C-6 epimer diol(L), as a mixture of diastereoisomers (D/L 8 : 1) in 90% yield (Scheme 2.7). Noteworthy,

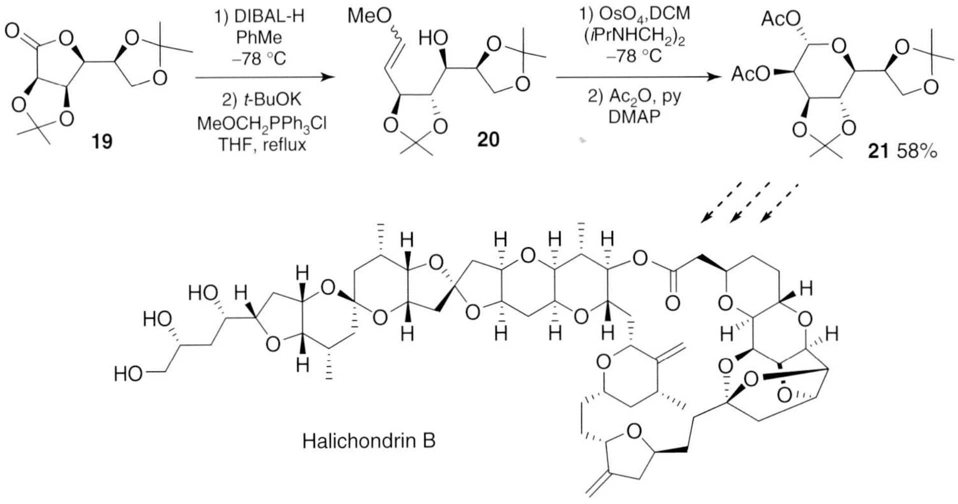

Scheme 2.6 Synthesis of protected L-*glycero*-L-*manno*-heptose derivative **21**.

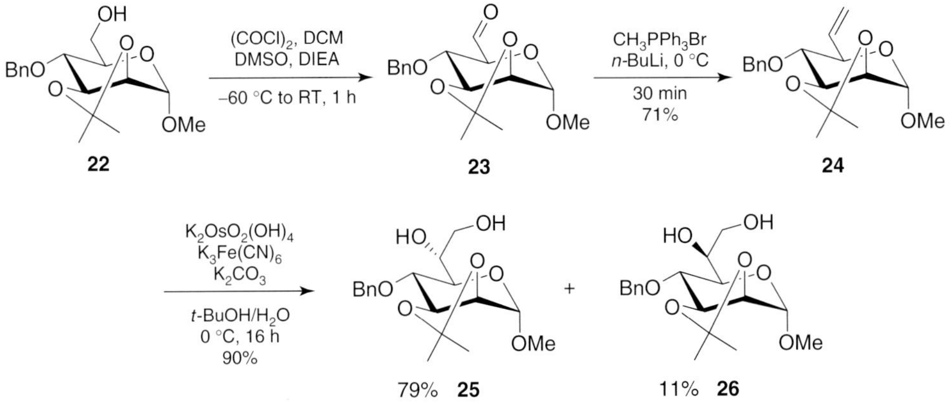

Scheme 2.7 Synthesis of diols (D,L) **25** and **26**.

Sharpless asymmetric dihydroxylation of **24** with AD-mix α or β did not improve the diastereoselectivity.

Later on, asymmetric nucleophilic acylation of various aldehydes was reported by Kirschning's group, using Horner–Wittig olefination followed by dihydroxylation. Thus, they extended this homologation method to D-*galacto*-dialdopyranose, affording both diastereomers of glycero-D-*galacto*-heptose derivatives [13]. Jarosz also described the synthesis of higher carbon sugars following a similar strategy. The subsequent ozonolysis of the resulting-*C*-disaccharide afforded two derivatives of D-*glycero*-D-*gluco*-heptose [14].

2.2.2
Homologation by Nucleophilic Additions

2.2.2.1 Elongation at C-6 of Hexoses

As detailed below, this strategy has been applied by many groups probably because of its robustness and the straightforward access of the direct hexose precursors.

Addition of Hydrogen Cyanide and 2-Methylfuran Additions of hydrogen cyanide or nitromethane to the carbonyl group of an aldose are classical methods for one-carbon elongations [15]. Zamojski [16] has elaborated two complementary methods for the synthesis of heptose derivatives starting from D-*manno*-hexodialdo-1,5-pyranoside **27**. The first one employs cyanuric acid that converts aldehyde **27** into the cyanohydrin **28** (Scheme 2.8). The subsequent reduction gave the corresponding amine, immediately converted with sodium nitrite into a derivative of L-*glycero*-D-*manno*-heptose **29** in modest yield (Scheme 2.8). The second route involved the condensation of **27** with 2-methylfuran that produced a mixture of two diastereoisomers at C-6 position. The ozonolysis of the major (L)-stereoisomer **30**, followed by reduction and then acetylation provided the peracetylated L-*glycero*-D-*manno*-heptose **32** in 11% yield (Scheme 2.8).

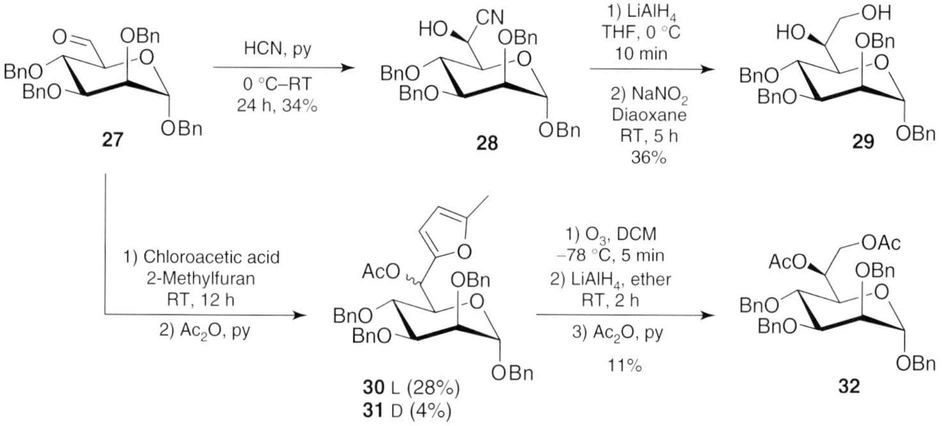

Scheme 2.8 Addition of hydrogen cyanide and 2-methylfuran for hexose elongation.

A similar homologation procedure was developed by Aspinall's group [17] as well as by Dondoni *et al.* [18] using 2-trimethylsilylthiazole as a nucleophile instead of 2-methylfuran. In 2008, Nemoto and coworkers described a general method for the synthesis of α-siloxy esters via one-carbon homologation of various aldehydes, using 2-[[(1,1-dimethylethyl)di-methylsilyl]oxy]-propanedinitrile as a masked acyl cyanide reagent. Thanks to this strategy, two protected D-*galacto*-heptose derivatives were synthesized in good yield and high stereoselectivity in favor of the L-isomer [19].

Addition of Ethynyl and Vinyl Grignard Reagents The conversion of dialdohexoses into heptoses via Grignard additions is very direct and is successfully applied in many instances. Vinyl, alkoxymethylene, as well as α-trisubstituted silylmethylene Grignard reagents are the most widely employed in this approach. Thus, alkoxymethylmagnesium chloride such as methoxy-, allyloxy-, and benzyloxy- directly gave 7-*O*-protected heptoside derivatives [16a]. It must be stressed that these Grignard reagents easily decompose and high yields of reactions depend strongly on the quality of the chloromethyl ethers used. Alternately, α-silylmethyl Grignard reagents are stable and the reactions can be performed under standard conditions [20]. Protecting groups previously installed at the hexose level must be properly selected to be compatible with the Grignard reaction conditions for further transformation into suitable heptosyl acceptors. This approach has been extensively described in the literature [16a, 21]. The one-carbon homologation via a Grignard approach from an aldehyde generated at C-6 of D-mannopyranosides was generally found to afford heptose derivatives in good yields and high stereoselectivity in favor of the L-*glycero*-isomers [4].

In 1972, Horton [22] reported the synthesis of D-*glycero*-D-*galacto*-heptitol (perseitol) and L-*glycero*-D-*galacto*-heptitol via ethynylation of known aldehyde **33** with ethynylmagnesium bromide to afford acetylenes **34** and **35** in 73% yield as a 2 : 3 ratio of epimers (Scheme 2.9). These propargyl alcohols were reduced to the corresponding alkenes by lithium aluminum hydride in 75% and 82% yields, respectively. Then, ozonolysis of allylic alcohol **37**, followed by borohydride reduction, provided the diisopropylidene acetal (D), which was deacetonated and the resulting heptose reduced with $NaBH_4$ to give the natural D-heptitol (perseitol, **39**) in 62% overall yield. The same sequence was repeated from **36** and led to the corresponding D-*glycero*-D-*galacto*-heptitol **38** in 55% overall yield (Scheme 2.9) [22b, 23].

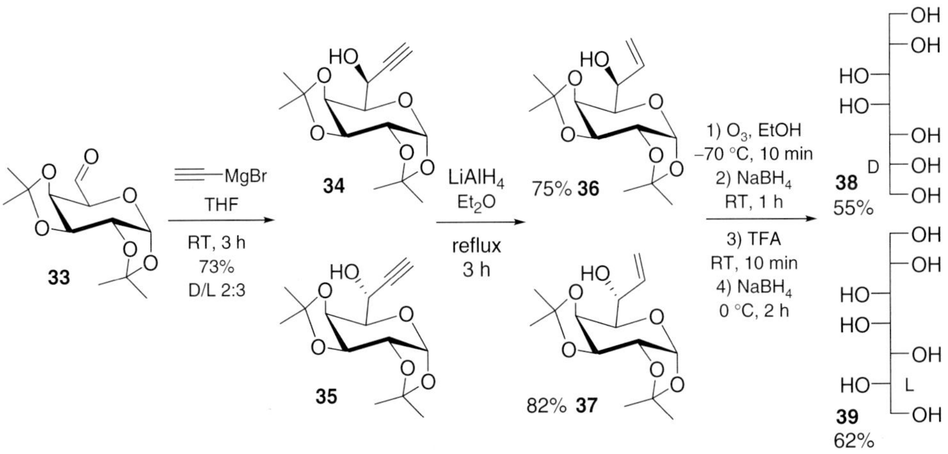

Scheme 2.9 Synthesis of L- and D-*glycero*-D-*galacto*-heptitol.

Chapleur *et al.* [24] described a concise synthesis of L-*glycero*-D-*manno*-heptose from readily available aldehyde **23** previously prepared from alcohol **22** (Scheme 2.7). The condensation of aldehyde **23** with commercial vinylmagnesium bromide afforded a single alcohol **40** in 83% yield (Scheme 2.10). The benzylation of allylic alcohol **40** followed by oxidative cleavage of the double bond and reductive workup gave alcohol **41** which was further elaborated into L-*glycero*-D-*manno*-heptose **11** in 38% overall yield, after deprotections.

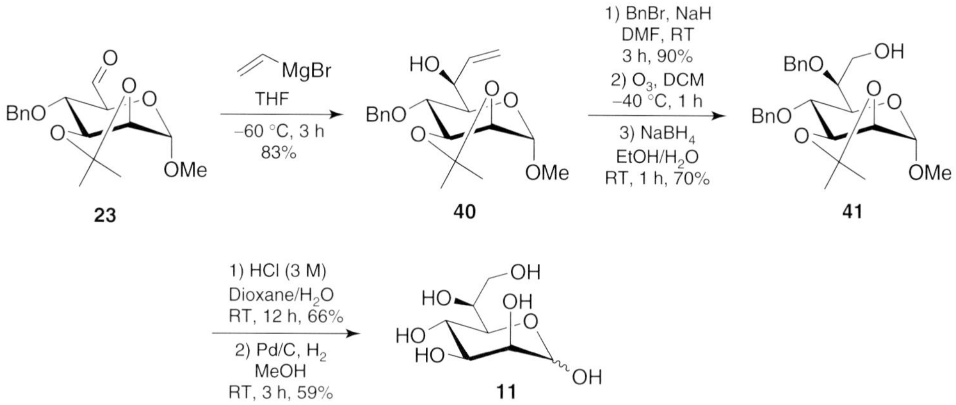

Scheme 2.10 Stereoselective synthesis of L-*glycero*-D-*manno*-heptose **11**.

This stereoselective vinylation at C-6 is achieved without the use of additives, and can be rationalized by the Cram chelate model [25]. In this model, the endocyclic oxygen atom and the oxygen of the aldehyde are chelated with magnesium as shown in Scheme 2.11. Then, the attack of the Grignard reagent occurs from the less hindered face. It should be noted that the diastereoselectivity of the Grignard addition is always high. As expected, the diastereomeric excess of the reaction depends on the structure of the Grignard reagent itself as well as the carbohydrate and its protecting group pattern [4, 16a, 21a, 23b, 26].

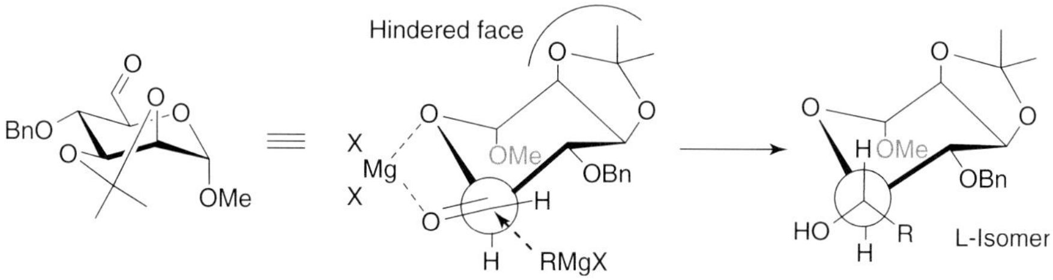

Scheme 2.11 Cram chelate model for the Grignard addition.

In 2011, Yamasaki *et al.* [27] adopted this vinylation approach for the preparation of methyl-L-*glycero*-D-*manno*-heptopyranoside and α-lactosyl-(1 → 3)-L-*glycero*-α-D-*manno*-heptopyranoside, a partial trisaccharide structure found in the lipooligosaccharide produced by *Neisseria gonorrhoeae* strain 15253. Moreover, following the same Grignard approach, Kiso [28] reported the first synthesis of the oligosaccharide skeleton of the *C. jejuni* lipooligosaccharide composed of a branched trisaccharide core structure.

Addition of Allyloxy- and Benzyloxy-Methyl Grignard Reagents In addition to vinyl and ethynyl Grignard reagents, the allyloxy- and benzyloxy-methylmagnesium chloride were developed by Zamojski [21h,i, 29] as reagents leading directly to the 7-*O*-protected heptoses. Thus, condensation of **27** with benzyloxy-methylmagnesium chloride afforded 65% of an inseparable mixture **42** in a 1 : 3 (D/L) ratio (Scheme 2.12). Interestingly, the use of allyloxymethylmagnesium chloride neither affected the diastereoselectivity nor the reaction yield, but gave a mixture of **43** and **44** readily separable by flash chromatography (Scheme 2.12). The Pd-/C-catalyzed deallylation of **43** and **44** provided the respective D-*glycero*-D-*manno*-heptose **10** and L-*glycero*-D-*manno*-heptose **11** (Scheme 2.12) [16a, 30].

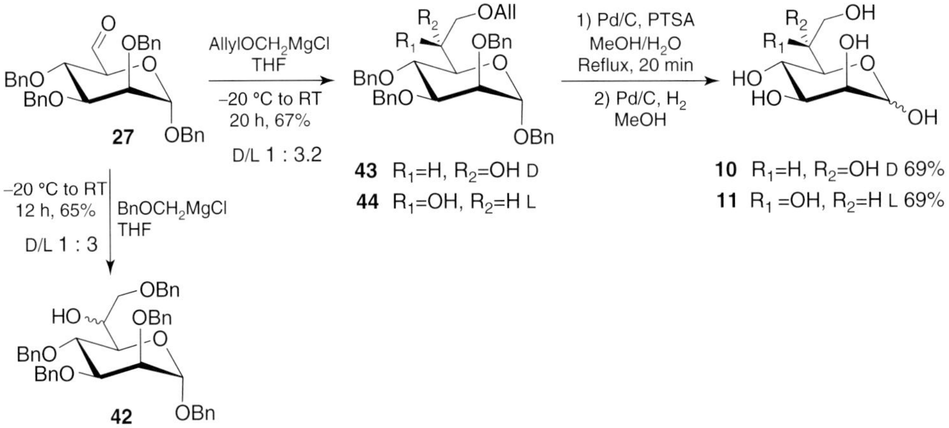

Scheme 2.12 Addition of allyloxy- and benzyloxy-methyl Grignard reagents.

The same authors applied this strategy to synthesize five L-*glycero*-D-*manno*-heptose acceptors displaying a single free hydroxyl group in C-2, C-3, C-4, C-6, and C-7 positions, after several protection–deprotection sequences [21a]. On the basis of the same strategy, the synthesis of a common tetrasaccharide core structure of *Haemophilus influenzae* LPSs was described by Oscarson's group [31]. In the latter study, the elongation was improved by performing the reaction under Barbier conditions. Starting from thioglycoside **45**, this reaction gave heptoside **46** in 59% yield as an inseparable mixture of L/D epimers (ratio 8 : 1). To resolve this problem (also observed by Zamojski [16a]), the 6-position of **46** was therefore trimethylsilylated to afford the readily separable L- and D-*glycero*-isomers, **47** and

48 in 83% and 8% yields, respectively. The syntheses of different heptose donors and acceptors were thus accomplished using a 1,6-anhydro-L-*glycero*-β-D-*manno*-heptopyranose intermediate **49**, which was effectively obtained in 85% yield through an internal glycosylation of a 6-*O*-trimethylsilylated ethyl thioheptoside **47** using NIS/TfOH (*N*-iodosuccinimide/triflic acid) as a promoter (Scheme 2.13) [31].

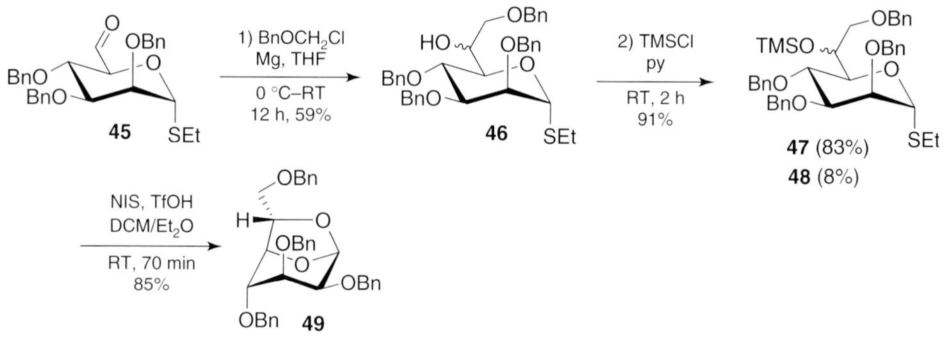

Scheme 2.13 Synthesis of 1,6 anhydro-heptose **49**.

Thanks to this methodology, several monophosphates of L- and D-*glycero*-D-*manno*-heptose **10** and **11** have been synthesized by Zamojski [21b,c] as well as by Oscarson [32] to perform studies with respect to migration and hydrolytic cleavage of the phosphate moiety. Importantly, this synthetic strategy was also adopted to the preparation of challenging oligosaccharides [33].

Addition of α-Silylmethyl Grignard Reagents In the late 1980s, a convenient diastereoselective hydroxymethylation of a suitably protected α-D-*manno*-hexodialdo-1,5-pyranoside with isopropoxydimethyl-, dimethyl(phenylthiomethyl)-, and dimethylphenyl-silylmethylmagnesium chloride was developed by van Boom [12a, 20a, 34]. As an example, treatment of perbenzylated α-D-*manno*-hexodialdo-1,5-pyranoside **50** with the Grignard reagent generated from dimethylphenylsilylmethyl chloride gave diastereoselectively the silane **51**, featuring a masked hydroxy group at C-7 position. Cleavage of the carbon–silicon bond in **51** by a Tamao–Fleming oxidation, in the presence of peracetic acid and potassium bromide, afforded the diol **52** displaying the L-configuration at C-6 (Scheme 2.14) [35].

Using the same protocol, Paulsen [36] reported the first synthesis of ADP D-*glycero*-α-D-*manno*-heptopyranose as well as several analogs containing amidophosphate and phosphodithioate as potential inhibitors of ADP heptose synthetase. Moreover, Oscarson *et al.* [37] disclosed the trisaccharide glycoside containing two L-*glycero*-D-*manno*-heptopyranose units, as a part of the core region of the LPS from *Salmonella*. Overall, among all Grignard reagents presented previously, the dimethylphenylsilylmethyl exhibited the highest L-selectivity.

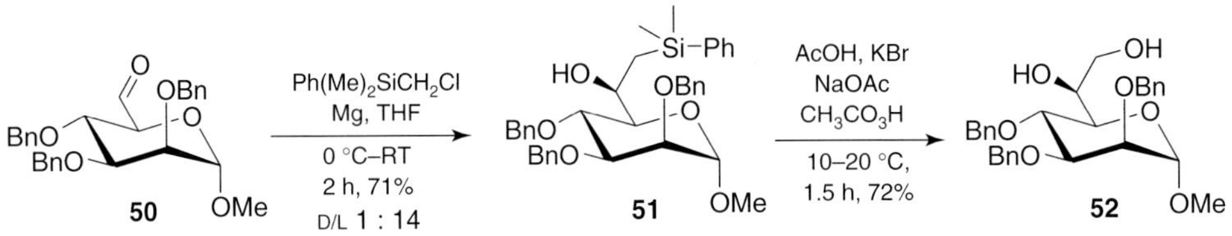

Scheme 2.14 Addition of dimethylphenylsilylmethyl Grignard followed by Tamao–Fleming oxidation.

Addition of Propargyltrimethylsilane A novel stereoselective carbon chain extension reaction at the C-6 position of 1,6-anhydromannose triacetate **53**, prepared from D-mannose in three steps [38], was reported by Nishikawa *et al.* [39]. Compound **54** was prepared by photobromination of **53** according to the Ferrier [40] procedure. The reaction of **54** with propargyltrimethylsilane **55** in the presence of AgOTf gave only one allenylated compound **56** (*exo*-stereochemistry, L) at C-6 position. Thus, ozonolysis of allene **56**, followed by reduction with $NaBH_4$, led to the alcohol **57**, which was efficiently transformed into L-D-Hep*p* hexaacetate **58** after acetylation and subsequent acetolysis (Scheme 2.15).

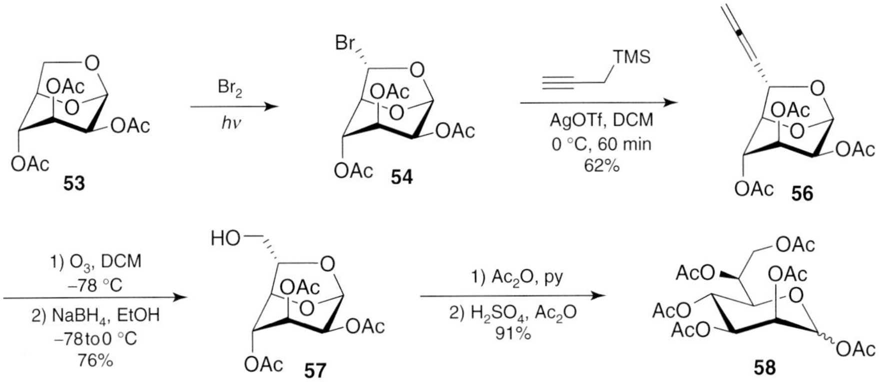

Scheme 2.15 Photochemical bromination/allenylation strategy.

Addition of Diazomethane In 2005, the Tanner group [41] developed a chemoenzymatic synthesis of ADP-D-*glycero*-β-D-*manno*-heptose **64** starting from primary alcohol **59**, readily available in four steps from D-mannose (Scheme 2.16). The primary alcohol was oxidized into acid **60** by potassium dichromate. This acid was activated with oxalyl chloride and subsequently added to a solution of diazomethane in ether to afford the α-diazoketone **61**. Nucleophilic displacement with dibenzylphosphoric acid gave an intermediate α-ketophosphate which was found unstable and decomposed on silica gel. Instead of isolating the compound, a reductive work-up using $NaBH_4$ yielded the corresponding dibenzyl phosphates **62** and **63** in 52% yield after two steps as a 1 : 1 ratio (Scheme 2.16). It should be noted that using $Zn(BH_4)_2$ as the reducing agent gave predominantly the L-isomer.

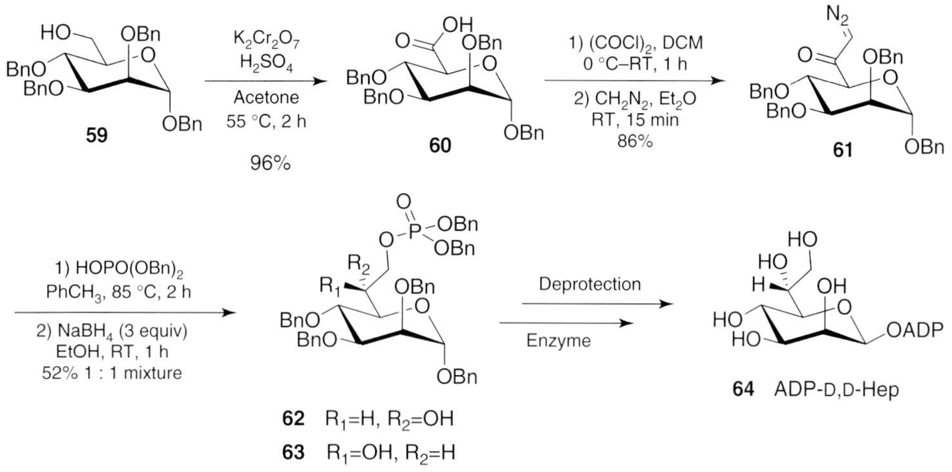

Scheme 2.16 Chemoenzymatic synthesis of ADP-D,D-heptose **64**.

2.2.2.2 Elongation at C-1 Position of Aldose

Addition of Cyanide – The Kiliani Reaction Probably the earliest method for the construction of heptoses consisted of the addition of one-carbon nucleophiles such as nitromethane and cyanide on the anomeric aldehyde of hexoses. Indeed, heptonic acids are readily available from the Kiliani reaction of hexoses as depicted in Scheme 2.17 [42].

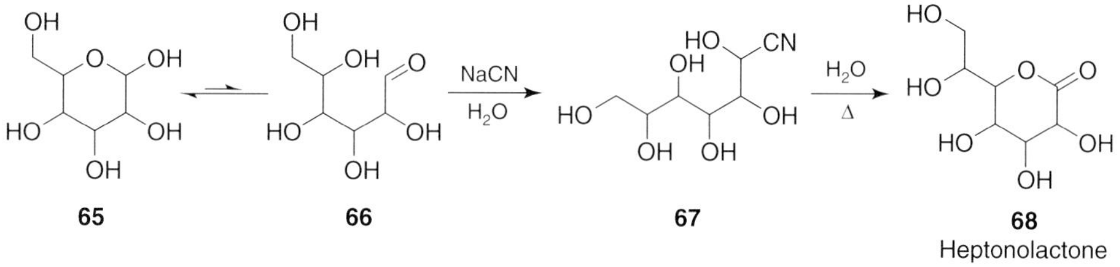

Scheme 2.17 Homologation of pyranoses via the Kiliani reaction.

The Kiliani reaction is amenable on a large scale, leading to γ- or δ-heptonolactones depending on the reaction conditions. The latter can further be protected and derivatized to give a wide range of chiral synthons.

This reaction has been efficiently exploited for the synthesis of eight homonojirimycin (HNJ) stereoisomers, a notorious class of glycosidase inhibitors [42]. It is noteworthy, that both the first steps (diacetonide formation from D-mannose and formation of the heptonolactone) have been performed on a kilogram scale (Scheme 2.18). Interestingly, α-HNJ **73** is an iminoheptitol that was synthesized before it was recognized as a natural product from *Omphalea diandra* [43]. Many of the stereoisomers of HNJ have then been extracted and characterized from a wide range of plants [44].

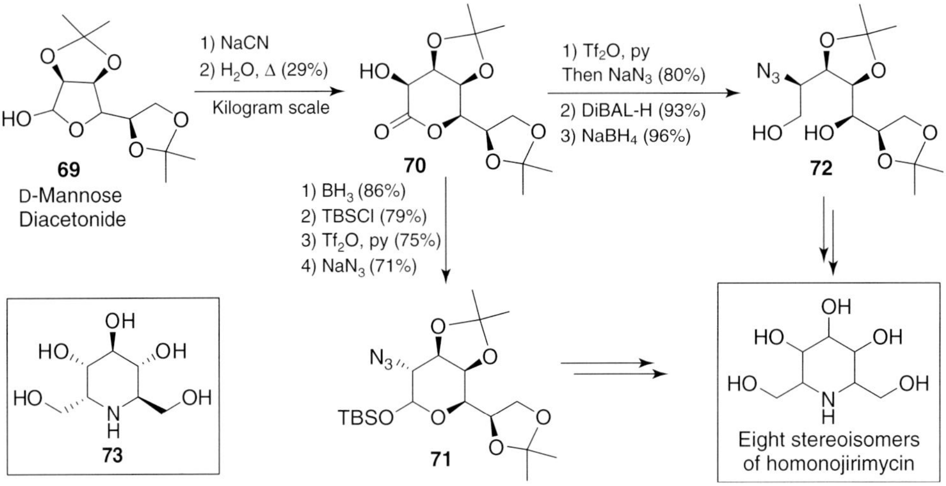

Scheme 2.18 Synthesis of eight homonojirimycins from D-mannose.

Recently, Rezanka *et al.* [45] described the transformation of D-allopyranose to D-*allo*-heptopyranose and D-*altro*-heptopyranose via Kiliani–Fischer reaction. Thus, the first step is to react D-allopyranose **74** with aqueous potassium cyanide to give a mixture of two epimeric cyanohydrins, which were reduced by palladium on barium sulfate under acidic conditions into the corresponding aldehydes that quickly cyclized into β-D-*glycero*-D-*allo*-heptopyranose **77** and D-*glycero*-D-*altro*-heptopyranose **78** in 84% yield after two steps (Scheme 2.19).

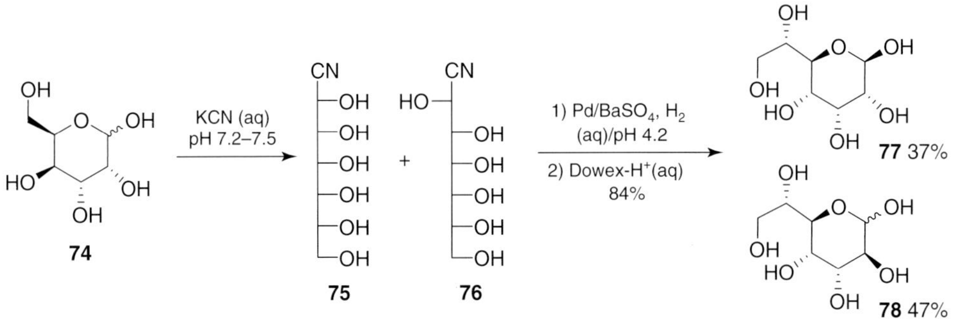

Scheme 2.19 Kiliani–Fischer synthesis of D-*allo*-heptopyranose and D-*altro*-heptopyranose.

Addition of Dithiane Although the most usual homologations were realized from aldehydes at C-6 or C-5 of hexoses and pentoses, Paulsen and coworkers [46] performed the first chain elongation at C-1 position of 2,3 : 5,6-*O*-isopropylidene-D-*manno*-furanose (**79**). Thus, condensation of this lactol with 2-lithio-1,3-dithiane provided a derivative of D-*glycero*-D-*galacto*-heptose **80** in good yield. The diastereoselectivity could be explained by a chelation of the lithium cation with the oxygens at C-1, C-2, and C-4, as depicted in Scheme 2.20. The nucleophilic attack from the less hindered face generates the D-isomer. The hydrolysis of dithioacetal **80** followed by deprotection of isopropylidene led to **81** but the yields and details of this sequence were not mentioned in this study (Scheme 2.20).

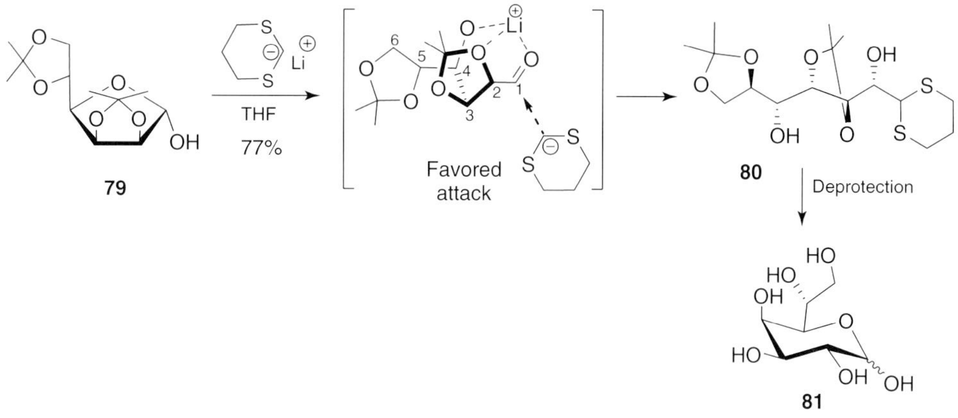

Scheme 2.20 Addition of 2-lithio-1,3-dithiane on lactol **79**.

Addition of Allyl Indium on Unprotected Sugars Schmid and Whitesides [47] described an extension by three carbon atoms of unprotected carbohydrates by a tin- or indium-mediated reaction with allyl bromide. This reaction was exploited in the synthesis of D-*glycero*-D-*galacto*-heptose **81** starting from D-arabinose (Scheme 2.21) [48].

In 2007, Kosma *et al.* [49] have adopted this indium-mediated chain elongation starting from L-lyxose to afford 2-deoxy-L-*galacto*-heptose, which was used as an intermediate in the synthesis of ADP 2-deoxy-L-*galacto*-heptopyranose.

Addition of 3-Bromopropenyl Esters Indium-mediated acyloxyallylation of unprotected D-ribose, D-arabinose, D-lyxose, and D-xylose were reported by Madsen *et al.* [50], in order to synthesize their corresponding heptosides. As an example, the treatment of D-ribose with 3-bromopropenyl acetate in ethanol at 50 °C in the presence of indium metal afforded the peracetylated polyol **85** in 82%. Only two out of the four possible diastereoisomers were observed with a 1.5 : 1 selectivity (Scheme 2.22). Deacetylation of the major D-isomer, isolated by flash chromatography, followed by ozonolysis gave the corresponding D-*glycero*-D-*manno*-heptose **10**.

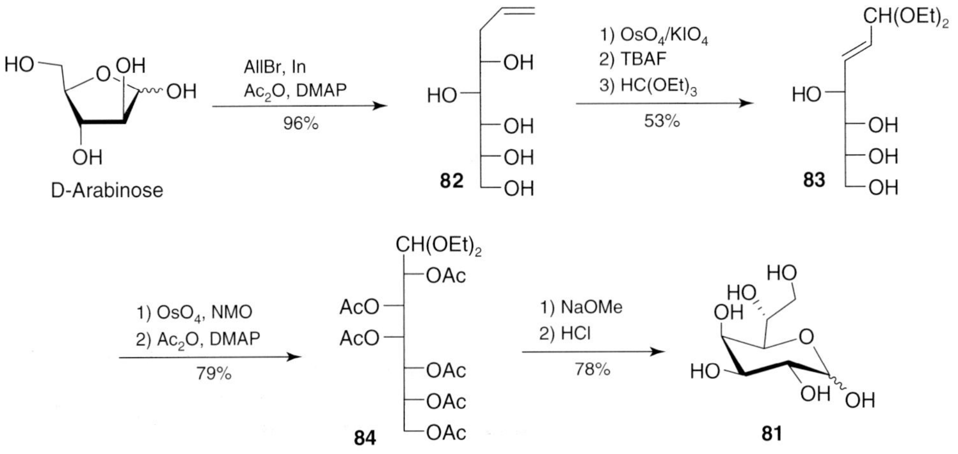

Scheme 2.21 Indium-mediated allylation of unprotected D-arabinose.

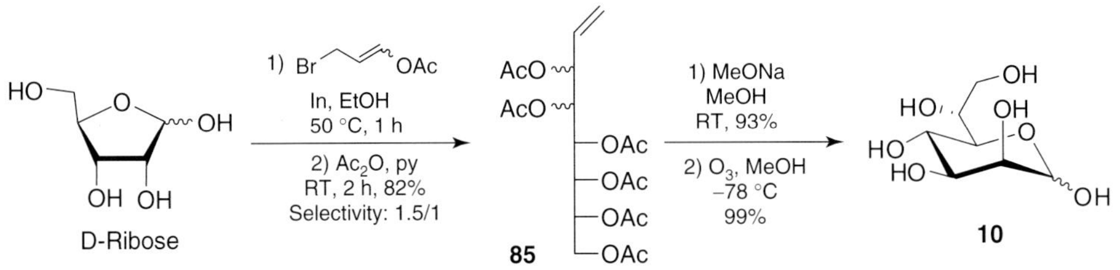

Scheme 2.22 Indium-mediated acyloxyallylation of unprotected D-ribose.

Addition of Divinyl Zinc to an Anomeric Lactol Recently, Lowary *et al.* [6a] described the first synthesis of the 6-*O*-methyl-D-*glycero*-α-L-*gluco*-heptopyranose moiety present in the capsular polysaccharide from *C. jejuni* CTC 11168. The key synthetic step was the two-carbon homologation of galactofuranose **86**, which was obtained from D-galactose via a series of protections and deprotections (Scheme 2.23). The addition of divinylzinc, generated *in situ* from vinylmagnesiumbromide and $ZnBr_2$, to **86** yielded adduct **87** in 91% yield as a single isomer, thanks to chelation control. It was observed that $ZnBr_2$ plays a very important role for the stereoselectivity of the reaction since the treatment of **86** with vinylmagnesium bromide alone afforded **87** with poor selectivity and in reduced yield. Ozonolysis of olefin **87** afforded the heptose derivative **88**, which was converted to the 8-aminooctyl glycoside **89** after several standard steps and then conjugated to bovine serum albumin (BSA) for the generation of antibodies.

2.2.3
Heptose de novo synthesis

In 2010, a short and enantioselective *de novo* synthesis of orthogonally protected L-*glycero*-D-*manno*-heptose building blocks was described by Seeberger and coworkers

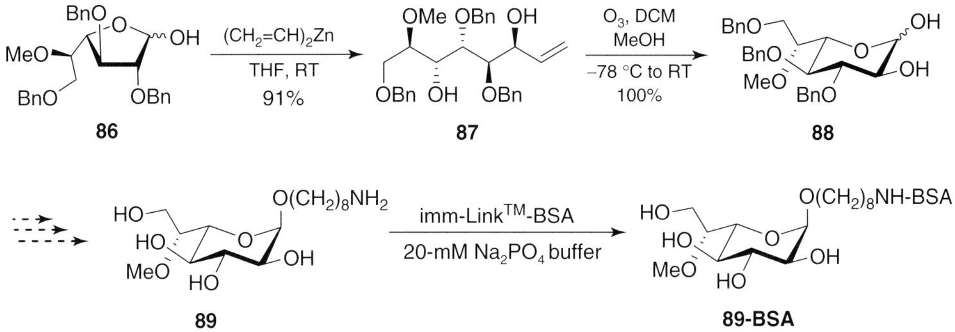

Scheme 2.23 Synthesis of heptose **89** and its bovine serum albumin conjugate **89-BSA**.

[51]. This strategy is based on the enantioselective construction of the seven-carbon skeleton using two aldol reactions (Scheme 2.24). Thus, the L-proline-catalyzed aldol reaction between **90** and **91** followed by silylation of the aldol product provided ketone **92**. Reduction of **92** with L-Selectride resulted in a 1,3-migration of the *tert*-butyldimethylsilyl (TBS) group that yielded the regioisomer **93** in 79%. The latter was converted to the linear aldehyde **94** after several protection/deprotection sequences. A Mukaiyama aldol reaction between **94** and silyl enolether **95** was then performed using $MgBr_2$:OEt_2 as a chelating activator, to afford aldol product **96** in moderate yield but in excellent 2,3-*anti*-3,4-*syn*-selectivity. Lactonization

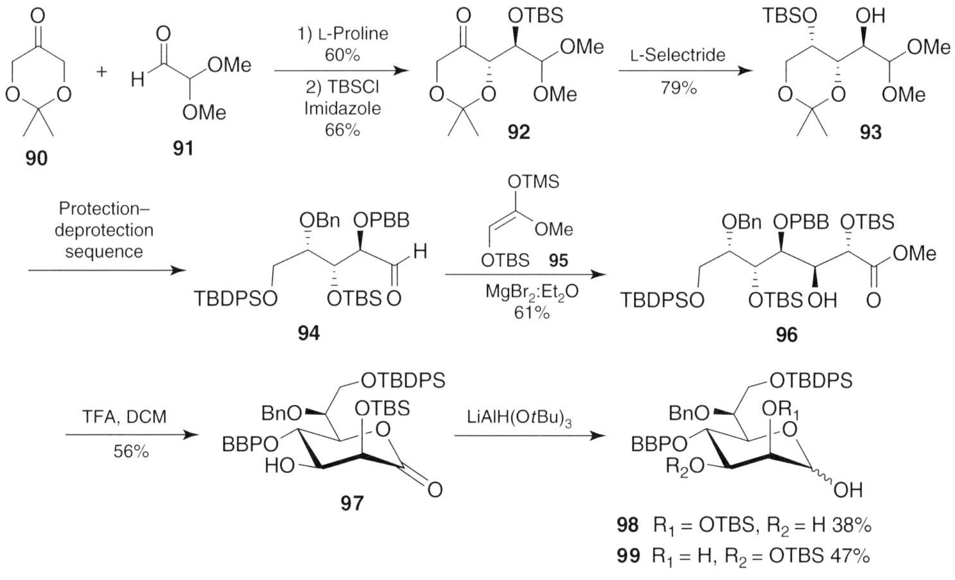

Scheme 2.24 Enantioselective *de novo* synthesis of L-*glycero*-D-*manno*-heptose building block.

of ester **96** with trifluoroacetic acid followed by reduction with lithium tri-*tert*-butoxyaluminum hydride led to both regioisomeric heptosides **98** and **99** resulting from the partial migration of the silyl group from C-2 to C-3 (Scheme 2.24).

Recently, Majewski *et al.* [6b] have applied this *de novo* strategy for the stereodivergent synthesis of D-*glycero*-β-D-*allo*-heptose and L-*glycero*-β-L-*allo*-heptose.

2.3 Synthesis of Heptosylated Oligosaccharides

The first syntheses of heptosylated oligosaccharides were performed in the 1980s, just after the development of the main "modern" glycosidation methods. The groups of Paulsen [52], Zamojski [29], and Garegg [37] were very active in this field: they used anomeric heptosyl trichloroacetimidates and halogenides as the heptosyl donor and various acceptors. Heptosylated oligosaccharides containing up to four sugar units were thus generated [53]. In a general manner, configuration of the newly formed heptosidic linkage can be confirmed based on the coupling constants between C-1 and H-1 [54].

To give an overview of the state of the art of this field, we selected five complementary approaches presented in their chronological order. A more exhaustive presentation has been published by Kosma and Oscarson [4, 5].

2.3.1 Synthesis of the Core Tetrasaccharide of *Neisseria meningitidis* Lipopolysaccharide

Neisseria meningitidis, a gram-negative bacterium, is a leading cause of bacterial meningitis, a severe infection of the meninges, which accounts for 1.2 million cases of the disease worldwide annually [55]. The LPS of *N. meningitidis* does not possess a polymeric O-antigen; thus, in this case, the core is the exposed and immunodominant part of LPS [56]. Well-defined, nontoxic LPS core structures produced by chemical synthesis constitute key tools to fully explore the effectiveness of LPS oligosaccharides as vaccine candidates against pathogenic bacteria and to identify the most immunogenic epitopes [33c, 57].

In 1989, the van Boom [34a] laboratory disclosed the total synthesis of an inner core-dephosphorylated trisaccharide **106** bearing a spacer suitable for bioconjugations (Scheme 2.25). This pioneering synthetic work was designed to generate chemical tools to study the antigenicity of the inner core structure and probe the specificity of the antibodies raised against this antigen. The two L,D-heptose building blocks **100** and **101** were constructed by addition of Grignard reagents on a 6-aldehydo-mannoside (Section 2.2.2), a highly diastereoselective methodology developed in the same laboratory. The two heptosides **100** and **101** were first assembled in 72% using silver triflate as promoter. Imidate **104** was then efficiently grafted to the bis-heptoside **105** in 92%. A remarkable feature of van Boom's approach is the very late transformation of the dimethylphenylsilyl methylene group into a primary alcohol under Tamao–Fleming oxidation conditions. This

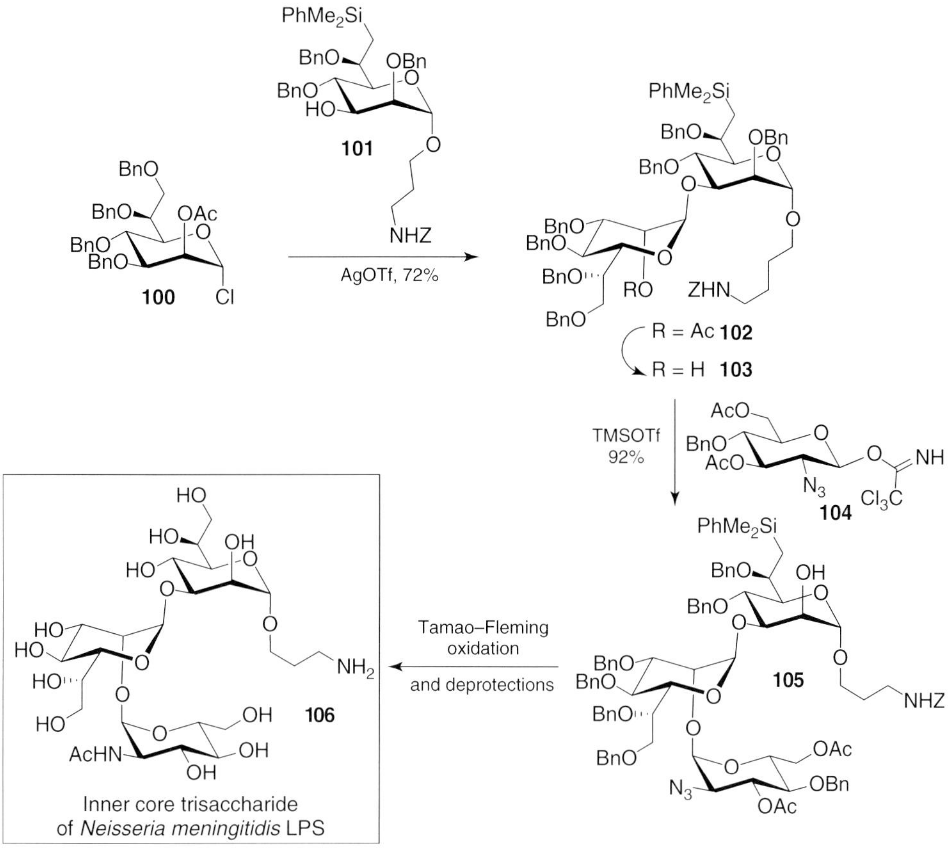

Scheme 2.25 van Boom's synthesis of *N. meningitidis* LPS inner core trisaccharide.

study showed that this group is not only at the origin of excellent L-stereoselectivities but it is also a convenient masking group of the heptose 7-position compatible with glycosidation conditions and protective group manipulations.

2.3.2
Synthesis of a Branched Heptose- and *Kdo*-Containing Common Tetrasaccharide Core Structure of *Haemophilus influenzae* Lipopolysaccharides

In 1998, Oscarson [31] described the syntheses of common tetrasaccharide core structures of *H. influenzae* LPSs. The oligosaccharides were synthesized as glycosides of a bifunctional spacer to allow the subsequent formation of immunogenic glycoconjugates. The syntheses of the 3,4-branched structures were successfully achieved, thanks to a strategic 1,6-anhydro-L-*glycero*-β-D-*manno*-heptopyranose intermediate **107** designed to decrease the steric crowding between the 3- and 4-substituents (Scheme 2.26). This intermediate was effectively prepared from a mannose precursor via a stereoselective one-carbon elongation at C-6 using a

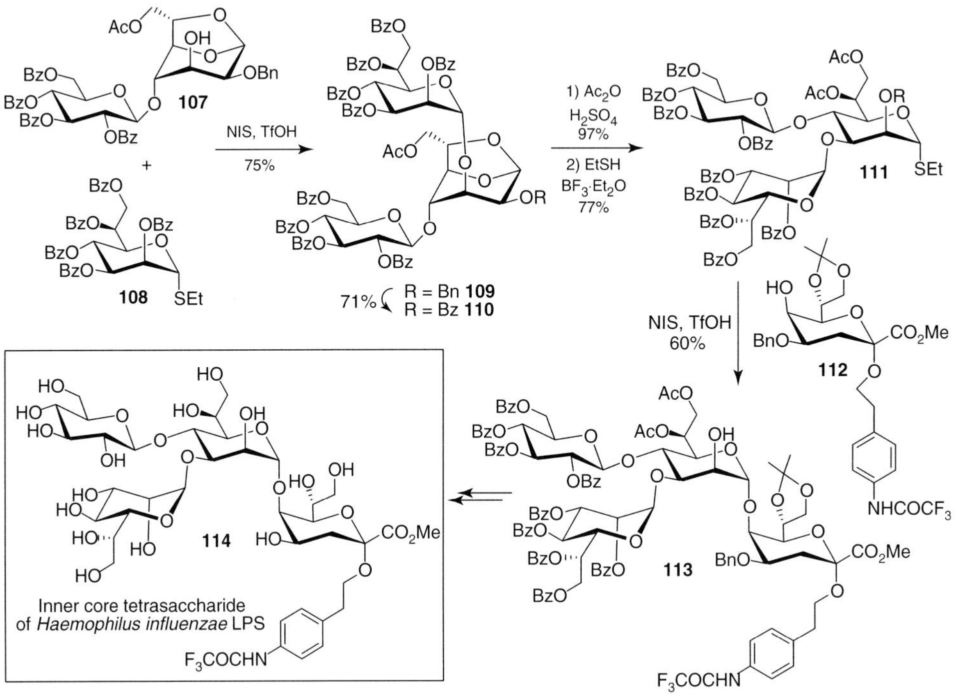

Scheme 2.26 Oscarson's synthesis of *Haemophilus influenzae* LPS inner core tetrasaccharide.

Barbier reaction [31, 33a]. The key anhydro bridge was realized through an internal glycosylation of a 6-*O*-trimethylsilylated ethyl thioheptoside using NIS/TfOH as a promoter. The 3- and 4-substituents were readily introduced into the 1,6-anhydro intermediate by glycosylation reactions using thioglycosides **108** and **111** as donors and NIS/TfOH as a promoter.

Interestingly, this transformation was not possible using acceptors with equatorial 3,4-substituents. Acetolysis of the anhydro bridge followed by conversion into the ethyl thioglycoside afforded a trisaccharide donor, which, in NIS/TfOH-promoted couplings to the spacer and to a Kdo acceptor followed by deprotection, efficiently gave the target compounds.

2.3.3
Synthesis of the Core Tetrasaccharide of *Neisseria gonorrhoeae* Lipopolysaccharide

In 2004, the Yamasaki group synthesized a tetrasaccharide containing a 3,4-dibranched L-*glycero*-D-*manno*-heptose (Hep), β-lactosyl-(1 → 4)-[L-α-D-Hep-(1 → 3)]-L-α-D-Hep **121**, using a central mannoside **116** derivative as an acceptor, and the L,D-heptose imidate [52f] as a donor [27c] (Scheme 2.27). After lactosylation and protective group manipulations, the central mannose residue was eventually homologated by a Swern oxidation, Grignard reaction, and oxidative cleavage

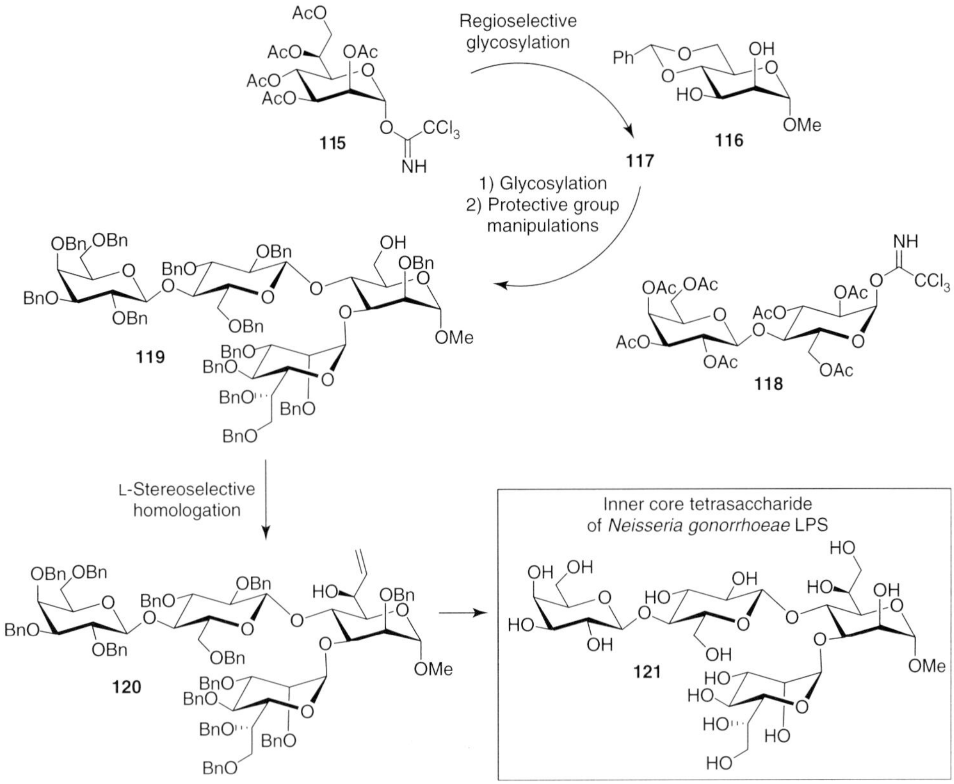

Scheme 2.27 Synthesis of the typical 3,4-dibranched heptosidic LPS substructure.

followed by reduction. This hexose-to-heptose elongation was performed, rather unusually, at a late stage of the tetrasaccharide synthesis, which was in reality a key feature of Yamasaki's strategy to resolve the challenging problem of the construction of 3,4-dibranched heptosidic structures. This approach complemented Oscarson's [31] strategy described above to synthesize the inner core oligosaccharide expressed in LPS produced by pathogenic gram-negative bacteria such as the *Neisseria* and *Haemophilus* species.

2.3.4
The Crich's Stereoselective β-Glycosylation Applied to the Synthesis of the Repeating Unit of the Lipopolysaccharide from *Plesimonas shigelloides*

The infections caused by *Plesimonas shigelloides*, a gram-negative bacterium, strongly correlate with the surface water contamination and are particularly common in tropical and subtropical habitats. The structure of the *O*-specific side chain of *P. shigelloides* LPS (strain CNCTC 113/92, serotype 54) has been recently elucidated and revealed a novel hexasaccharide repeating unit (Scheme 2.28) containing two unusual β-linked heptose units, one of which is 6-deoxy [58]. In

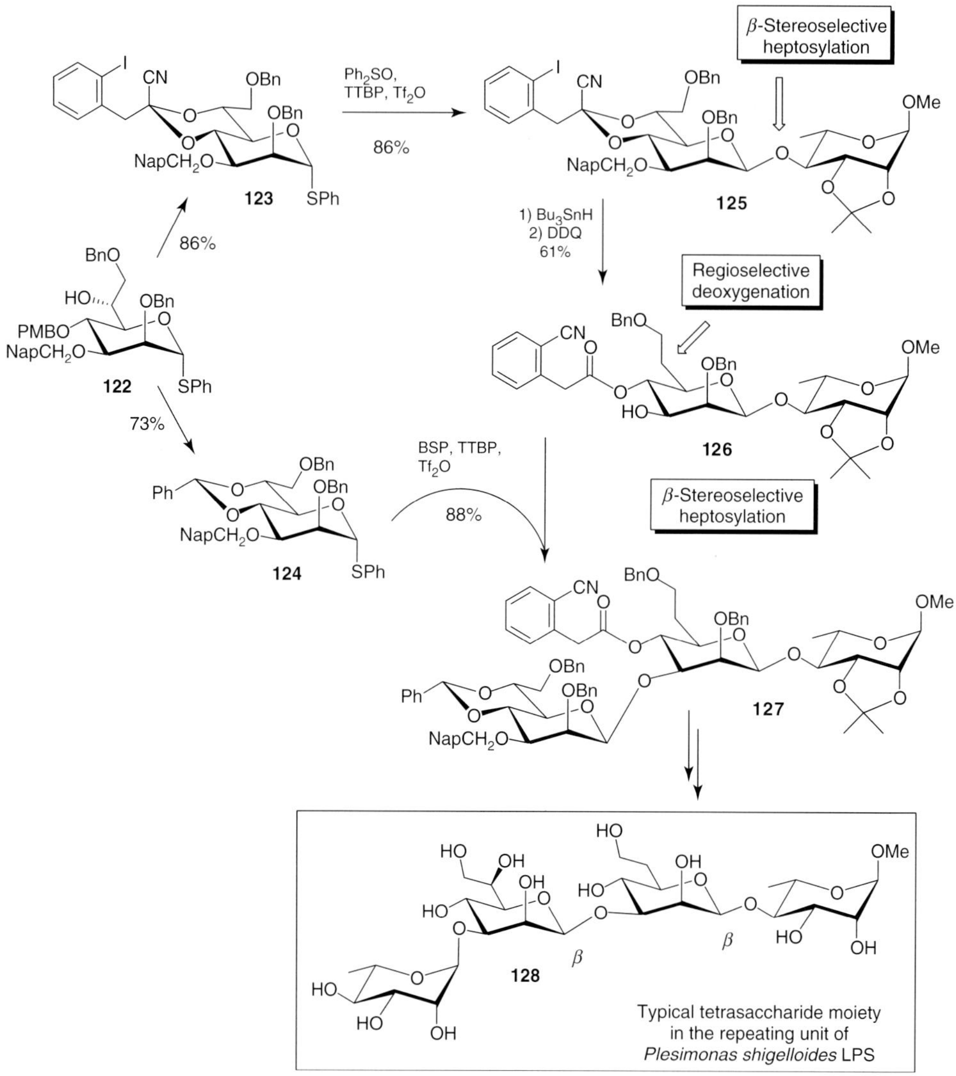

Scheme 2.28 Crich's β-stereoselective heptosylations.

2005, the Crich group [12b,c] developed an original strategy allowing the synthesis of tetrasaccharide **128**.

In their approach, two key issues were addressed: (i) stereoselective β-glycosidation in D,D- and L,D-heptoses and (ii) efficient generation of the 6-deoxy-β-Hep*p* unit with full regiocontrol at the 6-position and stereocontrol at the anomeric position. For the methodology study, various D,D- and L,D-heptoses were constructed mainly through the dihydroxylation of C6–C7 alkenes derived from D-mannose. This series of molecules allowed the study of the influence of both protecting groups and conformational restrictions on the course and the stereoselectivity of the heptosylation reactions.

For the synthesis of the tetrasaccharide **128**, two key building blocks **123** and **124** were constructed from the same heptoside **122** (Scheme 2.28). They were both protected with 4,6-*O*-alkylidene acetals to insure excellent β-stereoselectivities for the two heptosylation steps. The two glycosylations were performed under a preactivation protocol involving either diphenylsulfoxide (Ph_2SO) or 1-benzenesulfinyl piperidine (BSP), giving exclusively the β-anomer in both cases. Moreover, with a 4,6-*O*-[1-cyano-2-(2-iodophenyl)ethylidene] acetal-protected thioglycoside **123**, the subsequent treatment with tributyltin hydride and azoisobutyronitrile brought about clean fragmentation to the 6-deoxy-*glycero*-β-D-*manno*-heptopyranosides.

2.3.5
De Novo Approach Applied to the Synthesis of a Bisheptosylated Tetrasaccharide

In 2012, the Seeberger group [57] disclosed the total synthesis of the core tetrasaccharide α-GlcNAc-(1,2)-α-Hep-(1,3)-α-Hep-(1,5)-α-Kdo **136** of *N. meningitidis* LPS (Scheme 2.29).

The synthetic strategy was based on a convergent and stereocontrolled [2+2] approach. The heptose building blocks obtained by *de novo* synthesis [51] were fine-tuned for the effective assembly of the (1,5)-Hep*p*-Kdo disaccharide. The synthetic strategy incorporates an α-linked spacer at the reducing end of the tetrasaccharide for use as a handle during subsequent conjugation to a carrier protein. To realize the efficient coupling of heptose building blocks with other sugars, the protecting group pattern of heptose had to be adjusted to provide suitable reactivities. Thus, the two complementary heptosides **129** and **132** could be efficiently prepared using protocols easily scalable. It should be emphasized that the protecting groups SAr, Lev (levulinoyl), Ac (acetyl), and PBB (*p*-bromobenzyl) in **129** and/or **132** could be removed selectively. Thus the two building blocks **129** and **132** could be glycosylated at the O-1, O-2, O-3, and O-4, no matter the glycosylation sequence.

Coupling of the 2-azidoglucopyranosyl trichloroacetimidate **133** with heptose nucleophile **132** using a catalytic amount of TMSOTf as promoter in Et_2O, provided the desired α-linked GlcN-Hep disaccharide **134** in 68% yield. No β-anomer by-product was detected. On the other hand, identification of effective glycosylation conditions for the construction of the (1,5)-linked Hep-Kdo disaccharide proved extremely challenging because of a reactivity mismatch in the coupling of the heptose-glycosylating agent with the inert C-5 hydroxyl of the Kdo building block **130**, as already observed by Oscarson and coworkers [31, 59].

Eventually, heptose trichloroacetimidate **129**, activated by catalytic TMSOTf, proved to be superior for this glycosylation. The more reactive disaccharide *N*-phenyl trifluoroacetimidate **135** was then coupled with disaccharide **131** upon TMSOTf activation to give the desired α-linked tetrasaccharide **136** in a satisfactory yield of 72% (Scheme 2.29).

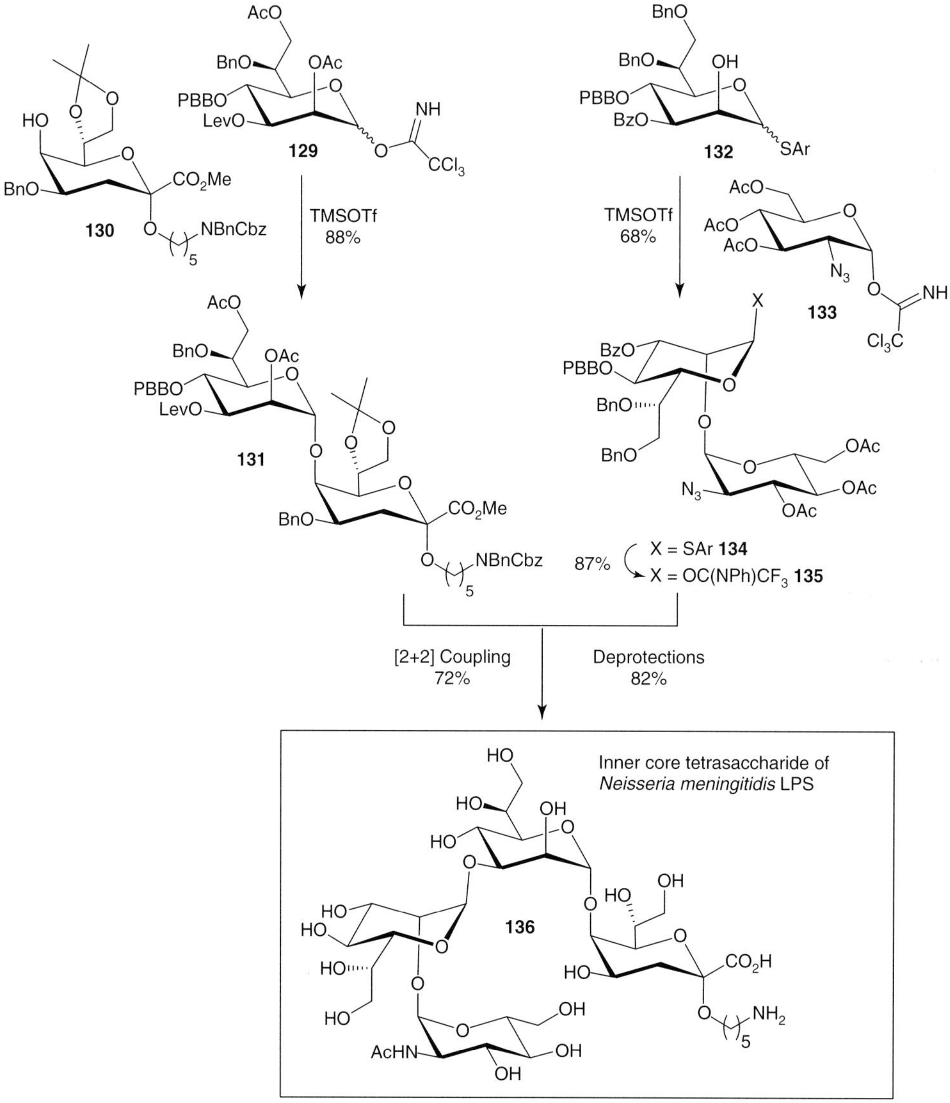

Scheme 2.29 Seeberger's synthesis of the tetrasaccharide of *N. meningitidis* LPS.

2.4 Synthesis of Heptosides as Biochemical Probes

Beyond the synthesis of heptosylated natural molecules, artificial heptosides have been designed and synthesized to address different biochemical problems such as the search of virulence factors [60]. As detailed above, the Kiliani reaction has allowed the synthesis of iminoheptitols, a validated class of glycosidase inhibitors

[42]. In 2008, Pohl *et al.* [12f] described an innovative technology based on fluorous-based carbohydrate microarrays applied to demonstrate that both diastereomers of glycero-D-*manno*-heptoses found in bacteria bind to ConA. These results showed that this plant lectin can accept modifications at the C-6 position of its natural mannose ligand.

However, the main biochemical applications of heptose synthesis have been directed toward the study or the inhibition of the bacterial heptose biosynthetic pathways.

2.4.1 Bacterial Heptose Biosynthetic Pathways

Because of their occurrence in the LPS of most gram-negative bacteria, and in particular, in major pathogenic strains, the biosynthesis of L,D-heptosides has attracted much attention [61]. These bacterial heptoses are constructed from sedoheptulose-7-phosphate (**137**) derived from the central metabolism (Scheme 2.30) [62]. A keto-aldose isomerase GmhA transforms **137** into D-*glycero*-D-mannoheptose 7-phosphate **138**, which is then phosphorylated by the kinase HldE. After hydrolysis of the terminal phosphate of **139**, intermediate **140** is transformed into nucleotide sugar **141**. Interestingly, bacterial strains such as *Escherichia coli* use the same HldE enzyme for two nonconsecutive steps. A regioselective D-to-L-epimerization is then catalyzed by HldD, yielding ADP-L-heptose **142**, the donor substrate of heptosyltransferases (WaaC, WaaF, and WaaQ) [63].

Recently, Tanner *et al.* [41, 64] have described a detailed mechanistic study on the HldD-catalyzed epimerization step. Crystal structures of HldD have been obtained and precious information about its catalytic pocket [41, 64b, 65] have been provided.

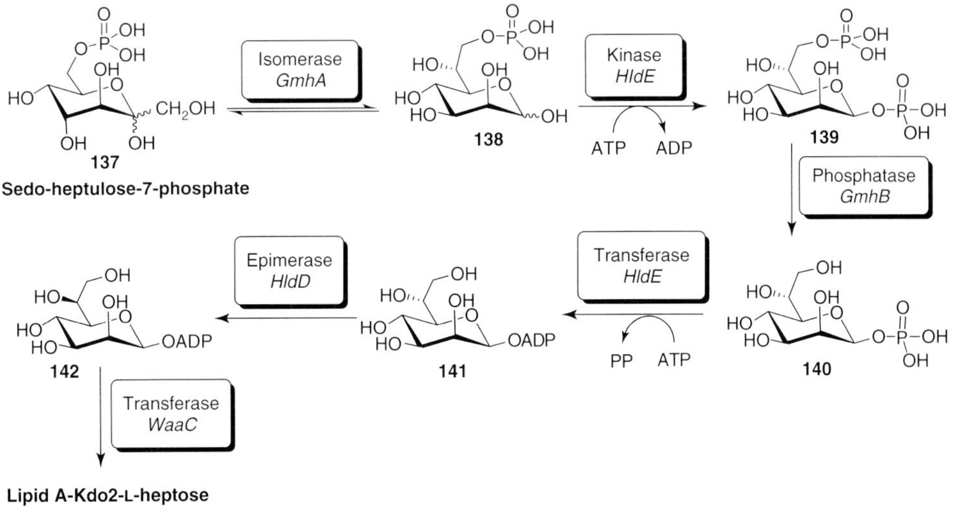

Scheme 2.30 Biosynthesis of bacterial L-heptosides found in LPS.

Despite this mechanistic and structural knowledge, no potent inhibitor of HldD has been described so far.

A similar pathway has been evidenced for the D,D-heptosylated oligosaccharides and the 6-deoxy-heptosides found in some bacterial O-antigens. The kinase HddA phosphorylates the same D,D-heptose **138** but this time to give the α-epimer **143** (Scheme 2.31). This intermediate 1,7-bisphosphate is then dephosphorylated by the phosphatase GmhB and transformed into GDP-heptose **145** that can be further processed.

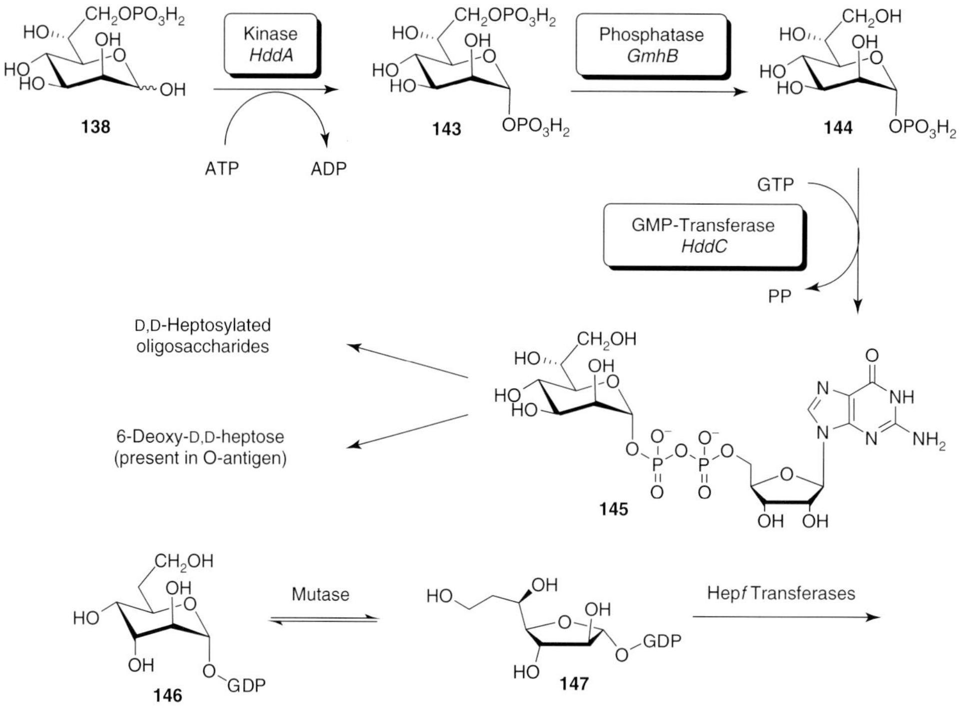

Scheme 2.31 Biosynthesis of bacterial D-heptosides and heptofuranosides.

Interestingly, Lowary and coworkers just discovered the biosynthetic origin of the heptofuranosides found, for instance, in capsular oligosaccharides of *C. jenuni*. The 6-deoxy-heptose-GDP **146** is isomerized into the corresponding GDP-heptofuranoside **147**, thanks to a mutase [66]. This rather unusual reaction is probably related, from a mechanistic point of view, to the more studied mycobacterial galactofuranose biosynthetic pathway [67].

2.4.2
Artificial D-Heptosides as Inhibitors of HldE and GmhA

Surprisingly, the inhibition of the first enzymes of the L,D-heptose biosynthetic pathway is an almost unexplored field, although its importance had already been acknowledged by Paulsen *et al.* [68] in 1994. They indeed described the synthesis of a series of D-*glycero*-D-mannoheptosides modified at the anomeric position, but, unfortunately, the biological evaluation has, never been published. In 2006, Wright *et al.* [69] developed an *in vitro* screening of bacterial LPS biosynthetic enzymes that allowed the identification of an inhibitor of HldE. It is noteworthy that several crystal structures of GmhB [70] and GmhA have been obtained, some of them in complex with its substrate that allowed a mapping of the interactions between the sugar–phosphate and the enzyme [71]. However, the very first GmhA inhibitors have been reported in 2011 (Figure 2.2) [72].

Derivatizations of a central scaffold allowed the preparation of heptosides with structural modifications at the 1-, 2-, 6-, and 7-positions [72]. The inhibition profile of the whole inhibitor family toward the first two enzymes of the heptose biosynthetic pathway clearly indicated that the two enzymes are extremely sensitive to structural modifications of the heptose scaffold at the 6- and 7-positions. Interestingly, it

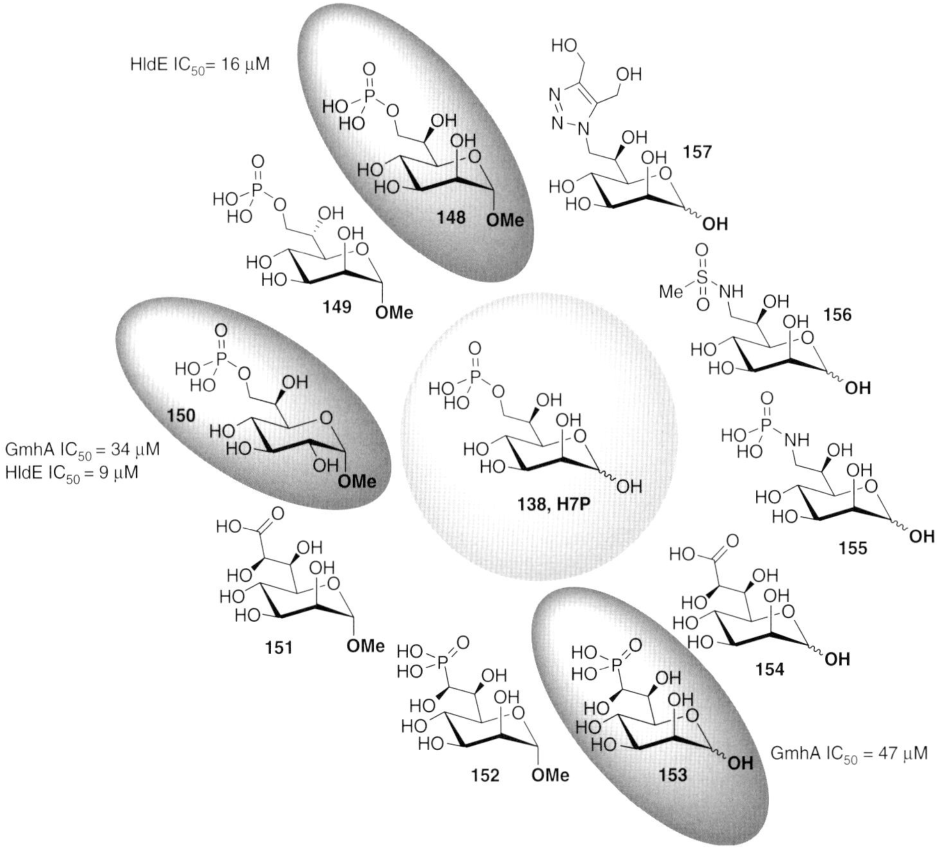

Figure 2.2 Synthetic analogs of heptose-7-phosphate **138** as inhibitors of GmhA and HldE.

could be shown that both enzymes tolerated an epimerization at the 2-position and maintain low-micromolar inhibition levels with glucoheptose analogs.

2.4.3
Inhibition Studies of Heptosyltransferase WaaC

To date, the most studied enzymes of this pathway are the epimerase HldD and heptosyltransferase WaaC. In 2000, Kosma *et al.* [73] described the first synthesis of the two anomers of ADP-L-heptose **142**, thus demonstrating the anomeric configuration of **142**. Later, the same team disclosed the synthesis of *C*-glycosidic analogs [74] of **142** for which no inhibition data is available to date. In 2008, the synthesis and the inhibition properties of the 2-fluoro analog **162** have been described by Dohi *et al.* [8] (Scheme 2.32). After the construction of the L-heptose-glycals **158** and **159**, the key step of this synthesis was the β-selective fluoroheptosylation mediated by Selectfluor [75]. Rewardingly, a 3D structure of WaaC in complex with **162** was also obtained [76]. The final molecule **162** was found to be a very potent competitive inhibitor of WaaC.

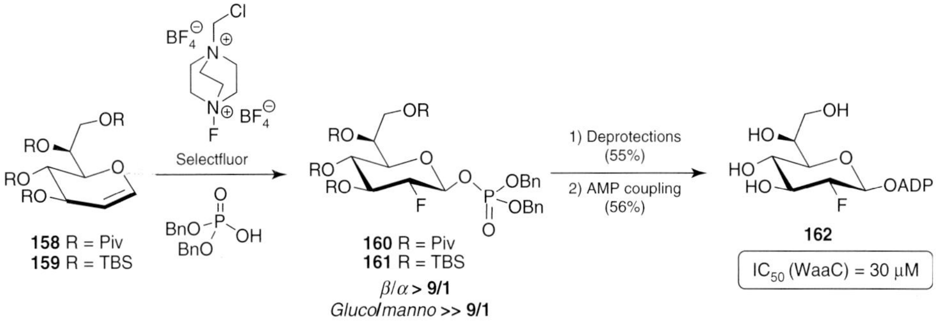

Scheme 2.32 Synthesis of ADP-2-fluoro-heptose **162**.

In 2012, the synthesis of a series of fullerene hexa-adducts bearing 12 copies of peripheral sugars displaying the mannopyranose core structure of bacterial L,D-heptoside was described [77]. The multimers were assembled through an efficient copper-catalyzed reaction as the final step. The final fullerene sugar balls were assayed as inhibitors of heptosyltransferase WaaC. Interestingly, the inhibition of the final molecules was found in the low micromolar range (IC_{50} 7–45 μM) while the corresponding monomeric glycosides **163** and **164** displayed high micromolar to low millimolar inhibition levels (IC_{50} always above 400 μM) (Figure 2.3).

When evaluated on a "per-sugar" basis, these inhibition data showed that, in each case, the average affinity of a single glycoside of the fullerenes toward WaaC was significantly enhanced when displayed as a multimer, thus demonstrating an unexpected multivalent effect. Such a multivalent mode of inhibition had not been previously evidenced with glycosyltransferases.

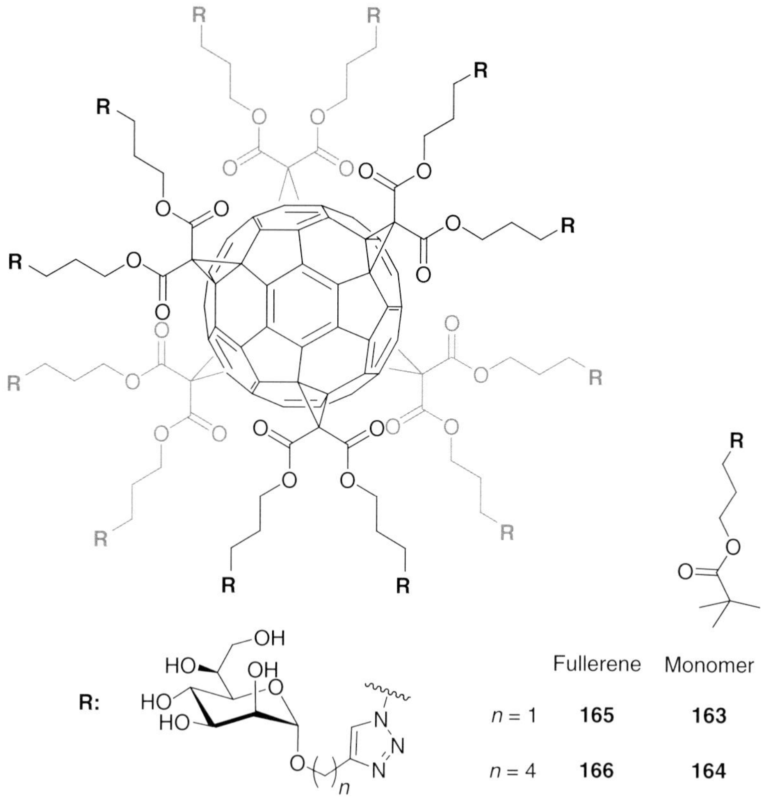

Figure 2.3 Glycofullerenes harboring 12 L,D-heptose copies as multivalent WaaC inhibitors.

2.5 Conclusions

This journey into the heptose chemistry and biochemistry nicely illustrated how this specific field of glycosciences accompanied the major twentieth century's breakthroughs of organic synthesis: from the Wittig reaction to organocatalysis. In parallel, the characterization of complex bacterial oligosaccharides has provided not only novel synthetic challenges but also new strategies to design synthetic vaccines. Moreover, thanks to the technologies recently developed in the genomic area, the biosynthesis of natural heptoses is much better understood. The availability of the corresponding genes now provides the opportunity to create synthetic heptose analogs targeting the virulence of pathogenic bacteria. Overall, major achievements have been accomplished in the field of aldoheptose chemistry and biochemistry but the recent literature indicates that significant advances are still to come.

2.6 Experimental Part

2.6.1 Typical Synthesis of a D-*glycero*-Heptoside by Dihydroxylation of a C6–C7 Alkene

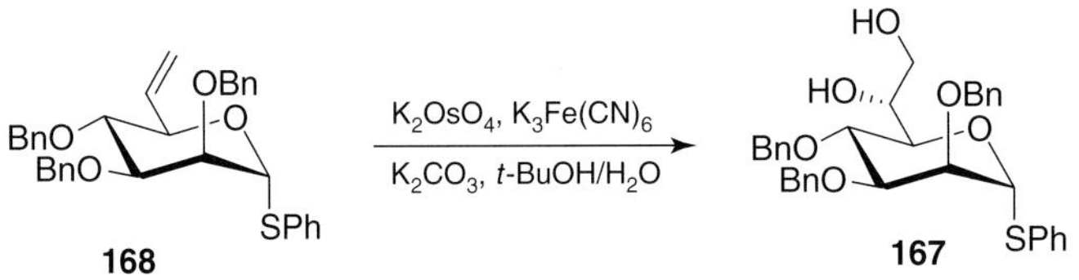

2.6.1.1 Phenyl 1-deoxy-2,3,4-tri-*O*-benzyl-1-thio-D-*glycero*-α-D-*manno*-heptopyranoside (167)

To a solution of $K_2OsO_4 \cdot 2H_2O$ (0.20 g, 0.55 mmol), $K_3Fe(CN)_6$ (16.94 g, 51.46 mmol), and K_2CO_3 (7.92 g, 56.94 mmol) in a mixture of water (94 ml) and *t*-BuOH (94 ml) at 0 °C was added a dropwise solution of phenyl 1-deoxy-2,3,4-tri-*O*-benzyl-1-thio-α-D-*manno*-hept-6-enopyranoside **168** (9.9 g, 18.38 mmol) in toluene (37 ml) [72]. The reaction was then allowed to warm to room temperature. After 48 h, $K_2OsO_4 \cdot 2H_2O$ (0.083 g, 0.23 mmol), $K_3Fe(CN)_6$ (2.54 g, 7.72 mmol), and K_2CO_3 (1.18 g, 8.49 mmol) were added and the mixture was stirred for 2 additional days. The reaction was then quenched by addition of Na_2SO_3 (32.2 g) and stirred for 1 h 30 min. The reaction mixture was extracted with EtOAc (4 × 100 ml). The organic layer was washed with an aqueous solution of 1 M KOH (100 ml), dried over $MgSO_4$, filtered, and concentrated *in vacuo*. The crude was purified by flash chromatography (cyclohexane/EtOAc 1 : 0 to 4 : 6) to afford the corresponding (D/L: 2/1) diol (7.74 g, 74%) as colorless oil. $[\alpha]^{20}_D$: +103.3 (*c* 1.00, $CHCl_3$). **^{1}H NMR** (400 Hz, $CDCl_3$): $\delta = 7.50$–7.22 (m, 20H, H^{arom}), 5.51 (d, $J_{1-2} = 1.8$ Hz, 1H, H-1), 5.09 (AB, $J_{AB} = 10.9$ Hz, 1H, CH_2^{Bn}), 4.76–4.67 (m, 3H, CH_2^{Bn}), 4.65 (s, 2H, CH_2^{Bn}), 4.25 (dd, $J_{4-5} = 9.2$ Hz, $J_{5-6} = 6.3$ Hz, 1H, H-5), 4.12 (dd, $J_{4-5} = 9.6$ Hz, $J_{3-4} = 8.9$ Hz, 1H, H-4), 4.04 (dd, $J_{2-3} = 2.8$ Hz, $J_{1-2} = 1.8$ Hz, 1H, H-2), 4.01–3.95 (m, 1H, H-6), 3.95 (dd, $J_{3-4} = 8.9$ Hz, $J_{2-3} = 3.0$ Hz, 1H, H-3), 3.73–3.53 (m, 2H, H-7), 3.15 (d, $J_{6\text{-}OH} = 4.7$ Hz, 1H, OH), 2.12 (dd, $J_{7b\text{-}OH} = 7.1$ Hz, $J_{7a\text{-}OH} = 5.9$ Hz, 1H, OH). **^{13}C NMR** (101 MHz, $CDCl_3$): $\delta = 137.7$, 137.6, 137.5, 133.4 (C_q^{arom}), 132.1, 129.2, 128.6, 128.5, 128.5, 128.2, 128.1, 128.0, 127.9, 127.8 (CH^{arom}) 85.8 (C-1), 80.2 (C-3), 76.5 (C-4), 75.9 (C-2), 75.0 (CH_2^{Bn}), 72.7 (C-5), 72.4 (C-6), 72.3, 71.9 (CH_2^{Bn}), 63.1 (C-7). **MS** (DCI-NH_3): *m*/*z*: 590 $[M + NH_4]^+$; **HRMS** calculated for $C_{34}H_{40}NO_6S$ $[M + NH_4]^+$: 590.2576; found: 590.2579.

2.6.2 Typical Synthesis of a L-*glycero*-Heptoside by Addition of Grignard Reagent Followed by a Tamao–Fleming Oxidation

This procedure was developed by van Boom *et al.* [26b, 34a,b] and recently improved by Durka *et al.* [72, 77].

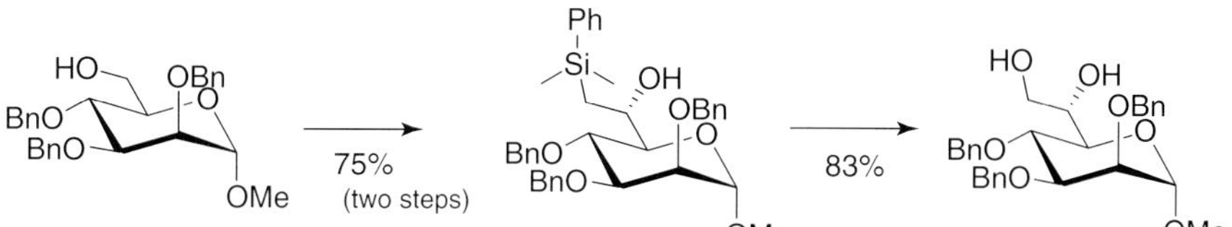

169 170 171

2.6.2.1 Methyl 2,3,4-tri-*O*-benzyl-7-(phenyldimethyl)silane-7-deoxy-L-*glycero*-α-D-*manno*-heptopyranoside (170)

Dimethylsulfoxide DMSO (123 µl, 1.72 mmol, 1.6 equiv) was added dropwise (one drop every ~15 s) in 30 min at −78 °C, under argon, to a solution of oxalyl chloride (127 µl, 1.51 mmol, 1.4 equiv), in dry CH_2Cl_2 (16 ml). After an additional 30 min, a solution of primary alcohol **169** [78] (500 mg, 1.08 mmol, 1equiv) in dry CH_2Cl_2 (5ml) was added dropwise in 12 min After 1 h at −78 °C, NEt_3 (450 µl, 3.23 mmol, 3 equiv) was added dropwise (over 2 min) and the reaction mixture was allowed to warm to room temperature. After 1 h, the solution was washed with saturated $NaHCO_3$ (10 ml), and the aqueous phases were extracted with CH_2Cl_2 (3 × 25 ml). The combined organic phases were washed with brine (2 × 75 ml), dried over $MgSO_4$, filtered and concentrated *in vacuo* and coevaporated with toluene (3 × 10 ml), and dried 2 h under vacuum. The resulting crude aldehyde was used without further purification and was stored at −18 °C under argon. ^{1}H-NMR (400 MHz, $CDCl_3$) $\delta = 9.73$ (s, 1H, H-6), 7.34–7.26 (m, 15H, H^{Ar}), 4.84 (d, $J = 2.7$ Hz, 1H, H-1), 4.83 (AB, $J = 10.7$ Hz, 1H, CH_2Ph), 4.70 (s, 2H, CH_2Ph), 4.64 (AB, $J = 10.7$ Hz, 1H, CH_2Ph), 4.60 (s, 2H, CH_2Ph), 4.08 (d, $J = 9.2$ Hz, 1H, H-5), 4.04 (t, $J = 8.9$ Hz, 1H, H-4), 3.93 (dd, $J = 3.0$ Hz, $J = 8.0$ Hz, 1H, H-3), 3.75 (t, $J = 2.8$ Hz, 1H, H-2), 3.36 (s, 3H, H^{Me}).

(Phenyldimethylsilyl)methyl chloride (482 µl, 2.69 mmol, 2.5 equiv) was added dropwise under argon to a suspension of dry magnesium turnings (69.7 mg, 2.91 mmol, 2.7 equiv) in dry tetrahydrofuran (THF; 12 ml). The mixture was refluxed for 40 min (magnesium turning almost totally disappeared). The Grignard reagent thus obtained was cooled to room temperature and transferred under argon by cannula to a dry flask and cooled to 0 °C. A solution of the aforementioned aldehyde (1.08 mmol, 1 equiv) in dry THF (5 ml) was added dropwise to the cold Grignard reagent. After stirring for 1 h at 0 °C, the mixture was allowed to warm to room temperature and stirred for 15 h. The reaction was cooled to 0 °C and quenched by slow addition of cold water (5 ml), and filtered through a Celite pad; the filtrate was diluted with a saturated solution of NH_4Cl (25 ml) and extracted with CH_2Cl_2 (3 × 30 ml). The organic layer was washed successively with a solution of diluted bleach (2%, 75 ml) and brine (2 × 75 ml). The organic layer was dried over $MgSO_4$, filtered, and concentrated under reduced pressure. Purification of the residue by flash chromatography (9 : 1 cyclohexane–EtOAc) afforded the product **170** (for similar NMR data, see literature [35]), (495 mg, 75%) as a colorless oil. ^{1}H-NMR (400 MHz, $CDCl_3$) $\delta = 7.59$-7.50 (m, 2H, H^{Ar}), 7.46–7.23 (m, 16H, H^{Ar}), 7.23–7.15 (m, 2H, H^{Ar}), 4.89 (d, $J = 10.5$ Hz, 1H, CH_2^{Bn}), 4.78 (d, $J = 12.3$ Hz, 1H,

CH_2^{Bn}), 4.71 (d, $J = 1.6$ Hz, 1H, H-1), 4.67 (d, $J = 12.4$ Hz, 1H, CH_2Ph), 4.63 (s, 2H, CH_2Ph), 4.60 (d, $J = 10.6$ Hz, 1H, CH_2Ph), 4.10 (dd, $J = 7.2$ Hz, $J = 14.0$ Hz, 1H, H-6), 4.05 (t, 1H, $J = 9.4$ Hz, H-4), 3.83 (dd, $J = 9.4$ Hz, $J = 3.0$ Hz, 1H, H-3), 3.80-3.73 (m, 1H, H-2), 3.32 (d, $J = 9.6$ Hz, 1H, H-5), 3.25 (s, 3H, H^{Me}), 1.37 (dd, $J = 14.8$ Hz, $J = 10.8$ Hz, 1H, H-7a), 0.92 (dd, $J = 14.8$ Hz, $J = 3.9$ Hz, 1H, H-7b), 0.36 and 0.35 (2 s, 6H, H^{SiMe}); ^{13}C-NMR (101 MHz, $CDCl_3$): $d = 139.4$, 138.7, 138.48, 138.46 ($4 \times C_q^{Ar}$), 133.8, 128.9, 128.5, 128.4, 128.3, 127.9, 127.8, 127.76, 127.7, 127.6 ($20 \times CH^{Ar}$), 99.6 (C-1), 80.4, 75.6, 75.3, 74.9 (C-2, C-3, C-4, C-5), 75.5, 72.9, 72.3 ($3 \times CH_2Ph$), 67.3 (C-6), 54.8 (C^{Me}), 21.9 (C-7), −1.9 (C^{SiMe}), −2.3(C^{SiMe}); α_D ($CHCl_3$, c = 1, 20 °C) + 13.1 (literature: +11.5 [35]); MS (ESI+): *m/z*: 635.3 (100%) $[M + Na]^+$; Elemental analysis: Calcd for $C_{37}H_{44}O_6Si$: C 72.52, H 7.24; found: C 72.49, H 7.36.

2.6.2.2 Methyl 2,3,4-tri-*O*-benzyl-L-*glycero*-α-D-*manno*-heptopyranoside (171)

AcOK (0.160 g, 1.63 mmol, 10 equiv) was added under argon to a solution of dry **170** (0.100 g, 0.163 mmol, 1 equiv) in 1 ml of glacial AcOH, and the solution was sonicated until all the solids dissolved (~5 min). Mercury trifluoroacetate (0.278 g, 0.653 mmol, 4 equiv) was added and the solution was sonicated for 5 min. The solution was cooled to 6 °C, and after 10 min at 6 °C, 40% peracetic acid in AcOH (135 µl, 0.816 mmol, 5 equiv) was added dropwise (one drop every 20 s). The solution was stirred at 6 °C for 30 min and for 4 h at room temperature. The solution was diluted with CH_2Cl_2 (15 ml) and washed successively with a saturated solution of $NaHCO_3$ (10 ml) and water (10 ml). Aqueous phases were extracted with CH_2Cl_2 (3×10 ml). Combined organic phases were dried over $MgSO_4$ and concentrated under reduced pressure. Chromatography (cyclohexane → 6 : 4 cyclohexane–EtOAc) afforded the desired product **171** [20b] as a colorless oil (67 mg, 83%).

^{1}H-NMR ($CDCl_3$, 400 MHz): δ 7.36–7.25 (m, 15H,H_{arom}), 4.97 (AB, $J_{AB} = 10.8$ Hz, 1H, CH_2Ph), 4.78 (AB, $J_{AB} = 12.4$ Hz, 1H, CH_2Ph), 4.72–4.63 (m, 5H, H-1, 4CH_2Ph), 4.14 (t, $J = 9.9$ Hz, 1H, H-4), 3.97 (m, 1H, H-6), 3.87 (dd, $J = 3.0$ Hz, $J = 9.4$ Hz, 1H, H-5), 3.82–3.77 (m, 2H, H-2, H-7a), 3.71-3.65 (m, 1H, H-7b), 3.60 (dd, $J = 1.4$, 9.6 Hz, 1H, H-3), 3.27 (s, 3H, OCH_3), 2.39 (d, $J = 9.9$ Hz, 1H, OH), 2.16 (dd, $J = 2.8$, 9.6 Hz, 1H, OH). ^{13}C-NMR ($CDCl_3$, 100 MHz): δ 138.47, 138.43, 138.26 ($3C_q^{arom}$), 128.51, 128.12, 127.93, 127.86, 127.69 ($15C_{arom}$), 99.65 (C-1), 80.15, 75.41 (CH_2Ph), 74.59, 74.41 (C-2, C-4), 73.08 (CH_2 Ph), 72.66 (C-3), 72.31 (CH_2 Ph), 69.46, 65.16 (C-5 and C-6), 54.97 (OCH_3). α_D ($CHCl_3$, $c = 1$, 20 °C): +24.5; MS (ESI+): *m/z*: 517.2 (100%) $[M + Na]^+$; HR-MS (ESI+): Calculated for $C_{29}H_{34}O_7Na$ 517.2197, found 517.2221.

List of Abbreviations

Ac	Acetyl
All	Allyl
Bn	Benzyl
BSA	bovine serum albumin
Bz	Benzoyl

n-Bu	*n*-Butyl
t-Bu	*tert*-Butyl
BSP	1-Benzenesulfinyl piperidine
Cbz or Z	Carbobenzoxy
DCM	Dichloromethane
DMAP	4-Dimethylaminopyridine
DMSO	Dimethylsulfoxide
DMF	Dimethylformamide
DIEA	*N*,*N*-Diisopropylethylamine
DIBAL-H	Diisobutylaluminum hydride
DDQ	2,3-Dichloro-5,6-dicyanobenzoquinone
HNJ	homonojirimycin
LPS	lipopolysaccharide
Kdo	3-Deoxy-D-*manno*-oct-2-ulosonic acid
Lev	Levulinoyl
Nap	Naphthyl
NMO	*N*-Methylmorpholine *N*-oxide
NIS	*N*-iodosuccinimide
PTSA	*p*-Toluenesulfonic acid
py	Pyridine
PBB	*p*-Bromobenzyl
PMB	*p*-Methoxybenzyl
RT	Room temperature
Selectfluor	1-Chloromethyl-4-fluoro-1,4-diazonia-bicyclo[2.2.2]octane bis(tetrafluoroborate)
THF	Tetrahydrofuran
TfOH	Triflic acid
TFA	Trifluoroacetic acid
TBAF	Tetrabutylammonium fluoride
TMS	Trimethylsilyl
TBS	*tert*-Butyldimethylsilyl
TBDPS	*tert*-Butyldiphenylsilyl
TTBP	2,4,6-tri-*tert*-Butylpyrimidine

Acknowledgments

The authors are grateful to FRS-FNRS, Marie Curie Actions, and Mutabilis S.A. (France) for financial supports to our own contributions to the synthesis of heptosides as biochemical probes.

References

1. (a) Suzuki, T. and Chida, N. (2003) *Chem. Lett.*, **32**, 190–191. (b) Suzuki, T., Suzuki, S.T., Yamada, I., Koashi, Y., Yamada, K., and Chida, N. (2002) *J. Org. Chem.*, **67**, 2874–2880.

2. (a) Marcelo, F., Jimenez-Barbero, J., Marrot, J., Rauter, A.P., Sinaÿ, P., and Blériot, Y. (2008) *Chem. Eur. J.*, **14**, 10066–10073. (b) Marcelo, F., Abou-Jneid, R., Sollogoub, M., Marrot, J., Rauter, A.P., and Blériot, Y. (2009) *Synlett*, 1269–1272.
3. Dong, L., Roosenberg, J.M., and Miller, M.J. (2002) *J. Am. Chem. Soc.*, **124**, 15001–15005.
4. Hansson, J. and Oscarson, S. (2000) *Curr. Org. Chem.*, **4**, 535–564.
5. Kosma, P. (2008) *Curr. Org. Chem.*, **12**, 1021–1039.
6. (a) Peng, W., Jayasuriya, A.B., Imamura, A., and Lowary, T.L. (2011) *Org. Lett.*, **13**, 5290–5293. (b) Palyam, N., Niewczas, I., and Majewski, M. (2012) *Synlett*, 2367–2370.
7. (a) Brimacombe, J.S. and Kabir, A.K.M.S. (1986) *Carbohydr. Res.*, **152**, 329–334. (b) Brimacombe, J.S. and Kabir, A.K.M.S. (1986) *Carbohydr. Res.*, **150**, 35–51.
8. Dohi, H., Perion, R., Durka, M., Bosco, M., Roué, Y., Moreau, F., Grizot, S., Ducruix, A., Escaich, S., and Vincent, S.P. (2008) *Chem. Eur. J.*, **14**, 9530–9539.
9. Guezlek, H., Graziani, A., and Kosma, P. (2005) *Carbohydr. Res.*, **340**, 2808–2811.
10. Artner, D., Stanetty, C., Mereiter, K., Zamyatina, A., and Kosma, P. (2011) *Carbohydr. Res.*, **346**, 1739–1746.
11. Duan, J.J. and Kishi, Y. (1993) *Tetrahedron Lett.*, **34**, 7541–7544.
12. (a) van Straten, N.C.R., Kriek, N.M.A.J., Timmers, C.M., Wigchert, S.C.M., van der Marel, G.A., and van Boom, J.H. (1997) *J. Carbohydr. Chem.*, **16**, 947–966. (b) Crich, D. and Banerjee, A. (2005) *Org. Lett.*, **7**, 1395–1398. (c) Crich, D. and Banerjee, A. (2006) *J. Am. Chem. Soc.*, **128**, 8078–8086. (d) Crich, D. and Li, M. (2008) *J. Org. Chem.*, **73**, 7003–7010. (e) Amigues, E.J., Greenberg, M.L., Ju, S., Chen, Y., and Migaud, M.E. (2007) *Tetrahedron*, **63**, 10042–10053. (f) Jaipuri, F.A., Collet, B.Y.M., and Pohl, N.L. (2008) *Angew. Chem. Int. Ed.*, **47**, 1707–1710.
13. Monenschein, H., Drager, G., Jung, A., and Kirschning, A. (1999) *Chem. Eur. J.*, **5**, 2270–2280.
14. (a) Jarosz, S., Gajewska, A., and Luboradzki, R. (2008) *Tetrahedron: Asymmetry*, **19**, 1385–1391. (b) Jarosz, S., Skora, S., and Kosciolowska, I. (2003) *Carbohydr. Res.*, **338**, 407–413.
15. (a) Gyoergydeak, Z. and Pelyvas, I.F. (1998) *Monosaccharide Sugars: Chemical Synthesis by Chain Elongation, Degradation, and Epimerization*, Academic Press, p. 508. (b) Kawauchi, N. and Hashimoto, H. (1987) *Bull. Chem. Soc. Jpn.*, **60**, 1441–1447. (c) Tronchet, J.M.J. and Zerelli, S. (1989) *J. Carbohydr. Chem.*, **8**, 217–232.
16. (a) Dziewiszek, K. and Zamojski, A. (1986) *Carbohydr. Res.*, **150**, 163–171. (b) Grzeszczyk, B., Dziewiszek, K., Jarosz, S., and Zamojski, A. (1985) *Carbohydr. Res.*, **145**, 145–151.
17. Khare, N.K., Sood, R.K., and Aspinall, G.O. (1994) *Can. J. Chem.*, **72**, 237–246.
18. Dondoni, A., Fantin, G., Fogagnolo, M., and Medici, A. (1987) *Tetrahedron*, **43**, 3533–3539.
19. Nemoto, H., Ma, R., Kawamura, T., Yatsuzuka, K., Kamiya, M., and Shibuya, M. (2008) *Synthesis*, 3819–3827.
20. (a) Boons, G.J.P.H., van der Klein, P.A.M., van der Marel, G.A., and van Boom, J.H. (1988) *Recl. Trav. Chim. Pays-Bas*, **107**, 507–508. (b) Boons, G.J.P.H., van der Marel, G.A., and van Boom, J.H. (1989) *Tetrahedron Lett.*, **30**, 229–232.
21. (a) Grzeszczyk, B. and Zamojski, A. (1994) *Carbohydr. Res.*, **262**, 49–57. (b) Grzeszczyk, B., Holst, O., and Zamojski, A. (1996) *Carbohydr. Res.*, **290**, 1–15. (c) Grzeszczyk, B., Holst, O., Muller-Loennies, S., and Zamojski, A. (1998) *Carbohydr. Res.*, **307**, 55–67. (d) Grzeszczyk, B. and Zamojski, A. (2000) *Collect. Czech. Chem. Commun.*, **65**, 610–620. (e) Spohr, U., Le, N., Ling, C.-C., and Lemieux, R.U. (2001) *Can. J. Chem.*, **79**, 238–255. (f) Brimacombe, J.S. and Kabir, A.K.M.S. (1988) *Carbohydr. Res.*, **174**, 37–45. (g) Jarosz, S. and Kozlowska, E. (1996) *Pol. J. Chem.*, **70**, 45–53. (h) Stepowska, H. and Zamojski, A. (2001) *Carbohydr. Res.*,

332, 429–438. (i) Grzeszczyk, B. and Zamojski, A. (2001) *Carbohydr. Res.*, **332**, 225–234.

22. (a) Horton, D., Nakadate, M., and Tronchet, J.M.J. (1968) *Carbohydr. Res.*, **7**, 56–65. (b) Hems, R., Horton, D., and Nakadate, M. (1972) *Carbohydr. Res.*, **25**, 205–216.
23. (a) Jarosz, S. (1992) *Pol. J. Chem.*, **66**, 1853–1858. (b) (2005) *Phys. Chem. News*, **22**, 107–112.
24. Dasser, M., Chretien, F., and Chapleur, Y. (1990) *J. Chem. Soc., Perkin Trans. 1*, 3091–3094.
25. Cram, D.J. and Wilson, D.R. (1963) *J. Am. Chem. Soc.*, **85**, 1245–1249.
26. (a) Kim, M., Grzeszczyk, B., and Zamojski, A. (2000) *Tetrahedron*, **56**, 9319–9337. (b) Boons, G.J.P.H., Overhand, M., van der Marel, G.A., and van Boom, J.H. (1989) *Carbohydr. Res.*, **192**, C1–C4.
27. (a) Yamasaki, R., Takajyo, A., Kubo, H., Matsui, T., Ishii, K., and Yoshida, M. (2001) *J. Carbohydr. Chem.*, **20**, 171–180. (b) Ishii, K., Kubo, H., and Yamasaki, R. (2002) *Carbohydr. Res.*, **337**, 11–20.(c) Ishii, K., Esumi, Y., Iwasaki, Y., and Yamasaki, R. (2004) *Eur. J. Org. Chem.*, 1214–1227.
28. Hori, K., Sawada, N., Ando, H., Ishida, H., and Kiso, M. (2003) *Eur. J. Org. Chem.*, 3752–3760.
29. Dziewiszek, K., Banaszek, A., and Zamojski, A. (1987) *Tetrahedron Lett.*, **28**, 1569–1572.
30. Kim, M., Grzeszczyk, B., and Zamojski, A. (2002) *Tetrahedron Lett.*, **43**, 1337–1340.
31. Bernlind, C. and Oscarson, S. (1998) *J. Org. Chem.*, **63**, 7780–7788.
32. Stewart, A., Bernlind, C., Martin, A., Oscarson, S., Richards, J.C., and Schweda, E.K.H. (1998) *Carbohydr. Res.*, **313**, 193–202.
33. (a) Bernlind, C., Bennett, S., and Oscarson, S. (2000) *Tetrahedron: Asymmetry*, **11**, 481–492. (b) Olsson, J.D.M. and Oscarson, S. (2010) *Carbohydr. Res.*, **345**, 1331–1338. (c) Oscarson, S. (2001) *Carbohydr. Polym.*, **44**, 305–311.
34. (a) Boons, G.J.P.H., Overhand, M., van der Marel, G.A., and van Boom, J.H. (1989) *Angew. Chem., Int. Ed. Engl.*, **28**, 1504–1506. (b) Boons, G.J.P.H., van der Marel, G.A., Poolman, J.T., and van Boom, J.H. (1989) *Recl. Trav. Chim. Pays-Bas*, **108**, 339–343. (c) Boons, G.J.P.H., Overhand, M., van der Marel, G.A., and van Boom, J.H. (1992) *Recl. Trav. Chim. Pays-Bas*, **111**, 144–148. (d) Smid, P., Schipper, F.J.M., Broxterman, H.J.G., Boons, G.J.P.H., van der Marel, G.A., and van Boom, J.H. (1993) *Recl. Trav. Chim. Pays-Bas*, **112**, 451–456. (e) van Delft, F.L., van der Marel, A., and van Boom, J.H. (1995) *Synlett*, 1069–1070. (f) Boons, G.J.P.H., Hoogerhout, P., Poolman, J.T., van der Marel, G.A., and van Boom, J.H. (1991) *Bioorg. Med. Chem. Lett.*, **1**, 303–308.
35. Boons, G.J.P.H., Steyger, R., Overhand, M., van der Marel, G.A., and van Boom, J.H. (1991) *J. Carbohydr. Chem.*, **10**, 995–1007.
36. Paulsen, H., Pries, M., and Lorentzen, J.P. (1994) *Liebigs Ann. Chem.*, **12**, 389–397.
37. Garegg, P.J., Oscarson, S., and Szoenyi, M. (1990) *Carbohydr. Res.*, **205**, 125–132.
38. Zottola, M.A., Alonso, R., Vite, G.D., and Fraser-Reid, B. (1989) *J. Org. Chem.*, **54**, 6123–6125.
39. Nishikawa, T., Mishima, Y., Ohyabu, N., and Isobe, M. (2004) *Tetrahedron Lett.*, **45**, 175–178.
40. Ferrier, R.J. and Furneaux, R.H. (1980) *Aust. J. Chem.*, **33**, 1025–1036.
41. Read, J.A., Ahmed, R.A., and Tanner, M.E. (2005) *Org. Lett.*, **7**, 2457–2460.
42. Lenagh-Snow, G.M.J., Jenkinson, S.F., Newberry, S.J., Kato, A., Nakagawa, S., Adachi, I., Wormald, M.R., Yoshihara, A., Morimoto, K., Akimitsu, K., Izumori, K., and Fleet, G.W.J. (2012) *Org. Lett.*, **14**, 2050–2053.
43. Kite, G.C., Fellows, L.E., Fleet, G.W.J., Liu, P.S., Scofield, A.M., and Smith, N.G. (1988) *Tetrahedron Lett.*, **29**, 6483–6486.
44. (a) Asano, N., Nishida, M., Kizu, H., Matsui, K., Watson, A.A., and Nash, R.J. (1997) *J. Nat. Prod.*, **60**, 98–101. (b) Kim, H.S., Kim, Y.H., Hong, Y.S., Paek, N.S., Lee, H.S., Kim, T.H., Kim, K.W., and Lee, J.J. (1999) *Planta Med.*, **65**, 437–439. (c) Asano, N., Kato, A., Miyauchi, M., Kizu, H., Kameda, Y.,

Watson, A.A., Nash, R.J., and Fleet, G.W.J. (1998) *J. Nat. Prod.*, **61**, 625–628. (d) Ikeda, K., Takahashi, M., Nishida, M., Miyauchi, M., Kizu, H., Kameda, Y., Arisawa, M., Watson, A.A., Nash, R.J., Fleet, G.W.J., and Asano, N. (2000) *Carbohydr. Res.*, **323**, 73–80. (e) Kato, A., Kato, N., Adachi, I., Hollinshead, J., Fleet, G.W.J., Kuriyama, C., Ikeda, K., Asano, N., and Nash, R.J. (2007) *J. Nat. Prod.*, **70**, 993–997. (f) Kite, G.C., Hoffmann, P., Lees, D.C., Wurdack, K.J., and Gillespie, L.J. (2005) *Biochem. Syst. Ecol.*, **33**, 1183–1186.

45. Rezanka, T., Siristova, L., Melzoch, K., and Sigler, K. (2011) *Lipids*, **46**, 249–261.
46. Paulsen, H., Schueller, M., Nashed, M.A., Heitmann, A., and Redlich, H. (1985) *Tetrahedron Lett.*, **26**, 3689–3692.
47. (a) Schmid, W. and Whitesides, G.M. (1991) *J. Am. Chem. Soc.*, **113**, 6674–6675. (b) Kim, E., Gordon, D.M., Schmid, W., and Whitesides, G.M. (1993) *J. Org. Chem.*, **58**, 5500–5507.
48. Prenner, R.H., Binder, W.H., and Schmid, W. (1994) *Liebigs Ann. Chem.*, **1994**, 73–78.
49. Balla, E., Zamyatina, A., Hofinger, A., and Kosma, P. (2007) *Carbohydr. Res.*, **342**, 2537–2545.
50. Palmelund, A. and Madsen, R. (2005) *J. Org. Chem.*, **70**, 8248–8251.
51. Ohara, T., Adibekian, A., Esposito, D., Stallforth, P., and Seeberger, P.H. (2010) *Chem. Commun.*, **46**, 4106–4108.
52. (a) Paulsen, H. and Brenken, M. (1991) *Liebigs Ann. Chem.*, **11**, 1113–1126. (b) Paulsen, H. and Heitmann, A.C. (1988) *Liebigs Ann. Chem.*, **1988**, 1061–1071. (c) Paulsen, H. and Heitmann, A.C. (1989) *Liebigs Ann. Chem.*, **1989**, 655–663. (d) Paulsen, H. and Hoeffgen, E.C. (1991) *Tetrahedron Lett.*, **32**, 2747–2750. (e) Paulsen, H., Schueller, M., Heitmann, A., Nashed, M.A., and Redlich, H. (1986) *Liebigs Ann. Chem.*, **1986**, 675–686. (f) Paulsen, H., Wulff, A., and Brenken, M. (1991) *Liebigs Ann. Chem.*, **1991**, 1127–1145. (g) Paulsen, H., Wulff, A., and Heitmann, A.C. (1988) *Liebigs Ann. Chem.*, **1988**, 1073–1078.
53. (a) Nepogod'ev, S.A., Backinowsky, L.V., Grzeszczyk, B., and Zamojski, A. (1994) *Carbohydr. Res.*, **254**, 43–60. (b) Nepogodev, S.A., Pakulski, Z., Zamojski, A., Holst, O., and Brade, H. (1992) *Carbohydr. Res.*, **232**, 33–45. (c) Pakulski, Z., Zamojski, A., Holst, O., and Zähringer, U. (1991) *Carbohydr. Res.*, **215**, 337–344.
54. Crich, D. and Sun, S. (1998) *Tetrahedron*, **54**, 8321–8348.
55. Morley, S.L. and Pollard, A.J. (2002) *Vaccine*, **20**, 666–687.
56. Mistretta, N., Seguin, D., Thiebaud, J., Vialle, S., Blanc, F., Brossaud, M., Talaga, P., Norheim, G., Moreau, M., and Rokbi, B. (2010) *J. Biol. Chem.*, **285**, 19874–19883.
57. Yang, Y., Martin, C.E., and Seeberger, P.H. (2012) *Chem. Sci.*, **3**, 896–899.
58. Niedziela, T., Lukasiewicz, J., Jachymek, W., Dzieciatkowska, M., Lugowski, C., and Kenne, L. (2002) *J. Biol. Chem.*, **277**, 11653–11663.
59. Olsson, J.D.M. and Oscarson, S. (2009) *Tetrahedron: Asymmetry*, **20**, 875–882.
60. Marchetti, R., Malinovska, L., Lameignere, E., Adamova, L., de Castro, C., Cioci, G., Stanetty, C., Kosma, P., Molinaro, A., Wimmerova, M., Imberty, A., and Silipo, A. (2012) *Glycobiology*, **22**, 1387–1398.
61. Raetz, C.R.H. and Whitfield, C. (2002) *Annu. Rev. Biochem.*, **71**, 635–700.
62. (a) Kneidinger, B., Graninger, M., Puchberger, M., Kosma, P., and Messner, P. (2001) *J. Biol. Chem.*, **276**, 20935–20944. (b) Kneidinger, B., Marolda, C., Graninger, M., Zamyatina, A., McArthur, F., Kosma, P., Valvano, M.A., and Messner, P. (2002) *J. Bacteriol.*, **184**, 363–369.
63. (a) Gronow, S., Brabetz, W., and Brade, H. (2000) *Eur. J. Biochem.*, **267**, 6602–6611. (b) Czyzyk, D.J., Liu, C., and Taylor, E.A. (2011) *Biochemistry*, **50**, 10570–10572.
64. (a) Mayer, A. and Tanner, M.E. (2007) *Biochemistry*, **46**, 6149–6155. (b) Kowatz, T., Morrison, J.P., Tanner, M.E., and Naismith, J.H. (2010) *Protein Sci.*, **19**, 1337–1343. (c) Morrison, J.P., Read, J.A., Coleman, W.G., and Tanner, M.E. (2005) *Biochemistry*, **44**, 5907–5915. (d) Morrison, J.P. and Tanner, M.E. (2007) *Biochemistry*, **46**, 3916–3924. (e) Read, J.A., Ahmed, R.A., Morrison, J.P., Coleman, W.G., and Tanner,

M.E. (2004) *J. Am. Chem. Soc.*, **126**, 8878–8879.

65. Deacon, A.M., Ni, Y.S., Coleman, W.G., and Ealick, S.E. (2000) *Structure*, **8**, 453–462.
66. Lowary, T.L. (2012) Furanoside biosynthesis in mycobacteria and campylobacters. International Carbohydrate Meeting ICS26 2012.
67. (a) Chlubnova, I., Legentil, L., Dureau, R., Pennec, A., Almendros, M., Daniellou, R., Nugier-Chauvin, C., and Ferrières, V. (2012) *Carbohydr. Res.*, **356**, 44–61. (b) Richards, M.R. and Lowary, T.L. (2009) *ChemBioChem*, **10**, 1920–1938.
68. Paulsen, H., Pries, M., and Lorentzen, J.P. (1994) *Liebigs Ann. Chem.*, **1994**, 389–397.
69. De Leon, G.P., Elowe, N.H., Koteva, K.P., Valvano, M.A., and Wright, G.D. (2006) *Chem. Biol.*, **13**, 437–441.
70. Nguyen, H.H., Wang, L.B., Huang, H., Peisach, E., Dunaway-Mariano, D., and Allen, K.N. (2010) *Biochemistry*, **49**, 1082–1092.
71. (a) Taylor, P.L., Blakely, K.M., De Leon, G.P., Walker, J.R., McArthur, F., Evdokimova, E., Zhang, K., Valvano, M.A., Wright, G.D., and Junop, M.S. (2008) *J. Biol. Chem.*, **283**, 2835–2845. (b) Harmer, N.J. (2010) *J. Mol. Biol.*, **400**, 379–392. (c) Seetharaman, J., Rajashankar, K.R., Solorzano, V., Kniewel, R., Lima, C.D., Bonanno, J.B., Burley, S.K., and Swaminathan, S. (2006) *Proteins Struct. Funct. Bioinf.*, **63**, 1092–1096.
72. Durka, M., Tikad, A., Périon, R., Bosco, M., Andaloussi, M., Floquet, S., Malacain, E., Moreau, F., Oxoby, M., Gerusz, V., and Vincent, S.P. (2011) *Chem. Eur. J.*, **17**, 11305–11313.
73. Zamyatina, A., Gronow, S., Oertelt, C., Puchberger, M., Brade, H., and Kosma, P. (2000) *Angew. Chem. Int. Ed.*, **39**, 4150–4153.
74. Graziani, A., Amer, H., Zamyatina, A., Hofinger, A., and Kosma, P. (2007) *Tetrahedron: Asymmetry*, **18**, 115–122.
75. (a) Caravano, A., Dohi, H., Sinaÿ, P., and Vincent, S.P. (2006) *Chem. Eur. J.*, **11**, 3114–3123. (b) Nyffeler, P.T., Duron, S.G., Burkart, M.D., Vincent, S.P., and Wong, C.-H. (2005) *Angew. Chem. Int. Ed.*, **44**, 192–212. (c) Vincent, S.P., Burkart, M.D., Tsai, C.Y., Zhang, Z., and Wong, C.H. (1999) *J. Org. Chem.*, **64**, 5264–5279.
76. Grizot, S., Salem, M., Vongsouthi, V., Durand, L., Moreau, F., Dohi, H., Vincent, S., Escaich, S., and Ducruix, A. (2006) *J. Mol. Biol.*, **363**, 383–394.
77. Durka, M., Buffet, K., Iehl, J., Holler, M., Nierengarten, J.-F., and Vincent, S.P. (2012) *Chem. Eur. J.*, **18**, 641–651.
78. Borén, H.B., Eklind, K., Garegg, P.J., Lindberg, B., and Pilotti, A. (1972) *Acta Chem. Scand.*, **26**, 4143.

3
Protecting-Group-Free Glycoconjugate Synthesis: Hydrazide and Oxyamine Derivatives in *N*-Glycoside Formation

Yoshiyuki A. Kwase, Melissa Cochran, and Mark Nitz

3.1
Introduction

The formation of *O*-glycosides is generally thought of as a challenging chemical transformation for which, despite well over a century of research, no general solutions exist. The formation of glycosyl hydrazides and *N*-glycosyl oxyamines offers numerous advantages over *O*-glycoside synthesis. The transformations are readily carried out, without the need for protecting groups, and can be achieved in aqueous solvents. This chapter covers the generation and use of hydrazide and hydroxylamine glycoconjugates, and the applications of these compounds in analysis, synthesis, medicinal chemistry, and glycobiology.

It was noted in an early review by Percival [1] that the first notable advance in the use of hydrazines in carbohydrate chemistry was by Fischer in 1884. Since Fischer's pioneering work, hydrazine derivatives of sugars have been used extensively to characterize mono- and disaccharides and have provided an effective method for derivatization of sugars prior to chromatographic or electrophoretic analysis [2]. Early work with glycosyl hydrazines was complicated by the equilibrium mixture of hydrazones as well as the ring-closed furanoses and pyranoses, which form upon incubation of sugars with hydrazines. Additionally, without access to nuclear magnetic resonance (NMR), these dynamic mixtures were difficult to analyze (Scheme 3.1). For example, treatment of glucose (Scheme 3.1, **1**) with phenylhydrazine (Scheme 3.1, **2**) leads to three isolable products that slowly interconvert [3]. However, the same treatment of galactose mainly yields the acylic hydrazone in solution [4]. An excellent review of this area as well as the reactions of sugar hydrazines has been written by El Khadem and Fatiadi [2]. Here, we focus on the glycosyl hydrazide (Scheme 3.1, **3–5a**) and *N*-alkyl-*N*-glycosyl oxyamine (Scheme 3.1, **3–5b**) derivatives that have proven to be readily adaptable to a wide range of conjugation reactions and exist predominantly as their cyclic tautomers.

Modern Synthetic Methods in Carbohydrate Chemistry: From Monosaccharides to Complex Glycoconjugates,
First Edition. Edited by Daniel B. Werz and Sébastien Vidal.

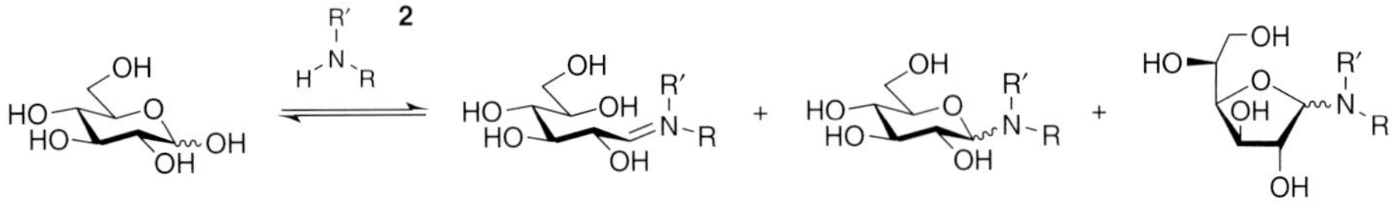

1a,b

a: R = NH(alkyl or aryl)
R′ = H
or
b: R = O(alkyl or aryl)
R′ = alkyl or aryl

3a,b

cis-trans-
hydrazones
or oximes

4a,b

α/β
Pyranosides

5a,b

α/β
Furanosides

Scheme 3.1 Tautomeric equilibrium present with hydrazine and oxyamine glucosyl condensation products.

3.2 Glycosyl Hydrazides (1-(Glycosyl)-2-acylhydrazines)

3.2.1 Formation, Tautomeric Preference, and Stability of Glycosyl Hydrazides

The first report of a condensation between a sugar and a hydrazide was published in 1895 by Wolff [5]. Although the product structure was not definitively determined, it was obvious that the condensation was a high-yielding transformation. Through the 1950s to 1970s, a series of reports were published, supporting Wolff's work. A variety of glycosyl hydrazides were explored, and the uses of glycosyl hydrazides in synthesis were evaluated; although often the tautomeric form of the glycosyl hydrazide formed was not definitively determined [6]. The advent of ^{1}H NMR characterization allowed the tautomeric forms of the hydrazide condensation products to be investigated in detail. Takeda [7] provided the first evidence that glycosyl hydrazides retain their pyranose structures in solution. Bendiak [8] carried out key ^{1}H NMR investigations on all of the 1-glycosyl-2-acetylhydrazines of the D-aldohexose series as well as five aldopentoses. The most important finding of these studies was that, with the exception of lyxose and idose, all of the glycosyl hydrazides existed predominantly (>85%) in the β-pyranose configuration in aqueous solution [8]. Additionally, crystal structures of glycosyl hydrazides support the cyclic pyranoses observed in solution [9]. Since these investigations, a wide range of glycosyl hydrazides have been characterized and the cyclic pyranoses usually dominate the aqueous equilibrium population [10]. Reports of glycosyl hydrazides having significant equilibrium populations as the open chain hydrazone have appeared only recently [11]. It is noteworthy that the reports of the acyclic hydrazone tautomers dominating the equilibrium population have relied on studies carried out in nonaqueous solvents (CD_3OD (deuterated methanol) or d_6-DMSO (deuterated dimethyl sulfoxide)), and the early report by Takeda [7] emphasizes that the nature of the solvent can affect the tautomeric equilibrium. Similarly to the acylhydrazides, sulfonylhydrazides of glucose, mannose, and galactose have also been shown to exist preferentially as the β-pyranosyl hydrazides [10g, 12].

In most cases, the formation of glycosyl hydrazides is reported to be a high-yielding transformation under a range of condensation conditions. The mechanism for glycosyl hydrazide formation has not been studied in detail, but is assumed to proceed through the acyclic sugar (Scheme 3.2, **6**), generating first the acylhydrazone (Scheme 3.2, **7**), which then ring closes to preferentially form the pyranose (Scheme 3.2, **8**). In some cases, the acyclic hydrazone is observed early in the reaction and converts slowly to the cyclic forms in solution [10c]. Much like hydrazone formation, the condensation proceeds under mildly acidic conditions, and alcohol solvents are generally used for reactions involving mono- or disaccharides. Aqueous conditions are employed when the solubility of larger oligosaccharides is limited in organic solutions. In aqueous conditions, the equilibrium constant for the condensation is low ($K \sim 10{-}100\,M^{-1}$) [10g]; thus, to achieve high yields in aqueous solvents, concentrated reaction conditions are necessary. Reports often focus on the number of equivalents of a hydrazide, or sugar, used in the condensation reaction. More important than an excess of either component is the overall concentration of the reagents, which controls the equilibrium, and concomitantly, the yield of the condensation product.

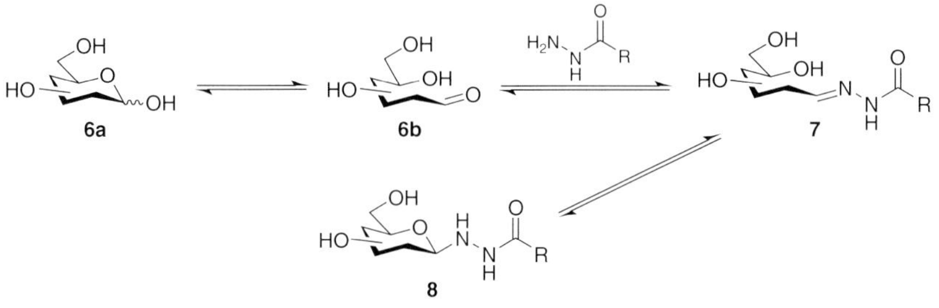

Scheme 3.2 Formation of glycosyl hydrazides.

Once formed, the glycosyl hydrazides (Table 3.1, **9**) can hydrolyze in aqueous solution, reverting to the free hemiacetal (Table 3.1, **10**) and hydrazide (Table 3.1, **11**) [8]. Kinetic evaluations of glycosyl hydrazide hydrolysis have shown the reaction to be first order, specific acid catalyzed, and to proceed with half-lives in the range of days to months at pH 7 [10f,g]. Thus, most glycosyl hydrazides certainly provide sufficient stability for labeling and chromatographic applications, but caution should be employed when using these derivatives in conditions below pH 7. Surprisingly, glycosyl hydrazides formed from oligogalacturonides are reported to be unstable in neutral water, suggesting that caution should also be employed in the case of uronic acids [13]. The rates of glycosyl hydrazide hydrolysis observed correlate well with the electron density at the primary amine of the hydrazide, with more electron-withdrawing acyl groups producing glycosyl hydrazides, which hydrolyze more slowly (Table 3.1, bottom left). The rates of hydrolysis also correlate with the relative electron density at the anomeric carbon of the glycosyl hydrazide, with more electron-withdrawn sugars, such as *N*-acetylglucosamine(GlcNAc), forming glycosyl hydrazides that are significantly

Table 3.1 Half-lives of glycosyl hydrazide hydrolysis.

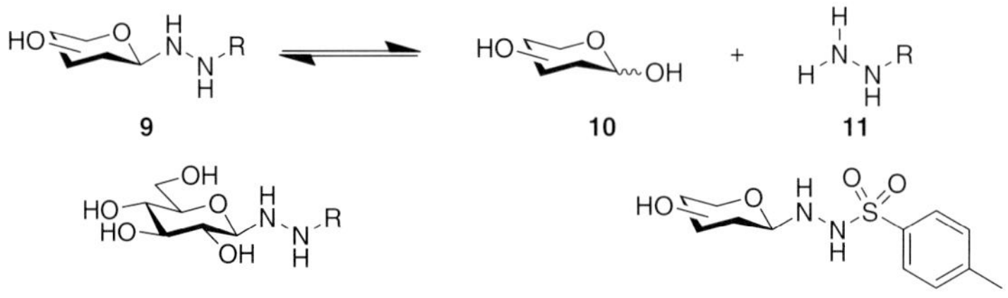

R	Hydrolysis half-life[a]		Hydrolysis half-life[b]
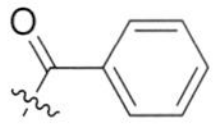	51 h	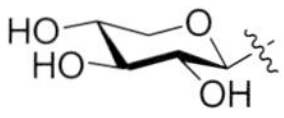	29 h
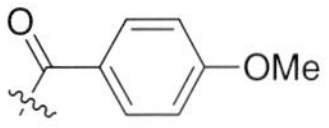	37 h	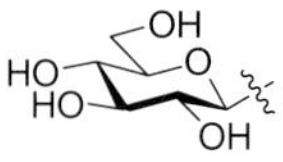	71 h
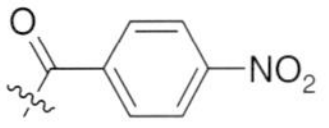	57 h	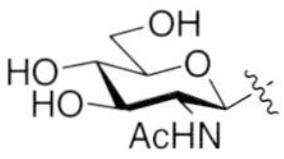	1100 h

[a]pH 6.0, 50 °C

[b]pH 6.0, 37 °C

more resistant to hydrolysis than glycosyl hydrazides formed from electron-rich sugars such as xylose (Table 3.1, bottom right) [10g].

3.2.2 Analytical Applications

The stability and facile-selective generation of glycosyl hydrazides have lead to an increased use of these conjugates in synthetic, analytical, and biochemical studies. Feizi and coworkers and Shinohara and coworkers [10a,b] were among the first to rigorously demonstrate the potential of glycosyl hydrazides through the development of bifunctional hydrazides bearing a chromophore and a biotin moiety. Both groups demonstrated that the hydrazide glycoconjugates could be formed efficiently with small quantities of oligosaccharides, and that these conjugates could be used for analytical studies. The naphthyl-functionalized hydrazide (Scheme 3.3, **12**), employed by Feizi and coworkers, could be used as a fluorescence tag for thin layer chromatography (TLC) and high-performance liquid chromatography (HPLC) analyses. Furthermore, they showed that the biotin-tagged carbohydrates could be immobilized in microtiter plates for enzyme-linked immunosorbent assay (ELISA) [14]. Shinohara and coworkers [10a, 14a,b] used their biotin-functionalized conjugates to immobilize the oligosaccharides for studies using surface plasmon resonance (SPR). Since these important contributions, the direct functionalization of glycans with biotin via hydrazide chemistry has proven its value in many applications. Recent examples include the generation of conjugates from muramyl

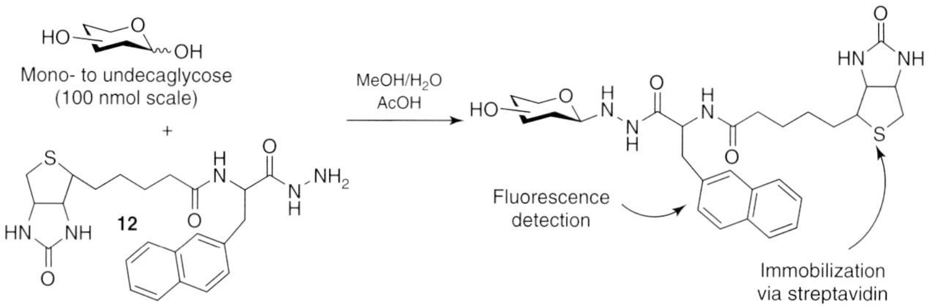

Scheme 3.3 Formation of bifunctional hydrazide conjugates for oligosaccharide analysis [10b].

peptides with fluorescent hydrazides or biotinyl hydrazides for use as microscopy probes of nodulation (Nod) signaling [15] and for the immobilization of synthetic disaccharides on streptavidin-coated SPR chips to investigate lectin-specific binding [16].

In addition to the use of biotinyl hydrazides as glycan analysis tools, the use of low-molecular weight bis-hydrazides provides a convenient and general approach to generate other functional glycoconjugates. Bis-hydrazide linkers are excellent substrates for the hydrazide condensations because of their high solubility in aqueous solution, allowing for the use of the highly concentrated conditions necessary to drive the condensation to completion. The low-molecular weight bis-hydrazides can also be readily separated from larger oligosaccharides by size exclusion chromatography or dialysis. Flinn and coworkers [10e] reported the use of adipic hydrazide which, when conjugated to the desired glycan, could be further functionalized with a salicylaldehyde derivative to form the hydrazone. Specifically, Flinn and coworkers showed that this two-step approach could be used to form stable neoglycoconjugates with proteins. Wei and coworkers [17] capitalized on the use of larger bis-hydrazides to form glycoconjugates designed for surface immobilization. After the initial condensation with their bis-hydrazides, the free remaining hydrazide was shown to be reactive toward active esters and isothiocyanates, allowing both immobilization on microspheres and fluorescent tagging. A photoreactive chromophore was also attached via the bis-hydrazide, allowing the photochemical functionalization of surfaces. These surfaces could be used for the detection of lectins and bacteria expressing specific adhesins.

Conjugation of glycans to surfaces, as demonstrated earlier with the biotin and bis-hydrazide examples, opens the possibility of a variety of immobilized analysis techniques such as ELISA, or using fluorescence, SPR, and quartz microbalances. Shin and coworkers [18] explored the direct functionalization of surfaces with hydrazides (Scheme 3.4, **14**) to allow for the printing of reducing oligosaccharides on glass slides (Scheme 3.4, **15**). These surfaces could be readily synthesized from amine-coated glass slides (Scheme 3.4, **13**) in three steps. The printed surfaces proved to be useful for rapid profiling of lectin binding and even binding of whole bacteria [19]. A similar approach modifying self-assembled monolayers on gold surfaces has also proved successful. The nonspecific adsorption and quenching

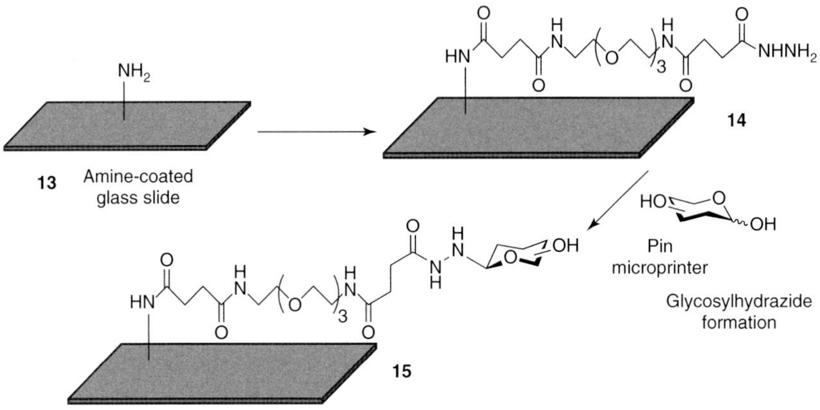

Scheme 3.4 Direct immobilization of oligosaccharides on glass slides via glycosyl hydrazides [18].

by the gold surface were overcome to give a robust surface functionalized with oligosaccharides including glycosaminoglycans [20].

The ease of formation of glycosyl hydrazides also lends itself to rapid labeling strategies to improve the ionization efficiency of oligosaccharides for mass spectrometric analysis. Oligosaccharides are not as efficiently ionized as peptides in matrix-assisted laser desorption/ionization mass spectrometry (MALDI-MS) likely because of the lack of basic sites for protonation in most carbohydrates. Improvements in the ionization by derivatizing reducing sugars through reductive amination with positively charged or basic groups have significantly improved (10–20-fold) the ionization efficiency. Kameyama and coworkers showed that greater improvements in ionization could be achieved with a simplified procedure using glycosyl hydrazide formation with cyanine hydrazides (Scheme 3.5, **17**). These conjugates (Scheme 3.5, **18**) showed sensitivities in the low- to sub-femtomolar range for the MALDI analysis of oligosaccharides [21]. This initial finding has been followed up by a series of studies that have analyzed different hydrazides,

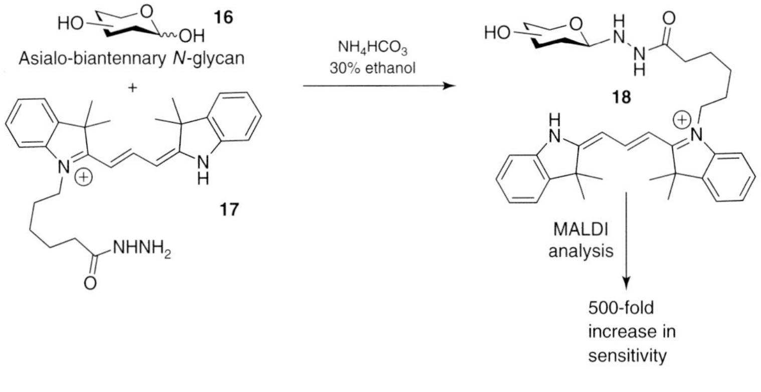

Scheme 3.5 Hydrazide labeling of oligosaccharides for MALDI analysis [21].

procedures for labeling, and purification protocols. The findings indicate that the best sensitivities are achieved with hydrazides that contain a positive ammonium group as well as a hydrophobic component [22]. The ammonium group provides a site to charge the carbohydrate, and the hydrophobic functional group is proposed to improve the incorporation of the glycosyl hydrazide in the MALDI matrix.

One of the most interesting analytical applications that has arisen from the use of glycosyl hydrazides is "glycoblotting" (Scheme 3.6) [23]. This is a powerful technique that circumvents major challenges in the analysis of complex oligosaccharides from heterogeneous biological samples, providing a useful tool for glycomics. Using solid-supported hydrazides (Scheme 3.6, **20**), reducing sugars can be selectively extracted by the formation of glycosyl hydrazides (Scheme 3.6, **21**), from a complex mixture such as that produced by cleavage of the *N*-glycans or *O*-glycans from a tryptic digest. After washing steps, and if required solid-phase esterification of the sialic acids [24], the oligosaccharides can be eluted from the solid support with a second hydroxylamine, hydrazide, or low pH treatment to allow for MALDI-MS analysis [25]. In its initial form, glycoblotting used hydroxylamine-functionalized beads as the oligosaccharide trapping matrix, but later modifications showed better recoveries with the use of hydrazide-functionalized beads in the oligosaccharide trapping step [26]. The glycoblotting solid-supported hydrazide beads are now commercially available (BlotGlyco H™). This approach has now been successfully applied to a wide variety of biological samples, allowing a rapid profiling of the glycome present [27].

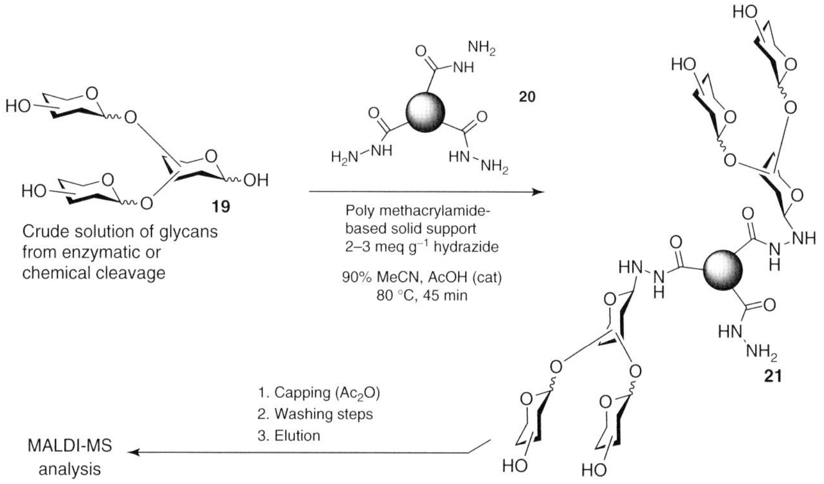

Scheme 3.6 Outline of glycoblotting procedure [23].

3.2.3
Hydrazides in Synthesis

Despite the ease of formation, the use of hydrazides in carbohydrate synthesis has thus far been limited. Early work often reported the acylation of glycosyl hydrazides as a method to confirm the structures of the condensation products. The use of glycosyl sulfonylhydrazides, with fully protected carbohydrates, was explored by Vasella and coworkers [28] as a convenient route to glycosyl carbenes, although this route was found to be lower yielding than with the use of glycosyl diazirines.

Recently, our lab has investigated the potential of glycosyl sulfonylhydrazides as a convenient route to protecting-group-free glycosidation reactions [12]. The use of protecting groups during *O*-glycoside formation has dominated the literature with the early exception of the work by Fischer, with his namesake glycosidation. Developing methods that forego protecting groups has the potential to dramatically improve the efficiency of carbohydrate synthesis.

Glycosyl sulfonylhydrazides glycosidation reactions are readily carried out in a series of two steps (Scheme 3.7). These reactions generally require smaller excesses of alcohol than are used with the Fischer glycosidations and can be carried out on oligosaccharide (Scheme 3.7, **22**) starting materials. First, like the formation of other glycosyl acylhydrazides, the condensation step is performed under mildly acidic conditions; we have had the most success in solutions of dimethylformamide (DMF) (for procedure, see Section 3.6.1). The glycosyl sulfonylhydrazide (Scheme 3.7, **24**) is then oxidized to activate the glycosyl donor with either a halogen or copper-based oxidant. Upon oxidation, the glycosyl sulfonylhydrazide is presumed to form a diazo-glycoside (Scheme 3.7, **25**) transiently prior to elimination of sulfinic acid and nitrogen gas, which is clearly observed evolving from the reaction mixture, resulting in the oxocarbenium ion (Scheme 3.7, **26**). The formed oxocarbenium ion can then be intercepted by the desired nucleophile. Generally, alcohols can be used in 10–20-fold molar excess without seeing significant self-polymerization. Other nucleophiles are also acceptable, provided they are tolerant to the oxidation

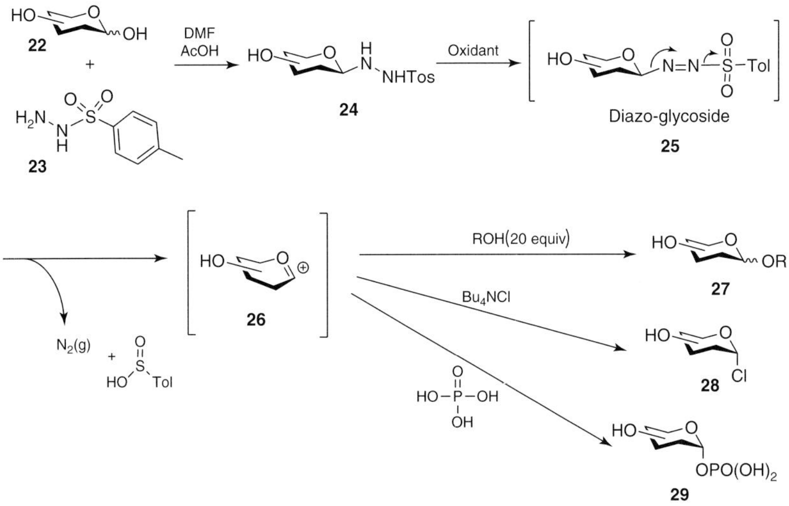

Scheme 3.7 Glycosides formed using glycosyl sulfonylhydrazides as glycosyl donors in protecting-group-free glycosidation reactions.

conditions. We have had success with chloride (Scheme 3.7, **28**) and phosphate (Scheme 3.7, **29**) among others (for procedure, see Section 3.6.3) [12, 29]. Surprisingly, the glycosyl chlorides are generated in good yields, as determined by ^{1}H NMR spectroscopy, although these species cannot be easily isolated because of their inherent lability. Alternatively, the glycosyl chloride generated *in situ* can be trapped with nucleophiles that are incompatible with the oxidation conditions necessary to activate the glycosyl sulfonylhydrazide such as thiourea and azide (for procedure, see Section 3.6.2) [18].

3.2.4
Biologically Active Glycoconjugates

In addition to the use of glycosyl hydrazides in oligosaccharide analysis, a variety of glycoconjugates have been used for applications in medicinal chemistry and glycobiology. In 2001, Peluso and Imperiali [30] synthesized an aspartyl hydrazide and incorporated this novel amino acid into peptides using standard solid-phase synthesis approaches. After deprotection, the hydrazide was condensed with GlcNAc to yield a mimic of an asparagine N-linked GlcNAc glycopeptide (Figure 3.1, **30**). This conjugate was further assayed as an inhibitor of oligosaccharyl transferase. Surprisingly, the GlcNAc hydrazide was a 20-fold more potent inhibitor than the corresponding natural glycosyl asparagine-containing peptide (Figure 3.1, **31**). The authors speculate that this may be due to the low barrier for rotation around the hydrazide in comparison with the amide linkage in the natural structure. These findings may explain why little product inhibition is observed in oligosaccharyl transferase-catalyzed glycosylations [10d]. Given the ease of synthesis of the glycosyl hydrazide linkages, other applications to combinatorially explore glycan drug conjugates will likely become more commonplace [11b, 31].

The high yielding nature of glycosyl hydrazide formation has allowed the late stage introduction of oligosaccharides on scaffolds for multivalent display. The first example of such was a mimic of a mucin produced by Godula and Bertozzi

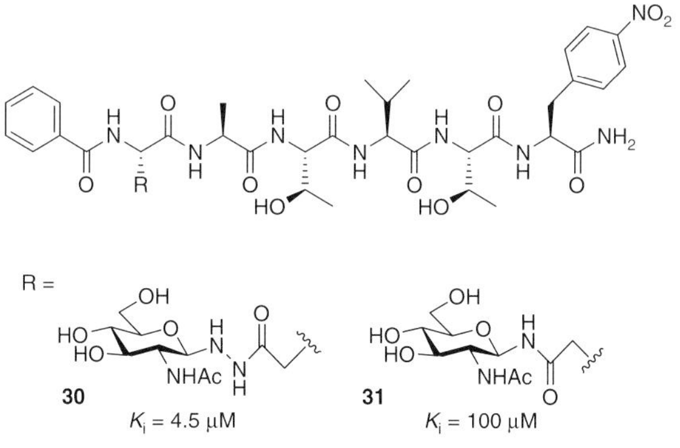

Figure 3.1 Inhibitors of yeast oligosaccharyl transferase show the preference for glycosyl hydrazides over natural linkage.

[32] (Scheme 3.8). From a poly(acryloyl hydrazide) bearing a terminal biotin (Scheme 3.8, **34**), they were able to condense a wide variety of oligosaccharides to form glycopolymers (Scheme 3.8, **35**). The advantage of this approach is that the reversible addition–fragmentation chain transfer (RAFT) polymerization used to synthesize the scaffold gives a narrow range of polymer lengths that can be readily controlled [33]. The functionalization step was done under aqueous conditions (>250 mM) and proceeded in good yield with high selectivity for the β-pyranoside. Aniline was used as a catalyst in these condensation reactions. Aniline has previously been shown to be a catalyst in hydrazone formation and in the formation of glycosyl oximes. Presumably, aniline plays a similar role in the formation of glycosyl hydrazides, although this remains to be definitively demonstrated [34]. The dense functionalization of the poly(acryloyl hydrazide) polymer scaffold mimics the dense display of glycans found in the mucins. These mucin mimics were readily immobilized on streptavidin-coated glass plates and probed with fluorescently labeled lectins.

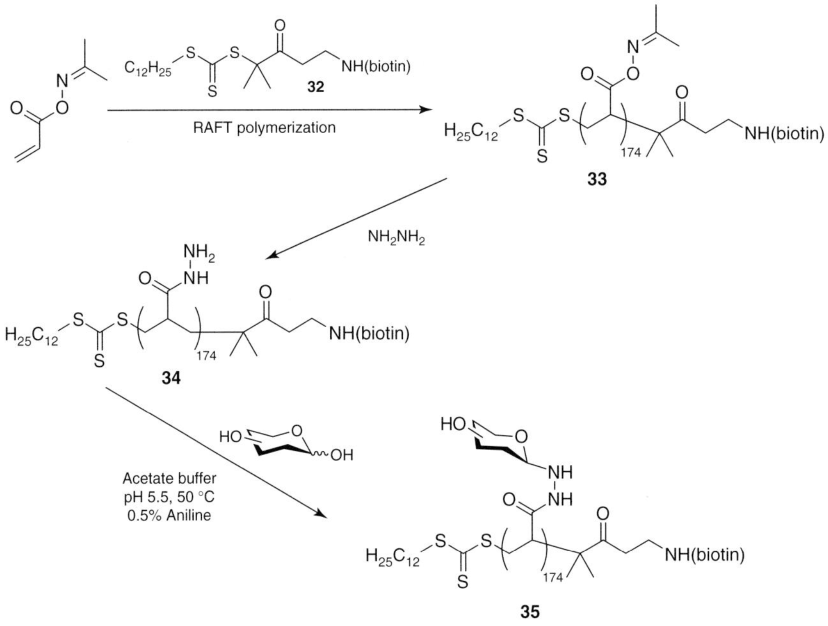

Scheme 3.8 Formation of mucin mimics.

Given the success of Bertozzi's example, it is perhaps surprising that not more multivalent scaffolds based on glycosyl hydrazide chemistry have been reported. One of the few examples has been described by Liu and coworkers, who have shown that oligosaccharides can readily be displayed on modified poly(amido amine) (PAMAM) dendrimers via a hydrazide linkage. These dendrimers were shown to be readily taken up by hepatocytes in a glycan-dependent fashion [35]. Similarly, dendrimers have been used to increase the density of carbohydrate display on solid surfaces by first coating a glass slide with PAMAM dendrimers and then modifying the surface with hydrazides [36].

3.2.5
Lectin-Labeling Strategies Using Glycosyl Hydrazides

One of the key challenges in glycobiology is the identification of protein-binding partners of a given oligosaccharide. This is difficult as affinities of the interaction are usually low (mM–μM K_D), and often only limited amounts of oligosaccharide are available, or the protein partner is not available in a purified form. Functional glycosyl hydrazides provide a tool to circumvent some of these challenges. The synthesis can be carried out in a single step on small amounts of oligosaccharide and these conjugates can then function as protein-labeling agents.

Hamachi and coworkers [37] gave the first elegant example of site specifically labeling the lectins using a catalytically active DMAP (4-dimethylaminopyridine)-based glycosyl hydrazide (Scheme 3.9, **38**). A straightforward synthesis of an *N,N*-dialkyl-4-aminopyridine hydrazide (Scheme 3.9, **36**) provided a compound that could be condensed with the desired glycan (Scheme 3.9, **37**). The lectin labeling chemistry is then achieved by addition of a phenylthioester (Scheme 3.9, **39**) to the solution of lectin and catalytic DMAP-functionalized glycoconjugate. The thioester is nucleophilically activated by the DMAP glycoconjugate, and acylpyridinium species is then held in close proximity to the lectin by the carbohydrates affinity. The approach works remarkably well as it is site selective, reacting with amino acids in close proximity to the carbohydrate-binding site, as well as protein selective working well in a mixture of proteins. Unfortunately, the hydrazide used in this

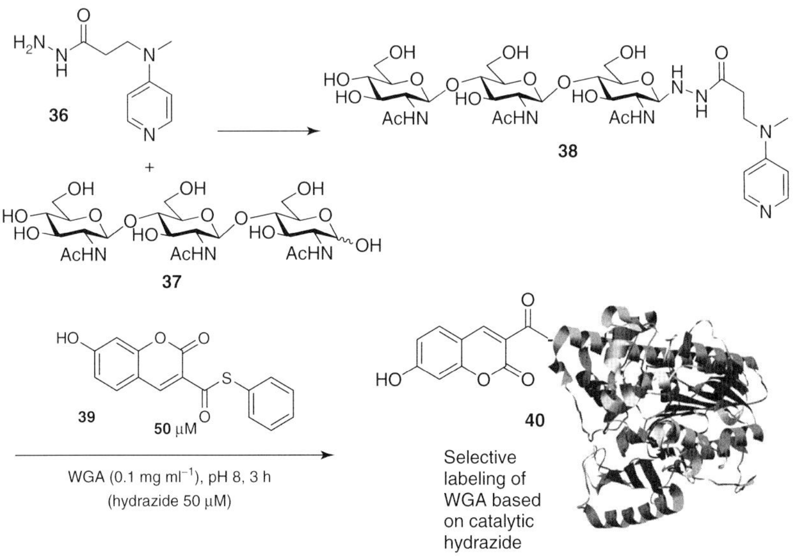

Scheme 3.9 Selective lectin labeling with catalytic glycosyl hydrazide [37].

report was monomeric, showing weak affinity for congerin II (Cong II), and more highly valent *O*-glycosides outperformed the glycosyl hydrazide. However, given the ease of hydrazide formation, it would be a logical extension to look at multivalent DMAP-bearing scaffolds.

An alternative strategy to the site-selective DMAP-catalyzed acylation is to prepare a single labeling construct based on an arylthioester-bearing glycosyl hydrazide (Scheme 3.10, **44**) [38]. These constructs can be prepared in two steps in aqueous solution. Initially, the desired reducing sugar (Scheme 3.10, **42**) is condensed with an arylthiol-bearing hydrazide (Scheme 3.10, **41**). The arylthiol-bearing glycoconjugate (Scheme 3.10, **43**) can then be exposed to thioester exchange conditions to allow formation of a thioester-bearing glycan (Scheme 3.10, **44**). It was found that sufficient acylation activity was realized with aryl thioesters, whereas alkyl thioesters were unreactive under the labeling conditions. The first example of this approach showed that site-specific labeling of maltose-binding protein (MBP) could be achieved with a maltosyl hydrazide construct. The labeling was selective for maltose, as lactose conjugates showed no background reaction. A single lysine residue, found at the mouth of the carbohydrate-binding groove in MBP, was labeled with the activated coumarin (Scheme 3.10, **45**). These hydrazide thioester

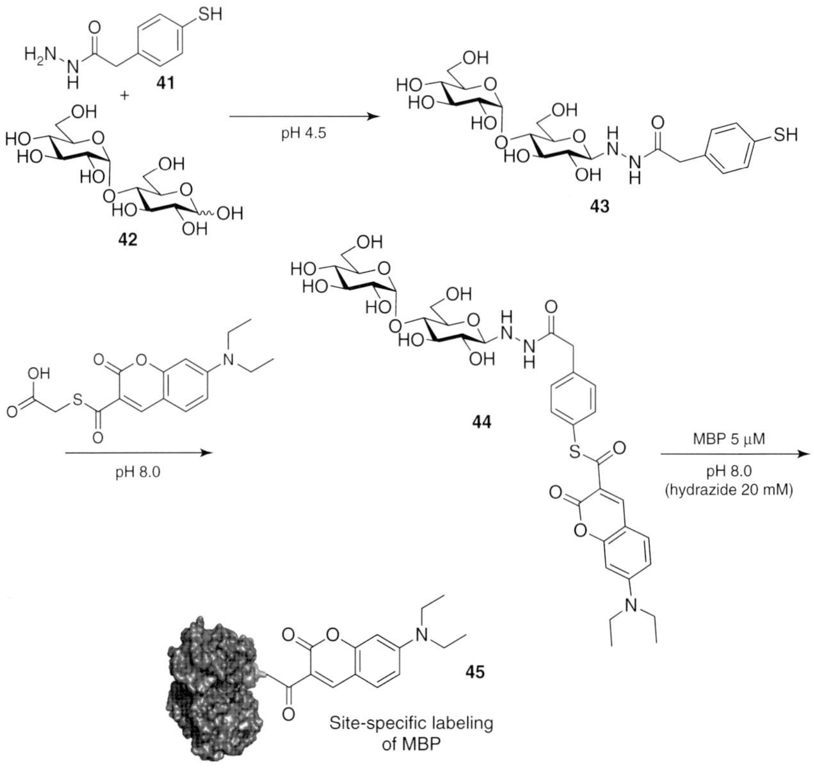

Scheme 3.10 Site-specific labeling of MBP with thioester-based glycosyl hydrazide.

conjugates also showed impressive selectivity, giving rise to site-selective labeling of MBP within live *Escherichia coli*.

3.2.6
Summary of Glycosyl Hydrazides

Glycosyl hydrazides represent an important group of glycoconjugates that are extremely versatile and readily synthesized in a predictable fashion. The ability to form glycosyl hydrazides in aqueous solutions makes them ideal interfaces between isolated or enzymatically synthesized oligosaccharides and analysis techniques, which often require immobilization or a functional handle to be installed. The key point to remember when forming these conjugates in aqueous solution is that the equilibrium constant for formation is small; thus, high concentrations (preferably >500 mM) of one of the components must be present in the reaction. The use of glycosyl hydrazides in synthesis provides an area that remains largely unexplored, but the early results suggest that they provide an excellent method to access reactive glycosylation intermediates, providing new routes to oligosaccharides or functional glycoconjugates.

3.3
O-Alkyl-*N*-Glycosyl Oxyamines

First accounts of the condensation of hydroxylamines with sugars can be found in the 1880s [39]. Owing to the lack of analytical methods to determine molecular structures, these early condensations were used for the characterization of sugars, the best known example of this chemistry is the Wohl degradation [40]. Here, we cover the more recent applications and characterization of *O*-alkyl-*N*-glycosyl oxyamines.

3.3.1
Formation, Configuration, and Stability of *O*-Alkyl-*N*-Glycosyloxyamines

Hydroxylamine condensations with reducing sugars proceed in high yield under mildly acidic conditions, but much like the hydrazines, a tautomeric mixture of condensation products is observed. The first NMR investigation of the condensation between glucose and hydroxylamines showed an equilibrium mixture made up of predominantly acyclic oximes and a small quantity of the pyranosyl conjugates [41]. More recent studies have confirmed these findings showing that, in the case of the glucose-configured hexoses, 20–30% of the equilibrium mixture exists in the pyranosyl form (Scheme 3.11, **48**) with the balance of the mixture made up of the diastereomeric oximes (Scheme 3.11, **46**, **47**) [42]. Further studies are required to elucidate the product distribution for other carbohydrate configurations, but it is likely that a significant portion of the population will exist as the open chain oxime.

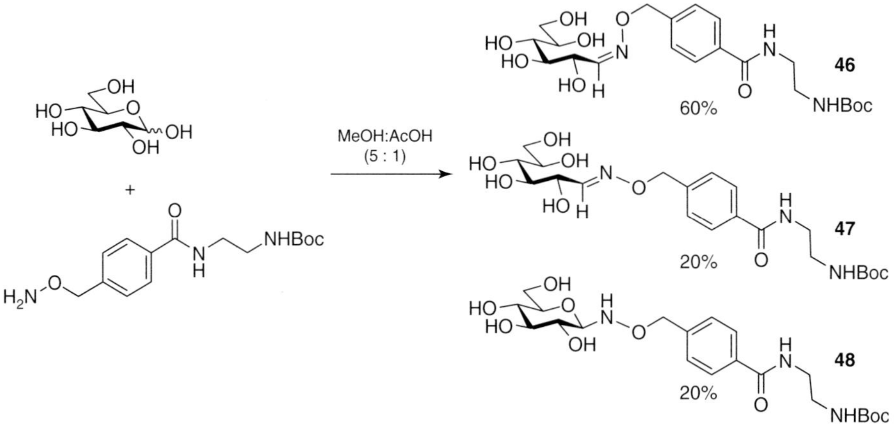

Scheme 3.11 Tautomeric equilibrium of glucose condensation products with *O*-alkyloxyamine [42c].

3.3.2
Uses of *O*-Alkyl-*N*-Glycosyl Oxyamines

Applications forming hydroxylamine conjugates such as glycopeptides have appeared despite the minor population of ring-closed products formed in these condensations. The condensation with hydroxylamines is more favorable than the formation of hydrazides or *N*-alkyl-*N*-glycosyloxyamines (see the subsequent text); thus, this reaction provides an excellent method for conjugating carbohydrates when the structure of the reducing sugar is not critical for function [42c, 43].

3.4
N,*O*-Alkyl-*N*-Glycosyl Oxyamines

In an insightful step, Dumy and coworkers [44] dramatically improved the utility of hydroxylamine-based glycoconjugates. With the introduction of an alkyl substituent at nitrogen, Dumy and coworkers demonstrated that the resulting *N*-alkyloxyamines can serve as robust coupling partners with reducing sugars under mildly acidic conditions. N-Alkylation of the oxyamine precludes oxime formation, and the resulting conjugation products adopt the ring-closed forms exclusively. Dumy and coworkers report the condensation to be diastereoselective. Using *N*-methylhydroxylamine derivatives with glucose and galactose mainly, the *N*-β-glycopyranosyloxyamines were produced, whereas mannose gave predominately the α-anomer. In the case of galactose and mannose, significant quantities of the furanosides were also observed. Later detailed studies using *N*-benzyloxyamines (Scheme 3.12, **49**) have observed similar mixtures of α- and β-pyranose anomers (Scheme 3.12, **50**, **51**) as well as the furanosyl oxyamines (Scheme 3.12, **52**) [45].

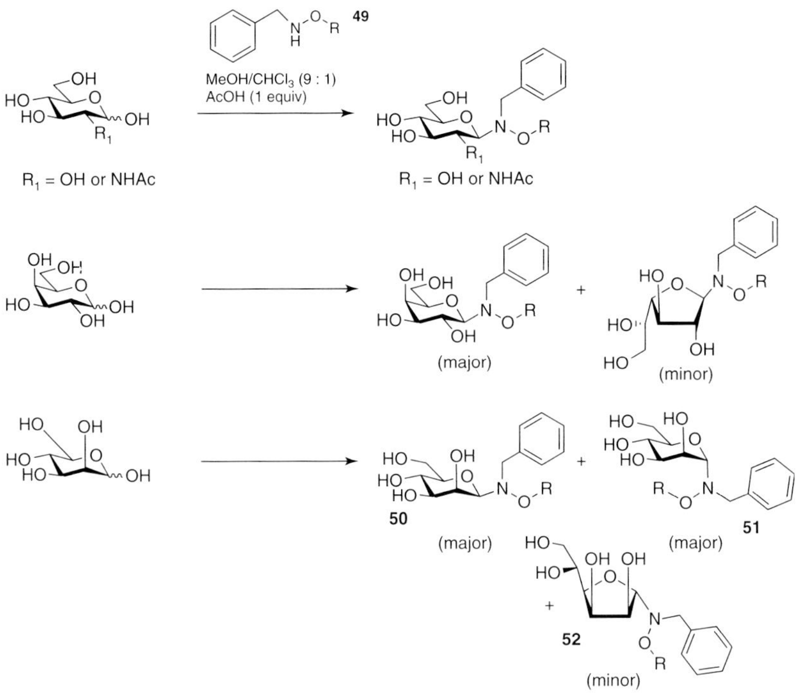

Scheme 3.12 Tautomeric mixture formed from condensation of monosaccharides with *N*-benzyl-*O*-alkyl-oxyamines [45].

Although, the selectivity of the reaction was initially proposed to be due to the selective *exo*-trig cyclization onto the oxyiminium ion, the reversibility of the reaction suggests that the relative product stabilities control the product distribution [45].

A range of *N*-alkyloxyamine conjugates have now been reported with applications ranging from medicinal chemistry to material chemistry. The conjugations have been performed under a wide range of conditions in both aqueous and organic solutions with good-to-excellent yield [45]. Much like in oxime formation, the reaction is catalyzed by weakly acidic solutions. The pK_a of *N*-alkylhydroxylamines is in the range of 5, and thus strongly acidic conditions should be avoided. The choice of reaction solvent is governed by the solubility of the coupling partners involved; many researchers have reported success in mixed solvents of DMF/H_2O/AcOH solutions. Thorson and coworkers have reported hundreds (see the subsequent text) of *N*-alkyloxyamine-based condensation reactions, and their conditions employ the coupling partners at concentrations >100 mM with acetic acid as the catalyst. In aqueous conditions, the concentrations of reagents are critical, as measurement of the apparent equilibrium constants for the reaction at pH 4.5 in D_2O for a series of *N*-methylhydroxylamines shows the equilibrium constants to be small (10–20 M) [10g]. Thus, to obtain high yields of the desired conjugates, the reactions must be run at high concentrations of one or both of the coupling partners. The small equilibrium constant for the reaction is likely a factor in the low yields reported for some condensation reactions.

In our own experiments on this matter, we could confirm the direct correlation between the concentration and equivalence of reagent to the yield of conversion. The ^{1}H NMR spectra from condensation reactions of *N*-acetyl glucosamine with *N*,*O*-dimethyl hydroxylamine (DMHA) showed a decrease of the free anomeric sugar proton signals (5.2 and 4.75 ppm) with increased amounts of reagent. The integral of this signal is equivalent to the percentage of unreacted sugar (Figure 3.2).

The stability of hydroxylamines to aqueous hydrolysis parallels that of glycosyl hydrazides. The hydrolysis is a specific acid-catalyzed process [10h, 42c, 45]. The half-lives of hydrolysis at pH values >6 are on the order of days to months. As with the glycosyl hydrazides, more electron-poor hydroxylamines and more electron-poor monosaccharides yield more hydrolytically stable conjugates [10g]. The difference in stabilities can be dramatic, as similar conjugates of xylose and

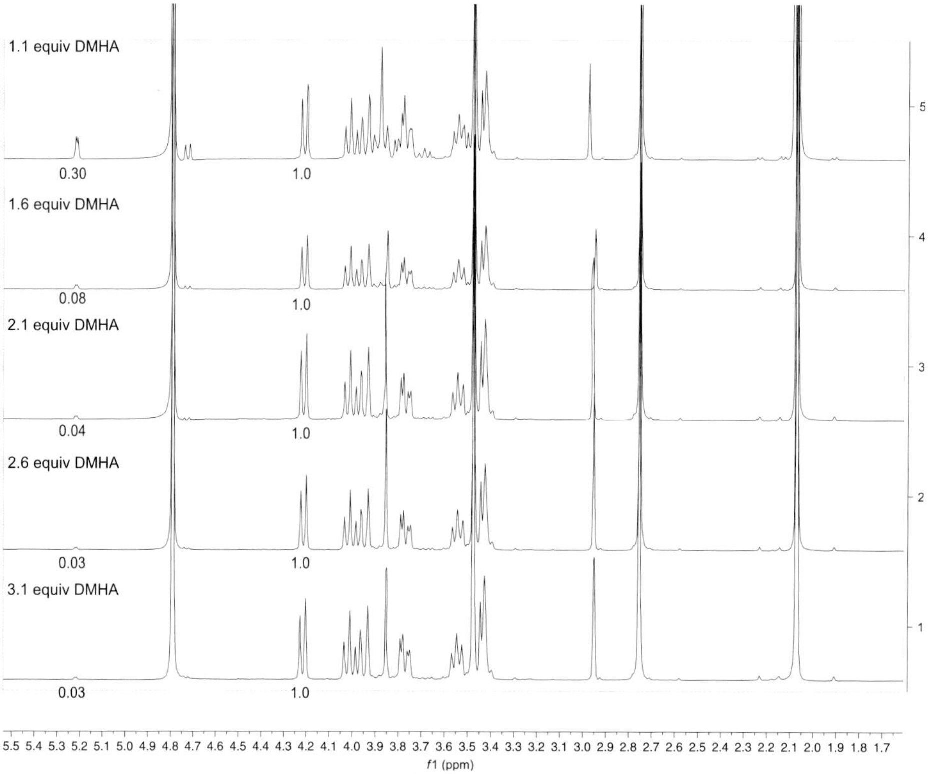

Figure 3.2 Crude ^{1}H NMR spectra in D_2O from condensation of GlcNAc with *N*,*O*-dimethyl hydroxylamine (DMHA) at different concentrations of reagent. Conditions: GlcNAc (100 mg, 1.0 equiv) in minimum amount of water (380 µl) combined with DMHA and sodium acetate (1.1, 1.6, 2.1, 2.6, or 3.1 equiv) in minimum amount of water (118, 172, 226, 280, or 334 µl) and stirred for 24 h at room temperature.

GlcNAc vary by a factor of 40 in hydrolysis half-life. Jensen and coworkers performed insightful studies on the effect of varying the *N*-alkyl group. Their comparisons, of *N*-methyl and *N*-benzyloxyamines, revealed *N*-benzyl conjugates to be five- to sixfold more stable than their methyl counterparts [42c]. This observation was explained by the aforementioned electronic effects, although steric effects could also be of importance. In addition, it is likely that electron-withdrawing substituents on the *N*-aryl ring can further decrease the rate of hydrolysis. As a rule, electron-withdrawing groups at the nitrogen of hydroxylamines seem to play the primary role in terms of hydrolysis stability, as Langenhan and coworkers [45] found only minor differences in hydrolysis rates for a series of *N*-benzyl-*O*-alkyloxyamine conjugates with varying *O*-alkyl substituents.

3.4.1
Uses of *N*-Alkyl-*N*-Glycosyloxyamines

Like the glycosyl hydrazides, *N*-alkyl-*N*-glycosyloxyamines have been used extensively to form functional glycoconjugates. Blixt and coworkers [46] were the first to exploit *N*-alkyloxyamine linkers for the attachment of oligosaccharides to *N*-hydroxysuccinimide(NHS)-activated glass slides. Later, introduction of an aryl substituent into the *N*-alkyloxyamine linker allowed for ease of HPLC purification of the conjugated oligosaccharides [42c]. Examples employing the fluorous phase using fluorous-functionalized *N*-alkyloxyamine linkers for efficient oligosaccharide immobilization have been reported [47]. The fluorous-functionalized linkers allowed for easy purification of the *N*-glycosyl oxyamines as well as their immobilization on fluorous-coated glass plates, making this approach especially conducive to automation. The formation of linkers to attach reducing sugars to proteins has also been reported [48]. The challenge associated with using *N*-alkyloxyamine linkers is that significant quantities of furanosyl conjugates are observed with nongluco-configured carbohydrates, complicating the conjugation chemistry. In comparison, hydrazides are observed to give cleaner conjugations with nongluco sugars, and despite glycosyl hydrazides being somewhat less stable than *N*-alkyl-*N*-glycosyloxyamines, they may be a superior choice for oligosaccharide immobilization.

3.4.2
Glycobiology

Dumy and coworkers [44] capitalized on the potential of using *N*-methyloxyamine-functionalized amino acids in the synthesis of glycopeptide mimics, reporting this work simultaneously with the formation of these useful conjugates. The proof of concept was illustrated through the acylation of a lysine residue with a protected *O*-oxyamine, and after deprotection and reductive amination with formaldehyde, the desired amino acid was formed (Scheme 3.13, **52**). Condensation of the functionalized peptide with a mono-, di-, or trisaccharide (pH 4.5) gave the desired glycopeptides mimics (Scheme 3.13, **53**) in 35–80% yield after HPLC purification.

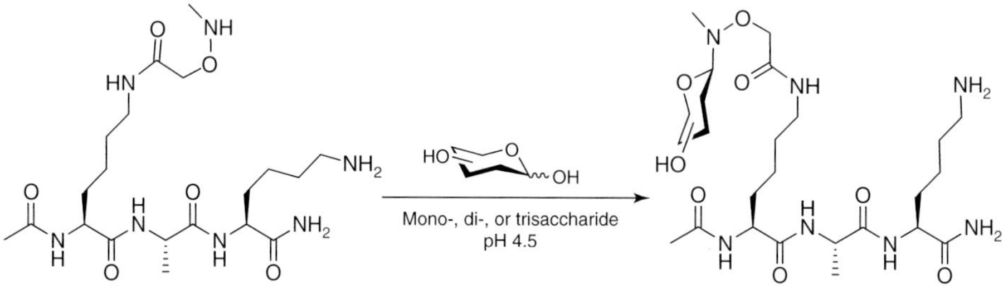

Scheme 3.13 Formation of glycopeptide conjugates.

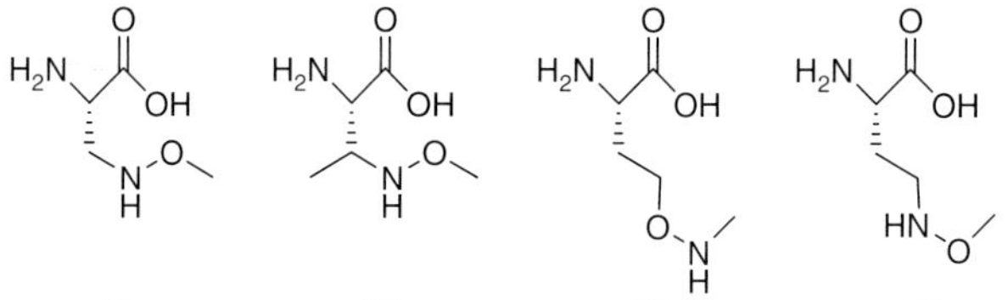

Figure 3.3 *N*-Alkylaminoxy-based amino acids for the synthesis of glycopeptides mimics.

Further work to improve *N*-alkyl-*N*-glycosyloxyamine peptide mimics has focused on preparing amino acids that are more structurally similar to asparagine (Figure 3.3, **57**), serine (Figure 3.3, **54**), homoserine (Figure 3.3, **56**), or threonine (Figure 3.3, **55**) in order to mimic naturally occurring *N*- or *O*-linked glycosides [49]. These amino acids have been prepared with either Boc or Fmoc protection and have been reported to be incorporated smoothly under standard solid-phase peptide synthesis conditions [50]. All of the amino acids with the exception of the threonine analog (Figure 3.3, **55**) could be glycosylated under aqueous conditions (pH 4.5) with glucose (Figure 3.3) [49]. The failure of the threonine residue is proposed to be due to steric bulk at the amine. Other branched carbon substituents have successfully been glycosylated (Section 3.4.3) albeit in low yield.

Peri and coworkers have explored the use of *N*-alkyl-*N*-glycosyloxyamine glycosides as *N*-linked mimics of interglycosidic *O*-glycosides in oligosaccharides. Given the ease of formation of the *N*-glycosyloxyamines and the diastereoselectivity of their formation, oligosaccharide mimics based on these linkages have the potential to provide an efficient route to good mimics of oligosaccharides. The *N*-glycosyloxyamine linkages are likely not recognized by glycosidase enzymes and as such these linkages may provide greater biological stability than their *O*-glycosidic counterparts. The approach has been explored for the formation of ($\beta1 \rightarrow 6$)-linkages of di- and trisaccharides (Scheme 3.14, **60**) [51]. Via a straightforward reductive amination strategy, it was possible to install a methoxyamino substituent at position 6 of glucose (Scheme 3.14, **58**), which can then be condensed with 6-methyloxime-substituted reducing sugars (Scheme 3.14, **59**). After oxime

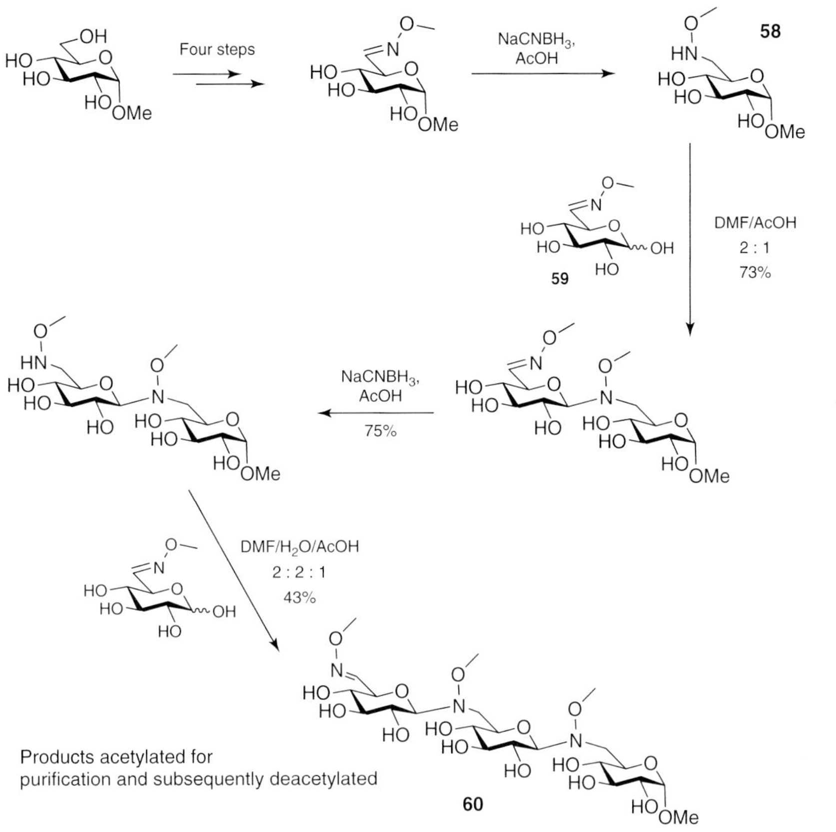

Scheme 3.14 Synthesis of *N*-linked oxyamine trisaccharide mimic.

reduction, the steps could be iterated. The glycosylations proceed with reasonable yield (35–90%) and allowed rapid generation of a trisaccharide (Scheme 3.14, **60**). The tolerance of the *N*-glycosyloxyamines to treatment with $NaCNBH_3$ (sodium cyanoborohydride) in glacial acetic acid supports the chemically inert nature of this linkage as the corresponding oxime is readily reduced under these conditions. The reduced glycosylation yield observed in the formation of the trisaccharide may be due to a small percentage of cleavage of the disaccharide under these acidic aqueous conditions where the *N*-glycosyloxyamines are labile. ^1H NMR conformational analysis of the *N*-linked disaccharide, which would be a mimic of gentiobiose, showed similar conformations to those found in the naturally occurring disaccharide, albeit with altered equilibrium populations [51]. Unfortunately, subsequent reports of the formation of ($\beta 1 \rightarrow 4$) oxyamine glycosidic linkages used glycosyl bromide donors rather than the simple condensation conditions, suggesting that the simple reactions were not successful perhaps because of a steric congestion around the secondary oxyamine substituent [52]. A similar problem was encountered in the synthesis of a lipid A analog [53].

3.4.3
Medicinal Chemistry

By far, the largest application of *N*-alkyl-*N*-glycosyl oxyamines has been in the area of medicinal chemistry. This is largely due to the work of the Thorson's group [54], which introduced these glycosides as an efficient and rapid method to "neoglycosylate" natural products to modulate their activities. The success of this approach lies in its general strategy (Scheme 3.15). Initially, a generally short synthesis from the drug or natural product to append an *N*-alkyloxyamine is established. The functionalized aglycon is then condensed with a library of reducing sugars to generate a library of neoglycosylated compounds. Often, mixtures of anomers and cyclic forms of the neoglycosides are produced in the condensation step. As this mixture represents an interconverting equilibrium population, they cannot be easily separated and thus these products are assayed as a mixture. The activity of these mixtures is then characterized in the desired biochemical assays.

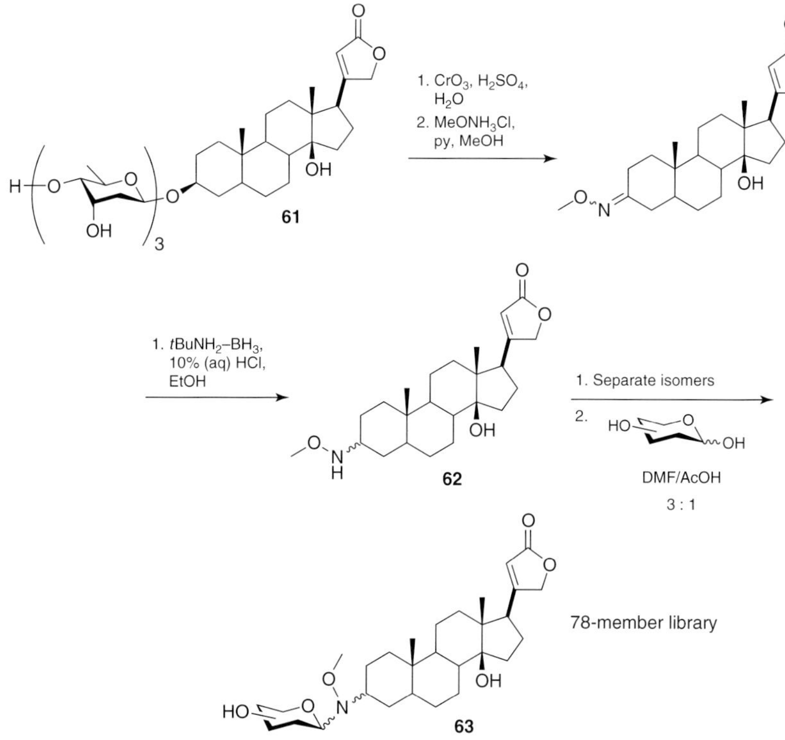

Scheme 3.15 Neoglycosylation of digitoxin [54].

The proof of concept example examined the cardiac glycoside digitoxin (Scheme 3.15, **61**). Digitoxin is an interesting target as it is a well-known inhibitor of the plasma membrane Na^+/K^+–ATPase (ATP, adenosine triphosphate), but also is known to have anticancer activity, activity against polyglutamine disorders, and to inhibit activation of the nuclear factor kappa-light-chain-enhancer of activated B cells (NF-κB) signaling pathway in cystic fibrosis (CF) cells. By neoglycosylating digitoxin, it provides a strategy to deconvolute these activities and potentially reduce off target effects of this promising molecule. In two steps, the glycoside of digitoxin was cleaved and a methoxyamine functional group was introduced by reductive amination (Scheme 3.15, **62**). The *N*-alkyloxyamine was then readily "neoglycosylated" in a 3 : 1 solution of DMF and acetic acid (Scheme 3.15, **63**). Interestingly, the crystal structure of the β-glucosyl neoglycoside of digitoxin was similar to that of the homologous *O*-glycoside. This suggests that if a promising lead was identified in a neoglycoside library, a specific *O*-glycoside that does not form tautomeric mixtures could be synthesized to supply a single compound for biological evaluations. Several of the 78-compound neoglycoside library synthesized showed significant improvements when analyzed for cytotoxicity against a panel of human cancer cell lines. Compounds with improved selectivity between cancer cell lines (three to fivefold over digitoxin) and/or compounds with improved potency (ninefold over digitoxin) were identified. Most notably, many of the cytotoxic compounds had significantly reduced Na^+/K^+–ATPase inhibition, suggesting that these compounds may have fewer off-target toxic effects when used in cancer treatment in comparison to digitoxin. The analysis of the library compounds indicated strong trends between glycoside stereochemistry and biological effects, suggesting that the glycoside is mediating the observed differences in activity.

Since reporting this groundbreaking study on digitoxin, a wide variety of drugs and natural products have been neoglycosylated including colchicine (Scheme 3.16, **65**) [55], vancomycin [56], betulinic acid [57], cyclopamine (Scheme 3.16, **64**) [58], warfarin [59], chlorambucil [11b], calicheamicin (Scheme 3.16, **66**) [60], and a Ras-inhibitor. In all of these cases, modulation of the neoglycosylated product's biological activities has been achieved in a glycan-dependent manner. The syntheses have illustrated that differences in the tautomeric mixture of glycosides obtained in the glycosylation vary with the aglycon, making it difficult to predict the products formed in the mixtures without detailed ^{1}H NMR investigations.

3.4.4 Carbohydrate Synthesis Using *N*-Alkyloxyamines

Given the ease of synthesis and cleavage of *N*-alkyl-*N*-glycosyloxyamines, they have been explored as tools for carbohydrate synthesis. Early reports demonstrated that these compounds are robust toward acetylation and deacetylation reactions, as these reactions have been used to facilitate purification of the *N*-alkyl-*N*-glycosyloxyamines [51, 61]. The use of *N*-alkyloxyamines has been further explored in our laboratory as persistent anomeric-protecting groups for oligosaccharide synthesis [62]. Given the tautomeric mixtures formed in the condensations with *N*-alkyloxyamines and reducing sugars, the protecting group is best suited to gluco-configured monosaccharides that generally form the β-pyranosyloxyamines exclusively. The formed conjugates are stable to many carbohydrate-protecting group manipulations such as benzylation, silylation, and benzylidene formation.

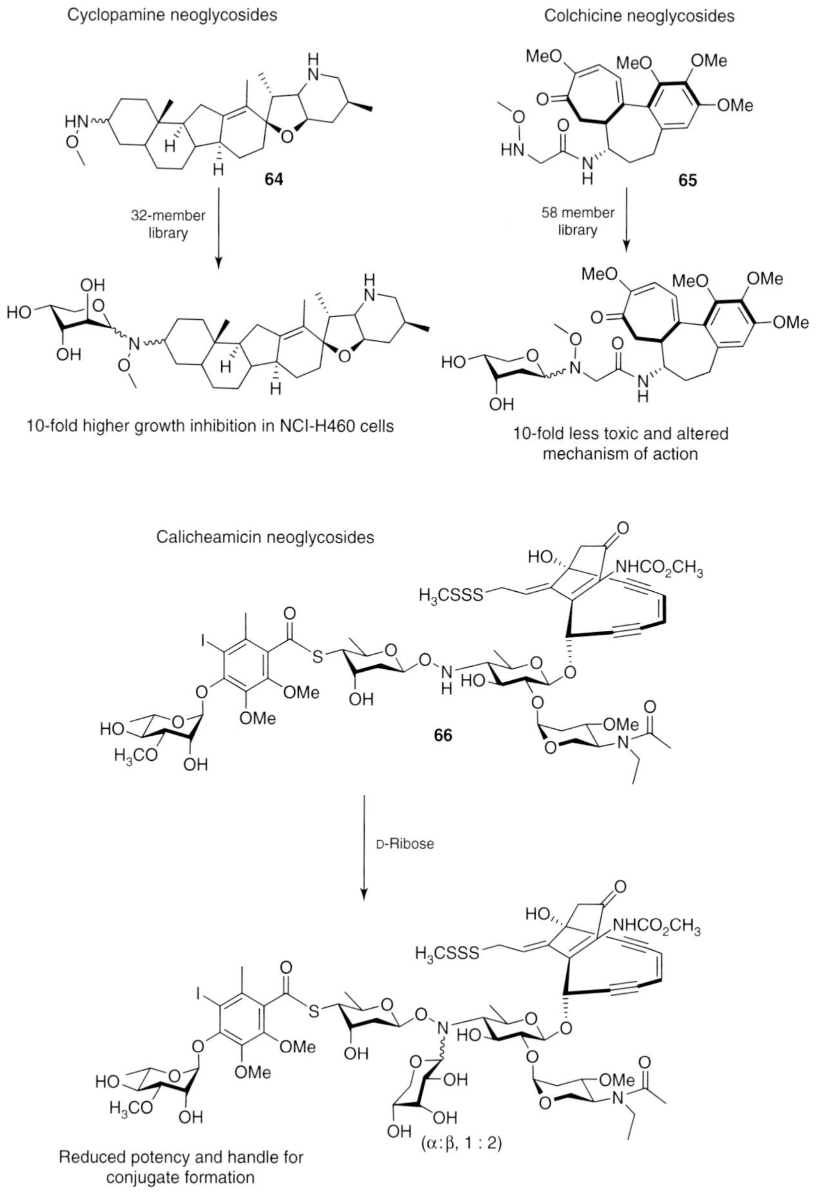

Scheme 3.16 Examples of neoglycosylation and resulting improved compounds.

Glycosylation conditions involving the activation of thioglycosides (Scheme 3.17, **68**) and trichloroacetimidates (Scheme 3.17, **69**) have also been successful in the presence of *N*-glycosyl-*N*,*O*-dimethyloxyamines (Scheme 3.17, **67a**, **67b**). Interestingly, the glycosyl oxyamine anomeric-protecting group can be cleaved in the presence of acid labile groups, such as benzylidenes, using *N*-chlorosuccinimide

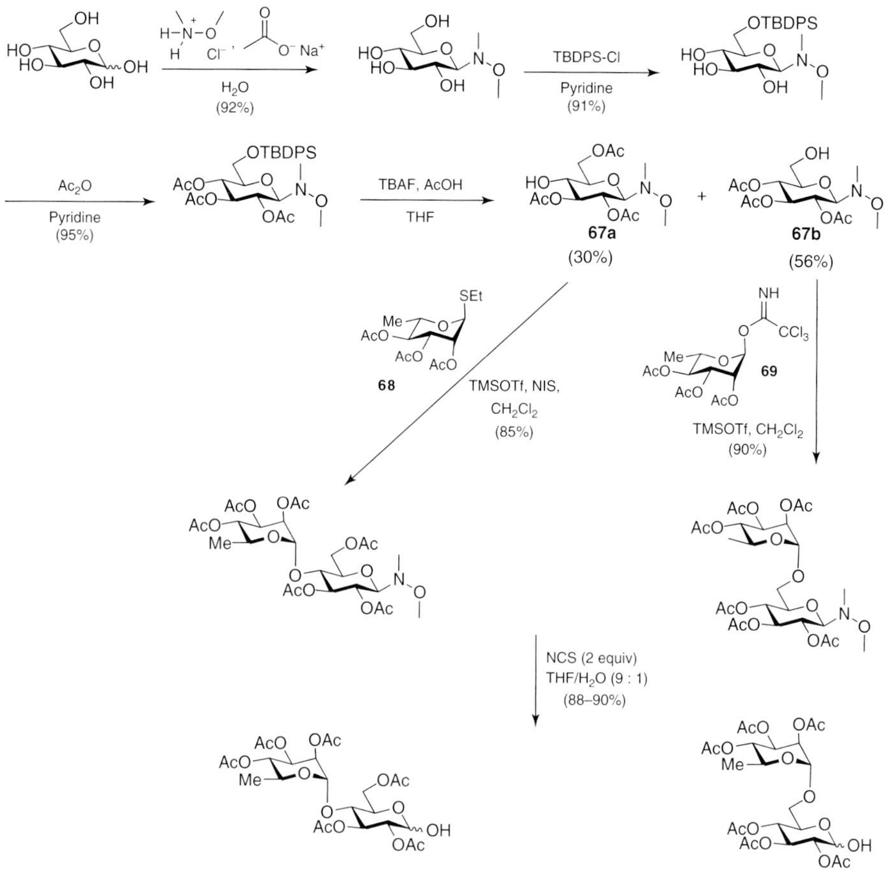

Scheme 3.17 Examples of disaccharide syntheses using alkyloxyamine glycoside-protecting groups.

as an oxidant. Unfortunately, to this point, efficient means of activating *N*-glycosyloxyamines as glycosyl donors has not been identified. An example of using *N*,*O*-dimethyloxyamine as a protecting group in oligosaccharide synthesis is given in Scheme 3.17. The major advantage of using this protecting group strategy is that the anomeric position can be directly protected, without intervening protecting group manipulations early in the synthesis. The free anomeric hemiacetal can then be readily liberated late in a synthesis via a mild aqueous acidic hydrolysis or relatively strong oxidizing conditions.

3.4.5 Summary of *N*-Alkyl-*N*-Glycosyl Oxyamines

The condensation reaction between free hemiacetals and *N*,*O*-alkyl or aryl glycosyl oximes developed by Dumy and coworkers allows the diastereoselective derivation of various sugars to ring-closed compounds, albeit in some cases as furanoside and pyranosides. Although reducing the number of possible equilibrium states during the product formation, these reaction conditions still required high concentrations to push the equilibrium toward a complete product formation. The increased hydrolytic stability of *N,O*-alkyl glycosyl oximes in comparison with glycosyl hydrazides makes these linkages more attractive for some conjugation reactions. The heightened stability, in particular of the *N*-benzylated species, can be explained by an electron-poor hydroxylamine system that disfavors the formation of the oxyiminium ion, which represents the first step of the acid-catalyzed hydrolysis.

3.5 Concluding Remarks and Unanswered Questions

The ease of formation of glycosides from hydrazide and oxyamines has provided useful strategies for carbohydrate analysis, binding studies, and glycodiversification. Key details of this process remain poorly understood and require further investigation. The controlling factors in determining the relative proportions of the cyclic, acyclic, and anomeric tautomers, observed in the condensations with reducing sugars and hydrazides as well as glycosyl oxyamines, remain unclear. The current available data suggests that the hydrazides form single condensation products in most situations and that mixtures are usually observed with oxyamine conjugates, with the exception being *gluco*-configured sugars. A detailed study of a range of glycoses and differentially substituted hydrazide and glycosyl oxyamines in aqueous as well as nonaqueous solvents should clarify this issue. It is likely that the balance between the strength of the anomeric effect and the steric interactions present will be the dominant factors. If conditions, or oxyamino aglycons that produce single isomers can be identified, this will dramatically increase the potential for the use of these conjugates in oligosaccharide synthesis.

From the hydrolysis rates of the hydrazide and aminoxy glycosides, it appears that the hydrazides form more easily and are hydrolyzed more easily. This is likely why the hydrazides have dominated the analytical applications to date. For biological applications and potentially medicinal chemistry applications, it is certainly desirable to form more stable conjugates. Work by the Raines's group [63] on the stability of oximes and hydrazones certainly suggests that increasing stability may be achieved by introducing a vicinal positive charged to destabilize any positive charges generated in the hydrolysis of the linkage. This would likely be achieved in a two-step protocol first forming the conjugate and then further functionalization, but this has not yet been explored.

The use of hydrazides and oxyamine condensation products in synthesis remains relatively uncultivated. The ability to form these compounds in the absence of protecting groups suggests that new protecting-group-free manipulations are feasible. Certainly, the recent results showing that metal-based oxidants such as copper chloride are able to activate glycosyl sulfonylhydrazides suggest that catalytic procedures amenable to mild conditions are within reach.

3.6 Procedures

3.6.1 Formation of the *p*-Toluylsulfonylhydrazide

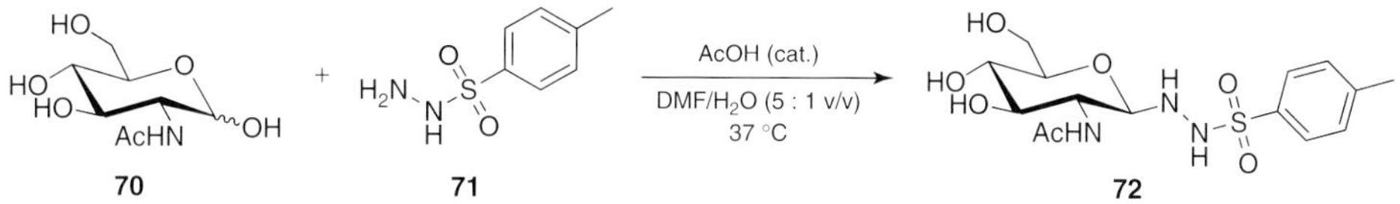

To a solution of 2-acetamido-2-deoxy-D-glucose (**70**) (5.0 g, 22.6 mmol) in DMF (10 ml) and water (5 ml), toluene sulfonyl hydrazide (**71**) (4.64 g, 24.9 mmol) and glacial acetic acid (1 ml, 18 mmol) are added. The suspension is incubated at 37 °C without stirring until all solids have dissolved (1.5 days). The solution is poured into diethylether (800 ml) and stirred vigorously for 18 h. After decanting the solvent, the solid is redissolved in water (20 ml) and freeze-dried to yield a white solid (**72**) in high purity (8.18 g, 21.0 mmol, 93%).

3.6.2 Formation of Azido-Glycosides

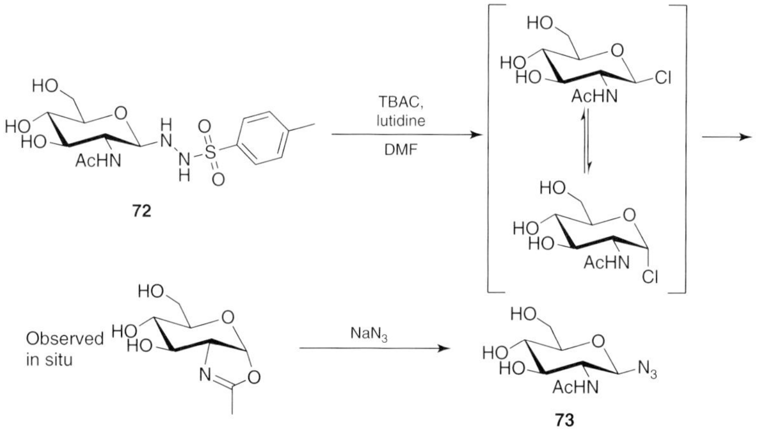

Tosylhydrazine donor (**72**) (100 mg, 257 μmol), tetrabutylammonium chloride (383 mg, 1.38 mmol), and 2,6-lutidine (149 μl, 1.28 mmol) were added to anhydrous DMF (2.0 ml), forming a homogeneous solution. *N*-Bromosuccinimide (110 mg, 618 μmol) was added to the solution at room temperature and after 15 min of stirring, nitrogen evolution had ceased, and NaN_3 (84 mg) was added. After stirring the reaction mixture overnight at room temperature, the reaction was quenched by the addition of water (10 ml). The aqueous layer washed with dichloromethane (DCM) (3 × 30 ml) and the aqueous solvent was removed *in vacuo*. The residue was purified by silica gel column chromatography (8% MeOH in DCM), yielding a white powder **73** (46 mg, 190 μmol, 73%).

3.6.3
Formation of Glycosyl Phosphate

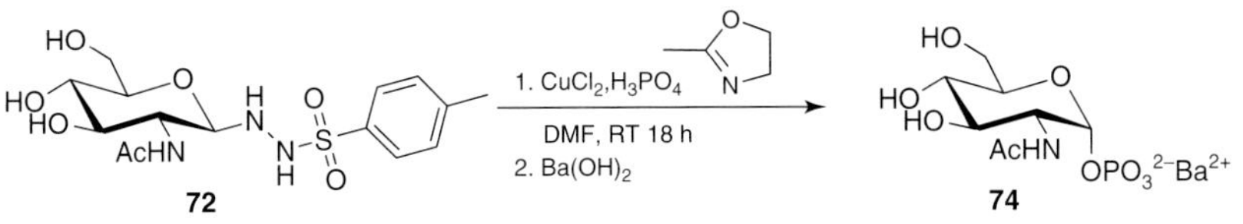

The reaction is conducted under a nitrogen atmosphere. Spherical 4 Å molecular sieves (~1 g) and the tosylhydrazine donor (**72**) (557 mg, 1.43 mmol) are added to anhydrous DMF (20 ml) and stirred for 15 min at room temperature. To a solution of cupric chloride (790 mg, 59.0 mmol) in dry DMF (5 ml), 2 methyl-2-oxazoline (0.50 ml, 5.9 mmol) is added and mixed by vigorous shaking. A solution of crystalline phosphoric acid (2.25 g, 23.0 mmol) in DMF (5 ml) is added to the cupric chloride and stirred for 30 s before addition to the solution of the tosylhydrazine donor. After stirring the reaction mixture for 18 h at room temperature, the mixture is poured into methylene chloride (150 ml) and filtered. The solid is dissolved in a minimal amount of DMF (~15 ml) and diluted with ethanol (75 ml). A saturated aqueous solution of barium hydroxide is added dropwise to a pH of 8–9. The suspension is filtered and the solid extracted with hot water (75 ml, 70 °C). The combined filtrates are reduced *in vacuo*. The solid is dissolved in water (5 ml), precipitated with ethanol (20 ml), pelleted by centrifugation (3700 g, 15 min), and the solvent decanted. This dissolution–precipitation was repeated three times, yielding the desired glycosyl-1-phosphate (**74**) (375 mg, 0.86 mmol, 60%) in 80% purity and an α/β ratio of 6.7 : 1.

3.6.4
Formation of *N*,*O*-Dialkyloxylamine Glycoside

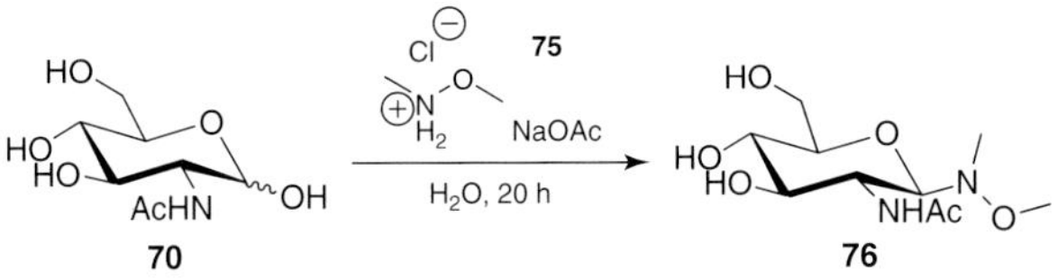

N,*O*-Dimethylhydroxylamine hydrochloride (**75**) (11.46 g, 117.5 mmol) and sodium acetate are dissolved in water (28 ml) and slowly added to a solution of 2-acetamido-2-deoxy-D-glucose (**70**) (10 g, 45.2 mmol) in water (38 ml) at 0 °C. The reaction mixture is stirred for 20 h at room temperature, after which the complete conversion can be observed by the appearance of a faster moving spot on a silica plate TLC (DCM:MeOH = 5 : 1, $R_f = 0.4$). The solvent is removed *in vacuo*, and the crude solid is purified by flash silica gel column chromatography (DCM:MeOH = 8 : 1), yielding a white crystalline solid (**76**) (11.35 g, 42.95 mmol, 95%).

List of Abbreviations

AcOH	Acetic acid
ATP	Adenosine triphosphate
Boc	*tert*-Butoxycarbonyl
CF	Cystic fibrosis
E. coli	*Escherichia coli*
K	Equilibrium constant
K_D	Dissociation constant
Cong II	Congerin II
DCM	dichloromethane
d_6-DMSO	Deuterated dimethyl sulfoxide
DMAP	4-Dimethylaminopyridine
DMF	Dimethylformamide
DMHA	*N*,*O*-Dimethyl hydroxylamine
ELISA	Enzyme-linked immunosorbent assay
Fmoc	Fluorenylmethyloxycarbonyl
GlcNAc	*N*-Acetylglucosamine
HPLC	High-performance liquid chromatography
MALDI	Matrix-assisted laser desorption/ionization
MBP	Maltose-binding protein
MS	Mass spectrometry
NF-κB	Nuclear factor kappa-light-chain-enhancer of activated B cells
NHS	*N*-Hydroxysuccinimide
NMR	Nuclear magnetic resonance
Nod	Nodulation
pK_a	Acid dissociation constant
PAMAM	Poly(amido amine)
RAFT	Reversible addition–fragmentation chain transfer
SPR	Surface plasmon resonance
TLC	Thin layer chromatography

Acknowledgment

The authors acknowledge NSERC Canada for the support of our work.

References

1. Percival, E.G.V. (1948) *Adv. Carbohydr. Chem.*, **3**, 23–44.
2. El Khadem, H.S. and Fatiadi, A.J. (2000) *Adv. Carbohydr. Chem. Biochem.*, **55**, 175–263.
3. Skraup, Z.H. (1889) *Monatsh. Chem.*, **10**, 401–410.
4. Compton, J. and Wolfrom, M.L. (1934) *J. Am. Chem. Soc.*, **56**, 1157–1162.
5. Wolff, H. (1895) *Chem. Ber.*, **28**, 160–163.
6. (a) Hirst, E.L., Jones, J.K.N., and Woods, E.A. (1947) *J. Chem. Soc.*, 1048–1051. (b) Helferich, B. and Schirp, H. (1951)

Chem. Ber., **84**, 469–471. (c) Helferich, B. and Schirp, H. (1953) *Chem. Ber.*, **86**, 547–556. (d) Zinner, H., Brock, J., Peter, B., and Schaukel, H. (1965) *J. Prakt. Chem.*, **29**, 101–112. (e) Stroh, H.H. and Tengler, H. (1968) *Chem. Ber.*, **101**, 751–755. (f) Bendiak, B. and Cumming, D.A. (1985) *Carbohydr. Res.*, **144**, 1–12. (g) Tang, P.W. and Williams, J.M. (1985) *Carbohydr. Res.*, **136**, 259–271.

7. Takeda, Y. (1979) *Carbohydr. Res.*, **77**, 9–23.
8. Bendiak, B. (1997) *Carbohydr. Res.*, **304**, 85–90.
9. (a) Ojala, W.H., Ojala, C.R., and Gleason, W.B. (1999) *J. Chem. Crystallogr.*, **29**, 19–26. (b) Ernholt, B.V., Thomsen, I.B., Lohse, A., Plesner, I.W., Jensen, K.B., Hazell, R.G., Liang, X.F., Jakobsen, A., and Bols, M. (2000) *Chem. Eur. J.*, **6**, 278–287. (c) Ojala, C.R., Ostman, J.M., and Ojala, W.H. (2002) *Carbohydr. Res.*, **337**, 21–29.
10. (a) Shinohara, Y., Sota, H., Gotoh, M., Hasebe, M., Tosu, M., Nakao, J., Hasegawa, Y., and Shiga, M. (1996) *Anal. Chem.*, **68**, 2573–2579. (b) Leteux, C., Childs, R.A., Chai, W.G., Stoll, M.S., Kogelberg, H., and Feizi, T. (1998) *Glycobiology*, **8**, 227–236. (c) Auge, J. and Lubin-Germain, N. (2000) *J. Carbohydr. Chem.*, **19**, 379–392. (d) Peluso, S., Ufret, M.D., O'Reilly, M.K., and Imperiali, B. (2002) *Chem. Biol.*, **9**, 1323–1328. (e) Flinn, N.S., Quibell, M., Monk, T.P., Ramjee, M.K., and Urch, C.J. (2005) *Bioconjugate Chem.*, **16**, 722–728. (f) Gudmundsdottir, A.V. and Nitz, M. (2007) *Carbohydr. Res.*, **342**, 749–752. (g) Gudmundsdottir, A.V., Paul, C.E., and Nitz, M. (2009) *Carbohydr. Res.*, **344**, 278–284.
11. (a) Guillaumie, F., Thomas, O.R.T., and Jensen, K.J. (2002) *Bioconjugate Chem.*, **13**, 285–294. (b) Goff, R.D. and Thorson, J.S. (2010) *J. Med. Chem.*, **53**, 8129–8139.
12. Gudmundsdottir, A.V. and Nitz, M. (2008) *Org. Lett.*, **10**, 3461–3463.
13. Ridley, B.L., Spiro, M.D., Glushka, J., Albersheim, P., Darvill, A., and Mohnen, D. (1997) *Anal. Biochem.*, **249**, 10–19.
14. (a) Shinohara, Y., Hasegawa, Y., Kaku, H., and Shibuya, N. (1997) *Glycobiology*, **7**, 1201–1208. (b) Kaneda, Y., Whittier, R.F., Yamanaka, H., Carredano, E., Gotoh, M., Sota, H., Hasegawa, Y., and Shinohara, Y. (2002) *J. Biol. Chem.*, **277**, 16928–16935.
15. Blanot, D., Lee, J., and Girardin, S.E. (2012) *Chem. Biol. Drug Des.*, **79**, 2–8.
16. Fais, M., Karamanska, R., Allman, S., Fairhurst, S.A., Innocenti, P., Fairbanks, A.J., Donohoe, T.J., Davis, B.G., Russell, D.A., and Field, R.A. (2011) *Chem. Sci.*, **2**, 1952–1959.
17. Adak, A.K., Leonov, A.P., Ding, N., Thundimadathil, J., Kularatne, S., Low, P.S., and Wei, A. (2010) *Bioconjugate Chem.*, **21**, 2065–2075.
18. Lee, M. and Shin, I. (2005) *Org. Lett.*, **7**, 4269–4272.
19. Park, S., Lee, M.R., and Shin, I. (2009) *Bioconjugate Chem.*, **20**, 155–162.
20. Zhi, Z.L., Powell, A.K., and Turnbull, J.E. (2006) *Anal. Chem.*, **78**, 4786–4793.
21. Kameyama, A., Kaneda, Y., Yamanaka, H., Yoshimine, H., Narimatsu, H., and Shinohara, Y. (2004) *Anal. Chem.*, **76**, 4537–4542.
22. (a) Kurzban, G.P., Bayer, E.A., Wilchek, M., and Horowitz, P.M. (1991) *J. Biol. Chem.*, **266**, 14470–14477. (b) Shinohara, Y., Furukawa, J., Niikura, K., Miura, N., and Nishimura, S.I. (2004) *Anal. Chem.*, **76**, 6989–6997. (c) Zhang, Y., Iwamoto, T., Radke, G., Kariya, Y., Suzuki, K., Conrad, A.H., Tomich, J.M., and Conrad, G.W. (2008) *J. Mass Spectrom.*, **43**, 765–772. (d) Kapkova, P. (2009) *Rapid Commun. Mass Spectrom.*, **23**, 2775–2784. (e) Chang, Y.L., Liao, S.K.S., Chen, Y.C., Hung, W.T., Yu, H.M., Yang, W.B., Fang, J.M., Chen, C.H., and Lee, Y.C. (2011) *J. Mass Spectrom.*, **46**, 247–255.
23. Nishimura, S.I., Niikura, K., Kurogochi, M., Matsushita, T., Fumoto, M., Hinou, H., Kamitani, R., Nakagawa, H., Deguchi, K., Miura, N., Monde, K., and Kondo, H. (2005) *Angew. Chem. Int. Ed.*, **44**, 91–96.
24. Miura, Y., Shinohara, Y., Furukawa, J.I., Nagahori, N., and Nishimurara, S. (2007) *Chem. Eur. J.*, **13**, 4797–4804.

25. Miura, Y., Hato, M., Shinohara, Y., Kuramoto, H., Furukawa, J.I., Kurogochi, M., Shimaoka, H., Tada, M., Nakanishi, K., Ozaki, M., Todo, S., and Nishimura, S.I. (2008) *Mol. Cell. Proteomics*, **7**, 370–377.

26. Furukawa, J.I., Shinohara, Y., Kuramoto, H., Miura, Y., Shimaokat, H., Kurogochi, M., Nakano, M., and Nishimura, S.I. (2008) *Anal. Chem.*, **80**, 1094–1101.

27. (a) Amano, M., Yamaguchi, M., Takegawa, Y., Yamashita, T., Terashima, M., Furukawa, J., Miura, Y., Shinohara, Y., Iwasaki, N., Minami, A., and Nishimura, S. (2010) *Mol. Cell. Proteomics*, **9**, 523–537. (b) Fujitani, N., Takegawa, Y., Ishibashr, Y., Araki, K., Furukawa, J., Mitsutake, S., Igarashi, Y., Ito, M., and Shinohara, Y. (2011) *J. Biol. Chem.*, **286**, 41669–41679.

28. (a) Mangholz, S.E. and Vasella, A. (1991) *Helv. Chim. Acta*, **74**, 2100–2111. (b) Somsak, L., Praly, J.P., and Descotes, G. (1992) *Synlett*, 119–120.

29. Edgar, L.J.G., Dasgupta, S., and Nitz, M. (2012) *Org. Lett.*, **14**, 4226–4229.

30. Peluso, S. and Imperiali, B. (2001) *Tetrahedron Lett.*, **42**, 2085–2087.

31. Lian, S., Su, H., Zhao, B.X., Liu, W.Y., Zheng, L.W., and Miao, J.Y. (2009) *Bioorg. Med. Chem.*, **17**, 7085–7092.

32. Godula, K. and Bertozzi, C.R. (2010) *J. Am. Chem. Soc.*, **132**, 9963–9965.

33. Chiefari, J., Chong, Y.K., Ercole, F., Krstina, J., Jeffery, J., Le, T.P.T., Mayadunne, R.T.A., Meijs, G.F., Moad, C.L., Moad, G., Rizzardo, E., and Thang, S.H. (1998) *Macromolecules*, **31**, 5559–5562.

34. (a) Dirksen, A., Dirksen, S., Hackeng, T.M., and Dawson, P.E. (2006) *J. Am. Chem. Soc.*, **128**, 15602–15603. (b) Thygesen, M.B., Munch, H., Sauer, J., Clo, E., Jorgensen, M.R., Hindsgaul, O., and Jensen, K.J. (2010) *J. Org. Chem.*, **75**, 1752–1755.

35. Liu, X.P., Liu, J., and Luo, Y. (2012) *Polym. Chem.*, **3**, 310–313.

36. Zhou, X.C., Turchi, C., and Wang, D.N. (2009) *J. Proteome Res.*, **8**, 5031–5040.

37. Koshi, Y., Nakata, E., Miyagawa, M., Tsukiji, S., Ogawa, T., and Hamachi, I. (2008) *J. Am. Chem. Soc.*, **130**, 245–251.

38. Rullo, A., Beharry, A.A., Gomez-Biagi, R.F., Zhao, X., and Nitz, M. (2012) *ChemBioChem*, **13**, 783–787.

39. (a) Meyer, V. and Schulze, E. (1884) *Ber. Dtsch. Chem. Ges.*, **17**, 1554–1558. (b) Rischbieth, P. (1887) *Ber. Dtsch. Chem. Ges.*, **20**, 2673–2674.

40. Wohl, A. (1893) *Ber. Dtsch. Chem. Ges.*, **26**, 730–744.

41. (a) Plenkiew, J. and Szarek, W.A. (1974) *Synthesis*, 56–58. (b) Finch, P. and Merchant, Z. (1975) *J. Chem. Soc., Perkin Trans. 1*, 1682–1686.

42. (a) Hatanaka, Y., Kempin, U., and Jong-Jip, P. (2000) *J. Org. Chem.*, **65**, 5639–5643. (b) Jimenez-Castells, C., Torre, B., Andreu, D., and Gutierrez-Gallego, R. (2008) *Glycoconj. J.*, **25**, 879–887. (c) Clo, E., Blixt, O., and Jensen, K.J. (2010) *Eur. J. Org. Chem.*, 540–554.

43. Vila-Perello, M., Gallego, R.G., and Andreu, D. (2005) *ChemBioChem*, **6**, 1831–1838.

44. Peri, F., Dumy, P., and Mutter, M. (1998) *Tetrahedron*, **54**, 12269–12278.

45. Langenhan, J.M., Endo, M.M., Engle, J.M., Fukumoto, L.L., Rogalsky, D.R., Slevin, L.K., Fay, L.R., Lucker, R.W., Rohlfing, J.R., Smith, K.R., Tjaden, A.E., and Werner, H.M. (2011) *Carbohydr. Res.*, **346**, 2663–2676.

46. Bohorov, O., Andersson-Sand, H., Hoffmann, J., and Blixt, O. (2006) *Glycobiology*, **16**, 21C–27C.

47. Pohl, N.L., Ko, K.S., and Zea, C.J. (2004) *J. Am. Chem. Soc.*, **126**, 13188–13189.

48. Leung, C., Chibba, A., Gomez-Biagi, R.F., and Nitz, M. (2009) *Carbohydr. Res.*, **344**, 570–575.

49. Carrasco, M.R. and Brown, R.T. (2003) *J. Org. Chem.*, **68**, 8853–8858.

50. (a) Carrasco, M.R., Nguyen, M.J., Burnell, D.R., MacLaren, M.D., and Hengel, S.M. (2002) *Tetrahedron Lett.*, **43**, 5727–5729. (b) Carrasco, M.R., Brown, R.T., Serafimova, I.M., and Silva, O. (2003) *J. Org. Chem.*, **68**, 195–197. (c) Carrasco, M.R., Brown, R.T., Doan, V.H., Kandel, S.M., and Lee, F.C. (2006) *Biopolymers*, **84**, 414–420.

51. Peri, F., Jimenez-Barbero, J., Garcia-Aparicio, V., Tvaroska, I., and

Nicotra, F. (2004) *Chem. Eur. J.*, **10**, 1433–1444.

52. Renaudet, O. and Dumy, P. (2002) *Tetrahedron*, **58**, 2127–2135.
53. Peri, F., Marinzi, C., Barath, M., Granucci, F., Urbano, M., and Nicotra, F. (2006) *Bioorg. Med. Chem.*, **14**, 190–199.
54. Langenhan, J.M., Peters, N.R., Guzei, I.A., Hoffmann, M., and Thorson, J.S. (2005) *Proc. Natl. Acad. Sci. U.S.A.*, **102**, 12305–12310.
55. Ahmed, A., Peters, N.R., Fitzgeraldo, M.K., Watson, J.A., Hoffmann, F.M., and Thorson, J.S. (2006) *J. Am. Chem. Soc.*, **128**, 14224–14225.
56. Griffith, B.R., Krepel, C., Fu, X., Blanchard, S., Ahmed, A., Edmiston, C.E., and Thorson, J.S. (2007) *J. Am. Chem. Soc.*, **129**, 8150–8155.
57. Goff, R.D. and Thorson, J.S. (2009) *Org. Lett.*, **11**, 461–464.
58. Goff, R.D. and Thorson, J.S. (2012) *Org. Lett.*, **14**, 2454–2457.
59. Peltier-Pain, P., Timmons, S.C., Grandemange, A., Benoit, E., and Thorson, J.S. (2011) *ChemMedChem*, **6**, 1347–1350.
60. Goff, R.D., Singh, S., and Thorson, J.S. (2011) *ChemMedChem*, **6**, 774–776.
61. Chen, G.S. and Pohl, N.L. (2008) *Org. Lett.*, **10**, 785–788.
62. Dasgupta, S. and Nitz, M. (2011) *J. Org. Chem.*, **76**, 1918–1921.
63. Kalia, J. and Raines, R.T. (2008) *Angew. Chem. Int. Ed.*, **47**, 7523–7526.

4
Recent Developments in the Construction of *cis*-Glycosidic Linkages

Alphert E. Christina, Gijsbert A. van der Marel, and Jeroen D. C. Codée

4.1
Introduction

Arguably the most important reaction in oligosaccharide synthesis is the union of two carbohydrate building blocks in a glycosylation reaction. Where the construction of 1,2-*trans*-linkages can, in general, reliably be achieved using neighboring group participation, the construction of 1,2-*cis*-glycosidic bonds can present a major challenge. Over the years, many elegant and efficient solutions for the construction of various 1,2-*cis*-glycosidic linkages have been developed. This chapter describes recently introduced strategies for the stereoselective construction of *cis*-glycosidic bonds by selected examples in the context of complex oligosaccharide synthesis. Attention is paid to the formation and reactivity of different reactive intermediates, such as anomeric triflates and specific oxocarbenium ion conformers. The chemistry of different glycosyl sulfonium ions is described as well as the possible remote participation by acyl functions. The presented overview is not intended to be exhaustive, it merely illustrates how contemporary insights into the reactivity of various reactive intermediates have paved the way to accomplish effective stereoselective syntheses of complex oligosaccharides. For an extensive treatise on different glycosylation methodologies, the reader is directed to recent reviews [1].

4.2
Cis-Glycosylation

Although the stereoselective introduction of the β-mannosidic linkage has traditionally been one of the biggest challenges in synthetic carbohydrate chemistry, β-mannoside formation now probably is the best-understood *cis*-glycosylation reaction of all. The construction of the β-mannosidic linkage can reliably be effected using the "benzylidene mannose" technology introduced by Crich and coworkers [2]. The serendipitous discovery of this methodology has provided a major impetus for studies into mechanistic pathways underlying the stereoselectivity

Modern Synthetic Methods in Carbohydrate Chemistry: From Monosaccharides to Complex Glycoconjugates,
First Edition. Edited by Daniel B. Werz and Sébastien Vidal.

in glycosylation reactions, and the identification of anomeric triflates as possible product-forming reactive intermediates now is one of the key pillars in these investigations [3]. The formation of β-mannosides can be rationalized with the kinetic scenario depicted in Scheme 4.1. Upon activation of a benzylidene mannose donor, an intermediate α-triflate (**1**) is stereoselectively formed. This species is significantly favored over its β-counterpart **5** because it benefits from a stabilizing anomeric effect, and lacks the destabilizing *gauche* interaction and the Δ2-effect (unfavorable interaction between the dipoles of O-1, O-2, and O-5), both of which are present in the β-triflate **5**. These anomeric triflates can interconvert through a series of equilibrium reactions involving contact ion pairs (CIPs) and solvent-separated ion pairs (SSIPs). Using both primary ^{13}C- and secondary ^{2}H-kinetic isotope effects [4], Crich and coworkers have been able to show that the formation of β-mannosides proceeds through an associative displacement of the anomeric triflate **1**. This S_N2-type displacement does take place with the development of significant oxocarbenium ion character at the anomeric [4b]. Collapse of the benzylidene mannosyl triflate **1** into the intermediate oxocarbenium ion **3** is prevented by the benzylidene acetal, which conformationally locks the mannopyranosyl ring, preventing it from flattening to accommodate the positive charge. In addition, it places the substituent at C-6 in a *tg*-position, which in case of an electronegative substituent represents the most *disarming* orientation [5]. Also the developing steric interactions between the substituents at C-2 and C-3 upon flattening of the mannopyranosyl ring disfavor the collapse of the covalent triflate **1** into the CIP **2** and SSIP **3** [6]. In contrast to the associative mechanism by which the β-mannosides are formed, α-mannoside formation from a benzylidene mannosyl donor was revealed to take place through a dissociative mechanism and therefore to proceed via an intermediate that approximates a discrete mannosyl oxocarbenium ion [4a]. Primary ^{13}C kinetic isotope effects further revealed that for the benzylidene glucose case, both the α- and the β-products originate from a S_N2-like substitution on an anomeric triflate [4a]. The difference in reaction pathways between the mannose and glucose systems can be rationalized by the fact that the formation of a β-mannosyl triflate is rather

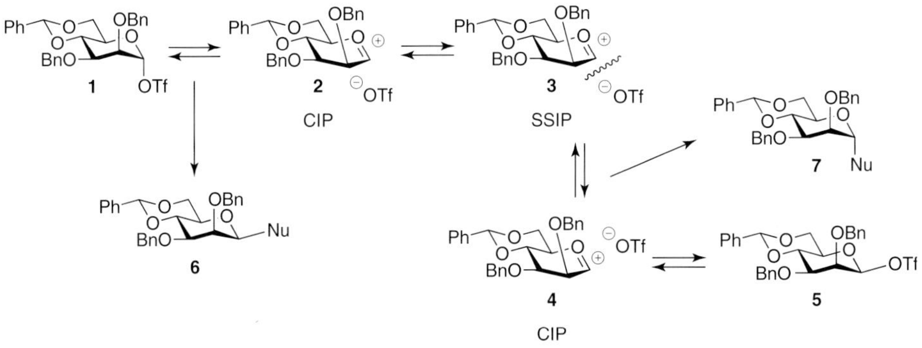

Scheme 4.1 The equilibrium of benzylidene mannosyl triflates, contact ion pairs (CIPs), and solvent-separated ion pairs (SSIPs).

unfavorable because of both steric and electronic reasons (*vide supra*). Although an α-glucosyl triflate is significantly more stable than its β-counterpart, the difference in stability of these intermediates is not large enough to exclude the β-triflate from the reaction pathway.

An example of the utility of the benzylidene methodology is provided by the synthesis of a tetrasaccharide subunit of the lipopolysaccharide from *Plesimonas shigelloides* depicted in Scheme 4.2 [7]. In this synthesis, the benzylidene β-directing principle was extended to allow for the introduction of two uncommon β-linked heptose units, one of which being a 6-deoxy moiety. Therefore, the 1-cyano-2-(2-iodophenyl)ethylidene group was introduced as a 4,6-*O*-benzylidene surrogate setup for deoxygenation by radical fragmentation. The assembly of the target tetrasaccharide started off with the glycosylation of methyl rhamnoside **9** with thioglycoside donor **8** following a preactivation protocol with the diphenylsulfoxide (Ph_2SO)/triflic anhydride (Tf_2O) promoter combination [8]. Because of the electron-withdrawing cyano-group on the benzylidene ketal in **8**, the use of this promoter system was required, because it generates a somewhat more reactive electrophile in comparison to the 1-benzenesulfinyl piperidine (BSP)/Tf_2O reagent system (Tf, trifluoromethanesulfonyl), originally developed by Crich and Smith [9]. Disaccharide **10** was obtained in 86% yield with complete β-selectivity. Treatment of this disaccharide with tributyltin hydride and azobis-*iso*-butyronitrile (AIBN) afforded a 6-deoxy-*manno*-heptopyranoside, which was transformed into acceptor **11** by 2,3-dichloro-5,6-dicyano-1,4-benzoquinone (DDQ) mediated removal of the 2-naphthylmethyl group. In the next preactivation-based glycosylation event, this time using mannoheptosyl donor **12** in combination with the BSP/Tf_2O promoter system, the second 1,2-*cis*-glycosidic bond was introduced to provide the trisaccharide **13**. Oxidative cleavage of the 2-naphthylmethyl group then set the stage for the final coupling. In this step, rhamnosyl donor **14** was preactivated with BSP, after which the addition of trisaccharide **13** stereoselectively led to the formation of tetrasaccharide **15** in 73% yield. Deprotection of the tetrasaccharide was accomplished by saponification and subsequent acid treatment, which gave a mixture of products consisting of tetrasaccharides **16** and **17** in 85% combined yield. Hydrogenolysis of the individual tetrasaccharides gave target **18** in 94% and 96% yield from **16** and **17**, respectively.

Mannuronic acid donors also proved very suitable for the stereoselective construction of 1,2-*cis*-mannosyl linkages, and these donors provide anomeric triflates upon activation with a triflate-based electrophile [10]. Interestingly, the anomeric triflates generated from mannuronic acid donors preferentially reside in a flipped $^1C_4$-chair conformation, placing the anomeric triflate in an unfavorable equatorial position (as in structure **21**, generated from either **19** or **20**, Scheme 4.3) [11]. In addition, this conformer places three of the five ring substituents in sterically unfavorable axial positions. A rationale for this unusual behavior has been forwarded, based on the influence of the orientation of the substituents on the pyranosyl core on the stability of a pyranosyl oxocarbenium ion [12]. In a half-chair oxocarbenium ion, the *O*-substituents at C-3 and C-4 prefer to take up a pseudo-axial position, which is less electron-withdrawing than the alternative pseudo-equatorial

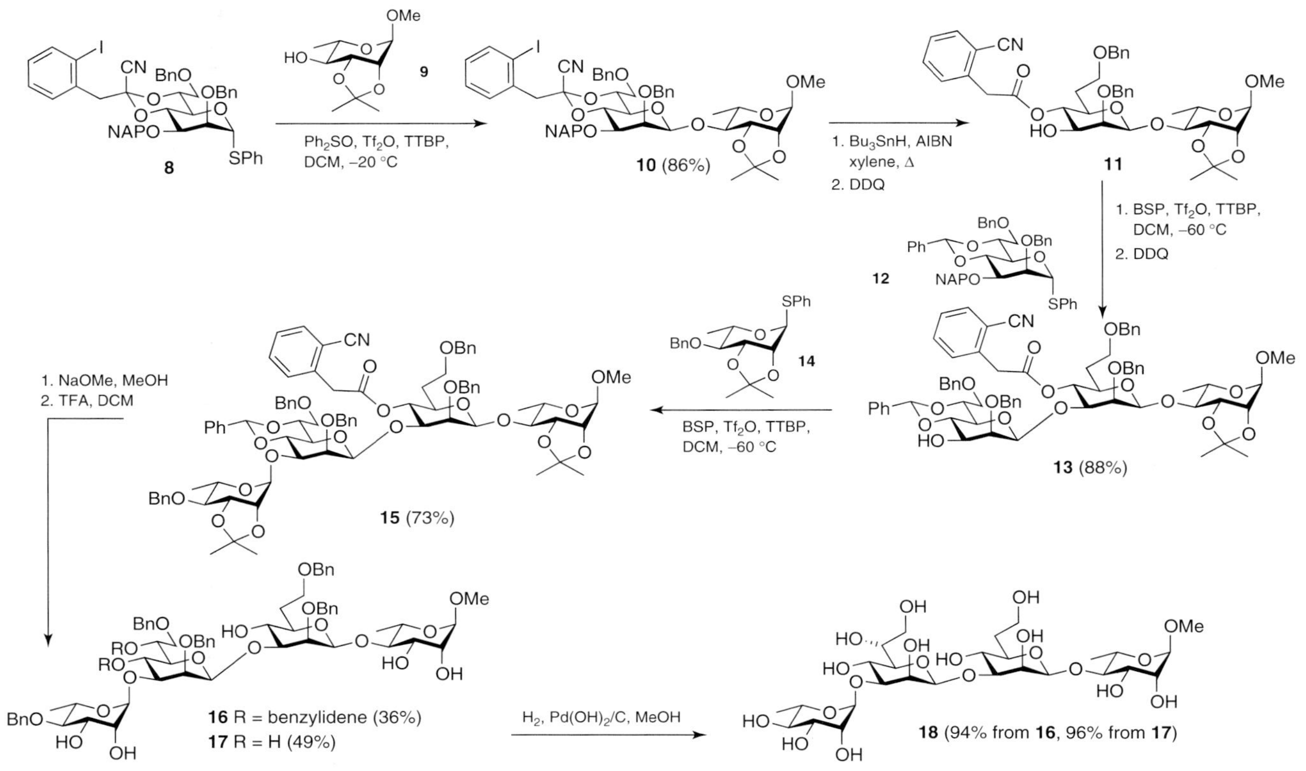

Scheme 4.2 Assembly of a *Plesimonas shigelloides* tetrasaccharide fragment.

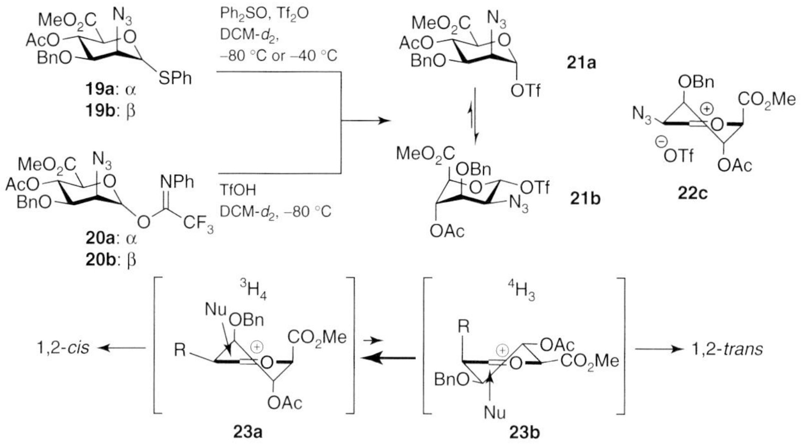

Scheme 4.3 Mannuronic acid triflates and oxocarbenium ion half-chair conformers.

orientation and which allows for the through-space stabilization of the positive charge. The C-2-*O*-substituent, on the other hand, prefers to occupy a pseudo-equatorial position, facilitating the hyperconjugative stabilization of the anomeric positive charge by the parallel σ_{C-H} orbital. Where the "common" C-5 methylene ether substituent prefers a pseudo-equatorial orientation because of steric reasons, a C-5 carboxylic acid ester preferentially occupies a pseudo-axial place for similar reasons as described for the C-3 and C-4-*O*-substituents. The mannuronic acid oxocarbenium ion can accommodate all the substituent preferences by taking up a $^3H_4$ half-chair conformation as depicted in Scheme 4.3 [12c]. Because the anomeric center of mannuronic acid triflate **21** bears significant carbocation character, it takes up a conformation close in conformational space to the favorable $^3H_4$ half-chair oxocarbenium ion **22**. The existence of this intermediate provides an indication that the formation of the β-mannosyl linkage from mannuronic acid donors can be accounted for with either an S_N2- or S_N1-type mechanism. In line with the benzylidene mannose case described earlier, an associative displacement of triflate **21** can be envisaged to explain the β-selectivity. On the other hand, nucleophilic attack on a $^3H_4$ half-chair-like oxocarbenium ion, such as **22**, which takes place on the diastereotopic face that leads directly to the chair product (i.e., the β-face), can also be at the basis of the observed stereoselectivity. The latter explanation is supported by the finding that mannuronic acid donors are not only more reactive than their benzylidene mannose counterparts but also significantly more reactive than what would be expected based on the presence of the electron-withdrawing C-5 carboxylic acid ester [13]. Also the decomposition temperature of the mannuronic acid triflate **21** is lower than the decomposition temperature of benzylidene mannosyl triflate **1** (−40 vs. −20 °C). A third difference between the mannuronic acid and benzylidene mannose systems is that the former seems to be more tolerant to changes in the substitution pattern at C-2 and C-3 than the latter [12c,d]. Where the β-selectivity in the benzylidene mannose system significantly erodes by replacing the C-2 or C-3 ether for a small azide or fluoride, the selectivity of the mannuronic acid system is not significantly influenced by these changes. Combined, these observations indicate that for the mannuronic acid case, a mechanism involving more oxocarbenium ion character in the product-forming intermediate is plausible. The excellent stereochemical behavior of mannuronic acid donors was exploited in the automated solid-phase synthesis of β-(1,4)-mannuronic acid alginate oligomers as depicted in Scheme 4.4 [14]. Using a "second-generation" carbohydrate synthesizer [15], Walvoort *et al.* assembled a mannuronic acid dodecamer **28**, featuring 12 *cis*-linkages using *N*-phenyl trifluoroimidate donor **25**. To this end, butenediol-functionalized Merrifield resin **24** [16] was treated with donor **25** and a catalytic amount of triflic acid (TfOH) at −40 °C (the decomposition temperature of the anomeric triflate). Next the C-4-OH was unmasked by hydrazinolysis of the levulinoyl ester. Repeating this procedure 12 times led, after releasing the products from the solid support in a cross-metathesis event, to a product mixture containing 42% of dodecamer **26** alongside deletion sequences, as indicated by LC-MS analysis. After saponification of the methyl esters, the product was purified and dodecamer **27** was obtained in 11% yield over 24 steps, equaling 90% per chemical step. Final hydrogenolysis led to synthetic alginate fragment **28** in 95% yield.

Hung and coworkers recently reported on another alginate fragment built up of L-guluronic acid (the C-5 epimer of D-mannuronic acid) monomers [17]. A bottleneck in this synthesis was presented by the poor nucleophilicity of the axially oriented C-4-OH. To overcome this drawback, 1,6-anhydro-gulopyranosyl building blocks, featuring an equatorial and accessible C-4-OH were introduced. As a result, a synthetic sequence was designed starting from the nonreducing end of the target tetramer as depicted in Scheme 4.5. Thus, coupling of donor **29** with acceptor **30** gave the desired α-linked disaccharide **31** in 70% yield along with 17% of its β-epimer. The stereoselectivity of this condensation reflects the intrinsic preference of gulosyl donors for the formation of 1,2-*cis*-products [18]. One of the factors at play here is the anomeric effect, which stabilizes the α-anomer with respect to its β-counterpart. However, this stabilizing anomeric effect should be largely offset by the unfavorable diaxial interaction with the substituent at C-3, and therefore the anomeric effect alone cannot be used to explain the unusual strong *cis* preference of gulosyl donors. Dinkelaar *et al.* have argued that also in this case the conformational preferences of an oxocarbenium ion intermediate can be at the basis of the observed selectivity. As depicted in Scheme 4.6, the gulosyl $^{3}H_{4}$ half-chair oxocarbenium **39a** places all ring substituents in electronically preferred orientations, thereby favoring this oxocarbenium ion over its $^{4}H_{3}$ counterpart **39b**. Notably the fact that the gulosyl oxocarbenium ion can adopt a conformation that is electronically most favorable is reflected in the reactivity of gulosides. In a series of methyl α-glycosides, methyl α-glucoside **42**, methyl α-galactoside **41**, and methyl α-guloside **40**, the latter was reported to hydrolyze fastest among the three monosaccharides (Scheme 4.6b) [19].

In donor **29**, the possible remote anchimeric assistance of the 6-*O*-acetyl group (*vide infra*) can also be of beneficial influence to the stereochemical outcome of the glycosylation at hand. To continue the synthesis of the guluronic acid tetramer, the 1,6-anhydro-bridge in disaccharide **31** was opened, followed by selective anomeric

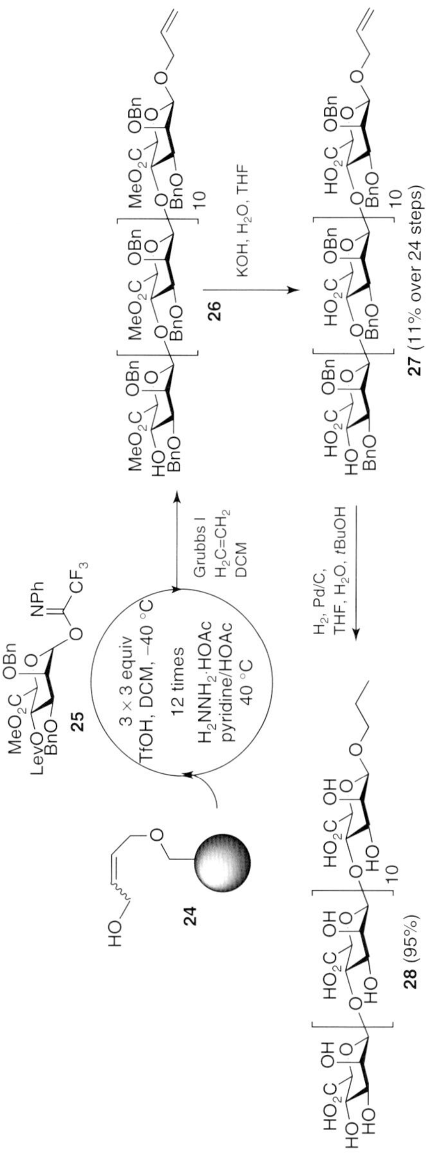

Scheme 4.4 Automated solid-phase assembly of a mannuronic acid dodecamer.

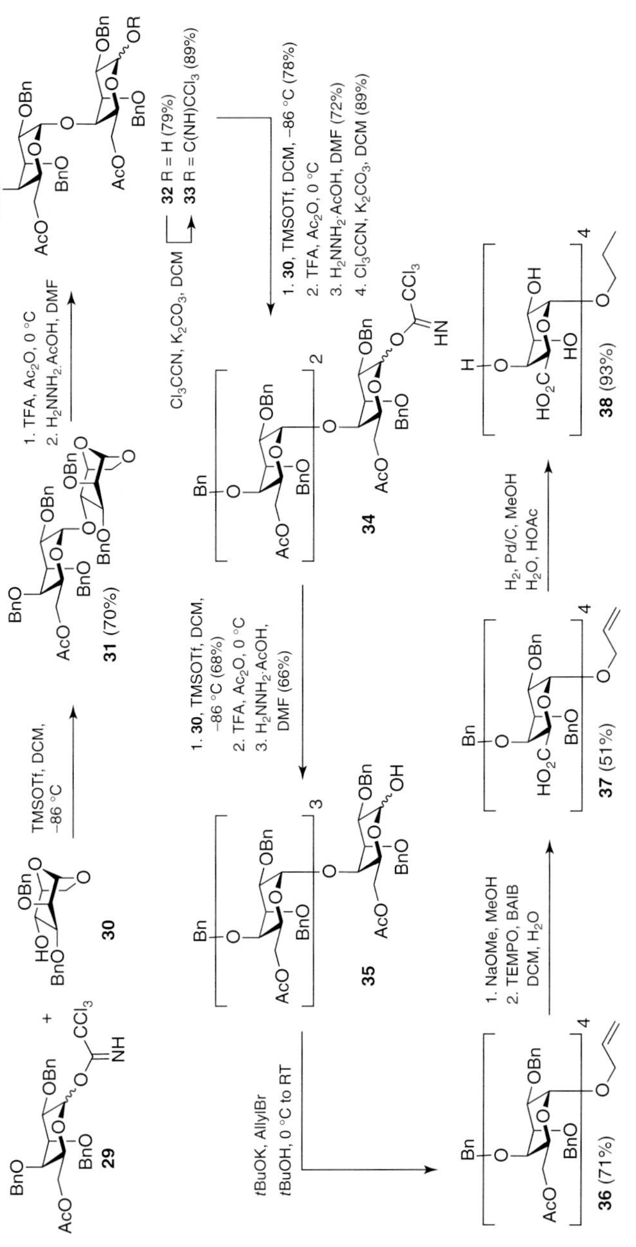

Scheme 4.5 Assembly of an L-guluronic acid alginate tetrasaccharide.

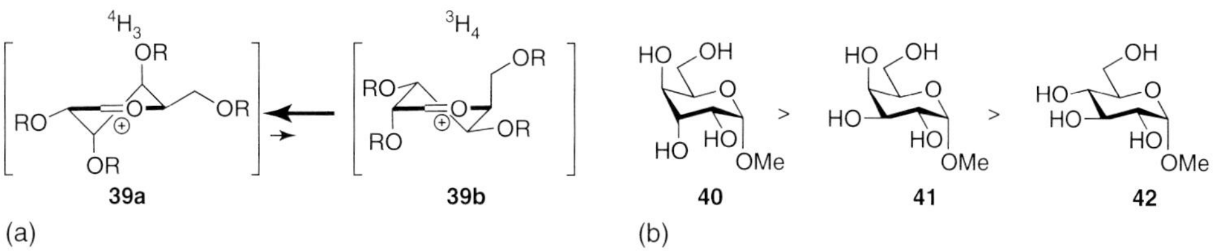

Scheme 4.6 (a) D-Gulosyl oxocarbenium ion half-chair conformers and (b) rate of hydrolysis of methyl gulose versus methylgalactose versus methyl glucose.

deacetylation using $H_2NNH_2{\cdot}AcOH$ to give lactol **32**. Conversion of this hemiacetal to the corresponding trichloroacetimidate **33** primed the dimer for another coupling reaction with **30**. This glycosylation proceeded with complete α-stereoselectivity in 78% yield. A repetition of the last four steps gave the fully protected tetramer (coupling 68%, α-anomer only). Again the bicyclic structure was disrupted under acidic conditions and the resulting anomeric acetate was selectively removed. Because initial attempts (probed in a disaccharide stage) to stereoselectively install an anomeric allyl functionality employing a TMSOTf-catalyzed coupling with an imidate donor failed, possibly because of the relatively high nucleophilicity of the acceptor used, an alternative procedure was developed. To this end, an anionic glycosylation was used, and a Williamson etherification using alcohol **35** and allyl bromide afforded the desired α-product **36** as a single isomer in good yield. It was suggested that the observed stereoselectivity originated from the α-alkoxide, which allows for chelation of the potassium counterion with the ring oxygen and the oxygen at C-2. Cleavage of the acetyl groups unmasked the primary alcohols, which were oxidized with TEMPO (2,2,6,6-tetra-methyl-1-piperidinyloxy) free radical employing BAIB ([(bis-acetyloxy)iodo]benzene) as a co-oxidant to furnish the carboxylic acid functions [20]. A hydrogenolysis step completed the synthesis of the all-*cis*-linked guluronic acid tetramer **38**.

The relatively poor nucleophilicity of the axial C-4-OH in a gulosyl acceptor is also encountered in glycosylations featuring galacturonic acid acceptors [21]. Christina *et al.* [22] have employed a similar trick as described earlier to circumvent the poor nucleophilicity of the galacturonic acid C-4-hydroxyl. By installing a cyclic lactone functionality bridging the C-3-OH and the C-5 carboxylic acid function in galacturonic acid building block **48**, the pyranosyl ring is flipped from its regular $^4C_1$ conformation to the alternative $^1C_4$ chair, placing the C-4-OH in an equatorial position [23]. Another important feature of the galacturonic acid lactone building blocks is their excellent *cis*-selectivity when used as a donor. Christina *et al.* [22] exploited both features in the assembly of a set of frame-shifted trimer repeating units of the zwitterionic polysaccharide Sp1. A selected synthesis is presented in Scheme 4.7. Galacturonic acid lactone **43** was preactivated with Ph_2SO/Tf_2O and subsequently treated with diaminofucosyl acceptor **45** to give disaccharide **46** in 75% as the sole anomer. The stereoselectivity in this glycosylation can be explained with anomeric triflate **45** as a product-forming intermediate. This species was identified by low-temperature NMR studies, and its intermediacy was shown to be

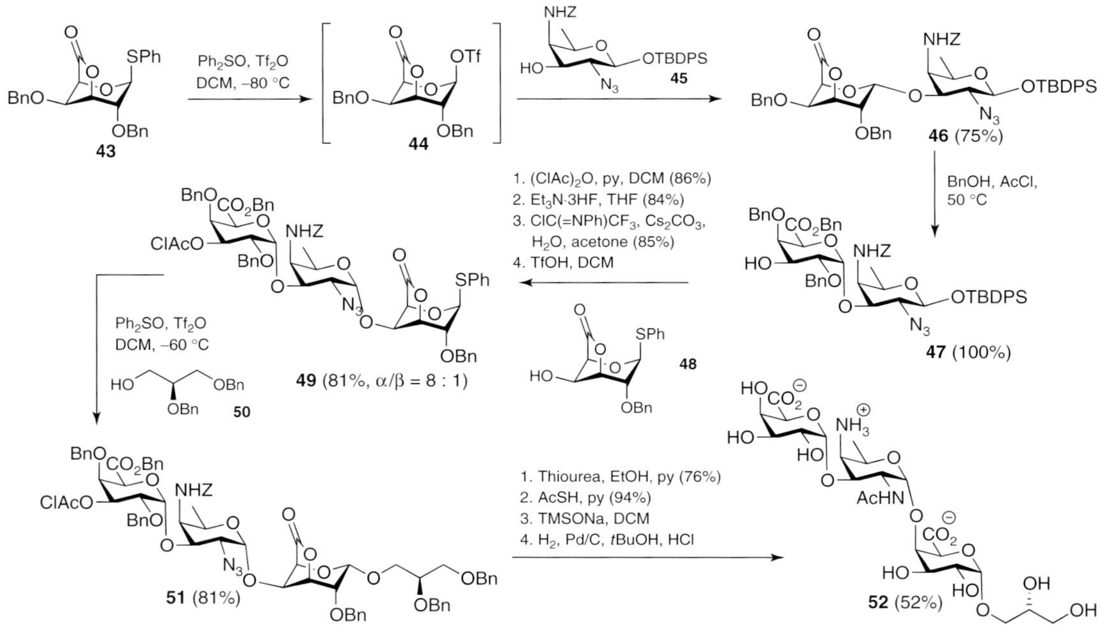

Scheme 4.7 Galacturonic acid lactones in the construction of an Sp1 trisaccharide.

crucial for the observed α-selectivity. Next the lactone bridge in **46** was opened and dimer **47** was transformed into an *N*-phenyltrifluoroacetimidate donor, which was condensed with lactone acceptor **48** under the *aegis* of a catalytic amount of TfOH. Trisaccharide **49** was formed in 81% yield showcasing the apt nucleophilicity of the lactone alcohol **48**. In the final glycosylation event, the trisaccharide was functionalized with a glycerol spacer to provide trimer **51** as the sole anomer. Deprotection of this trisaccharide was accomplished by removal of the chloroacetyl ester and ensuing transformation of the azide functionality into an acetamide by treatment with thioacetic acid. Next the lactone bridge of the reducing end sugar was opened with TMSONa, and the benzyl ethers and benzyloxycarbonate were removed by hydrogenation.

In an alternative synthesis of a fragment of the Sp1 polysaccharide, Bundle and coworkers [24] chose to employ galactosyl donors having a C-6-*O*-acetyl function as precursor for the α-galacturonic acid moieties, to circumvent the difficult reactivity of the galacturonic acid residues (Scheme 4.9). The 6-*O*-acyl function also served to aid in the stereoselective formation of the *cis*-galactosyl linkages. The issue of remote participation by acyl functions grafted on the C-3-OH, C-4-OH, and/or C-6-OH has been the subject of considerable debate, and although there is ample empiric evidence (especially in the gluco-, manno-, and galacto-series), to substantiate the profitable effect of an acyl function at these positions, direct evidence for actual participation is scarce. (Also see Chapter 5 of this volume for a more elaborate discussion on the subject.) [1] Participation from the C-3-position has recently been investigated by the groups of Crich [25] and Kim [26]. The former group was inspired by the striking α-selectivity in condensation reactions of benzylidene mannosyl donors featuring a C-3-acyl protecting group [27] and initially postulated that remote participation could be at the basis of this selectivity (Scheme 4.8). To probe the possible neighboring group participation, they designed several model donors to identify putative bicyclic intermediates, such as probes **53** and **56**, having an axially and equatorially oriented C-3-*O*-*tert*-butyl carbonate. Activation of allosyl donor **53**, using the BSP/Tf_2O reagent combination, in the absence of an acceptor led to the isolation of bicyclic product **54**, indicating that the axial C-3-*O*-*tert*-butyl carbonate can effectively trap the anomeric cation. Inclusion of cyclohexanol as a model acceptor led to the formation of allosylated cyclohexanol in 61% yield as an 11 : 1 β/α mixture, alongside 7% of the bicyclic carbonate. When the *tert*-butyl carbonate was mounted on a benzylidene mannosyl donor, as in **56**, no proof in the form of a tricyclic carbonate could be found for the participation of the carbonate, while the sole product from the condensation of this donor with cyclohexanol was the α-configured **57**. Kim and coworkers performed a series of glycosylation reactions of benzylated mannosyl donors equipped with different electron-withdrawing groups at C-3, C-4, and C-6 and noted that with increasing electron-withdrawing capacity of the substituents, the β-selectivity increased, the striking exceptions being the 3-*O*-acetyl/benzoyl and 6-*O*-acetyl donors. To account for these results, the C-3 and C-6-*O*-trichloroacetimidate probes **58** and **60** were investigated. While probe **58** cleanly furnished the bicycle **59** upon activation, the seven-membered ring oxazepine **61** could not be isolated and the only evidence for its presence in the reaction

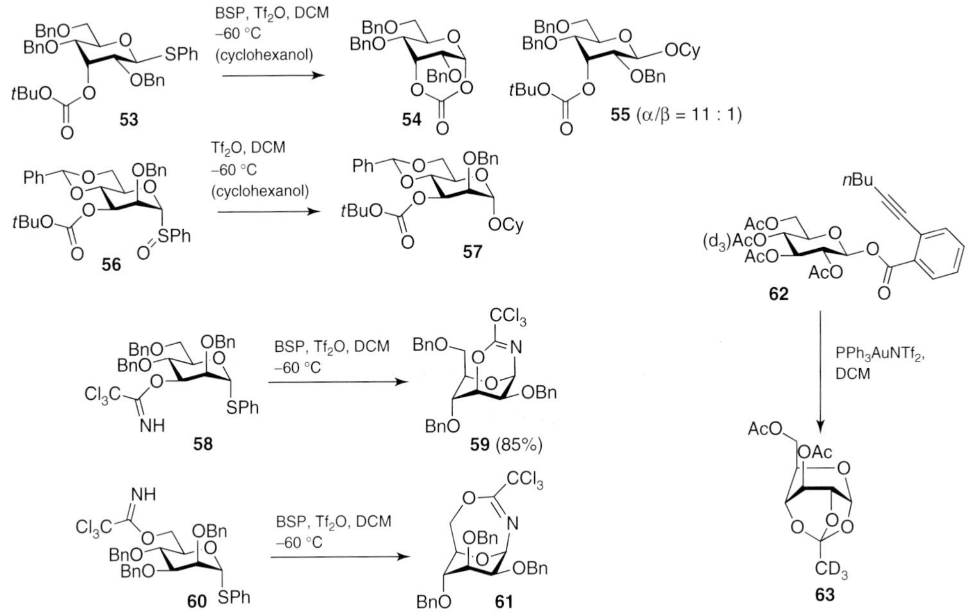

Scheme 4.8 Various probes to study remote participation by acyl functions.

mixture came from mass spectrometry. The apparent contradiction between the Boc- and imidate probes can be the result of the difference in nucleophilicities of the carbonate and imidate functions and the conformational constraint posed by the benzylidene acetal in **56**. In 1999, Boons and coworkers revealed that the α-selectivity in a series of galactosylation reactions, using NIS/TMSOTf (NIS, *N*-iodosuccinimide) as a promoter system in a 1,4-dioxane/toluene solvent system (an "α-directing" solvent system) critically depended on the nature of the C-4 substituent [28]. Where the benzyl and methyl ethers were moderately α-selective, the presence of especially electron-rich esters, such as the *para*-methoxybenzoyl, led to the formation of the 1,2-*cis*-products with great selectivity. Crich and coworkers revisited this system with, among others, a C-4-*O*-*tert*-butyl carbonate donor and again found no direct evidence for remote participation [25]. Probably the most compelling evidence for the participatory behavior of C-4-acetates has come from a study by Yu and coworkers [29], who reported on the gold(I)-catalyzed condensation of *ortho*-hexynylbenzoate glucosyl donor **62** with pentenyl alcohol. From the complex reaction mixture that arose during this reaction, tricyclic ortoester **63** was isolated in 12% yield. Control experiments with deuterium-labeled acetates at either the C-2-OH or the C-4-OH revealed that the methyl group in this orthoester originated from the C-4-*O*-acetate, clearly showcasing that this acetyl group can reach over to participate in the reaction, even at the expense of the formation of an unfavorable $B_{1,4}$ boat conformation. Thus, while there is evidence that acetyl esters are capable of remote participation, it is probably an oversimplification to attribute the enhanced α-selectivity of acetyl-containing donor glycosides solely to this effect, and the seemingly contradictory results that are reported warrant further investigation and sophistication of the current mechanistic models.

Returning to the assembly of the Sp1 hexamer fragment by the Bundle lab (Scheme 4.9), the TMSOTf-mediated condensation of diacetyl glucosazide imidate donor **66** and galactosyl acceptor **64** proceeded with excellent α-selectivity to give disaccharide **68**. For the construction of the very similar disaccharide **69**, a trichloroacetimidate-based glycosylation proved ineffective and the authors switched to a dehydrative coupling strategy [30] using lactol **67** and thexyldimethylsilyl (TDS) galactoside **65**. In this event, dimer **67** was formed in 73% as a single anomer. Now, the glucosazide residues of both dimers **66** and **67** were converted to a diamino fucose residue in seven steps: azide reduction, *N*-acetylation, *O*-deacetylation, 6-*O*-mesylation, $NaBH_4$ reduction, triflation, and azide substitution. Deprotection of the PMB (*para*-methoxybenzyl) group, acetylation, and deallylation then gave alcohols **70** and **71** in 10% and 20% over 10 steps, respectively. Dimers **70** and **71** were glycosylated with C-6-acetyl galactosyl donors **72** and **73**, respectively, to provide trisaccharides **74** and **75**, with excellent stereoselectivity. Next trimer **74** was transformed into acceptor building block **76** by treatment with DDQ (73%), while trisaccharide **75** was turned into donor **77** by removal of the anomeric TDS group and subsequent installation of an trichloroacetimidate function. The union of the two trisaccharide parts required careful tuning of the reaction conditions (temperature, donor equivalents, and amount of Lewis acid activator) and was finally stereoselectively accomplished in 85% yield. The fully protected hexasaccharide **78** was deacetylated to give the tetraol, which was oxidized in a two-step procedure to provide the tetracarboxylate. Immediate benzylation then gave hexamer **79** in 52% over the last steps. Hydrogenolysis of all benzyl ethers and esters and the two azide groups gave the zwitterionic target compound **80**.

A conceptually different approach to stereoselectively access 1,2-*cis*-glucosyl and 1,2-*cis*-galactosyl linkages was recently introduced by Boons and coworkers [31, 32]. The stereoselective introduction of the 1,2-*cis*-linkages was accomplished by the exploitation of an (*S*)-(phenylthiomethyl)benzyl chiral auxiliary mounted at the C-2-OH (Scheme 4.10). Upon activation of the anomeric-leaving group, the auxiliary traps the intermediate oxocarbenium ion to form a *trans*-decalin β-sulfonium ion having the phenyl substituent in an equatorial position. The alternative α-sulfonium ion would possess an axially oriented phenyl group that suffers from a steric clash with the glucosyl H-3. The intermediate β-sulfonium ion **83** can be substituted in a S_N2-like manner to provide the 1,2-*cis*-linked products. It was revealed that an electron-withdrawing acyl functionality at the C-3-OH was needed to effectively suppress the participation of oxocarbenium ion intermediate in the reaction. Notably the Boons laboratory also reported that external sulfides can be used to generate intermediate β-sulfonium ions, allowing for the construction of α-glucosyl linkages, but this methodology seems to be less robust [33].

In Scheme 4.11, the use of the chiral auxiliary is illustrated in the solid-phase assembly of an α-glucan pentasaccharide repeating unit found in *Aconitum carmichaeli* [34]. The synthesis commenced with the TMSOTf-catalyzed union of polystyrene resin-bound alcohol **85** and donor **86**. The glycosylation was carried out

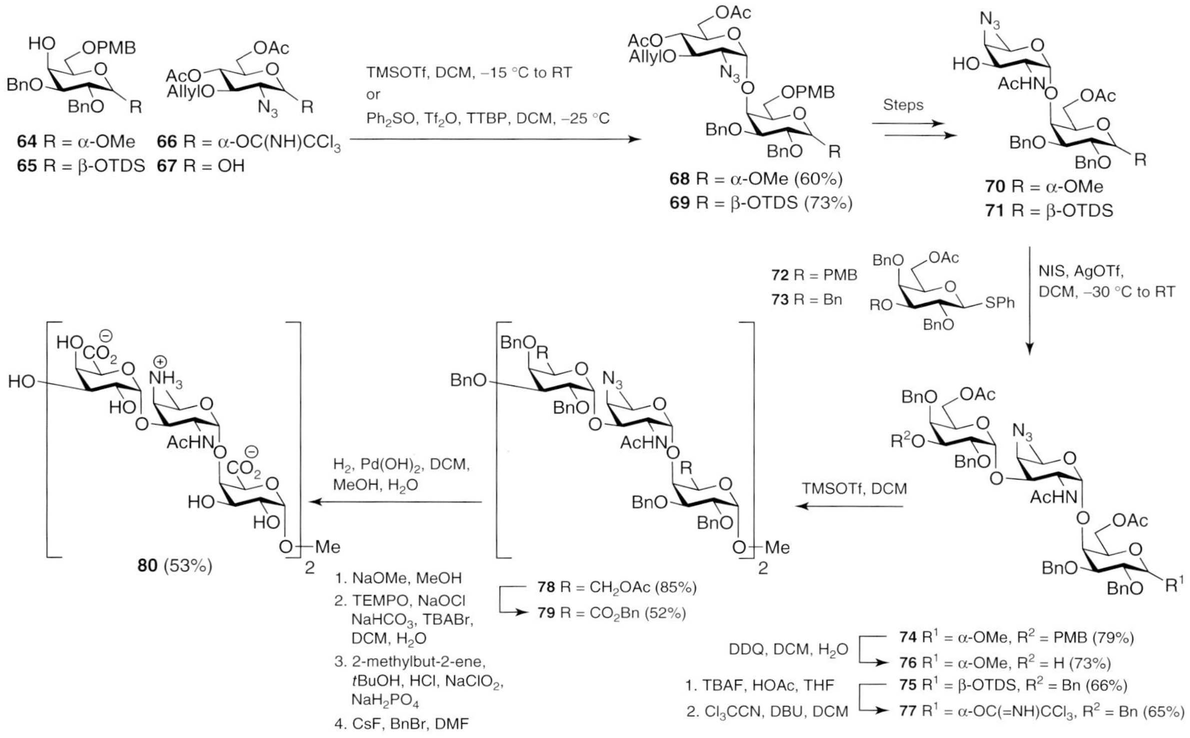

Scheme 4.9 Assembly of an Sp1 hexasaccharide.

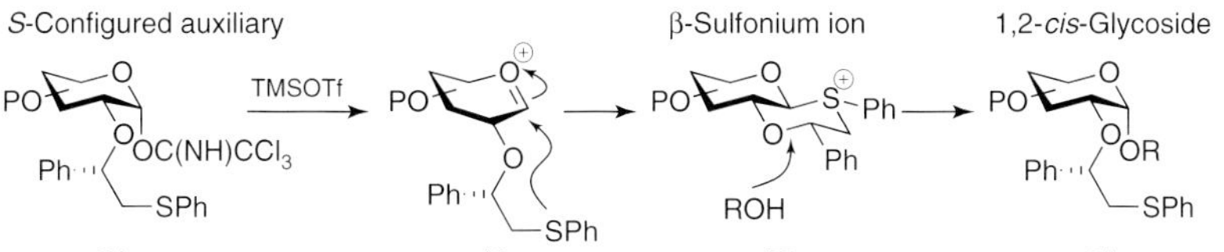

Scheme 4.10 Anchimeric assistance by the (*S*)-(phenylthiomethyl)benzyl chiral auxiliary.

using a preactivation protocol involving treatment of the donor with a stoichiometric amount of TMSOTf at −40 °C, and subsequent addition of the so-formed sulfonium ion to the resin-bound acceptor. Because model studies had revealed that the C-3″-OH of this disaccharide was a rather unreactive nucleophile as a result of the steric shielding by the auxiliary, the (*S*)-(phenylthiomethyl)benzyl group in **87** was converted to an acetyl function using Ac_2O and $BF_3{\cdot}OEt_2$ prior to the removal of the allyl carbonate. Liberation of the C-3″-OH was followed by the second glycosylation event, in which preactivated donor **89** was condensed with resin-bound dimer **88**. After Fmoc deprotection, the same glycosylation protocol was followed using donor **92** to construct the resin-bound tetrasaccharide **93**. Removal of the Fmoc in **93** preceded the final coupling toward the fully protected branched pentaglucan. Conversion of the chiral auxiliaries to acetyl functions, Fmoc deprotection, release of the glucan from the resin under Zemplén conditions, and reacetylation gave, after size-exclusion chromatography, a pentasaccharide as the major product with its mono-debenzylated counterpart as a side-product. From this mixture, the stereochemical integrity of the introduced glycosidic linkages could be determined and no anomeric epimers were detected. After 13 steps on resin, the overall yield was 25%, corresponding to a yield of 90% per step. Finally, deacetylation and hydrogenation gave the target pentamer **96**. In addition, the authors showed that the same methodology could be used to construct an α-galactosyl-branched analog of pentasaccharide **96**.

Another example of a sulfur-based "auxiliary" has recently been reported by Christina *et al.* [35], who described that a C-6-thioether can be used as a stereocontrolling element for the synthesis of β-mannosides (Scheme 4.12). Upon reduction of the C-6-thioether, β-rhamnosides can be obtained in a straightforward manner [36]. It was shown that activation of imidate donor **97**, having a C-6-thiophenyl ether, led to the formation of bicyclic sulfonium ion **98**, which slowly reacted at −80 °C to give the *cis*-linked products. Higher reaction temperatures led to erosion of stereoselectivity. To account for the observed stereoselectivity and reaction kinetics, a reaction pathway based on a Curtin–Hammett kinetic scheme was invoked. In this scenario, the bridged sulfonium ion serves as a reservoir for the more reactive $^3H_4$ oxocarbenium ion **99a**, which is close in conformational space to the parent sulfonium ion and which can be attacked by an incoming nucleophile to furnish the β-linked product. The $^3H_4$ oxocarbenium **99a** should be significantly favored over its $^4H_3$ counterpart **99b**, because the former places all its substituents in electronically favorable orientations. The C-6-thiophenyl donors were exploited in the synthesis of a tetrarhamnoside fragment of the capsular polysaccharide of the

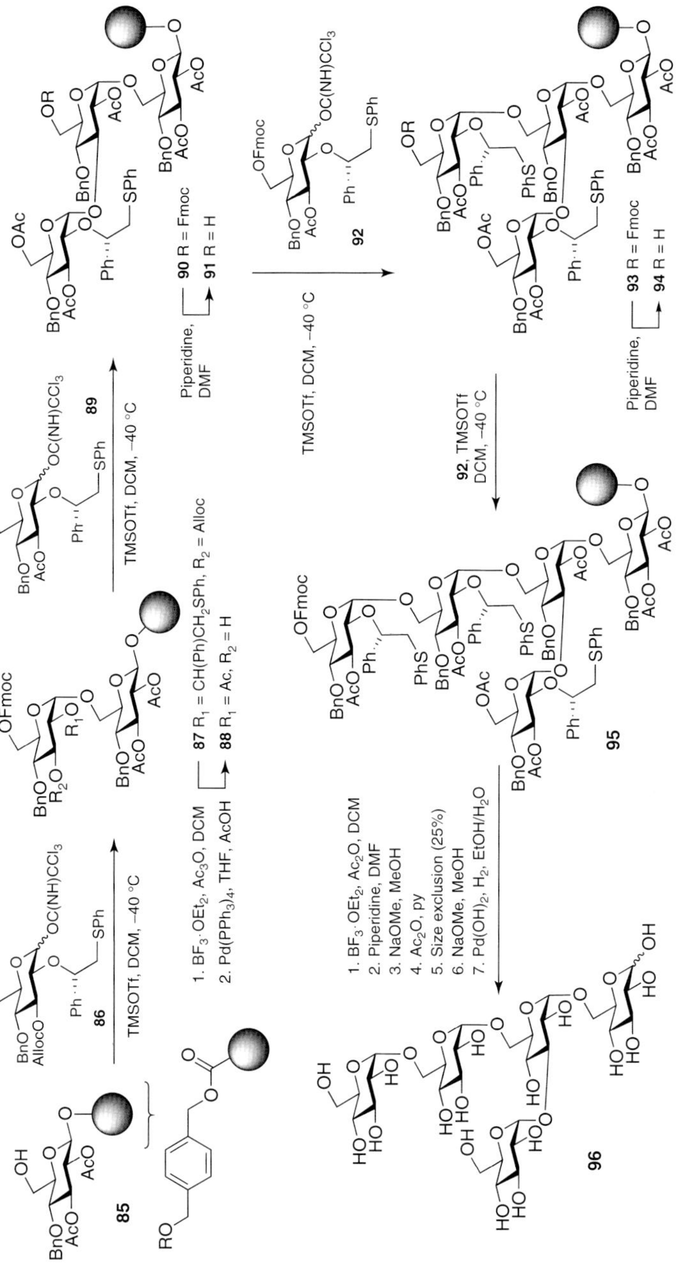

Scheme 4.11 Solid-phase assembly of an α-glucan pentamer **94**.

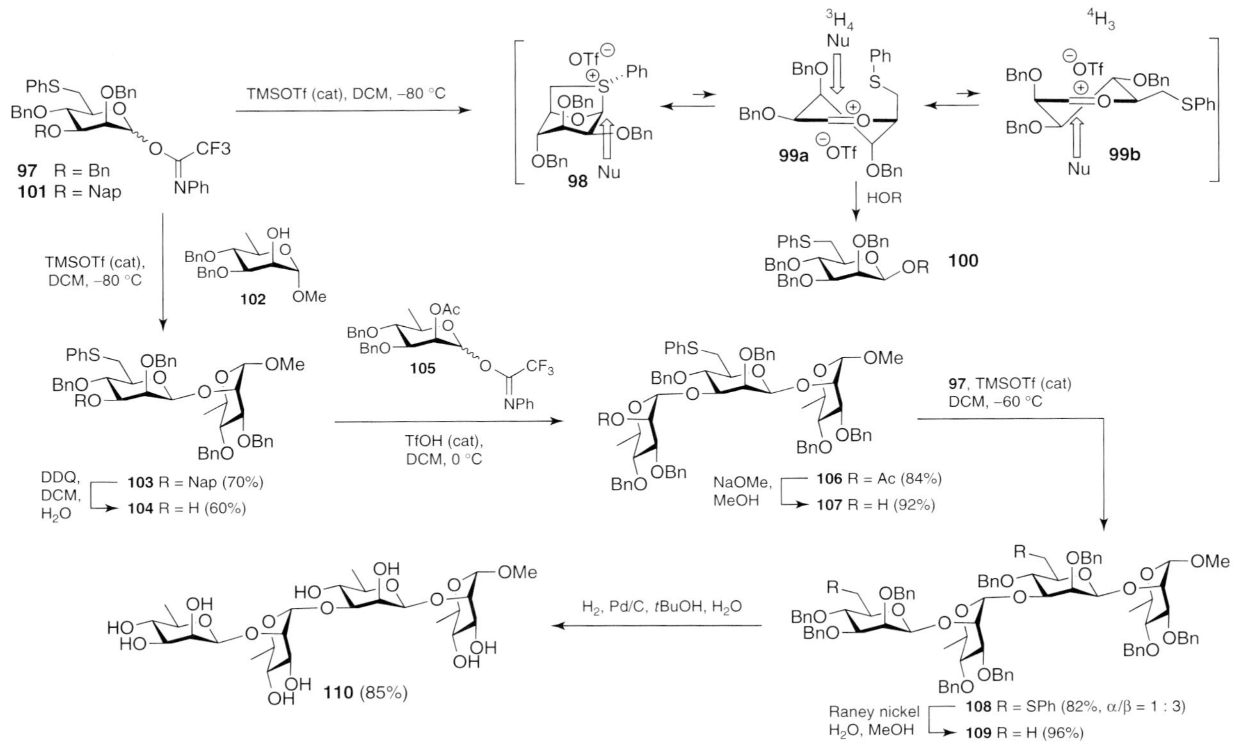

Scheme 4.12 Stereoselective synthesis of a *Xanthomonas campestris pathovar campestris* tetrasaccharide using C-6-(*S*)-phenyl mannosyl donors to construct the β-rhamnosidic linkages.

phytopathogen *Xanthomonas campestris pathovar campestris*, featuring alternating *cis*- and *trans*-glycosidic linkages. Thus, C-6-thiophenyl donor **101** and rhamnosyl acceptor **102** were condensed to provide disaccharide **103** in 72% yield. The 2-naphthylmethyl ether in this dimer was removed to set the stage for the condensation with rhamnosyl donor **105**, equipped with a participating 2-*O*-acetyl function. In the event, trisaccharide **106** was uneventfully formed. Deacetylation was then followed by the third condensation in which C-6-thiophenyl mannosyl donor **97** was used to provide the tetrasaccharide **108**. Although the second β-mannosyl linkage was formed with somewhat reduced stereoselectivity, this experiment showed that the *cis*-mannosyl linkage can also be introduced on more complex acceptors. It is not clear whether the thioether in the acceptor influences the erosion of stereoselectivity in this case. To complete the assembly of the target structure, tetramer **108** was desulfurized by reduction with Raney nickel to give tetrarhamnoside **109** in 96% yield. Global debenzylation then completed the synthesis of tetramer **110**.

cis-Linked glucosamine residues are regularly encountered in biologically relevant oligo- and polysaccharides [37]. In 2001, Kerns and coworkers [38] introduced cyclic *N*,*O*-carbamate-protected glucosamine donors to stereoselectively construct this type of glycosidic linkage. In-depth studies revealed a unique mechanism, underlying the stereoselectivity. It was discovered that the stereoselectivity in condensations of *N*,*O*-carbamate-protected glucosamine donors critically depends on the reactivity of the acceptor and both amount and strength of Lewis acid used for the activation. Using mild conditions and reactive nucleophiles led to the predominant formation of the β-glucosamine products, whereas less reactive nucleophiles and more forcing conditions provide the α-linked glycosides. Several groups, including those of Ito [39], Oscarson [40], and Ye [41] have explored the mechanistic details of this reaction, both experimentally and computationally, and the mechanistic picture that has emerged is summarized in Scheme 4.13. Activation of a *N*,*O*-carbamate-protected glucosamine donor can lead to an α-triflate intermediate **112**, which can be substituted by the incoming nucleophile to provide the initial β-product **113**. This disaccharide can then isomerize through an endocyclic ring cleavage pathway to provide the α-linked product **114**. Two factors

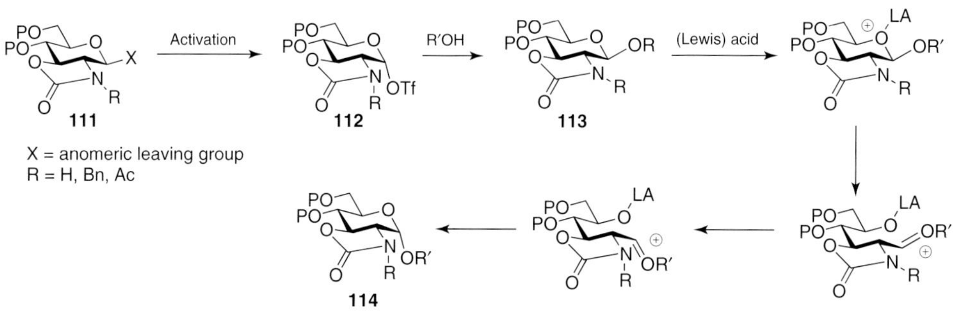

Scheme 4.13 Mechanistic pathway leading to the formation of α-glucosamine bonds from cyclic *N*,*O*-carbamate-protected glucosamine donors.

contribute to the efficient equilibration of the β- into the α-products. The first, and most important, is the strain present in the bicyclic system originating from the *trans* fusion of the two rings. Secondly, the $^4C_1$ conformation of the glucosamine ring allows for the efficient overlap of the nonbonding electron pair on the exocyclic substituent and the antibonding orbital of the C-1-O-5 bond, the exo-anomeric effect. To prevent side reactions on the carbamate nitrogen, a second *N*-protecting group, such as a benzyl or acetyl moiety, has been used. The methodology can also be used for the stereoselective construction of α-galactosamine bonds and recently the mechanistic principle has been extended to cyclic carbonate-protected glucosyl donors [42]. It should be noted, however, that the ease of isomerization seems to be more general for the glucosamine case than for the glucose system.

The cyclic *N*,*O*-carbamate methodology has been applied by Manabe and coworkers [43] in the synthesis of an hexasaccharide antibiotic, active against *Helicobacter pylori*, which is characterized by the presence of two terminal α-GlcNAc residues (Scheme 4.14). The hexasaccharide has two similar disaccharide-reducing end moieties and therefore a convergent strategy was used to assemble the target structure **126**. The synthesis started with the construction of the α-glucosamine linkage using *N*-benzyl oxazolidinone bromide donor **115** and thioglycoside

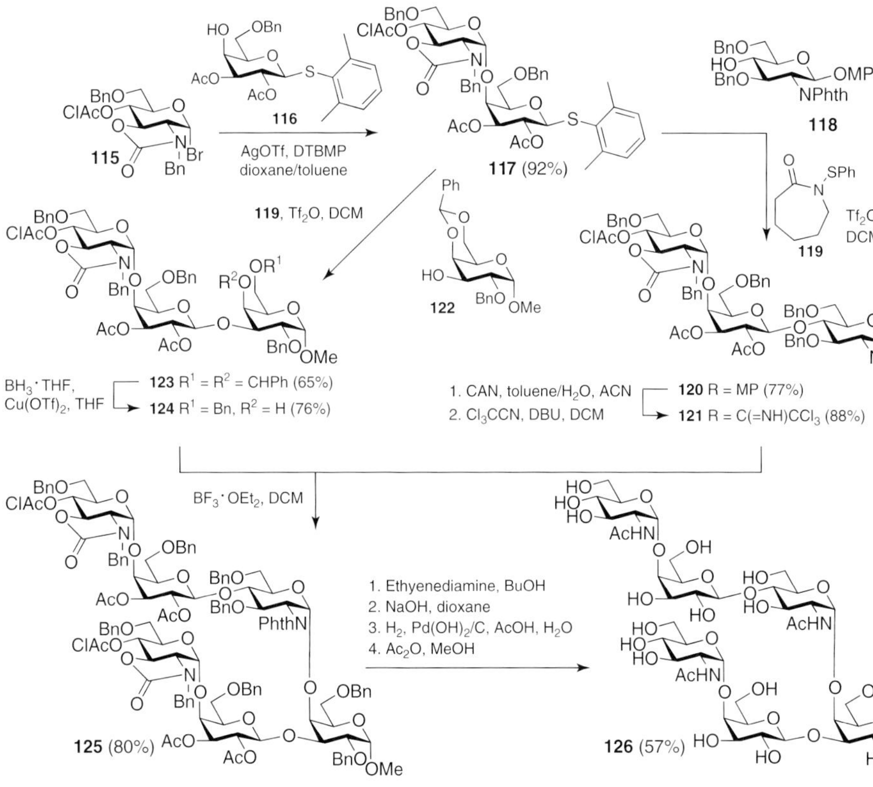

Scheme 4.14 Assembly of an anti-*Helicobacter pylori* hexasaccharide.

acceptor **116** in an AgOTf-mediated condensation reaction in a mixture of toluene and dioxane (an "α-directing" solvent as noted above) near room temperature. The desired disaccharide **117** was obtained in 92% as the sole anomer. Notably when the analogous *S*-phenyl acceptor was used, the yield of the reaction was significantly compromised because of a competing aglycon transfer process [44]. Next the disaccharide was elongated either with acceptor **118** to give trisaccharide **120** or monomer **122** to furnish trimer **123**. The former trisaccharide was transformed into a donor by removal of the anomeric *para*-methoxyphenyl group and installation of a trichloroacetimidate functionality. The latter trimer was subjected to a regioselective benzylidene ring opening to procure the trimer acceptor **124**. Both trisaccharides were united in the final glycosylation event which employed $BF_3{\cdot}OEt_2$ as a promoter. Deprotection of the resulting hexamer **125** started with removal of the phthalimide group using ethylenediamine in butanol. The cyclic carbamate proved to be stable to these conditions, and an alkali treatment of the hexamer was required to remove the oxazoline moiety. Global debenzylation and chemoselective acetylation completed the synthesis of target hexamer **126**.

The cyclic *N*,*O*-carbamate-type protection has also found application in the stereoselective synthesis of α-sialic acids, which has been reviewed [45]. Sialic acid donors with a 4-*O*-5-*N*-carbamate, bearing an additional *N*-acetyl protecting group are now among the most selective sialylating agents, and an interesting feature of these donors is that the formation of 2,3-glycal side products is suppressed [46]. Possibly the ring strain induced by the cyclic protecting group is an important factor here. Finally, the cyclic *N*,*O*-carbamate also has a beneficial effect on the reactivity of the proximal 4-OH in glucosamine acceptors [47]. The carbamate ties back the C-3 substituent, thereby making the 4-OH more accessible to engage in a nucleophilic attack.

Also for the stereoselective construction of furanosidic linkages, effective methodologies have been recently developed [48], exploiting specific oxocarbenium ion conformers or by directing the mechanism toward the transposition of a single anomeric triflate. For example, the group of Lowary [49] has found that 2,3-anhydropentosyl thioglycosides and sulfoxides can be used as donors, which favor the formation of 1,2-*cis*-products (Scheme 4.15). The mechanistic principle behind the selectivity was addressed through computational chemistry and low-temperature NMR spectroscopy, and anomeric triflates were identified as glycosylating species. The triflates take a position *trans* with respect to the 2,3-epoxide ring. The epoxide ring rigidifies the otherwise flexible furanose system, thereby stabilizing the triflate with respect to the associated ion pair. The preferred orientation of the anomeric triflate can possibly be explained from the more favorable dipole interactions between the epoxide and the triflate in the case of the *trans*-isomer. Following the glycosylation event, the epoxide can be regioselectively opened to provide the desired *cis*-furanoside. A recent and illustrative example of this approach is the synthesis of trisaccharide **134** from 2,3-anhydro-D-gulofuranosyl glycosyl sulfoxides, as depicted in Scheme 4.15 [50]. This trisaccharide is structurally related to an antigenic polysaccharide from *Eubacterium saburreum* strain T19. By varying the protective group pattern on the key 2,3-anhydro-D-gulofuranosyl sulfoxide donor,

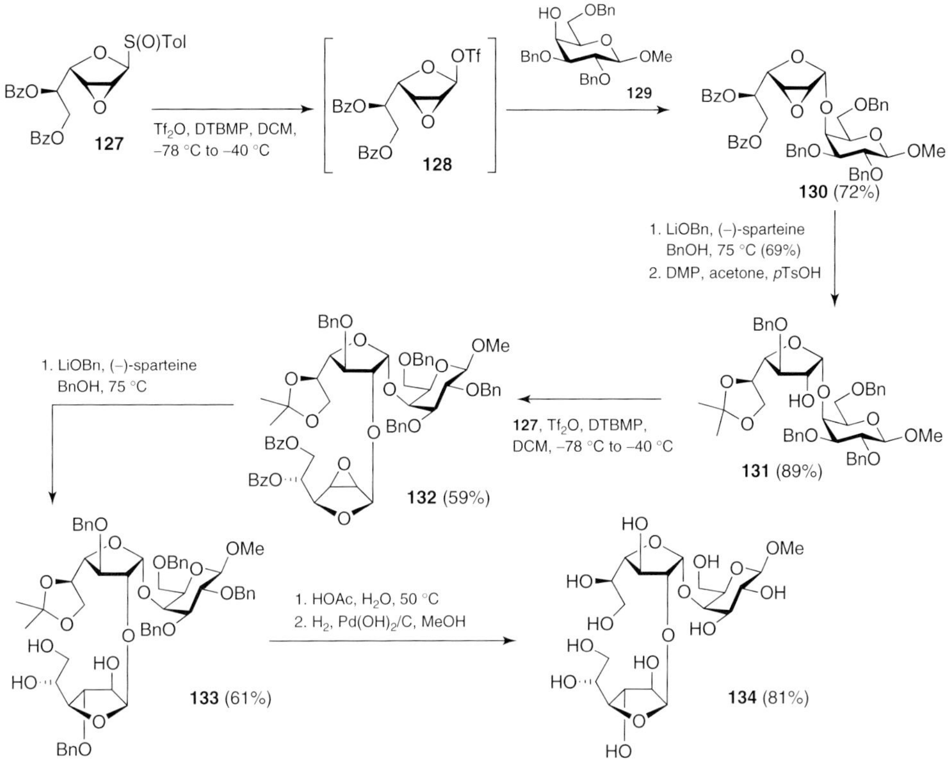

Scheme 4.15 Synthesis of a *Eubacterium saburreum* -related trisaccharide.

it was found that a benzoyl-protected building block gave the best results in terms of stereoselective coupling and subsequent epoxide opening. Thus, the assembly of the target trisaccharide started with the coupling of 2,3-anhydro-donor **127** with acceptor **129**, using conditions that ensured the intermediacy of triflate **128**. Upon S_N2-like substitution of this intermediate, dimer **130** was formed as the sole anomer. Opening of the epoxide with LiOBn and (−)-sparteine and concomitant benzoyl deprotection gave triol **131** in 69% yield along with 5% of its regioisomer. Although the mechanistic details underlying the observed regioselectivity have not been determined, a specific lithium–sparteine–epoxide complex has been proposed to account for the regiochemistry in the epoxide-opening step. Acid-mediated installation of an isopropylidene acetal then gave alcohol **131**, which was subjected to the next cis furanosylation event. Prolonged reaction times were necessary for the glycosylation of acceptor **131**, which most probably is due to the sterically hindered nature of the acceptor. Opening of the epoxide in **132** gave triol **133** in 61% alongside 13% of the regioisomeric product. Acidic hydrolysis of the acetal and catalytic hydrogenation took place uneventfully to afford the desired target trisaccharide **134**.

Besides stereospecific reactions on anomeric triflate intermediates, furanosyl oxocarbenium ions have also been exploited for the introduction of *cis*-furanosidic

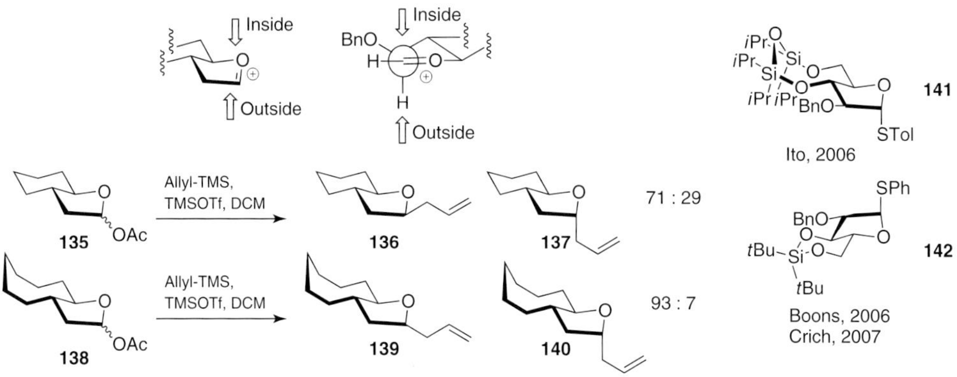

Scheme 4.16 Stereoselectivity of conformationally locked (*arabino*) furanosides.

linkages. Almost simultaneously the groups of Boons [51], Ito [52], and Crich [53] reported on the glycosylation behavior of arabinofuranosides that were conformationally locked using silyl-based cyclic protecting groups (Scheme 4.16). Boons and coworkers and the group of Crich reported on the use of 3,5-*O*-silylidene-protected arabinosides **141**, where Ito described the use of the eight-membered tetra-*iso*-propyldisiloxanylylidene (TIPS) ring to span the C-3 and C-5 hydroxyls (as in **142**). Both cyclic groups served to lock the arabinose ring in a single envelope conformation (the ^{3}E for the D-Ara system and the E_3 for the L-Ara system). Woerpel and coworkers [54] have developed a model to explain the diastereoselective substitution of furanosyl oxocarbenium ions, stipulating that attack on an envelope conformer takes place from the inside to minimize steric interactions with the substituent at the C-2 position. Outside attack would lead to a staggered interaction with this substituent and staggered interactions between the C-1 and C-2 substituents in the product. In the arabinose systems, the inside attack model translates to a preferential 1,2-*cis*-attack, as observed with these donors. Woerpel and coworkers also described that the overall conformational change of the furanosyl system in the course of the reaction has to be accommodated for. Two bicyclic models were studied by the Woerpel laboratory, which comprised either a *trans*-5,6-bicyclic system or a *trans*-5,8-ring system (Scheme 4.16) [55]. It was observed that the 5,8-fused furanosyl system **138** reacted with excellent diastereoselectivity, while the 5,6-fused system **135** was less stereoselective. It was argued that the developing ring strain in the 5,6-system upon rehybridization from a trigonal to a tetrahedral system is at the basis of the observed erosion in selectivity. In this light, the stereoselectivity of the 3,5-silylidene arabinosyl is intriguing.[1)]

1) Double stereodifferentiation also plays a role in these condensation reactions, as the glycosylations of 3,5-silylidene D-arabinosyl donors proceed with significantly less stereoselectivity than the condensation of their L-configured counterparts. See Ref. [56]. Also of note is the high *cis*-stereoselectivity described by Kim and coworkers for condensations involving perbenzylated D-arabinosyl carboxybenxyl donors. See Ref. [57].

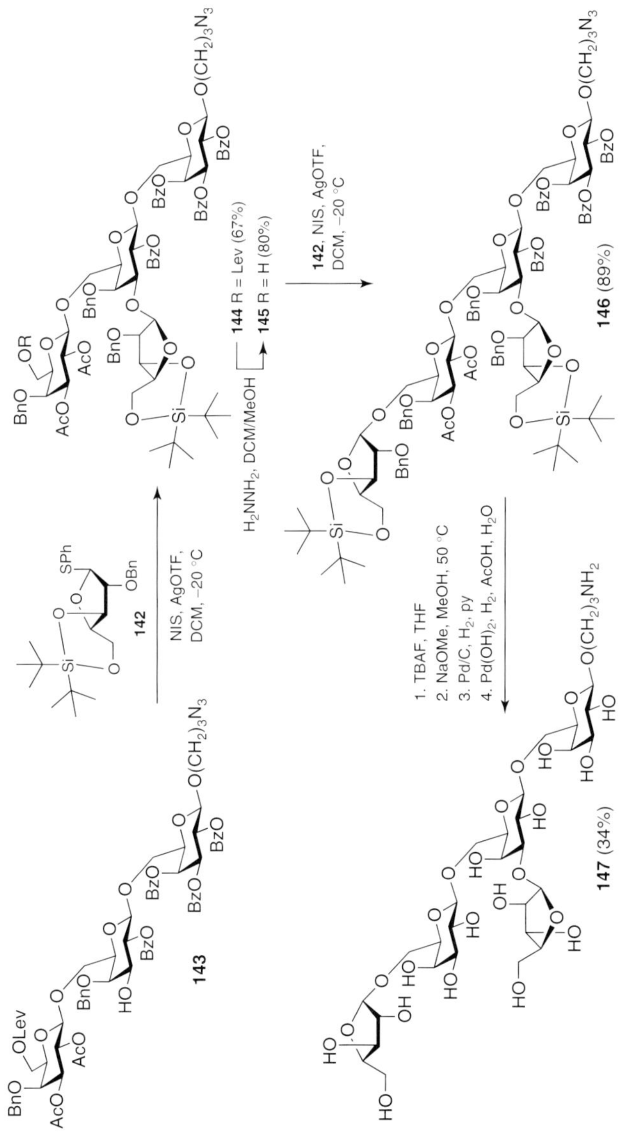

Scheme 4.17 Synthesis of a β-L-arabinofuranose-containing pentasaccharide.

In the following, the synthesis of an arabinogalactan fragment, a constituent of the primary plant cell wall, is depicted to illustrate the applicability of the silylidene methodology (Scheme 4.17) [51]. Attachment of the arabinose moiety to trisaccharide **143** was accomplished by a NIS/AgOTf-mediated coupling, employing silylidene donor **142**. The tetrasaccharide **144** was obtained in 67% yield with complete *cis*-selectivity. Liberation of the nonreducing end C-6-OH by levulinoyl deprotection afforded acceptor **145**. The second arabinose residue was coupled to this acceptor using the same promoter system, and pentamer **146** was obtained as a single diastereomer. Global deprotection of this product was achieved in four steps, entailing removal of the silylidene groups with tetrabutylammonium fluoride (TBAF), saponification of the acetyl and benzoyl esters under Zemplén conditions, reduction of the azide moiety to an amine, and final catalytic hydrogenolysis of the benzyl ethers.

4.3 Conclusion

Glycosylation reactions can proceed via a multitude of pathways, passing through a variety of reactive intermediates. Because all these intermediates have their specific reactivity and associated selectivity, predicting and controlling the stereochemical course of glycosylation reaction can be a precarious undertaking. Although our understanding of the stereoelectronic effects, controlling the stereochemistry in glycosidic bond formation, is continuously growing, optimization of a glycosylation reaction is often still a game of trial and error. This chapter has described some recent developments that effectuate stereoselective glycosylations in the context of complex carbohydrate synthesis, and from the presented examples it becomes clear that there is a broad pallet of reaction intermediates that can be summoned to achieve this goal. The key to success in these approaches is to promote one reactive intermediate over another and through the judicious tuning of the carbohydrate core and covalent anomeric triflates, oxocarbenium ion-like intermediates and anomeric sulfonium ion intermediates have all been used to this end.

Acknowledgments

The Netherlands Organisation for Scientific Research (NWO) is kindly acknowledged for financial support.

List of Abbreviations

Ac	Acetyl
ACN	Acetonitrile
AIBN	Azobis-*iso*-butyronitrile
BAIB	[(Bis-acetyloxy)iodo]benzene
Bn	Benzyl

BSP	Benzenesulfinyl piperidine
Bu	Butyl
Bz	Benzoyl
CAN	Ceric(IV) ammonium nitrate
CIP	Contact ion pair
Cy	Cyclohexyl
DBU	1,8-Diazabicyclo-[5.4.0]undec-7-ene
DCM	Dichloromethane
DDQ	2,3-Dichloro-5,6-dicyano-1,4-benzoquinone
DMF	Dimethylformamide
DMP	Dimethoxypropane
DTBMP	Di-*tert*-butylmethylpyridine
Lev	Levulinoyl
Me	Methyl
NAP	Naphthyl-2-methyl
NIS	*N*-Iodosuccinimide
Ph	Phenyl
PMB	*para*-Methoxybenzyl
py	Pyridine
SSIP	Solvent-separated ion pair
TBAF	Tetrabutylammonium fluoride
TBDPS	*tert*-Butyldiphenylsilyl
TDS	Thexyldimethylsilyl
TEMPO	2,2,6,6-tetra-Methyl-1-piperidinyloxy
THF	Tetrahydrofuran
TIPS	Tri-*iso*-propylsilyl
TMS	Trimethylsilyl
Tf	Trifluoromethanesulfonyl
Tol	Tolyl
TTBP	Tri-*tert*-butylpyrimidine
Z	Benzyloxycarbonyl

References

1. (a) Boltje, T.J., Buskas, T., and Boons, G.-J. (2009) *Nat. Chem.*, **1**, 611–622; (b) Zhu, X.M. and Schmidt, R.R. (2009) *Angew. Chem. Int. Ed.*, **48**, 1900–1934; (c) Levy, D.E. and Fügedi, P. (2006) *The Organic Chemistry of Sugars*, CRC Press, Boca Raton, FL. (d) Demchenko, A. (2008) *Handbook of Chemical Glycosylation*, Wiley-VCH Verlag GmbH, Weinheim.
2. (a) Crich, D. and Sun, S. (1998) *Tetrahedron*, **54**, 8321–8348; (b) Crich, D. (2010) *Acc. Chem. Res.*, **43**, 1144–1153.
3. Aubry, S., Sasaki, K., Sharma, I., and Crich, D. (2011) *Top. Curr. Chem.*, **301**, 141–188.
4. (a) Huang, M., Garrett, G.E., Bohé, L., Pratt, D.A., and Crich, D. (2012) *Nat. Chem.*, **4**, 663–667; (b) Crich, D. and Chandrasekera, N.S. (2004) *Angew. Chem. Int. Ed.*, **43**, 5386–5389.
5. Jensen, H.H., Nordstrøm, M., and Bols, M. (2004) *J. Am. Chem. Soc.*, **126**, 9205–9213.
6. Crich, D. and Vinogradova, O. (2006) *J. Org. Chem.*, **71**, 8473–8480.

7. Crich, D. and Banerjee, A. (2006) *J. Am. Chem. Soc.*, **128**, 8078–8086.
8. (a) Codée, J.D.C., Litjens, R.E.J.N., van den Bos, L.J., Overkleeft, H.S., van Boom, J.H., and van der Marel, G.A. (2003) *Org. Lett.*, **5**, 1947–1950; (b) Codée, J.D.C., van den Bos, L.J., Litjens, R.E.J.N., Overkleeft, H.S., van Boeckel, C.A.A., van Boom, J.H., and van der Marel, G.A. (2004) *Tetrahedron*, **60**, 1057–1064; (c) Codée, J.D.C., Boltje, T.J., and van der Marel, G.A. (2011) *Carbohydrate Chemistry: Proven Synthetic Methods*, Vol. 1 Chapter 6, CRC Press, Taylor & Francis Group, pp. 67–72.
9. Crich, D. and Smith, M. (2001) *J. Am. Chem. Soc.*, **123**, 9015–9019.
10. (a) van den Bos, L.J., Dinkelaar, J., Overkleeft, H.S., and van der Marel, G.A. (2006) *J. Am. Chem. Soc.*, **128**, 13066–13067; (b) Codée, J.D.C., van den Bos, L.J., de Jong, A.R., Dinkelaar, J., Lodder, G., Overkleeft, H.S., and van der Marel, G.A. (2009) *J. Org. Chem.*, **74**, 38–47; (c) Walvoort, M.T.C., Lodder, G., Overkleeft, H.S., Codée, J.D.C., and van der Marel, G.A. (2010) *J. Org. Chem.*, **75**, 7990–8002; (d) Walvoort, M.T.C., Moggré, G.-J., Lodder, G., Overkleeft, H.S., Codée, J.D.C., and van der Marel, G.A. (2011) *J. Org. Chem.*, **76**, 7301–7315; (e) Codée, J.D.C., Walvoort, M.T.C., de Jong, A.R., Dinkelaar, J., Lodder, G., Overkleeft, H.S., and van der Marel, G.A. (2011) *J. Carbohydr. Chem.*, **30**, 438–457.
11. Walvoort, M.T.C., Lodder, G., Mazurek, J., Overkleeft, H.S., Codée, J.D.C., and van der Marel, G.A. (2009) *J. Am. Chem. Soc.*, **131**, 12080–12081.
12. (a) Romero, J.A.C., Tabacco, S.A., and Woerpel, K.A. (2000) *J. Am. Chem. Soc.*, **122**, 168–169; (b) Ayala, L., Lucero, C.G., Romero, J.A.C., Tabacco, S.A., and Woerpel, K.A. (2003) *J. Am. Chem. Soc.*, **125**, 15521–15528; (c) Lucero, C.G. and Woerpel, K.A. (2006) *J. Org. Chem.*, **71**, 2641–2647.
13. Walvoort, M.T.C., de Witte, W., van Dijk, J., Dinkelaar, J., Lodder, G., Overkleeft, H.S., Codée, J.D.C., and van der Marel, G.A. (2011) *Org. Lett.*, **13**, 4360–4363.
14. Walvoort, M.T.C., van den Elst, H., Plante, O.J., Kröck, L., Seeberger, P.H., Overkleeft, H.S., van der Marel, G.A., and Codée, J.D.C. (2012) *Angew. Chem. Int. Ed.*, **51**, 4393–4396.
15. Kröck, L., Esposito, D., Castagner, B., Wang, C.-C., Bindschädler, P., and Seeberger, P.H. (2012) *Chem. Sci.*, **3**, 1617–1622.
16. Plante, O.J., Palmacci, E.R., and Seeberger, P.H. (2001) *Science*, **291**, 1523–1527.
17. Chi, F.-C., Kulkarni, S.S., Zulueta, M.M.L., and Hung, S.-C. (2009) *Chem. Asian J.*, **4**, 386–390.
18. (a) Dinkelaar, J., van den Bos, L.J., Hogendorf, W.F.J., Lodder, G., Overkleeft, H.S., Codée, J.D.C., and van der Marel, G.A. (2008) *Chem. Eur. J.*, **14**, 9400–9411; (b) Dinkelaar, J., De Jong, A.R., van Meer, R., Lodder, G., Overkleeft, H.S., Codée, J.D.C., and van der Marel, G.A. (2009) *J. Org. Chem.*, **74**, 4982–4991.
19. (a) Woods, R.J., Andrews, C.W., and Bowen, J.P. (1992) *J. Am. Chem. Soc.*, **114**, 850–858; (b) Dudley, T.J., Smoliakova, I.P., and Hoffmann, M.R. (1999) *J. Org. Chem.*, **64**, 1247–1253.
20. (a) De Mico, A., Margarita, R., Parlanti, L., Vescovi, A., and Piancatelli, G. (1997) *J. Org. Chem.*, **62**, 6974–6977; (b) van den Bos, L.J., Codée, J.D.C., van der Toorn, J.C., Boltje, T.J., van Boom, J.H., Overkleeft, H.S., and van der Marel, G.A. (2004) *Org. Lett.*, **6**, 2165–2168.
21. Codée, J.D.C., Christina, A.E., Walvoort, M.T.C., Overkleeft, H.S., and van der Marel, G.A. (2011) *Top. Curr. Chem.*, **301**, 253–289.
22. Christina, A.E., van den Bos, L.J., Overkleeft, H.S., van der Marel, G.A., and Codée, J.D.C. (2011) *J.Org. Chem.*, **76**, 1692–1706.
23. van den Bos, L.J., Litjens, R.E.J.N., van den Berg, R.J.B.H.N., Overkleeft, H.S., and van der Marel, G.A. (2005) *Org. Lett.*, **7**, 2007–2010.
24. Wu, X.Y., Cui, L.N., Lipinski, T., and Bundle, D.R. (2010) *Chem. Eur. J.*, **16**, 3476–3488.
25. Crich, D., Huy, T., and Cai, F. (2008) *J. Org. Chem.*, **73**, 8942–8953.

26. Baek, J.Y., Lee, B.-Y., Jo, M.G., and Kim, K.S. (2009) *J. Am. Chem. Soc.*, **131**, 17705–17713.
27. Crich, D., Cai, W., and Dai, Z. (2000) *J. Org. Chem.*, **65**, 1291–1297.
28. Demchenko, A.V., Rousson, E., and Boons, G.-J. (1999) *Tetrahedron Lett.*, **40**, 6523–6526.
29. Ma, T., Lian, G., Li, Y., and Yu, B. (2011) *Chem. Commun. (Camb.)*, **47**, 7515–7517.
30. (a) Garcia, B.A., Poole, J.L., and Gin, D.A. (1997) *J. Am. Chem. Soc.*, **119**, 7597–7598; (b) Garcia, B.A. and Gin, D.A. (2000) *J. Am. Chem. Soc.*, **122**, 4269–4279.
31. (a) Kim, J.-H., Yang, H., Park, J., and Boons, G.-J. (2005) *J. Am. Chem. Soc.*, **127**, 12090–12097; (b) Kim, J.-H., Yang, H., and Boons, G.-J. (2005) *Angew. Chem. Int. Ed.*, **44**, 947–949; (c) Kim, J.-H., Yang, H., Khot, V., Whitfield, D., and Boons, G.-J. (2006) *Eur. J. Org. Chem.*, **22**, 5007–5028; (d) Fang, T., Mo, K.-F., and Boons, G.-J. (2012) *J. Am. Chem. Soc.*, **134**, 7545–7552.
32. See for related mechanistic work: (a) Fascione, M.A., Adshead, S.J., Stalford, S.A., Kilner, C.A., Leach, A.G., and Turnbull, W.B. (2009) *Chem. Commun. (Camb.)*, 5841–5843; (b) Fascione, M.A., Kilner, C.A., Leach, A.G., and Turnbull, W.B. (2012) *Chem. Eur. J.*, **18**, 321–333.
33. Park, J., Kawathar, S., Kim, J.-H., and Boons, G.-J. (2007) *Org. Lett.*, **9**, 1959–1962.
34. Boltje, T.J., Kim, J.-H., Park, J., and Boons, G.-J. (2010) *Nat. Chem.*, **2**, 552–557.
35. Christina, A.E., van der Es, D., Dinkelaar, J., Overkleeft, H.S., van der Marel, G.A., and Codée, J.D.C. (2012) *Chem. Commun. (Camb.)*, **48**, 2686–2688.
36. For an alternative approach using a 6-S-benzylidene mannosyl donor approach, see: (a) Crich, D. and Li, L. (2009) *J. Org. Chem.*, **74**, 773–781; (b) Picard, S. and Crich, D. (2011) *Chimia*, **65**, 59–64.
37. Bongat, A.F.G. and Demchenko, A.V. (2007) *Carbohydr. Res.*, **342**, 374–406.
38. (a) Benakli, K., Zha, C., and Kerns, R.J. (2001) *J. Am. Chem. Soc.*, **123**, 9461–9462; (b) Wei, P. and Kerns, R.J. (2004) *J. Org. Chem.*, **70**, 4195–4198.
39. (a) Manabe, S., Ishii, K., and Ito, Y. (2006) *J. Am. Chem. Soc.*, **128**, 10666–10667; (b) Manabe, S., Ishii, K., Hashizume, D., Koshino, H., and Ito, Y. (2009) *Chem. Eur. J.*, **15**, 6894–6901; (c) Satoh, H., Manabe, S., Ito, Y., Lüthi, H.P., Laino, T., and Hutter, J. (2011) *J. Am. Chem. Soc.*, **133**, 5610–5619.
40. (a) Boysen, M., Gemma, E., Lahmann, M., and Oscarson, S. (2005) *Chem. Commun. (Camb.)*, 3044–3046; (b) Olson, J.D.M., Eriksson, L., Lahmann, M., and Oscarson, S. (2008) *J. Org. Chem.*, **73**, 7181–7188.
41. (a) Geng, Y.Q., Zhang, L.H., and Ye, X.S. (2008) *Tetrahedron*, **64**, 4949–4958; (b) Geng, Y.Q., Zhang, L.H., and Ye, X.S. (2005) *Chem. Commun. (Camb.)*, 597–599.
42. (a) Geng, Y.Q. and Ye, X.S. (2012) *J. Org. Chem.*, **77**, 5255–5270; (b) Zhu, T. and Boons, G.-J. (2001) *Org. Lett.*, **3**, 4201–4203.
43. Manabe, S., Ishii, K., and Ito, Y. (2007) *J. Org. Chem.*, **72**, 6107–6115.
44. Li, Z. and Gildersleeve, J.C. (2006) *J. Am. Chem. Soc.*, **128**, 11612–11619.
45. (a) Boons, G.-J. and Demchenko, A.V. (2000) *Chem. Rev.*, **100**, 4539–4565; (b) De Meo, C. (2008) *Carbohydr. Res.*, **343**, 1540–1552.
46. (a) Tanaka, H., Nishiura, Y., and Takahashi, T. (2006) *J. Am. Chem. Soc.*, **128**, 7124–7125; (b) Crich, D. and Li, W. (2007) *J. Org. Chem.*, **72**, 2387–2391; (c) Crich, D. and Wu, B. (2008) *Org. Lett.*, **10**, 4033–4035; (d) Farris, M.D. and De Meo, C. (2007) *Tetrahedron Lett.*, **48**, 1225–1227; (e) Hsu, C.-H., Chu, K.-C., Lin, Y.-S., Han, J.L., Peng, Y.-S., Ren, C.-T., Wu, C.-Y., and Wong, C.-H. (2010) *Chem. Eur. J.*, **16**, 1754–1760.
47. Crich, D. and Vinod, A. (2003) *Org. Lett.*, **5**, 1297–1300.
48. Imamura, A. and Lowary, T.L. (2011) *Trends Glycosci. Glycotechnol.*, **23**, 134–152.
49. Callam, C.S., Gadikota, R.R., Krein, D.M., and Lowary, T.L. (2003) *J. Am. Chem. Soc.*, **125**, 13112–13119.
50. Bai, Y. and Lowary, T.L. (2006) *J. Org. Chem.*, **71**, 9658–9671.

51. Zhu, X.M., Kawatkar, S., Rao, Y., and Boons, G.-J. (2006) *J. Am. Chem. Soc.*, **128**, 11948–11957.
52. (a) Ishiwata, A., Akao, H., and Ito, Y. (2006) *Org. Lett.*, **8**, 5525–5528; (b) Ishiwata, A. and Ito, Y. (2011) *J. Am. Chem. Soc.*, **133**, 2275–2291.
53. Crich, D., Pedersen, C.M., Bowers, A.A., and Wink, D.J. (2007) *J. Org. Chem.*, **72**, 1553–1565.
54. Bear, T.J., Shaw, J.T., and Woerpel, K.A. (2002) *J. Org. Chem.*, **67**, 2056–2064.
55. Smith, D.M., Tran, M.B., and Woerpel, K.A. (2003) *J. Am. Chem. Soc.*, **125**, 14149–14152.
56. Wang, Y., Maguire-Boyle, S., Dere, R.T., and Zhu, X. (2008) *Carbohydr. Res.*, **343**, 3100–3106.
57. Lee, Y.L., Lee, K., Jung, E.H., Jeon, H.B., and Kim, K.S. (2005) *Org. Lett.*, **7**, 3263–3266.

5
Stereocontrol of 1,2-*cis*-Glycosylation by Remote *O*-Acyl Protecting Groups

Bozhena S. Komarova, Nadezhda E. Ustyuzhanina, Yury E. Tsvetkov, and Nikolay E. Nifantiev

5.1
Introduction

The formation of a glycoside bond is the key step in any oligosaccharide synthesis. As a rule, only one of two possible anomeric products is desired. For half a century, a great experience in the field of stereoselective glycosylation has been acquired. The most prominent finding was the possibility to prepare stereoselectively 1,2-*trans*-glycopyranosides by glycosylation with 2-*O*-acylated pyranosyl donors. 1,2-*trans*-Selectivity is governed by the anchimeric assistance of the vicinal acyl group (Scheme 5.1a). The need to access 1,2-*cis*- or 1,2-*trans*-glycopyranosides bearing no vicinal participating group posed a question whether nonvicinal, for example, attached to either O-3, O-4, or O-6, acyl groups can similarly influence the stereoselectivity of glycosylations because of the remote anchimeric assistance (Scheme 5.1b,c) and to which extent.

On the one hand, enough data has been accumulated up to date to affirm that remote acyl groups affect the stereoselectivity of glycosylation. On the other hand, the degree of their stereodirecting influence is not equal for different pyranoses, and this absence of generality makes the concept of the remote anchimeric assistance questionable. In this review, we present data concerning the stereocontrol of glycosylation by remote *O*-acyl groups separately for each position of acyl group attachment and for each relative configuration of the pyranose to give an array of options that offers the introduction of a remote acyl group into a donor. Different points of view on the mechanism of the stereocontrol are discussed as well.

5.2
Stereodirecting Influence of Acyl Groups at Axial and Equatorial *O*-3: Opposite Stereoselectivity Proves Anchimeric Assistance

The review aims at providing a general overview of stereocontrolling properties of remote potentially participating groups. However, in spite of plenty of published

Modern Synthetic Methods in Carbohydrate Chemistry: From Monosaccharides to Complex Glycoconjugates, First Edition. Edited by Daniel B. Werz and Sébastien Vidal.

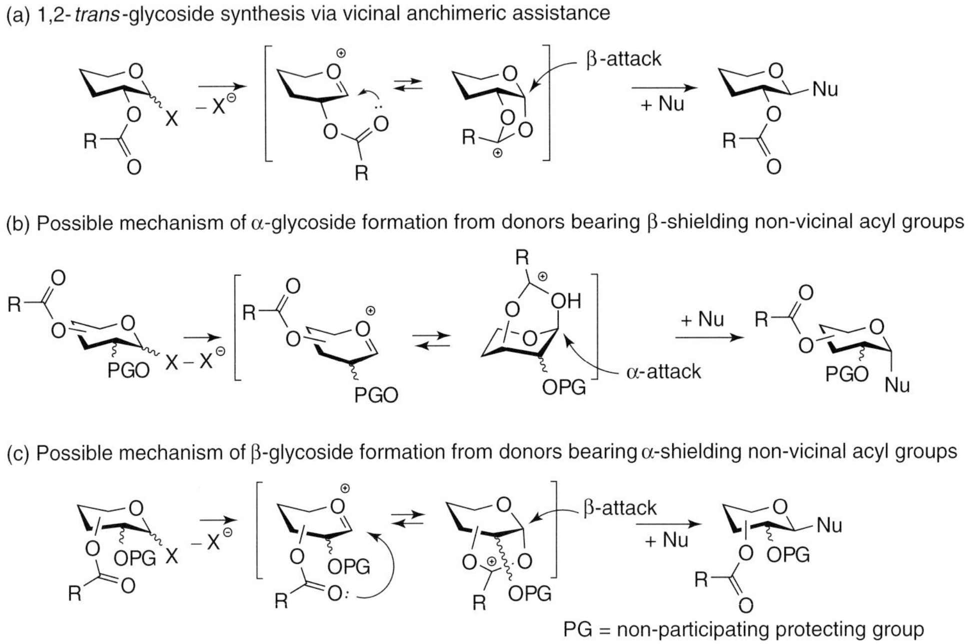

Scheme 5.1 Anchimeric assistance of vicinal acyl groups and possible stereoselective synthesis of α- and β-glycopyranosides bearing nonvicinal acyl groups. (a) 1,2-*trans*-Glycoside synthesis via vicinal anchimeric assistance, (b) possible mechanism of α-glycoside formation from donors bearing β-shielding nonvicinal acyl groups, and (c) possible mechanism of β-glycoside formation from donors bearing α-shielding nonvicinal acyl groups.

experimental data on glycosylation, a generalization of the results turns out to be questionable, as the reaction partners and conditions are very different. To make the analysis of the data reliable, we comply with some restrictions: (i) the considered data include results of glycosylation of the same acceptor with a donor bearing the same leaving group and under the same conditions; (ii) basic factors, apart from the anchimeric assistance, that determine that the stereoselectivity are thoroughly taken into account. The first of these basic factors is the type of reaction mechanism. There are glycosylations that definitely go through the S_N2 mechanism, like Lemieux glycosylation [1]. These reactions are excluded from the consideration, as the stereocontrolling effect of acyl substituents implies the S_N1 mechanism. Second, the solvent medium affects the stereochemistry of glycosylations. β-Directing effect of nitrile solvents [2–6] and α-directing effect of ether solvents [7, 8] are well known. Given that the most neutral solvent is CH_2Cl_2, only minimal attention is paid to reactions carried out in other solvents. Third, the reaction temperature is an important parameter, but it should be admitted that it is very difficult to take it into account.

Acyl groups at O-3 in pyranosyl donors change the stereoselectivity of glycosylation as compared to a nonparticipating alkyl-protecting group at the same O-3:

in $^4C_1$ conformation, an axially oriented acyl group favors the formation of a β-product, while an equatorial one ensures the predominance of an α-product. For example, β-stereodirecting influence of a potentially participating acyl group at axial O-3 is proved for altro donors [9]. Altrosyl fluoride **1a** bearing only nonparticipating protecting groups at O-2, O-3, and O-4 gave less proportion of the β-isomer **2** than fluorides **1b–1d** with potentially participating groups at O-3 (Scheme 5.2). It should be underlined that here the β-selectivity manifests itself even in Et_2O, which is considered favoring the α-product formation [9].

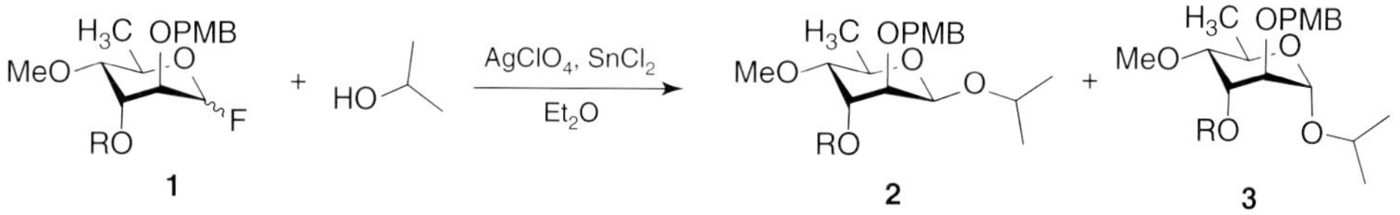

1 **2** **3**

Donor	R	2 : 3 ratio (yield, %)
1a	TBDMS	1.5 : 1 (80)
1b	Bz	2.5 : 1 (92)
1c	*p*-MeOBz	3 : 1 (92)
1d	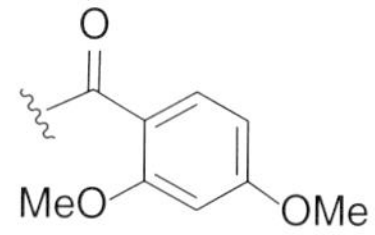	2.5 : 1 (69)

Scheme 5.2 Stereodirecting effect of acyl groups at axial O-3.

Donor **1c** demonstrated its utility in the synthesis of sordarine **5**, in which complex tetracyclic aglycon **4** is β-glycosylated with 6-deoxyaltrose (Scheme 5.3) [9]. In this synthesis, a *p*-methoxybenzoyl group fulfilled the role of β-stereodirecting group.

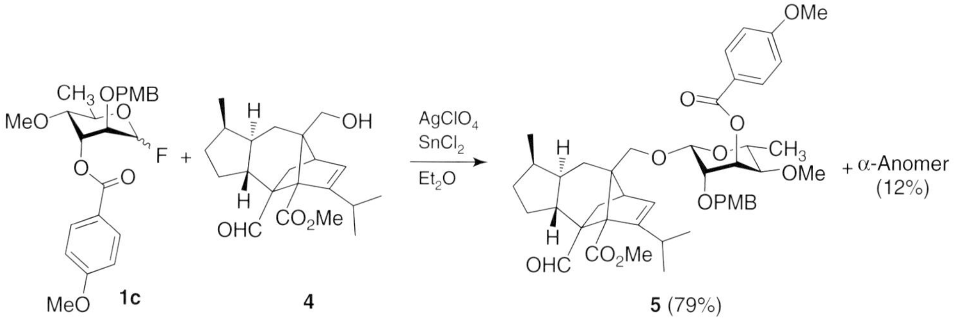

1c **4** **5** (79%)

Scheme 5.3 Stereoselective synthesis of sordarine.

Likewise, reaction of donor **6** with acceptor **7** resulted in the highly stereoselective formation of β-glycoside **8** (Scheme 5.4) [10].

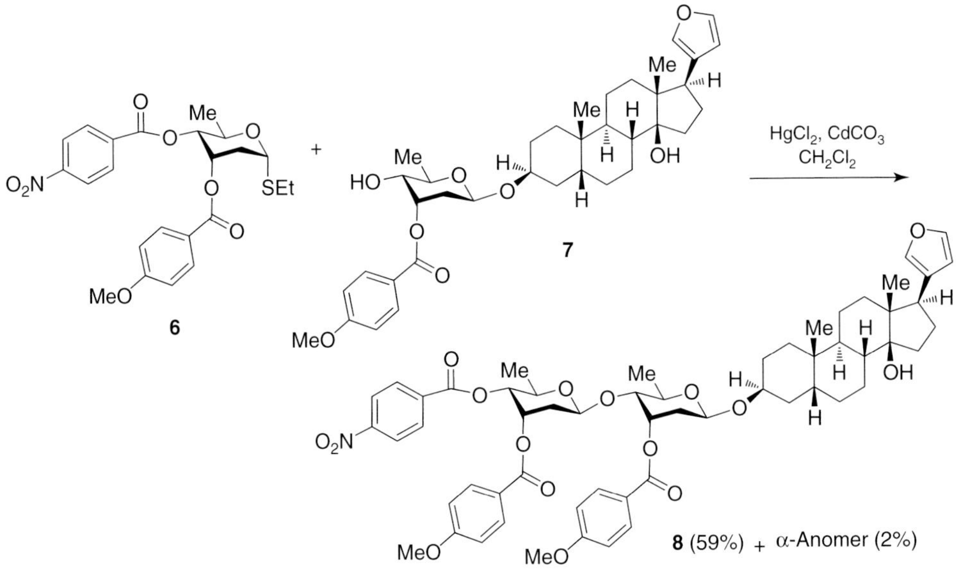

Scheme 5.4 Stereodirecting effect of the *p*-methoxybenzoyl group at axial O-3.

Crich *et al.* [11] attempted to prove directly the possibility of the remote anchimeric assistance by trapping bicyclic stabilized cations corresponding to this mechanism. To this aim, glycosyl donors of different configurations bearing a *tert*-butoxycarbonyl protecting group at one of the oxygen atoms (except O-2) were activated under glycosylation conditions in the absence of an acceptor. Allosyl donor **9** with the axial Boc group gave upon activation the bicyclic carbonate **10** (Scheme 5.5a). Besides, it is known that 2-deoxy donor **11** with *N*-methylcarbamoyl group at O-3 also provided similar bicycle **12** (Scheme 5.5b) [10]. But all the other tested donors with the Boc-protecting group at O-3, O-4, or O-6 did not form such cyclic carbonates. From these results, the authors concluded that the remote anchimeric assistance is possible only from the position 3 of allose. This conclusion is apparently correct, if the remote anchimeric assistance is considered in the strict sense of the term as the formation of a true bicyclic intermediate (Scheme 5.1b,c). However, one may treat the remote anchimeric assistance more broadly, namely as an interaction of carbonyl oxygen with cationic C-1, which is able to govern the access of a nucleophile but does not lead to the formation of a covalent bond between them. In this interpretation, the lack of formation of the corresponding cyclic carbonates does not exclude the possibility of the remote anchimeric assistance of acyl groups, to greater or lesser extent, from other positions of pyranoses of other configurations. The results given in the following seem to confirm this statement.

First, another research group has directly demonstrated a possibility of the remote participation of an acyl group at equatorial O-3 by conversion of mannosyl donor **13** into 1,3-bicycle **14** (Scheme 5.5c) [12]. Acyl groups at equatorial O-3 in glucosyl donors also display an evident stereodirecting effect despite the

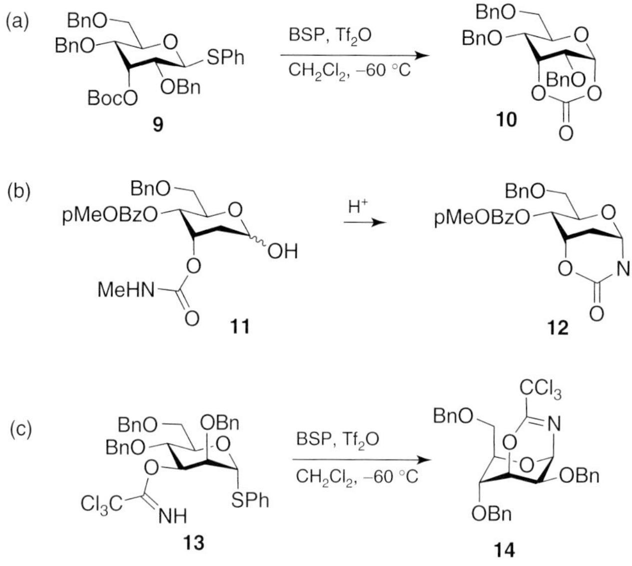

Scheme 5.5 Formation of the (a) bicyclic carbonate, (b) carbamate, and (c) imidate as a result of the remote anchimeric participation.

fact that the 3-O-Boc-protected glucosyl donor failed to produce corresponding 1,3-bicyclic carbonate under glycosylation conditions [11]. Thus, 3-*O*-acetylated glucosyl trifluoroacetimidate **15b** provided upon activation with AgOTf in CH_2Cl_2 essentially higher α:β ratio than fully benzylated counterpart **15a** (Scheme 5.6) [13, 14].

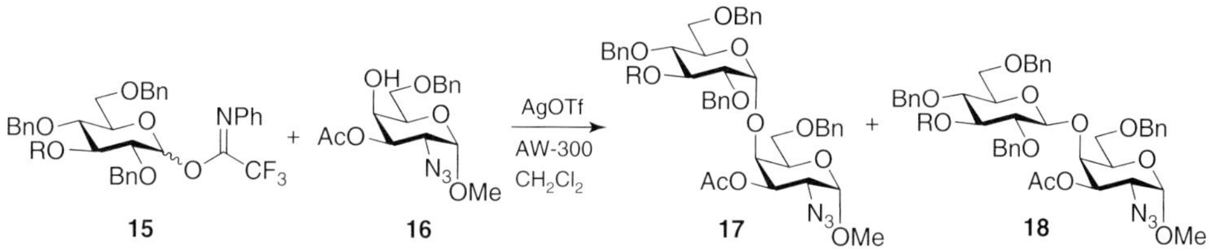

Donor	R	**17** : **18** ratio
15a	Bn	2 : 1
15b	Ac	4 : 1

Scheme 5.6 Stereodirecting effect of the acetyl group at equatorial O-3.

The same trend was demonstrated in the Mukaiyama glycosylation [15] with 6-deoxyglucosyl donor **19** (Scheme 5.7). Although this type of glycosylation is considered β-stereoselective, reaction of acceptor **20** with compound **19** gave unexpectedly a very high proportion of α-isomer **21**. Model glycosylation of cholesterol **24** with 3-*O*-acetylated donor **19** and its fully benzylated analog **23** under the same conditions showed that the replacement of the benzyl group at O-3 by acetyl one resulted in the reversal of the reaction stereoselectivity [16]. Thus, 3-*O*-acyl groups

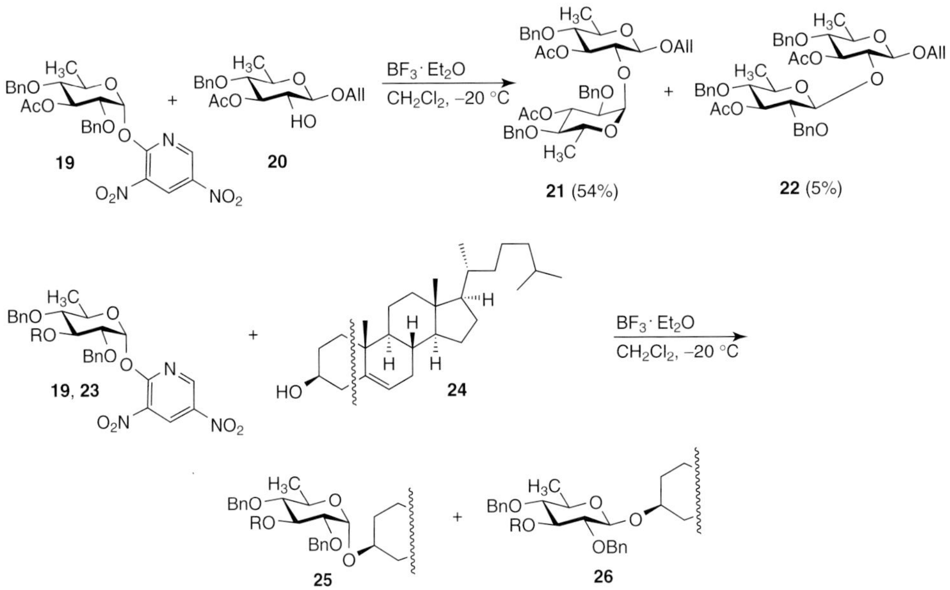

Donor	R	**25** : **26** ratio
19	Ac	2 : 1 (60)
23	Bn	1 : 2 (56)

Scheme 5.7 Stereodirecting effect of the acetyl group at equatorial O-3 in Mukaiyama glycosylation.

in glucosyl donors favor the formation of α-glucosides irrespective of the type of the donor and reaction conditions.

There are numerous additional illustrations of the α-directing effect of acyl groups at equatorial O-3 in different types of donors of manno, gluco, and galacto series. An acyl group at O-3 of 4,6-*O*-benzylidene-protected mannosyl donors is able to affect the stereoselectivity dramatically. The presence of a 4,6-*O*-benzylidene protecting group is a prerequisite for the efficient synthesis of β-mannosides [17–19], which are otherwise difficult to access [20]. But acylation of O-3 leads to the complete loss of the 4,6-*O*-benzylidene-dependent β-selectivity [21, 22]. Thus, mannosylation of acceptor **31** with 3-O-benzoylated thiomannoside **30** gives only α-glycoside **32** (Scheme 5.8b) [22] under the same conditions, in which thiomannoside **27**, bearing no participating group, affords almost exclusively β-mannoside **29** (Scheme 5.8a) [17].

The same reversal of the stereoselectivity in the manno series was observed in the case of an electron-withdrawing 3-*O*-monochloroacetyl group [23]. Obviously, the α-stereoselectivity of mannosylation associated with the presence of 3-*O*-acyl groups is a consequence of the anchimeric assistance. An alternative assumption that the cause of the α-selectivity is electron-withdrawing nature of substituents at O-3 is disproved by next several examples, which reveal the effect of electron-withdrawing groups at O-3 on the stereochemical result.

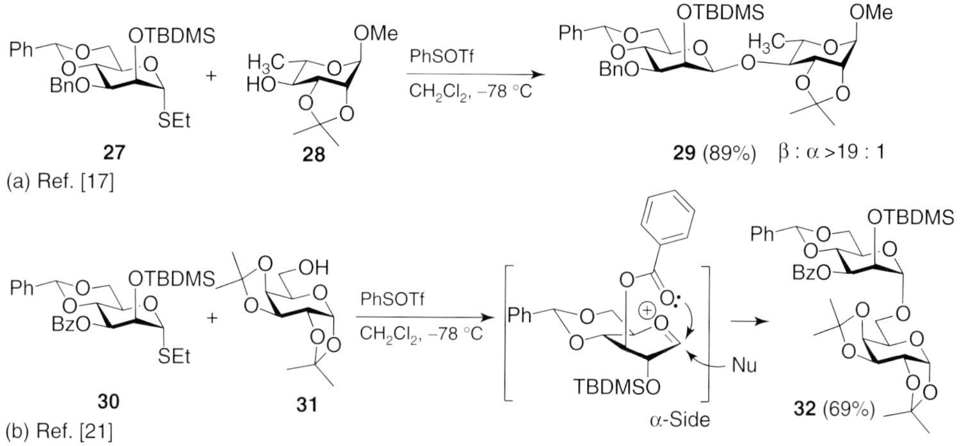

Scheme 5.8 Acyl group at O-3 eliminates the β-directing effect of the 4,6-*O*-benzylidene group in mannosyl donors.

The next series of experiments helps to evaluate the properties of potentially participating acyl groups. Mannosyl donors **33** bearing at O-3 benzyl, acyl and electron-withdrawing groups of the different degrees of electronegativity and nucleophilicity were tested in glycosylation of acceptor **34** (Scheme 5.9) [12]. Donor **33a** having neither electron-withdrawing nor participating benzyl group at O-3 affords a mixture of disaccharides **35** and **36** in a ratio of 1 : 2.7. This result was taken as a reference point.

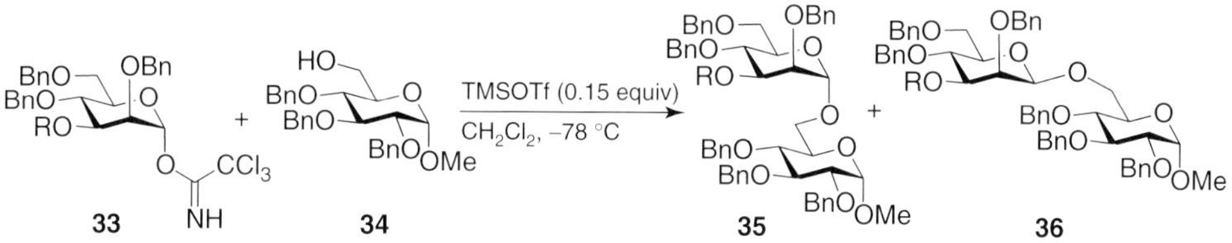

Donor	R	**36** : **35** ratio (yield, %)
33a	Bn	2.7 : 1 (91)
33b	SO_2Bn	15.9 : 1 (95)
33c	Bz*p*-NO_2	1 : 1.7 (92)
33d	Bz	1 : 9.8 (81)
33e	Ac	1 : 25.9 (91)

Scheme 5.9 Effect of potentially participating acyl and electron-withdrawing groups at O-3 on the stereoselectivity of mannosylation.

Donor **33b** having a strongly electron-withdrawing 3-*O*-benzylsulfonyl group provided an excellent β-selectivity. That is, the electronegativity of a substituent at O-3 shifts the selectivity of mannosylation toward β-anomer **36**, and the value of the shift depends on the degree of the electronegativity. Less electronegative *p*-nitrobenzoate, which additionally possesses nucleophilic properties, ensured only slight increase of the α-product formation with regard to fully benzylated donor **33a**. Further decrease of the electronegativity and increase of the nucleophilicity of the substituent at O-3 (benzoylated donor **33d**) resulted in further improvement of α-selectivity. Finally, almost complete α-selectivity was achieved with 3-O-acetylated donor **33e**. These results lead to a conclusion that changes in the stereoselectivity of glycosylation caused by remote acyl groups cannot be explained only by electron-withdrawing properties of those groups as van Boeckel [24] did. These examples also show that electron-withdrawing and nucleophilic properties of substituents should be distinguished.

Glycosylation of acceptor **38** with 3-*O*-acetylated glucuronyl bromide **37b** afforded pure α-linked disaccharide **39** (Scheme 5.10), while fully benzylated counterpart **37a** gave an α:β-mixture of the corresponding disaccharides in a ratio of 2 : 1 [13] under the same conditions. However, such a high stereoselectivity of 3-*O*-acetylated donor **37b** may be partly accounted for by the assistance of the methoxycarbonyl group at C-5.

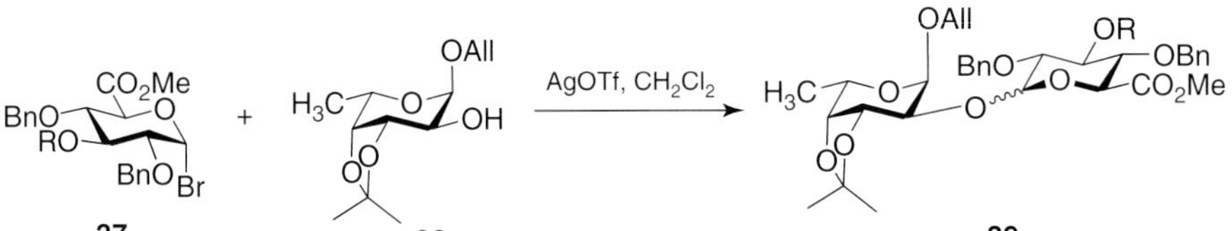

Donor	R	α : β ratio (yield, %)
37a	Bn	2 : 1 (89)
37b	Ac	Only α (90)

Scheme 5.10 Effect of the 3-*O*-acetyl group on the stereoselectivity of glucuronosylation.

A very similar effect of 3-*O*-acetylation was observed in the case of xylosylation of acceptor **41** with imidates **40** [13]. Glycosylation with fully benzylated donor **40a** leads to an anomeric mixture of disaccharides **42**, while the presence of an acetyl group at O-3 in **40b** ensures the complete α-stereoselectivity (Scheme 5.11).

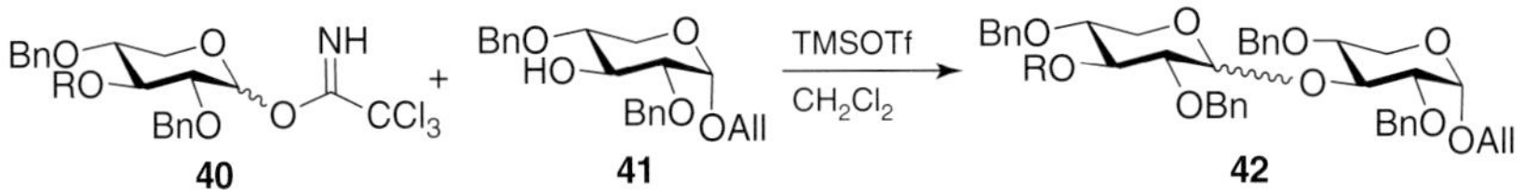

Donor	R	α : β ratio (yield, %)
40a	Bn	2.5 : 1 (85)
40b	Ac	Only α (80)

Scheme 5.11 α-Stereodirecting effect of an acetyl group at O-3 in xylosyl donors.

Highly α-stereoselective glycosylation with donor **40b** was successfully applied for the synthesis of di- and trisaccharides **43** and **44** comprising one and two α-xylosyl residues, respectively (Scheme 5.12) [25]. These oligosaccharides served as starting compounds for the preparation of neoglycolipid acceptors for investigation of carbohydrate specificity of α-xylosyltransferases acting on *O*-glucosylated epidermal growth factor repeats of Notch [26, 27].

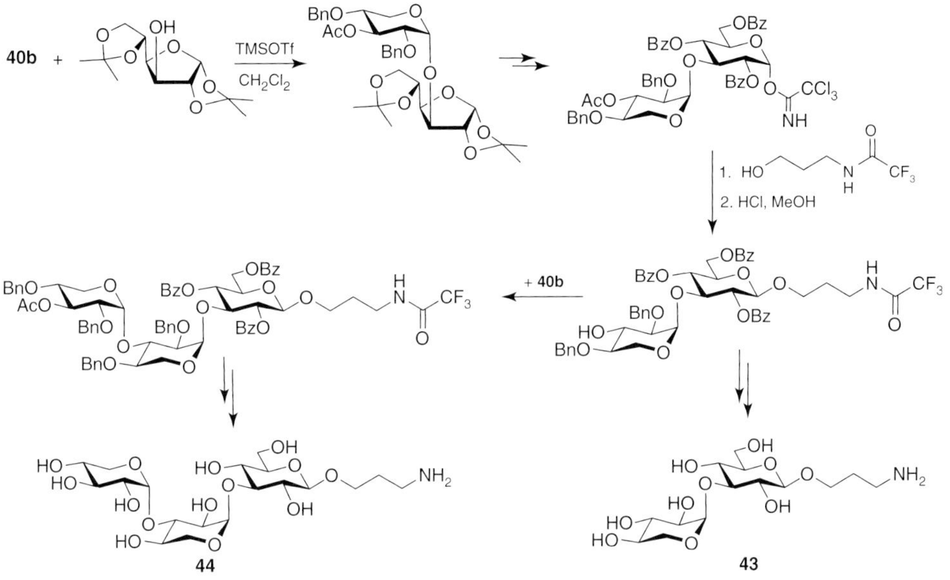

Scheme 5.12 Application of 3-*O*-acetylated donor **40b** for the synthesis of di- and trisaccharides bearing α-xylopyranosyl units.

The stereodirecting effect of acyl groups at O-3 is also manifested in the glycosylation with fucosyl donors. In contrast to fully benzylated donor **45a**, which demonstrated no selectivity in the reaction with acceptor **38**, 3-*O*-benzoyl imidate **45b** provided highly α-stereoselective formation of disaccharide **46** (Scheme 5.13) [28].

Thus, the stereodirecting effect of acyl groups at O-3 of various pyranosyl donors is a well-documented fact. However, some other factors, which are able to affect the stereochemistry of glycosylation, can interfere with this effect. For example, the influence of an acyl group at O-3 in 2-azido-2-deoxy donors of the gluco and galacto configurations is not always noticeable. While 3-*O*-acyl groups in 2-azido-2-deoxygalactosyl donors provide an increase of the α-selectivity [29], little or no changes in the stereoselectivity was observed upon replacement of a nonparticipating group at O-3 with an acyl group in 2-azido-2-deoxyglucosyl donors [30, 31]. Partly it may be explained by good intrinsic O-selectivity of 2-azido-2-deoxyglucosyl donors even in the absence of any participating protecting groups.

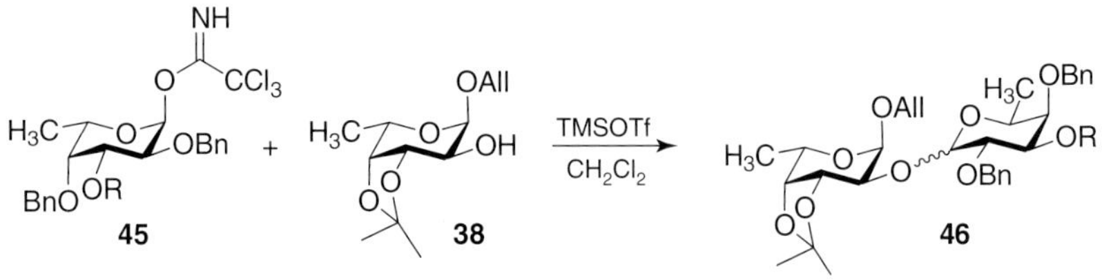

Donor	R	$\alpha:\beta$ ratio (yield, %)
45a	Bn	1.4 : 1 (83%)
45b	Bz	12 : 1 (85%)

Scheme 5.13 Different stereoselectivities of glycosylations with the 3-*O*-benzoylated fucosyl donor and the donor bearing no participating groups.

The β-directing effect of nitrile solvents is also able to decrease the influence of a participating acyl substituent at O-3. For example, glycosylation of primary acceptor **48** with 3-*O*-acetylated imidate **47** in CH_2Cl_2 resulted in the formation of an anomeric mixture of disaccharides **49** with a slight predominance of the α-anomer (Scheme 5.14a) [29]. However, the reaction of closely related imidate **50** with acceptor **51** under similar conditions but in acetonitrile provided only β-isomer **52** (Scheme 5.14b) [32].

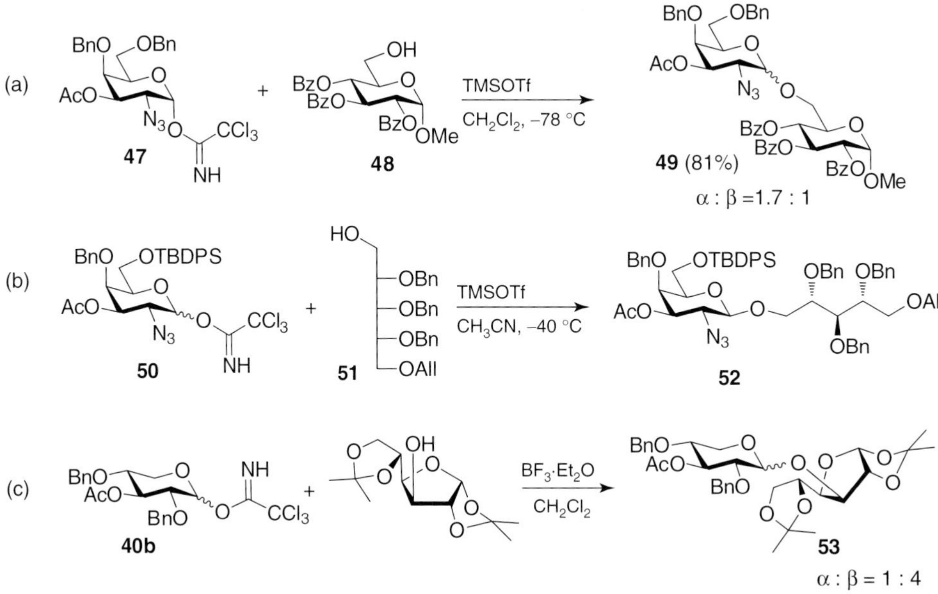

Scheme 5.14 Different cases of the loss of the stereocontrol by acyl groups at O-3.

Another factor, which can eliminate the effect of remote acyl groups, is the choice of a promoter. Thus, trimethylsilyl trifluoromethanesulfonate (TMSOTf)-promoted xylosylation of di-isopropylidene-D-glucose with imidate **40b** afforded the α-disaccharide (Scheme 5.12), whereas the same reaction in the presence of $BF_3 \cdot Et_2O$ yields β-isomer of **53** (Scheme 5.14c) [25] predominantly. The role of $BF_3 \cdot Et_2O$ can also be evaluated from glycosylation with a derivative of 2-azido-2-deoxygalactose bearing a levulinoyl group at O-3: reaction in the presence of TMSOTf led to a 1 : 2 α:β-mixture, while $BF_3 \cdot Et_2O$ provided complete β-selectivity [33].

5.3
Acyl Groups at O-4 in the galacto Series: Practical Synthesis of α-Glycosides: Complete Stereoselectivity

Natural oligosaccharides containing α-linked galacto-configurated pyranosides include blood group antigens [34, 35], selectin ligands, glycolipids [36], and fucoidans [37], while pectins are composed of α-galacturonic acid [38]. Numerous syntheses of such oligosaccharides provide extensive data that allow examination of the influence of acyl groups linked to axial O-4.

A series of 2,3-di-*O*-benzyl-protected fucosyl bromides **54** bearing different participating acyl groups at O-4 or nonparticipating benzyl one were tested in glycosylation of acceptor **38** in dichloromethane under Helferich conditions (Scheme 5.15) [39]. No stereoselectivity of the formation of disaccharide **56** was observed for glycosylation with benzylated bromide **54a**. Introduction of a benzoyl group at O-4 (donor **54b**) improved the α-stereoselectivity to α : β = 3.5 : 1. The degree of the stereodirecting effect depends on the nucleophilicity of carbonyl

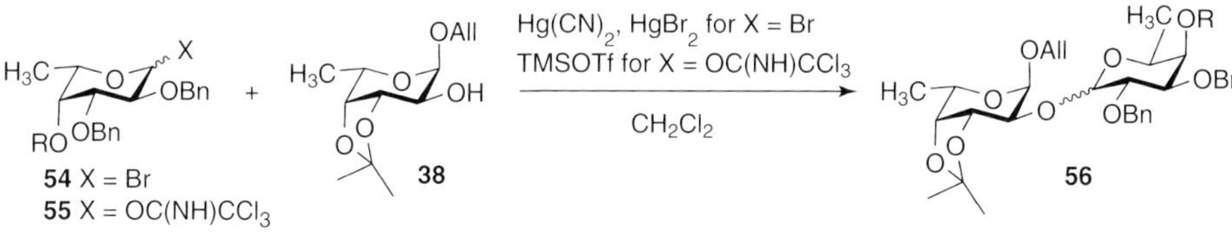

Donor	R	α : β ratio X = Br	α : β ratio X = OC(NH)CCl$_3$	ΔE (kcal mol^{-1})
54a, 55a	Bn	1 : 1	1.4 : 1	0
54b, 55b	Bz	3.5 : 1	4 : 1	−3.6
54c	*p*-NO$_2$Bz	1.9 : 1	—	−2.1
54d	*p*-MeOBz	5 : 1	—	−4.7

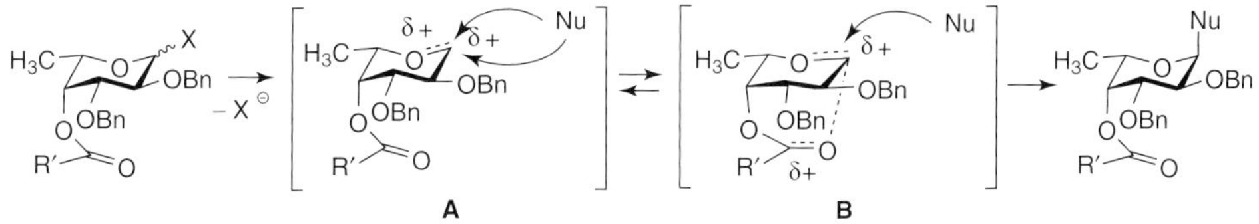

Scheme 5.15 Effect of the nucleophilicity of stereodirecting groups on the stereoselectivity. Correlation with the calculated energies of stabilization.

oxygen of the acyl group involved in the carbocation stabilization. Donor **54c** with a less nucleophilic *p*-nitrobenzoyl group at O-4 turned out to be less α-selective than 4-*O*-benzoylated donor **54b**. On the contrary, donor **54d** with a more nucleophilic *p*-methoxybenzoyl group furnished the highest α-selectivity within this series. Imidates **55** displayed a similar tendency. This evidence in favor of the mechanism of the remote anchimeric assistance was supported by theoretical calculations.

Molecular mechanics calculations by employing the MM+ molecular mechanics force field have been performed (Scheme 5.15). The observed stereoselectivity pattern correlates with calculated energies of stabilization ("stabilization energies") of carbocations with the acyl groups. The largest estimated value ($-4.7\,\mathrm{kcal\,mol^{-1}}$) of carbocation stabilization corresponds to the highest α-stereoselectivity (α : β = 5 : 1). The fact that the *p*-NO_2Bz group exhibited smaller α-directing influence than the benzoyl one is supported by calculated values of stabilization energies being 1.5 times smaller for *p*-NO_2Bz than for Bz.

However, contrary to the tendency with substituted benzoyl groups, the less nucleophilic chloroacetyl group has surprisingly the higher stereodirecting effect than the acetyl one (Scheme 5.16) [40]. That is, fucosylation of acceptor **58** with 4-*O*-chloroacetylated donor **57c** provided α-anomer of disaccharide **59** exclusively, whereas 4-*O*-acetylated donor **57b** produced a 5 : 1 α,β-mixture. Allylated counterpart **57a** displayed the lowest α-stereoselectivity.

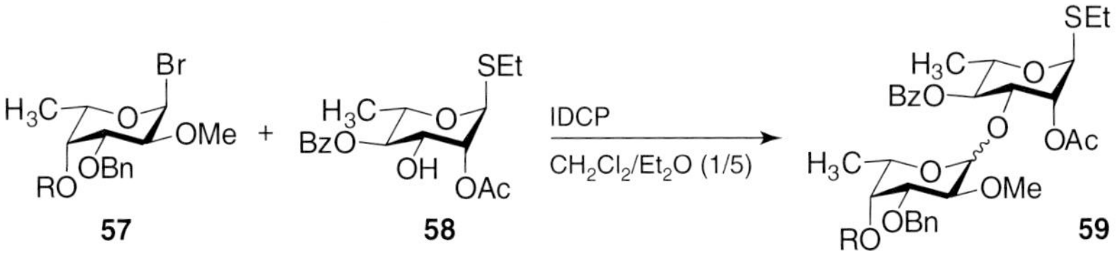

Donor	R	α : β ratio (yield, %)
57a	All	2.5 : 1 (80)
57b	Ac	5 : 1 (61)
57c	C(O)CH_2Cl	Only α (65)

Scheme 5.16 Stereodirecting properties of acetyl and monochloroacetyl groups.

The influence of the 4-*O*-chloroacetyl group in fucose is similar to that of pentafluoropropionyl (PFP) group at O-4 in 2-azido-2-deoxygalactose. The reaction of 2-azido-2-deoxygalactosyl fluoride **60a** having a PFP group with acceptor **61** provided as high α-stereoselectivity of the disaccharide **62** formation as donor **60b** with a participating 4-*O*-acetyl group did (Scheme 5.17) [41]. That is, an electron-withdrawing fluorinated acyl group at O-4, which is much less nucleophilic than an acetyl group, in 2-azido-2-deoxy-galacto donor ensures, nevertheless, highly α-stereoselective glycosylation. An attempt was made to interpret the stereodirecting properties of the PFP group as an alternative mechanism of stereodirecting influence [41, 42]. In authors' opinion, the access of a nucleophile from the α-face of the glycosyl cation is much more preferable, as it allows neutralization of the

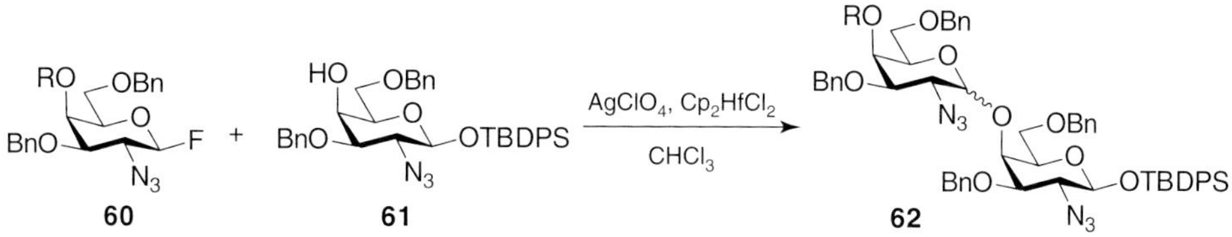

Donor	R	α : β ratio (yield, %)
60a	PFP	94 : 6 (92)
60b	Ac	96 : 4 (80)

Scheme 5.17 Effect of acetyl and pentafluoropropionyl groups at O-4 of 2-azido-2-deoxygalactosyl donors.

strong dipole moment caused by the PFP group at axial O-4. Hence, a combination of two properties, namely, the ability to the anchimeric assistance and a relatively high electronegativity, may be responsible for the strong α-stereodirecting effect of the chloroacetyl group.

However, another example showed that the strong α-stereodirecting effect of a fluorinated acyl group at O-4 of galactosyl donors is not general. A homolog of the PFP group, trifluoroacetyl one, at O-4 in thiogalactoside **63e** provide only a modest increase of the α-selectivity (Scheme 5.18) compared to perbenzylated donor **63a** [43].

In the search for the best conditions for the stereoselective synthesis of α-galactosides, two solvent mixtures, both containing coordinating solvent, were examined (Scheme 5.18) [43]. In both of them, 4-*O*-acetylated thiogalactoside **63b** is more stereoselective than the fully benzylated counterpart **63a**, and the benzoyl

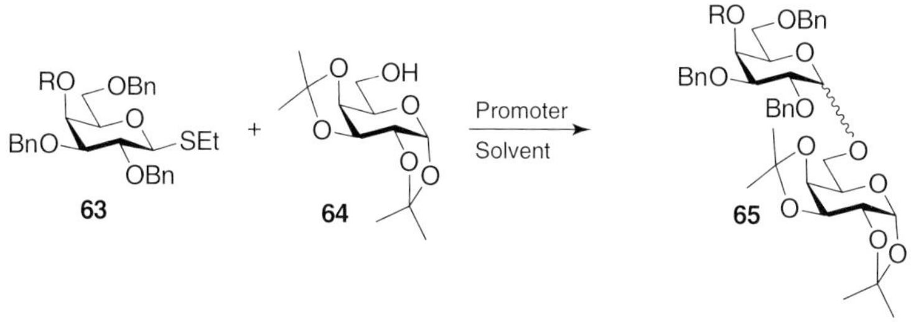

Glycosyl donor	R	α : β ratio (yield, %)			
		$Et_2O/(ClCH_2)_2$ (1/5)		1,4-Dioxane/toluene (1/3)	
		NIS/TMSOTf	IDCP	NIS/TMSOTf	IDCP
63a	Bn	1.2 : 1 (83)	2.1 : 1 (77)	2.2 : 1 (91)	2.6 : 1 (74)
63b	Ac	3.6 : 1 (74)	6.0 : 1 (62)	7.2 : 1 (76)	14 : 1 (83)
63c	Bz	9.5 : 1 (79)	12.5 : 1 (66)	17 : 1 (72)	32 : 1 (74)
63d	*p*-MeOBz			33 : 1 (85)	Only α (75)
63e	CF_3CO			3.0 : 1	

Scheme 5.18 Stereocontrolling effect of acyl groups at O-4 of the galactosyl donor in different solvent mixtures.

group at O-4 (**63c**) has a more powerful α-stereodirecting influence than the acetyl one. The dioxane–toluene mixture provided the best stereoselectivity especially with iodine(I) dicollidine perchlorate (IDCP) as a promoter. Pure α-disaccharide was obtained under these conditions upon glycosylation with donor **63d** bearing the *p*-methoxybenzoyl group. The latter example demonstrates that the combined effect of the participating group, the solvent composition, and the promoter enabled the achievement of the complete α-selectivity. However, some lowering of the coupling yields occurred when IDCP was used as the promoter.

The alkoxycarbonyl group at C-5 of galacturonosyl donors seem to enhance the α-stereodirecting properties of acyl substituents at O-4 (Scheme 5.19) [44]. It is difficult to compare the stereochemical results with those from the Scheme 5.18 because of a great difference in the properties of acceptors **64** and **67** and conditions, but it is noticeable that glycosylation with 4-*O*-acetylated donor **66b** produced pure α-isomer of disaccharide **68**, whereas coupling of 4-*O*-acetylated galactosyl donor **63b** led to α : β-mixtures in ratios from 3.6 : 1 to 14 : 1 even in coordinating solvents.

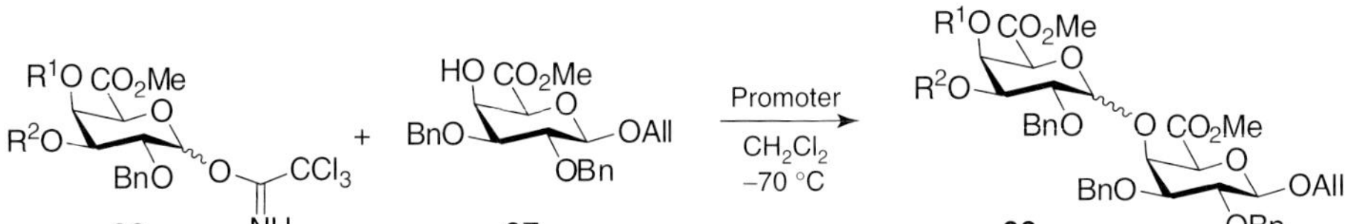

66 **67** **68**

Donor	R^1	R^2	Promoter	α : β ratio (yield, %)
66a	Bn	Bn	TMSOTf	1 : 1 (52)
66b	Ac	Bn	TMSOTf	Only α (59)
66c	Ac	Ac	TMSOTf	13.4 : 1 (72)
66b	Ac	Bn	$BF_3{\cdot}Et_2O$	1.6 : 1 (53)

Scheme 5.19 Stereoselectivity of glycosylation with galacturonosyl donors bearing acetyl substituents at O-3 and O-4.

It can be seen also from the Scheme 5.19 that the use of $BF_3{\cdot}Et_2O$ as a promoter eliminates the α-directing influence of the acetyl group at O-4 in galacto series as it took place with xylosyl donor **40b** (Scheme 5.14). It is interesting that glycosylation with di-*O*-acetylated donor **66c** occurred less stereoselective than with donor **66b** having the only acetyl group at O-4, although both acetyl groups are capable, in principle, of the remote anchimeric participation.

Relative value of the α-stereodirecting effect of acyl groups attached to O-3 and O-4 in sugars of galacto configuration was evaluated on fucosyl donors **69** (Scheme 5.20) [28]. 4-*O*-Benzoylated bromide **69b** demonstrated a lower α-stereoselectivity in the reaction with acceptor **38** than 3-*O*-benzoyl isomer **69c**. These data are supported by calculated values of stabilization energy of the cations **C** and **B** (the latter is shown in Scheme 5.15). The largest stabilization energy was revealed for the cation **C** corresponding to the most stereoselective donor **69c**. Two benzoyl groups at O-3

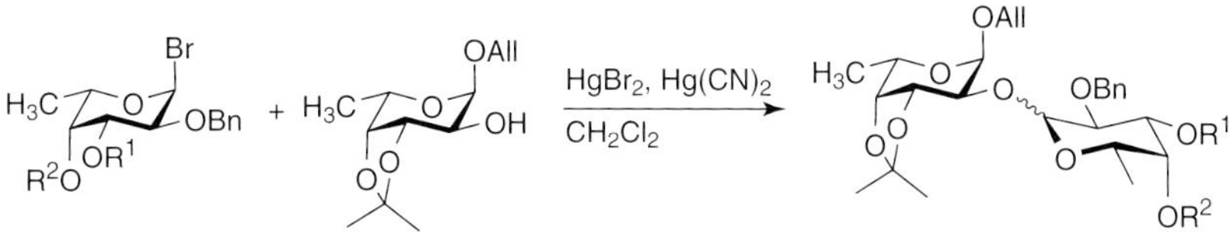

69 **38** **70**

Donor	R^1	R^2	α : β ratio (yield, %)	ΔE (kcal mol^{-1})	Cation
69a	Bn	Bn	1 : 1 (75)	0	**A**
69b	Bn	Bz	3.5 : 1 (74)	−3.6	**B**[a]
69c	Bz	Bn	13 : 1 (82)	−8.9	**C**
69d	Bz	Bz	20 : 1 (81)	−3.6	**C**
				−8.9	**B**[a]

[a]See Scheme 5.15

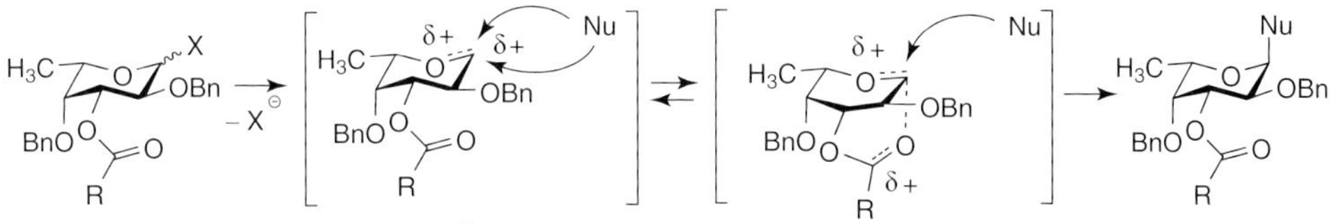

A **C**

Scheme 5.20 Comparison of the efficiency of the remote anchimeric participation from the positions 3 and 4 in fucosyl donors.

and O-4 in donor **69d** displayed a cooperative effect leading to almost complete α-selectivity of glycosylation.

The cooperative effect of participating substituents at O-4 and O-3 in fucosyl donors was also shown for *p*-NO_2Bz groups [45–48]. Historically, these works were the first where donors with remote potentially participating groups were used.

In 2-azido-2-deoxygalactosyl imidates **71**, acetyl groups at both O-3 and O-4 also facilitated the formation of α-products **73** upon glycosylation of acceptor **72**; however, the difference in the stereoselectivity between 3-*O*- and 4-*O*-acetyl groups (donors **71b** and **71c**) was negligible and could be revealed only at a low temperature (Scheme 5.21) [29]. The presence of acetyl groups at both oxygen atoms (donor **71d**) led to complete α-selectivity only at room temperature, while at −78 °C the combined influence of two acyl groups resulted in a marginal change of stereochemistry.

Calculated energies of stabilized cations corresponding to the anchimeric assistance from the positions 3, 4, and 6 were compared with the energies of two possible conformations of the nonstabilized oxocarbenium cation (Scheme 5.21). Two most stable conformations of the nonstabilized cation, $^3H_4$ and $^4H_3$, undergo a nucleophilic attack preferentially from β- and α-sides, respectively, giving products with the opposite stereochemistry [49–55]. Calculations showed that α-selective $^4H_3$ conformation of the oxocarbenium ion is by almost 3 kcal mol^{-1} more stable than β-selective $^3H_4$ one, that is, the α-selectivity is intrinsic to this

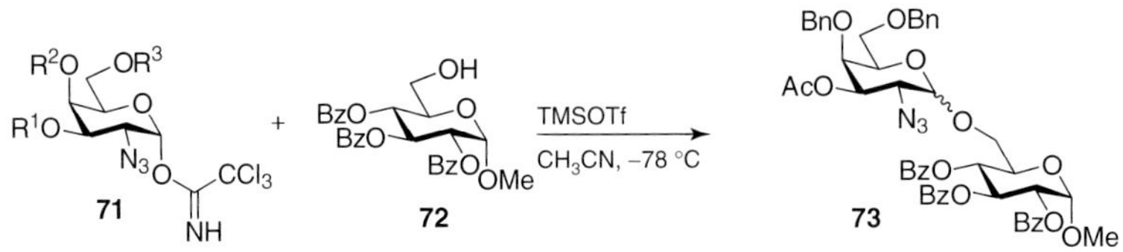

Donor	R^1	R^2	R^3	α : β ratio (yield, %)	
				At −78 °C	at RT
71a	Bn	Bn	Bn	1 : 3 (74)	3 : 1 (66)
71b	Ac	Bn	Bn	1.7 : 1 (81)	6 : 1 (66)
71c	Bn	Ac	Bn	1 : 1.1 (51)	6 : 1 (53)
71d	Ac	Ac	Bn	1.5 : 1 (67)	Only α (50)
71e	Bn	Bn	Ac	1 : 5 (52)	1.5 : 1 (75)

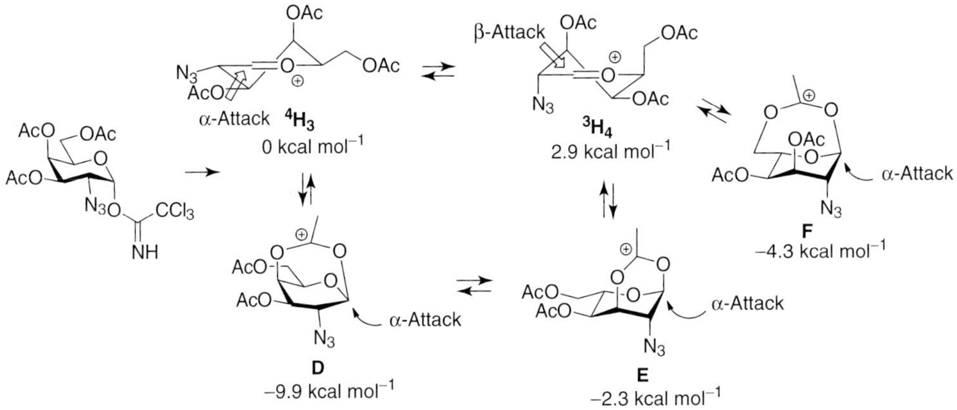

Scheme 5.21 Comparison of the α-stereodirecting effects of acetyl groups at O-3, O-4, and O-6 in 2-azido-2-deoxygalactosyl donors.

donor. This is in accordance with the result of glycosylation with fully benzylated imidate **71a**. But the intermediates **E** and **D**, corresponding to the anchimeric participation from the positions 3 and 4, are more stable than both conformations of the nonstabilized oxocarbenium ion. Thus, acetyl groups at O-3 and O-4 can also add the α-selectivity, what was really observed in glycosylations with 3-*O*- and 4-*O*-acetylated donors **71b** and **71c**.

On the other hand, there are also some contradictions between experimental results and calculations. First, computer calculations predicted (compare energies of stabilization for **E** and **D**) much more effective participation from the position 4 than that from the position 3, whereas the experiments showed almost no difference in the stereoselectivity provided by donors **71b** and **71c**. Second, relatively high calculated stabilization energy of the intermediate **F**, corresponding to the anchimeric participation from the position 6, was not corroborated experimentally: donor **71e** showed the preferential formation of the β-product at −78 °C and only very low α-stereoselectivity at room temperature. This result is somewhat surprising, as a number of other 6-*O*-acylated donors do demonstrate α-selectivity.

This example shows that computer modeling still cannot serve as a reliable instrument for prediction of chemical behavior of donors with different patterns of protecting groups.

Calculated relative energies of the intermediates corresponding to the anchimeric participation from the positions 3, 4, and 6 compared to those of oxocarbenium ions are as illustrated in Scheme 5.21.

Some examples of application of galacto-configurated donors with acyl groups at O-4 and/or O-3 for the synthesis of complex natural oligosaccharides are given later. Particularly, the α-fucosylation of galactal diol **74** was the key step in the synthesis of selectin ligands. Glycosylation with fully benzylated donor **75** proceeded with low stereoselectivity and led to a 2.7 : 1 α : β-mixture **76**, while 4-*O*-benzoylated donor **77** provided an α-linked disaccharide **78** (Scheme 5.22) exclusively [56].

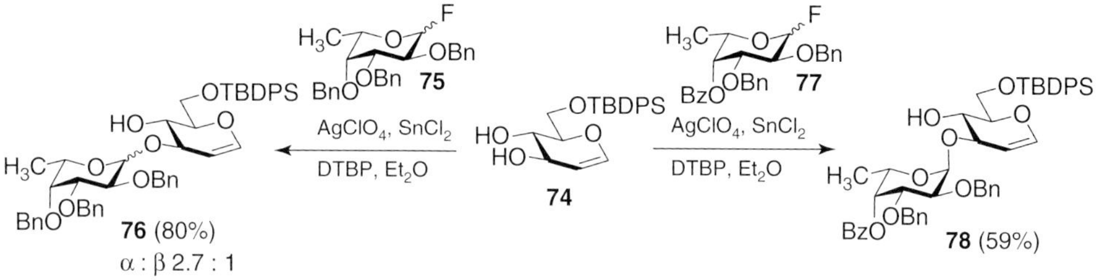

Scheme 5.22 Key step of Danishefsky's synthesis of selectin ligands.

Fucosyl donor **77** was also used in the syntheses of octasaccharide tumor-related antigens N3 [57]. Bis-glycosylation of acceptors **79** and **80** at the last step of the assembly of the oligosaccharide backbone was highly effective and stereoselective and produced octasaccharides **81** and **82** in very good yields (Scheme 5.23).

Effective syntheses of fucoidan fragments were carried out using 3,4-di-*O*-acylated trichloroacetimidate **83** (Scheme 5.24) [58]. Its TMSOTf-promoted reaction with acceptor **84** gave exclusively α-linked disaccharide **85** in 81% yield.

Moreover, 4-*O*-benzoylated donor **87** generated from disaccharide **85** was also demonstrated to be an efficient α-glycosylating agent (Scheme 5.25). Coupling of donor **87** with acceptor **86** resulted in the stereospecific formation of α-linked tetrafucoside **88** in 78% yield. This made a convergent scheme for the oligofucoside assembly reliable [58, 59].

Thus, the application of the reaction sequence depicted in the Scheme 5.25 to tetrasaccharide **88** enabled the preparation of linear (1 → 3)-α-linked octafucoside **89** in good yield of 75% (Scheme 5.26).

A tetrasaccharide fragment of the glycopeptidolipid of *Mycobacterium avium* serotype 4 was synthesized using α-fucosyl-rhamnoside disaccharide block **92** (Scheme 5.27) [60]. The latter was constructed stereoselectively by IDCP-promoted coupling of thiorhamnoside acceptor **90** with thiofucoside **91** bearing a chloroacetyl group at O-4. Further elongation of the oligosaccharide chain led to the target tetrasaccharide.

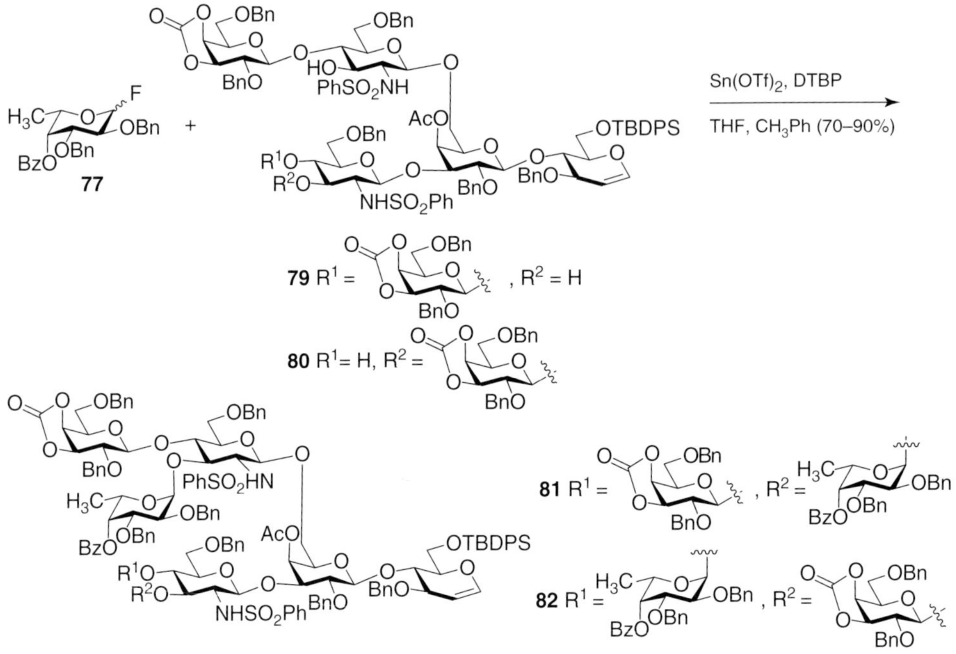

Scheme 5.23 Efficient bis-α-fucosylation in the synthesis of the octasaccharide tumor-related antigens N3.

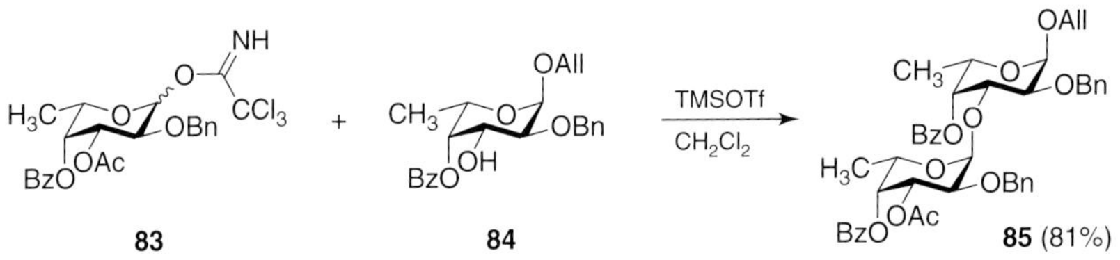

Scheme 5.24 Stereospecific glycosylation with 3,4-di-*O*-acylated fucosyl donor.

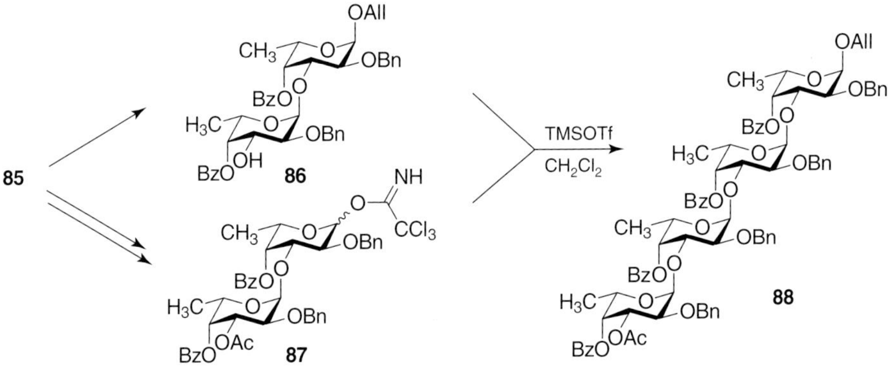

Scheme 5.25 Preparation of linear tetrafucoside **88**.

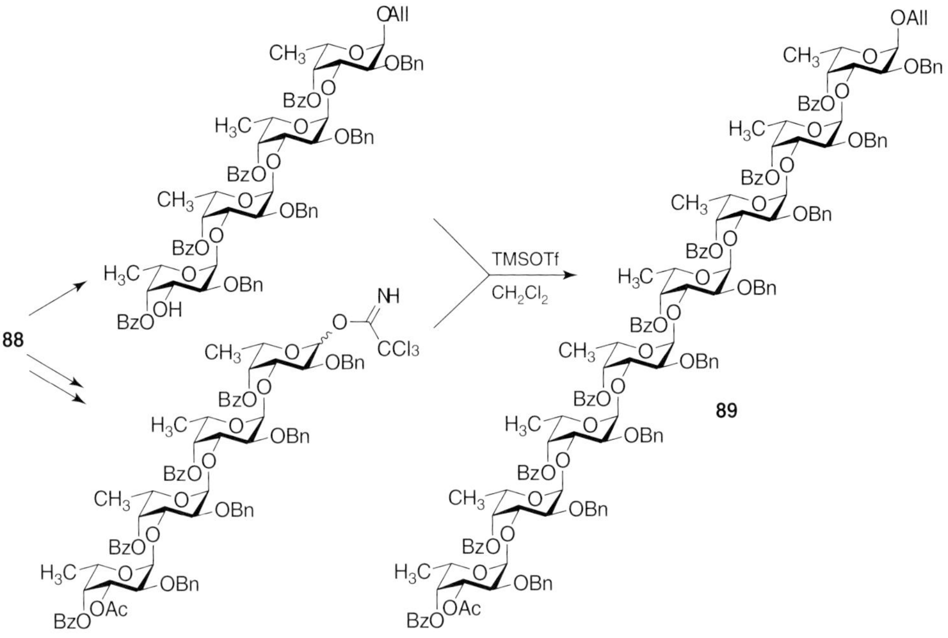

Scheme 5.26 Preparation of linear octafucoside **89**.

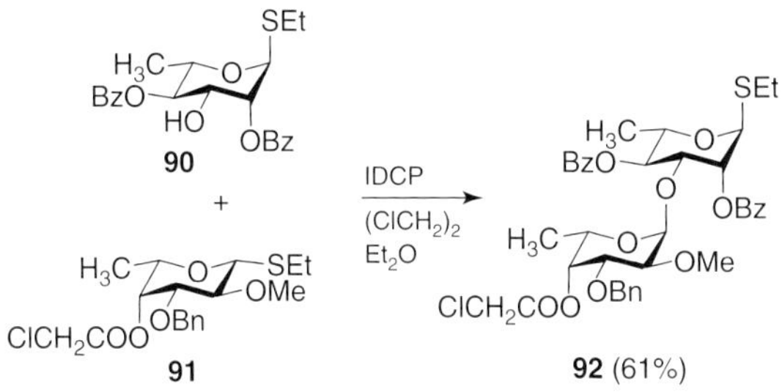

Scheme 5.27 Synthesis of a fragment of glycopeptidolipid of *Mycobacterium avium* through α-stereoselective fucosylation with 4-*O*-chloroacetylated donor.

5.4
Lack of Stereocontrolling Effect of Acyl Groups at Equatorial O-4 in $^4C_1$ Conformation

Mannosyl and glucosyl donors bearing a trichloroacetyl group at O-4 demonstrated in heterogeneously promoted glycosylation higher β-selectivity compared to per-alkylated counterparts [24, 61]. But, to the best of our knowledge, this is the only manifestation of the stereocontrolling effect of an acyl group at equatorial O-4, as other examples of glycosylation with glucosyl and mannosyl donors bearing acyl groups at O-4 revealed no influence.

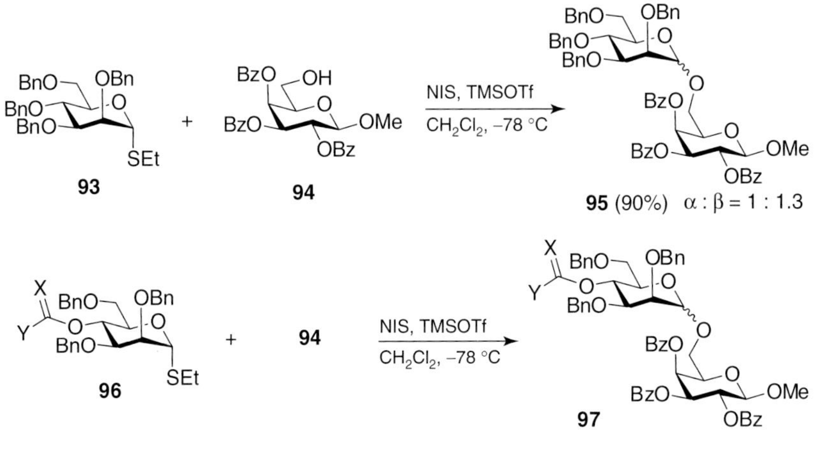

Donor	X	Y	α : β ratio (yield, %)
96a	O	*p*-MeOPh	1 : 1.5 (99)
96b	S	NHEt	1 : 1.5 (64)

Scheme 5.28 The absence of β-directing influence of potentially participating groups at O-4 in mannosyl donors.

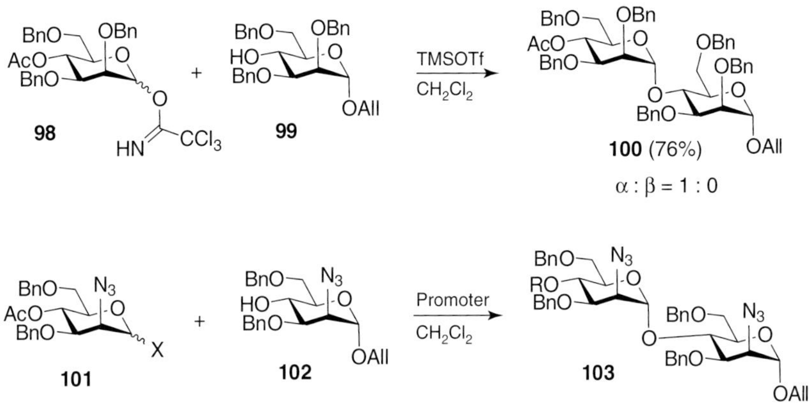

X	Promoter	α : β ratio (yield, %)
α-OC(NH)CCl_3	TMSOTf	Only α (60)
OH	Ph_2SO, Tf_2O	Only α (78)

Scheme 5.29 Acetyl group at O-4 in mannosyl and 2-azido-2-deoxymannosyl donors displays no β-directing effect.

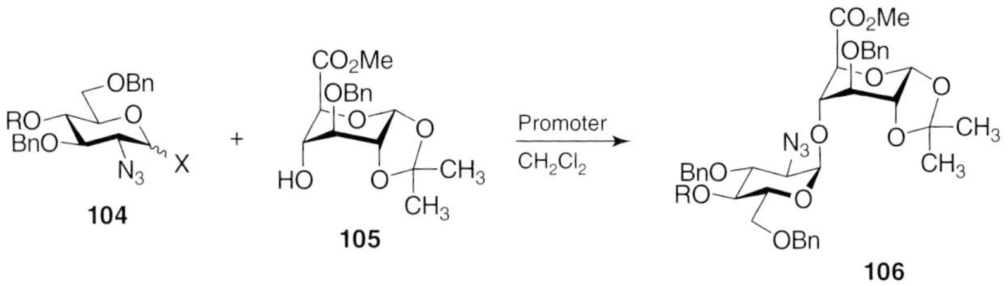

Donor	R	X	Promoter, conditions	Ref.
104a	TBDMS	$OC(NH)CCl_3$	TMSOTf, −25 °C→RT	[31]
104b	Ac	$OC(NH)CCl_3$	TMSOTf, −25 °C→RT	[31]
104c	PMB	SEt	AgOTf, NIS, 0 °C	[65]
104d	Lev	SEt	AgOTf, NIS, 0 °C	[65]

Scheme 5.30 Acyl groups at O-4 of 2-azido-2-deoxyglucosyl donors display no stereodirecting effect.

Demchenko *et al.* [62] investigated different potentially participating groups, either carbonyl or thiocarbonyl, at O-4 of thiomannosides **96** by comparison with fully benzylated thioglycoside **93** but found almost no influence (Scheme 5.28).

Complete α-selectivity was even observed on glycosylation with 4-*O*-acetylated mannosyl and 2-azido-2-deoxymannosyl donors **98** and **101** (Scheme 5.29) [63, 64].

Coupling of uronic acid derivative **105** with 2-azido-2-deoxyglucosyl donors **104** bearing different groups at O-4 resulted in the exclusive formation of α-products **106** irrespective of the type of the protecting group at O-4 (Scheme 5.30) [31, 65].

5.5 Effect of Substituents at O-6

A series of experiments outlined later demonstrate the α-directing effect of acyl substituents at O-6. Comparative glycosylation of acceptors **108** and **110**, differing in protecting group pattern and configuration, with 2-azido-2-deoxyglucosyl trichloroacetimidates **107** were carried out [30]. The results of only four experiments are shown in Scheme 5.31, but all others demonstrated the same tendency. Namely, perbenzylated trichloroacetimidate **107a** provided a modest predominance of the α-isomer in reactions with both acceptors. But application of 6-*O*-benzoylated glucosyl trichloroacetimidate **107b** increased the proportion of the α-isomer in the case of the conformationally strained acceptor **110** and ensured the formation of pure α-disaccharide with methyl glucoside **108**. The above-mentioned examples also demonstrate that the stereoselectivity depends on the acceptor structure.

Stereochemistry of glycosylations with partially benzylated glucosyl donors is also sensitive to the presence of an acyl group at O-6. The results of glycosylation of acceptor **113** with a series of glucosyl donors **112** showed that the presence of an

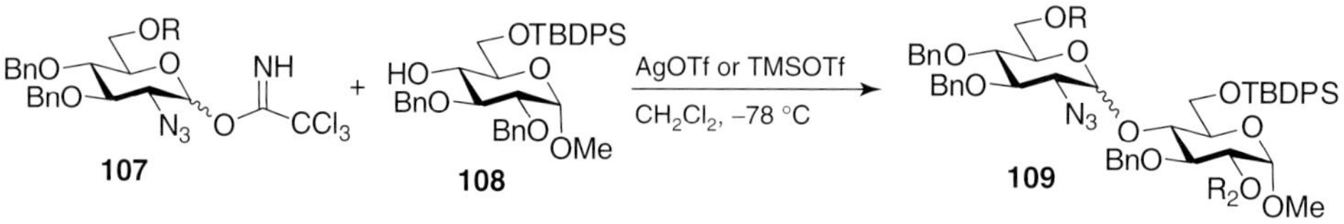

Donor	R	α : β ratio (yield, %)
107a	Bn	1.9 : 1 (45)
107b	Bz	Only α (57)

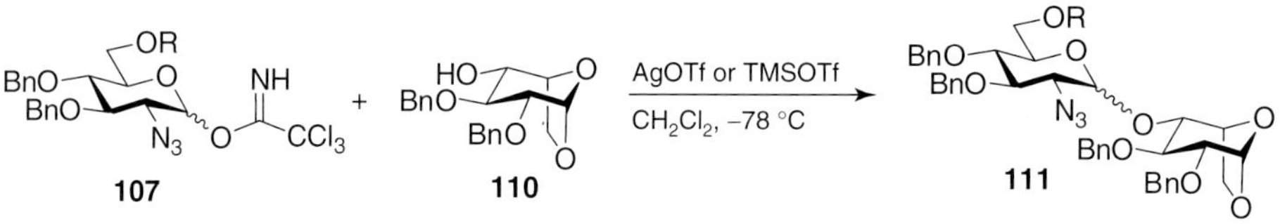

Donor	R	α : β ratio (yield, %)
107a	Bn	2.7 : 1 (82)
107b	Bz	3.9 : 1 (68)

Scheme 5.31 Effect of the 6-*O*-benzoyl group on stereoselectivity of glycosylation with 2-azido-2-deoxyglucosyl donors.

acyl group at O-6 improved the α-stereoselectivity independently from the leaving group and the promoter (Scheme 5.32) [14].

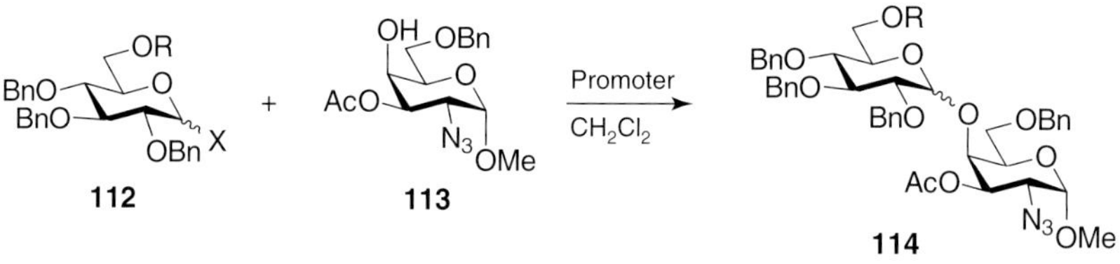

Donor	R	X	Promoter	α : β ratio (yield, %)
112a	Bn	SEt	NIS, TfOH	2 : 3 (90)
112b	Bz	SEt	NIS, TfOH	Only α (47)
112c	Ac	$OC(NH)CCl_3$	AgOTf	4 : 1 (48)
112d	Bz	$OC(NH)CCl_3$	AgOTf	5 : 1 (50)
112e	Bn	$OC(NPh)CF_3$	AgOTf	2 : 1(95)
112f	Ac	$OC(NPh)CF_3$	AgOTf	5 : 1 (90)
112g	Bz	$OC(NPh)CF_3$	AgOTf	6 : 1 (84)

Scheme 5.32 Influence of substituents at O-6 on the stereoselectivity of the glucosylation.

Glucosyl donors with acyl groups at either O-6 (**112g**) or at both O-6 and O-3 (**118**) were successively applied for the synthesis of glycoforms of the outer core region of the *Pseudomonas aeruginosa* lipopolysaccharide as it is illustrated by the synthesis of pentasaccharide **119** (Scheme 5.33) [66, 67].

In the next example, an acyl group at O-6 of a glucosyl donor predetermined the undesired disturbance of the β-selectivity. The synthetic plan toward macroviracins, macrocyclic bis-lactones containing two β-linked glucose residues [68–71], dictated the necessity to use both 6-*O*-acetylated and 6-*O*-benzylated glucosyl donors. To provide higher β-selectivity, the reactions were carried out in acetonitrile. Glucosylation of acceptor **120** with perbenzylated donor **121a** gave β-isomer **122** stereospecifically, whereas in the case of 6-*O*-acetylated donor **121b**, in spite of all efforts to synthesize pure β-glucoside, a substantial amount of α-isomer was formed (Scheme 5.34).

This example also gives an opportunity to compare α-directing effects of the participating acetyl group and the bulky *tert*-butyldiphenylsilyl (TBDPS) substituent attached to O-6 [68]. Coupling of TBDPS-protected donor **121c** with acceptor **120** afforded the corresponding glycoside **122** with almost the same yield and stereoselectivity as with acetylated donor **121b**. This result is in accordance with the known role of TBDPS at O-6 as a bulky substituent having an α-directing influence for steric reasons [72].

6-*O*-Acyl groups are also able to impede heterogeneously promoted β-glycosylation that was developed by Paulsen, Garegg, and van Boeckel [24, 61, 73, 74] specially for the synthesis of β-mannosides. Under these conditions, perbenzylated donor **123a** gave predominantly β-products because of S_N2 character of the substitution of α-bromide, while 6-*O*-acetylated donor **123b** provided a moderate α-selectivity (Scheme 5.35) [61].

Surprisingly, donor **123c** with a 6-*O*-trichloroacetyl group, which is substantially less nucleophilic than the acetyl one [75, 76], provided a higher α-stereoselectivity with acceptor **126**. Obviously, this result cannot be explained by remote anchimeric assistance. Instead, the authors interpreted it in terms of the molecular orbitals [61]. Interaction of the lone electron pair of O-6 with $C{=}O^+$ in the oxocarbenium ion is affected by an electron-withdrawing group at O-6 thus, the oxocarbenium ion becomes more stable. As a result, the mechanism of substitution becomes more S_N1-like providing more α-product.

However, the explanation of the stereodirecting effect of the trichloroacetyl group by only its electron-withdrawing properties contradicts the results of mannosylation with donors **126** (Scheme 5.36) [12]. Indeed, donor **126b** containing a nonparticipating electron-withdrawing sulfonate group at O-6 gave a mixture of disaccharides with a significant predominance of β-isomer **127b**, while perbenzylated donor **126a** displayed only a moderate β-selectivity and 6-*O*-acetylated donor **126c** provided exclusively α-isomer **127a**. Thus, the effect of nonparticipating electron-withdrawing group (EWG) at O-6 of mannosyl donor **126b** is opposite to that of the trichloroacetyl group in glucosyl donor **123c**. The relations between the acetyl and trichloroacetyl groups in donors **123** (Scheme 5.35) are similar to those observed for the acetyl and PFP groups at O-4 in galactosyl donors **60** (Scheme 5.17). Thus, so-called "non-nucleophilic" acyl groups, such as PFP or trichloroacetyl, which are usually regarded as nonparticipating [41, 42, 76], should be considered as a third

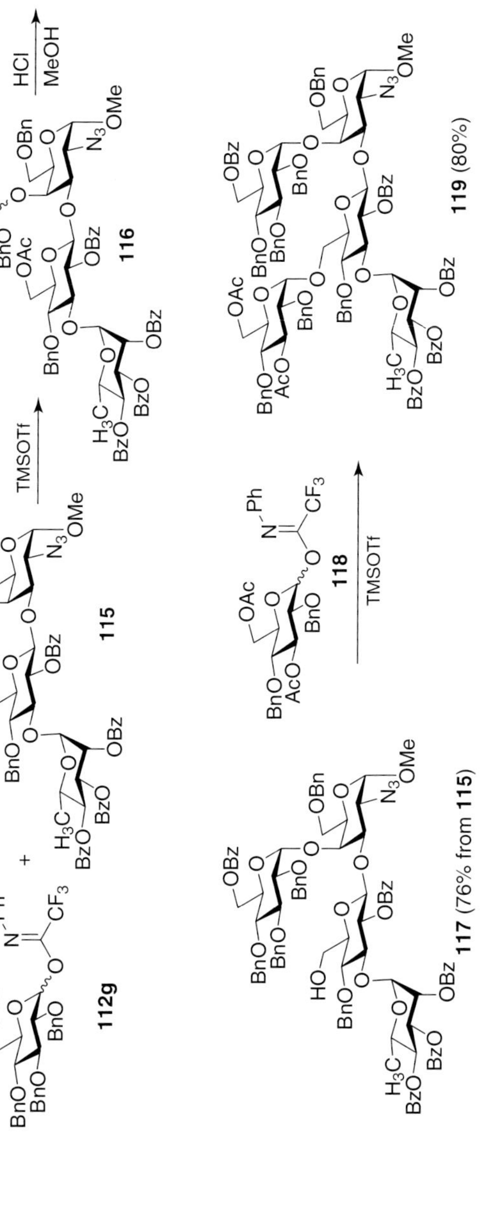

Scheme 5.33 Synthesis of pentasaccharide glycoform of outer core of the lipopolysaccharide of *P. aeruginosa*.

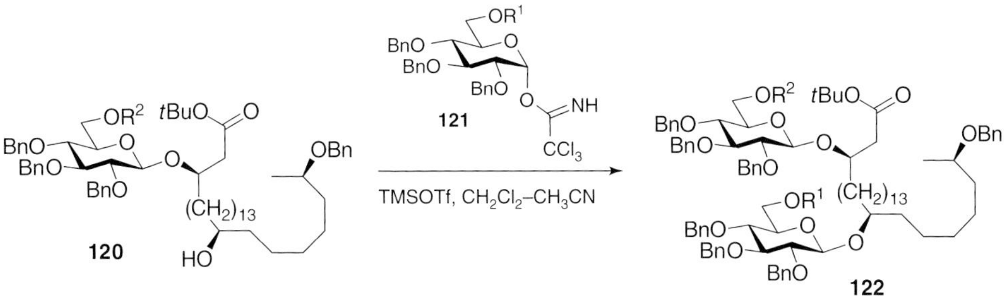

Donor	R^1	R^2	α : β ratio (yield, %)
121a	Bn	Bn	Only β (88)
121b	Ac	Bn	1 : 2.6 (80)
121c	TBDPS	Ac	1 : 2.8 (79)

Scheme 5.34 Demonstration of α-directing effect of an acetyl group at O-6 in reactions carried out in acetonitrile.

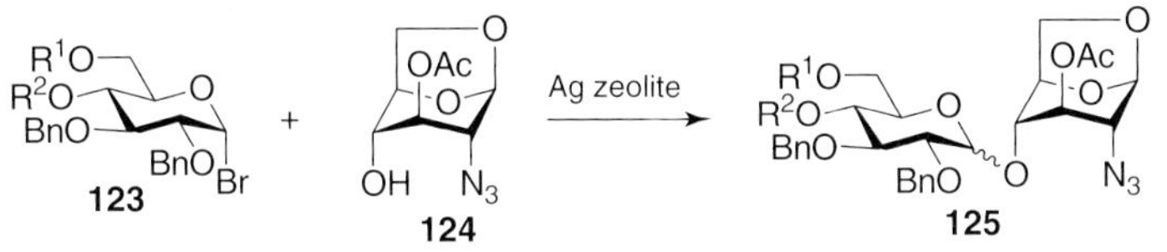

Donor	R^1	R^2	α : β ratio (yield, %)
123a	Bn	Bn	1 : 2.8 (50)
123b	Ac	All	2.2 : 1 (59)
123c	$Cl_3CC(O)$	All	3.6 : 1 (56)

Scheme 5.35 Effect of 6-*O*-acyl groups on the stereoselectivity of heterogeneously promoted β-glycosylation.

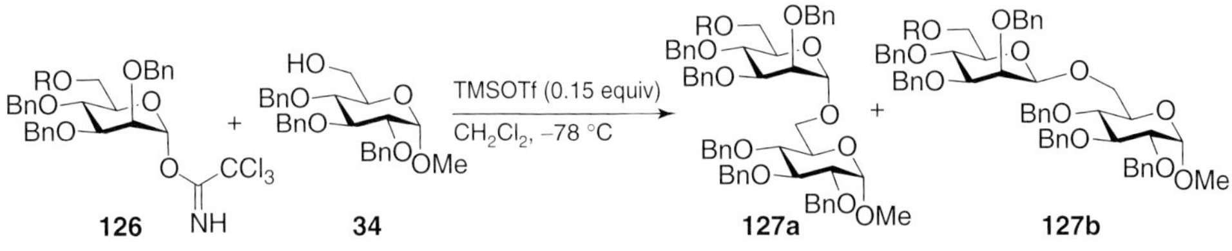

Donor	R	**127b** : **127a** ratio (yield, %)
126a	Bn	2.7 : 1 (91)
126b	SO_2Bn	13.8 : 1(76)
126c	Ac	Only **127a**

Scheme 5.36 Effect of the electron-withdrawing sulfonate group at O-6 at the stereoselectivity of mannosylation.

type of stereocontrolling groups. Their stereodirecting properties resemble those of nucleophilic acyl groups, but electronic properties are similar to the properties of EWGs such as sulfonate esters.

Acceleration of substitution reactions is considered as kinetic evidence for the neighboring group participation. The rate of solvolysis of 6-*O*-(*N*-phenylcarbamoyl) tribenzylated glucosyl tosylate in methanol or isopropanol is three times higher than that of perbenzylated glucosyl tosylate [77]. It can be regarded [78] as evidence in favor of the mechanism of the remote anchimeric assistance at least for some potentially participating groups. However, according to van Boeckel, this acceleration may also be a consequence of stabilization of the oxocarbenium ion by EWG at O-6 through a molecular orbital interaction [61] described earlier (Scheme 5.35).

To conclude, the stereodirecting effect of 4-*O*-acyl substituents is manifested only in the case of their axial orientation, that is, mainly in the galacto series, while equatorial acyl groups at O-4 in manno and gluco series give rise to almost no effect. The impact of acyl groups at O-3 seems to be almost equal for both axial and equatorial orientations; however, it is not particularly strong. The above-mentioned generalizations are in accordance with the intrinsic stereoselectivity of pyranoses. This phenomenon and its reasons are discussed in detail by Cumpstey [79].

Sometimes the intrinsic stereoselectivity is explained by the anomeric effect, which is, for example, stronger for mannose than for galactose and glucose. But more likely [79], the anomeric effect, as it was defined by Edward and Lemieux [80, 81], has no influence on the stereoselectivity of the majority of glycosylation reactions. Most of them proceed through either S_N1 mechanism or a mechanism intermediate between S_N1 and S_N2. Conformational analysis showed [79] that in both these cases the orbital interaction responsible for the anomeric effect can be realized for neither of the most stable conformations of the oxocarbenium ion. Thus, the failure to control the stereoselectivity of glycosylation by acyl groups at O-4 of gluco and manno series may be explained if the conformation of the oxocarbenium ions is considered according to Woerpel [52, 53, 82, 83] and van der Marel [84–86].

5.6
Interplay of Stabilized Bicyclic Carbocation and Two *H* Conformations of Oxocarbenium Ions

The energy needed for an acylated oxocarbenium ion to change its conformation and to form a stabilized bicyclic cation, corresponding to the anchimeric assistance, is not usually discussed. The comparative reactivity of the bicyclic cation and the oxocarbenium ion is not considered either. However, as it follows from the examples given later, both factors should not be neglected [79]. An alkoxycarbonyl group at C-5 of uronic acids shares the characteristics of acyl groups at O-6 but not stereodirecting properties. Location of the alkoxycarbonyl group and nucleophilic properties of carbonyl oxygen allow it to be referred to potentially participating.

However, stereochemistry of glycosylations with such donors does not reflect any participation by the alkoxycarbonyl group.

Glycosylation with various 2,3,4-tri-*O*-benzylated hexuronopyranosyl donors exhibits usually very low or no stereoselectivity at all [13, 44]. An exception is glycosylation with perbenzylated mannuronosyl donors [84, 85], as their coupling with different acceptors are completely β-stereoselective.

The comparative study of glycosylation with different perbenzylated hexopyranosyl and hexuronopyranosyl donors **128–137** demonstrated no rise of the α-selectivity for the uronic donors of gluco, galacto, allo, and gulo configurations (**130**, **132**, **134**, **136**) as compared to the corresponding hexopyranosyl analogs (**131**, **133**, **135**, **137**) and proved the high β-selectivity of mannuronosyl donor **128** (Table 5.1) [85].

Table 5.1 Effect of the substituent at C-5 in glycosyl donors of different configurations on the stereoselectivity.

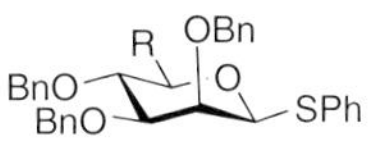

128 R = CO_2Me
129 R = CH_2OBn

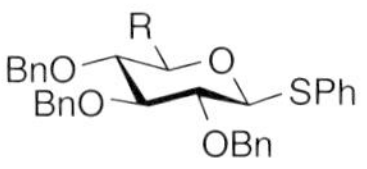

130 R = CO_2Me
131 R = CH_2OBn

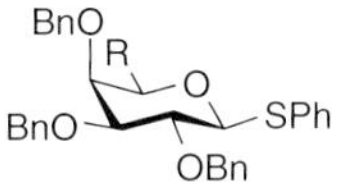

132 R = CO_2Me
133 R = CH_2OBn

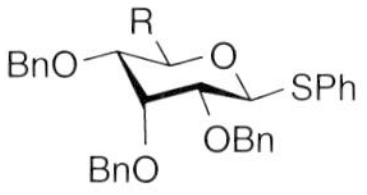

134 R = CO_2Me
135 R = CH_2OBn

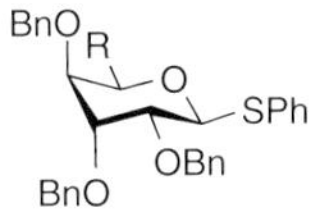

136 R = CO_2Me
137 R = CH_2OBn

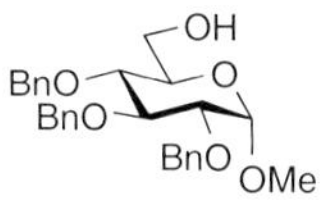

138

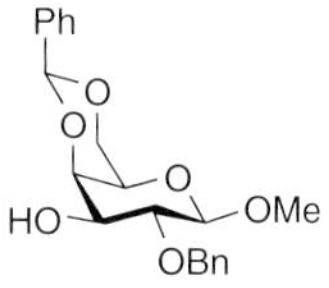

139

Donors Configuration	Acceptor	Substituent R at C-5 in the glycosyl donor	
		CO_2Me α : β Ratio in the mixture of glycosylation products (yield, %)	CH_2OBn
128, **129** manno	138	0 : 1 (77)	1 : 2 (71)
	139	0 : 1 (58)	1 : 1.5 (52)
130, **131** gluco	138	1 : 1.4 (68)	1 : 1.4 (75)
	139	1 :0 .6 (86)	1 : 1.7 (89)
132, **133** galacto	138	1 : 2.3 (49)	1 : 3 (67)
	139	1 : 0.4 (86)	1 : 0.1 (72)
134, **135** allo	138	1 : 0.4 (91)	1 : 0.5 (92)
	139	1 : 0 (52)	1 : 0.6 (65)
136, **137** gulo	138	1 : 0.33 (86)	1 : 0.10 (76)
	139	1 : 0.17 (63)	1 : 0.12 (70)

High β-selectivity of mannuronosyl donors draws a special attention, as it demonstrates the importance of the stability and reactivity of the intermediate oxocarbenium ion for the stereochemical outcome of glycosylation. Two of the possible conformations of the oxocarbenium ion are most stable, namely $^3H_4$ and $^4H_3$ (Figure 5.1) [49–55]. Each of them can be attacked with a nucleophile from only one side [52, 55]. Several factors are considered to determine the conformation of the oxocarbenium ion [52]. These are (i) alkoxy substituents at C-3 and C-4 of the oxocarbenium ion that tend to be axial; (ii) substituents at C-2 that prefer to be pseudoequatorial to provide the possibility of hyperconjugation between the pseudoaxial C–H bond at C-2 and the adjacent 2p orbital on positively charged C-1; (iii) the alkoxycarbonyl group at C-5 that adopts preferentially the pseudoaxial orientation. All these factors act concerted in the case of the manno configuration, thus ensuring the pro-β $^3H_4$ conformation of mannuronic oxocarbenium ion and the further formation of β-mannuronosides. Among other configurations, such a synchronous action of the ring substituents can be found only for the gulo oxocarbenium ion that exists preferentially in the pro-α $^4H_3$ conformation. This consideration is supported by a good α-selectivity achieved upon glycosylation with glycosyl donors **136** and **137** (Table 5.1).

Thus, the conformation of the oxocarbenium ion, which is much influenced by dipole–charge interaction between the alkoxycarbonyl group at C-5 and the positive charge of the oxocarbenium ion, is more important for the stereochemistry of glycosylation with perbenzylated uronosyl donors than the stabilization of the positive charge by the remote anchimeric assistance of the alkoxycarbonyl group. More detailed this question was evaluated in experiments with 6-thiomannosyl donor **140b** (Scheme 5.37) [87]. Thiophenyl group at C-6 increased the proportion of β-isomers significantly as compared to the perbenzylated donor **140a** in reactions with acceptor **141–143**. Other groups such as STol, SEt, and SePh demonstrated a weaker β-directing effect.

A sulfonium species **I** corresponding to anchimeric assistance was even observed per se in an NMR experiment, and its properties were thoroughly investigated. A temperature dependence of the selectivity of the reaction of **I** with acceptor **141** was examined. At −60 °C, the α : β ratio was 1 : 4, while it changed to 1 : 1 at room temperature. It shows that the sulfonium ion **I** is much less reactive than

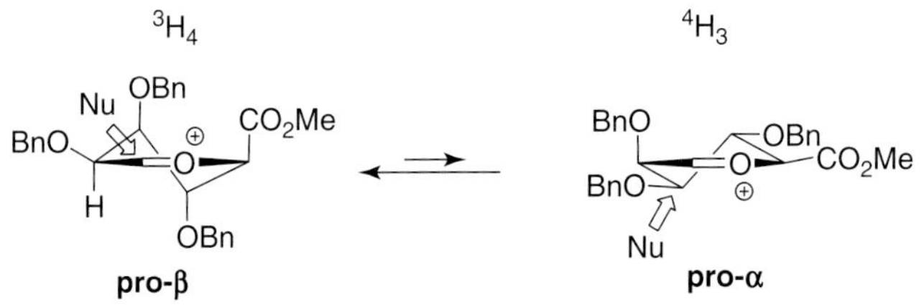

Figure 5.1 Conformations of the oxocarbenium ion and the stereoselectivity of glycosylation with the mannuronosyl donor.

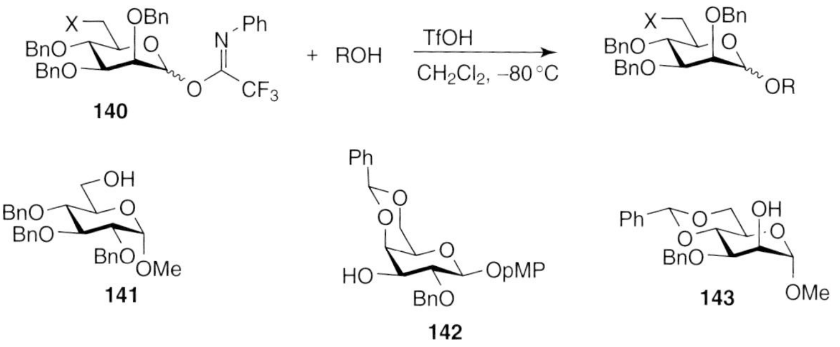

Donor	X	Acceptor		
		141	142	143
		α : β ratio (yield, %)		
140a	OBn	1 : 3.5 (94)	1 : 1 (88)	1 : 4 (99)
140b	SPh	1 : 7 (90)	1 : 11 (89)	1 : 5 (87)
140c	STol	1 : 5 (86)	1 : 8 (56)	1 : 3.5 (89)
140d	SEt	1 : 4 (80)	1 : 3.5 (86)	1 : 4 (90)
140e	SePh	1 : 7 (99)	1 : 10 (96)	1 : 3 (92)

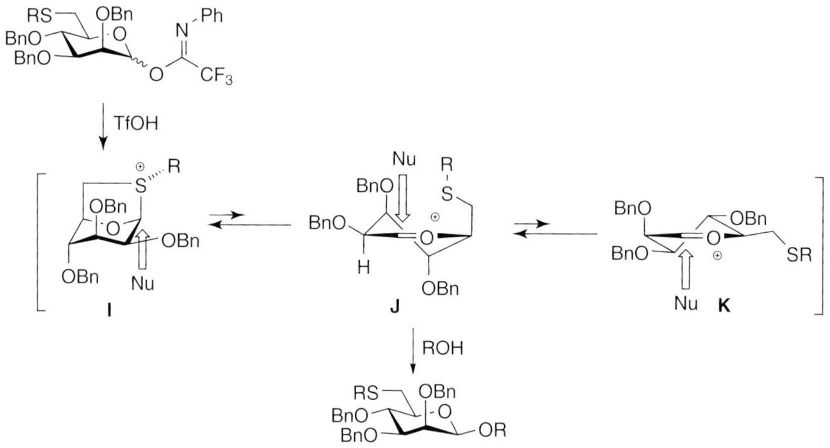

Scheme 5.37 Stereoselectivity of glycosylation with 6-thiomannosyl donors and interplay of sulfonium and oxocarbenium ions.

oxocarbenium ions **J** and **K** at a low temperature, but their reactivity becomes comparable at room temperature.

The fact that stabilized bicylic cation corresponding to the anchimeric assistance can be less reactive than the "opened" oxocarbenium ion is supported by another example (Scheme 5.38) [88]. Reactions of acceptors **145** and **147** with furanoid sulfonium salt **144** gave only a moderate preponderance of the α-anomers of disaccharides **146** and **148**, although the structure of **144** is "preorganized" for the selective α-attack.

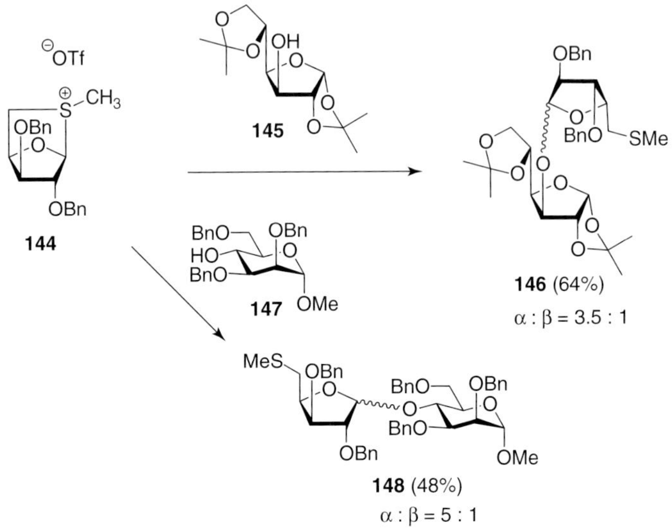

Scheme 5.38 Stereochemistry of glycosylations with the furanoid sulfonium salt.

5.7 Conclusion

Remote acyl groups at O-3, O-4, or O-6 affect the stereochemistry of glycosylation in a manner that can be considered as evidence for the mechanism of the remote anchimeric assistance. In other words, acyl substituents that are located above the β-side of a ring usually facilitate the formation of α-products, while acyl groups at the α-side sometimes favor the formation of β-products. Additionally to the stereochemical evidences supporting the mechanism of the anchimeric assistance, some bicyclic products, corresponding to the stabilized bicyclic cation, were obtained, and there is also a single example of acceleration of the solvolysis reaction as a consequence of introduction of the acyl group. However, the stereochemical effect of remote acyl groups is not general and depends on the position of the acyl group, the configuration of the pyranose, and the type of substituents at C-2 and C-5. Thus, acyl groups at O-4 in *galacto*-configurated donors are able to provide the absolute α-stereoselectivity of glycosylation, and acyl groups at O-3 in mannuronosyl and 4,6-*O*-benzylidene-protected mannosyl donors can disrupt completely β-directing effect of 5-C-alkoxycarbonyl or 4,6-*O*-benzylidene groups. At the same time, acyl groups at O-4 in donors of gluco and manno configurations demonstrate no β-directing effect. These facts evidence that the anchimeric assistance does take place, but some other factors can significantly disturb its influence on stereochemistry of glycosylation products.

The degree of the stereodirecting effect of remote acyl substituents correlates with the value of the anomeric effect inherent in pyranoses of different configurations. However, the difference in the stereodirecting effect of remote acyl substituents cannot be explained by anomeric effect itself, but rather by relative stabilities of the intermediate species, such as the oxocarbenium ion in different conformations and the bicyclic carbocation stabilized by the anchimeric assistance. The factor of the relative stability is manifested especially clear in glycosylation with mannuronic acid donors, 4,6-*O*-benzylidene protected, and 6-thio mannosides. Carefully taking into account the above-mentioned factors will allow an adequate explanation of the effect of remote acyl groups on stereochemistry of glycosylation. Meanwhile, glycosyl donors with an accurately designed protecting group pattern, including remote acyl groups, are being successfully used for the stereoselective total synthesis of complex oligosaccharides.

5.8 Key Experimental Procedures

5.8.1 Example of Stereocontrolled α-Fucosylation: Synthesis of Allyl 3-*O*-acetyl-4-*O*-benzoyl-2-*O*-benzyl-α-L-fucopyranosyl-(1 → 3)-4-*O*-benzoyl-2-*O*-benzyl-α-L-fucopyranoside (85)

To a solution of glycosyl donor **83** (900 mg, 1.65 mmol) and glycosyl acceptor **84** (600 mg, 1.50 mmol) in dry CH_2Cl_2 (19 ml), a 0.1 M solution of TMSOTf in CH_2Cl_2 (50 μl) was added portionwise at −30 °C under argon, and the reaction mixture was stirred for 15 min [58]. The mixture was neutralized with Et_3N and the solvent was evaporated. Column chromatography of the residue on a silica gel column gave disaccharide **85** (1.04 g, 89%): $[\alpha]_D$ −236 (*c* 1, EtOAc), R_f 0.66 (toluene–EtOAc, 5 : 1).

5.8.2 Example of Stereocontrolled α-Glucosylation: Synthesis of Methyl 2,3,4-tri-*O*-benzoyl-α-L-rhamnopyranosyl-(1 → 3)-[3,6-di-*O*-acetyl-2,4-di-*O*-benzyl-α-D-glucopyranosyl-(1 → 6)]-2-*O*-benzoyl-4-*O*-benzyl-β-D-glucopyranosyl)-(1 → 3)-[6-*O*-benzoyl-2,3,4-tri-*O*-benzyl-α-D-glucopyranosyl-(1 → 4)]-2-azido-6-*O*-benzyl-2-deoxy-α-D-galactopyranoside (119)

A solution of donor **118** (404 mg, 0.66 mmol) and acceptor **117** (794 mg, 0.48 mmol) in CH_2Cl_2 (17 ml) was added to powdered molecular sieve AW-300 (2.3 g), the resulting mixture was stirred at room temperature for 1 h, and then cooled to −20 °C [67]. A solution of TMSOTf (25 μl, 0.07 mmol), prepared by mixing of 25 μl TMSOTf and 0.5 ml of CH_2Cl_2, was added. The same solution was added three times (each time 10 μl) within next 3 h. When thin layer chromatography (TLC) revealed full conversion of acceptor **117**, the reaction mixture was diluted with CH_2Cl_2 (150 ml) and filtered through a pad of Celite. The filtrate was washed with saturated aqueous $NaHCO_3$ solution and the organic phase was concentrated. The residue was purified by silica gel column chromatography (toluene–CH_3CN, 16 : 1). The isolated mixture (0.83 g) of pentasaccharide anomers was separated by preparative high-performance liquid chromatography (HPLC) on a silica gel column (particle size 5 μm, 250 × 21.2 mm) in toluene–CH_3CN (18 : 1) to yield pure α-isomer **119** (0.80 g, 80%) as a syrup: $[\alpha]_D$ 82.1 (*c* 1, $CHCl_3$), R_f 0.16 (toluene–CH_3CN, 15 : 1).

List of Abbreviations

Ac	Acetyl
All	Allyl
Bn	Benzyl
Boc	*tert*-Butoxycarbonyl
BSP	1-Benzenesulfinyl piperidine
Bz	Benzoyl
Cp	Cyclopentadienyl
DTBP	2,6-Di-*tert*-butylpyridine
IDCP	Iodine(I) dicollidine perchlorate
Lev	Levulinoyl
p-MeOBz	*p*-Methoxybenzoyl
NIS	*N*-Iodosuccinimide
p-NO_2Bz	*p*-Nitrobenzoyl
PFP	Pentafluoropropionyl
PMB	*p*-Methoxybenzyl
TBDMS	*tert*-Butyldimethylsilyl
TBDPS	*tert*-Butyldiphenylsilyl
Tf	Trifluoromethylsulfonyl
Tf_2O	Trifluoromethanesulfonic anhydride
TMS	Trimethylsilyl

References

1. Lemieux, R.U., Hendriks, K.B., Stick, R.V., and James, K. (1975) *J. Am. Chem. Soc.*, **97**, 4056–4062.
2. Pougny, J. and Sinaÿ, P. (1976) *Tetrahedron Lett.*, **17**, 4073–4076.
3. Lemieux, R.U. and Ratcliffe, R.M. (1979) *Can. J. Chem.*, **57**, 1244–1251.
4. Schmidt, R.R. and Michel, J. (1985) *J. Carbohydr. Chem.*, **4**, 141–169.
5. Ito, Y. and Ogawa, T. (1987) *Tetrahedron Lett.*, **28**, 4701–4704.
6. Rattcliffe, A.J. and Fraser-Reid, B. (1990) *J. Chem. Soc., Perkin Trans. 1*, 747–750.
7. Wolff, G. and Röhle, G. (1974) *Angew. Chem. Int. Ed. Engl.*, **86**, 173–1871974, **13**, 157–170.
8. Demchenko, A., Stauch, T., and Boons, G.-J. (1997) *Synlett*, 818–820.
9. Chiba, S., Kitamura, M., and Narasaka, K. (2006) *J. Am. Chem. Soc.*, **128**, 6931–6937.
10. Wiesner, K., Tsai, T.Y.R., and Jin, H. (1985) *Helv. Chim. Acta*, **68**, 300–314.
11. Crich, D., Hu, T., and Cai, F. (2008) *J. Org. Chem.*, **73**, 8942–8953.
12. Baek, J.Y., Lee, B.-Y., Jo, M.G., and Kim, K.S. (2009) *J. Am. Chem. Soc.*, **131**, 17705–17713.
13. Ustyuzhanina, N., Komarova, B., Zlotina, N., Krylov, V., Gerbst, A., Tsvetkov, Y., and Nifantiev, N. (2006) *Synlett*, 921–923.

14. Komarova, B.S., Tsvetkov, Y.E., Knirel, Y.A., Zähringer, U., Pier, G.B., and Nifantiev, N.E. (2006) *Tetrahedron Lett.*, **47**, 3583–3587.
15. Mukaiyama, T., Murai, Y., and Shoda, S. (1981) *Chem. Lett.*, 431–432.
16. Smith, A.B. III,, Rivero, R.A., Hale, K.J., and Vaccaro, H.A. (1991) *J. Am. Chem. Soc.*, **113**, 2092–2112.
17. Crich, D. and Sun, S. (1998) *J. Am. Chem. Soc.*, **120**, 435–436.
18. Crich, D. and Sun, S. (1996) *J. Org. Chem.*, **61**, 4506–4507.
19. Crich, D. and Rahaman, M.Y. (2011) *J. Org. Chem.*, **76**, 8611–8620.
20. Gridley, J.J. and Osborn, H.M.I. (2000) *J. Chem. Soc., Perkin Trans. 1*, 1471–1491.
21. Crich, D., Cai, W., and Dai, Z. (2000) *J. Org. Chem.*, **65**, 1291–1297.
22. Crich, D. and Sharma, I. (2010) *J. Org. Chem.*, **75**, 8383–8391and references cited therein.
23. Crich, D. and Yao, Q. (2004) *J. Am. Chem. Soc.*, **126**, 8232–8236.
24. van Boeckel, C.A.A. and Beetz, T. (1985) *Recl. Trav. Chim. Pays-Bas*, **104**, 171–173.
25. Krylov, V., Ustyuzhanina, N., Bakker, H., and Nifantiev, N. (2007) *Synthesis*, **20**, 3147–3154.
26. Sethi, M.K., Buettner, F.F.R., Krylov, V., Takeuchi, H., Nifantiev, N., Haltiwanger, R.S., Gerardy-Schahn, R., and Bakker, H. (2010) *J. Biol. Chem.*, **285**, 1582–1586.
27. Sethi, M.K., Buettner, F.F.R., Ashikov, A., Krylov, V.B., Takeuchi, H., Nifantiev, N.E., Haltiwanger, R.S., Gerardy-Schahn, R., and Bakker, H. (2012) *J. Biol. Chem.*, **287**, 2739–2748.
28. Gerbst, A.G., Ustuzhanina, N.E., Grachev, A.A., Khatuntseva, E.A., Tsvetkov, D.E., Whitfield, D.M., Berces, A., and Nifantiev, N.E. (2001) *J. Carbohydr. Chem.*, **20**, 821–831.
29. Kalikanda, J. and Li, Z. (2011) *J. Org. Chem.*, **76**, 5207–5218.
30. Lu, L.-D., Shie, C.-R., Kulkarni, S.S., Pan, G.-R., Lu, X.-A., and Hung, S.-C. (2006) *Org. Lett.*, **8**, 5995–5998.
31. Orgueira, H.A., Bartolozzi, A., Schell, P., Litjens, R.E.J.N., Palmacci, E.R., and Seeberger, P.H. (2003) *Chem. Eur. J.*, **9**, 140–169.
32. Pedersen, C.M., Figueroa-Perez, I., Ulmer, A.J., Zähringer, U., and Schmidt, R.R. (2012) *Tetrahedron*, **68**, 1052–1061.
33. Tamura, J.-I., Neumann, K.W., and Ogawa, T. (1996) *Liebigs Ann.*, **1996**, 1239–1257.
34. Morgan, W.T. and Watkins, W.M. (1969) *Br. Med. Bull.*, **25**, 30–34.
35. Watkins, W.M. (1980) *Adv. Hum. Genet.*, **10**, 379–385.
36. Makita, A. and Taniguchi, N. (1985) in *Glycolipids*, Vol. 10 (ed H. Wiegandt), Elsevier, Amsterdam, pp. 1–82.
37. Fitton, J.H. (2011) *Mar. Drugs*, **9**, 1731–1760.
38. Mohnen, D. (1999) in *Comprehensive Natural Products Chemistry* (eds D. Barton, K. Nakanishi, and O. Meth-Cohn), Elsevier, Oxford, pp. 497–527.
39. Gerbst, A.G., Ustuzhanina, N.E., Grachev, A.A., Tsvetkov, D.E., Khatuntseva, E.A., and Nifant'ev, N.E. (1999) *Mendeleev Commun.*, **9**, 114–116.
40. Zuurmond, H.M., van der Laan, S.C., van der Marel, G.A., and van Boom, J.H. (1991) *Carbohydr. Res.*, **215**, C1–C3.
41. Ishiwata, A., Ohta, S., and Ito, Y. (2006) *Carbohydr. Res.*, **341**, 1557–1573.
42. Takatani, M., Matsuo, I., and Ito, Y. (2003) *Carbohydr. Res.*, **338**, 1073–1082.
43. Demchenko, A.V., Rousson, E., and Boons, G.-J. (1999) *Tetrahedron Lett.*, **40**, 6523–6526.
44. Nolting, B., Boye, H., and Vogel, C. (2001) *J. Carbohydr. Chem.*, **20**, 585–610.
45. Dejter-Juszynski, M. and Flowers, H.M. (1971) *Carbohydr. Res.*, **18**, 219–226.
46. Dejter-Juszynski, M. and Flowers, H.M. (1975) *Carbohydr. Res.*, **41**, 308–312.
47. Dejter-Juszynski, M. and Flowers, H.M. (1972) *Carbohydr. Res.*, **23**, 41–45.
48. Dejter-Juszynski, M. and Flowers, H.M. (1973) *Carbohydr. Res.*, **28**, 61–74.
49. Stevens, R.V. and Lee, A.W.M. (1979) *J. Am. Chem. Soc.*, **101**, 7032–7035.
50. Stevens, R.V. (1984) *Acc. Chem. Res.*, **17**, 289–296.
51. Hayashi, M., Sugiyama, M., Toba, T., and Oguni, N. (1990) *J. Chem. Soc., Chem. Commun.*, 767–768.

52. Ayala, L., Lucero, C.G., Romero, J.A.C., Tabacco, S.A., and Woerpel, K.A. (2003) *J. Am. Chem. Soc.*, **125**, 15521–15528.
53. Romero, J.A.C., Tabacco, S.A., and Woerpel, K.A. (2000) *J. Am. Chem. Soc.*, **122**, 168–169.
54. Woods, R.J., Andrews, C.W., and Bowen, J.P. (1992) *J. Am. Chem. Soc.*, **114**, 859–864.
55. Lucero, C.G. and Woerpel, K.A. (2006) *J. Org. Chem.*, **71**, 2641–2647.
56. Danishefsky, S.J., Gervay, J., Peterson, J.M., McDonald, F.E., Koseki, K., Griffith, D.A., Oriyama, T., and Marsden, S.P. (1995) *J. Am. Chem. Soc.*, **117**, 1940–1953.
57. Kim, H.M., Kim, I.J., and Danishefsky, S.J. (2001) *J. Am. Chem. Soc.*, **123**, 35–48.
58. Ustyuzhanina, N.E., Krylov, V.B., Grachev, A.A., Gerbst, A.G., and Nifantiev, N.E. (2006) *Synthesis*, **23**, 4017–4031.
59. Krylov, V.B., Kaskova, Z.M., Vinnitskiy, D.Z., Ustyuzhanina, N.E., Grachev, A.A., Chizhov, A.O., and Nifantiev, N.E. (2011) *Carbohydr. Res.*, **346**, 540–550.
60. Zuurmond, H.M., Veeneman, G.H., van der Marel, G.A., and van Boom, J.H. (1993) *Carbohydr. Res.*, **241**, 153–164.
61. van Boeckel, C.A.A., Beetz, T., and van Aelst, S.F. (1984) *Tetrahedron*, **40**, 4097–4107.
62. De Meo, C., Kamat, M.N., and Demchenko, A.V. (2005) *Eur. J. Org. Chem.*, 706–711.
63. Liao, W. and Lu, D. (1996) *Carbohydr. Res.*, **296**, 171–182.
64. Turský, M., Veselý, J., Tišlerová, I., Trnka, T., and Ledvina, M. (2008) *Synthesis*, **16**, 2610–2616.
65. Hansen, S.U., Baráth, M., Salameh, B.A.B., Pritchard, R.G., Stimpson, W.T., Gardiner, J.M., and Jayson, G.C. (2009) *Org. Lett.*, **11**, 4528–4531.
66. Komarova, B.S., Tsvetkov, Y.E., Pier, G.B., and Nifantiev, N.E. (2008) *J. Org. Chem.*, **73**, 8411–8421.
67. Komarova, B.S., Tsvetkov, Y.E., Pier, G.B., and Nifantiev, N.E. (2012) *Carbohydr. Res.*, **360**, 56–68.
68. Mlynarski, J., Ruiz-Caro, J., and Fürstner, A. (2004) *Chem. Eur. J.*, **10**, 2214–2222.
69. Takahashi, S., Souma, K., Hashimoto, R., Koshino, H., and Nakata, T. (2004) *J. Org. Chem.*, **69**, 4509–4515.
70. Fürstner, A., Ruiz-Caro, J., Prinz, H., and Waldmann, H. (2004) *J. Org. Chem.*, **69**, 459–467.
71. Fürstner, A., Albert, M., Mlynarski, J., Matheu, M., and DeClercq, E. (2003) *J. Am. Chem. Soc.*, **125**, 13132–13142.
72. Adinolfi, M., Iadonisi, A., and Schiattarella, M. (2003) *Tetrahedron Lett.*, **44**, 6479–6482.
73. (a) Paulsen, H. and Kutschker, W. (1983) *Liebigs Ann. Chem.*, **1983**, 557–569. (b) Paulsen, H. (1982) *Angew. Chem., Int. Ed. Engl.*, **94**, 184–201. 1982) *Angew. Chem., Int. Ed. Engl.*, **21**, 155–173.
74. Garegg, P.J. and Ossowski, P. (1983) *Acta Chem. Scand. Ser. B*, **37**, 249–250.
75. Köpper, S. and Zehavi, U. (1989) *Carbohydr. Res.*, **193**, 296–302.
76. Lemieux, R.U., Brice, C., and Huber, G. (1955) *Can. J. Chem.*, **33**, 134–147.
77. Eby, R. and Schuerch, C. (1974) *Carbohydr. Res.*, **34**, 79–90.
78. Winstein, S., Grunwald, E., and Ingraham, L.L. (1948) *J. Am. Chem. Soc.*, **70**, 821–828.
79. Cumpstey, I. (2012) *Org. Biomol. Chem.*, **10**, 2503–2508.
80. Edward, J. (1955) *Chem. Ind.*, 1102–1104.
81. Lemieux, R.U. and Chu, P. (1958) Abstracts of Papers 133th National Meeting American Chemical Society, San Francisco, California, p. 31N.
82. Beaver, M.G., Billings, S.B., and Woerpel, K.A. (2008) *J. Am. Chem. Soc.*, **130**, 2082–2086.
83. Shenoy, S.R., Smith, D.M., and Woerpel, K.A. (2006) *J. Am. Chem. Soc.*, **128**, 8671–8677.
84. Walvoort, M.T.C., Dinkelaar, J., van den Bos, L.J., Lodder, G., Overkleeft, H.S., Codée, J.D.C., and van der Marel, G.A. (2010) *Carbohydr. Res.*, **345**, 1252–1263.
85. Dinkelaar, J., de Jong, A.R., van Meer, R., Somers, M., Lodder, G., Overkleeft, H.S., Codée, J.D.C., and van der Marel, G.A. (2009) *J. Org. Chem.*, **74**, 4982–4991.

86. Codée, J.D.C., van den Bos, L.J., de Jong, A.-R., Dinkelaar, J., Lodder, G., Overkleeft, H.S., and van der Marel, G.A. (2009) *J. Org. Chem.*, **74**, 38–47.
87. Christina, A.E., van der Es, D., Dinkelaar, J., Overkleeft, H.S., van der Marel, G.A., and Codée, J.D.C. (2012) *Chem. Commun.*, **48**, 2686–2688.
88. Stalford, S.A., Kilner, C.A., Leach, A.G., and Turnbull, W.B. (2009) *Org. Biomol. Chem.*, **7**, 4842–4852.

6
Synthesis of Aminoglycosides

Yifat Berkov-Zrihen and Micha Fridman

6.1
Introduction

Aminoglycosides (AGs) are broad-spectrum antibiotics commonly used for the treatment of serious bacterial infections. These antibacterial agents target the prokaryotic ribosome by binding to the decoding A-site of the 16S ribosomal RNA (rRNA) and interfering with the protein translation process [1–5]. Following the isolation of the first known AG, streptomycin, from *Streptomyces griseus* by Schatz and Waksman [6] in 1944, multiple natural AGs with high structural diversity have been discovered (Figure 6.1). All AGs are composed of one or more carbohydrate units and contain a single pseudo-sugar aminocyclitol ring.

The substitution pattern of aminocyclitol rings vary between different AGs (Figure 6.1): For example, 2-deoxystreptamine (DOS) ring is found in common AGs such as neomycin B (**2**) and kanamycin A (**6**); the streptidine aminocyclitol ring is found in streptomycin (**16**), the actinamine ring is part of the AG spectinomycin (**17**), and the fortamine is part of the pseudo-disaccharide AG astromicin (**20**). Over eight decades of intensive clinical use of AGs have led to the emergence of bacterial resistance to this family of antibiotics [7]. Three major pathways have evolved to cause resistance to AGs in various bacterial strains: enzymatic modification by aminoglycoside-modifying enzymes (AMEs) [2, 8], reduction of the drug–target affinity through structural changes to the A-site rRNA [9], or active removal of the drug from the bacterial cell through efflux pumps [10]. An additional drawback to the clinical use of AGs is their nephro- and oto-toxic side effects; the molecular mechanisms that lead to the toxic side effects remain unclear and several studies have explored directions to reduce these undesired toxic side effects through chemical modifications of the parent AG [11, 12].

AGs and their analogs have been utilized in a wide variety of biological applications. In addition to binding to the bacterial 16S rRNA [13], AGs bind to several other RNA structures. For example, AGs inhibit processes such as human immunodeficiency virus (HIV) replication *in vivo* by inhibiting the function of the regulator of virion expression (Rev) protein through binding to the rev response element (RRE) RNA sequence and blocking Rev-RRE interaction [14]. Additional RNA-catalyzed

Modern Synthetic Methods in Carbohydrate Chemistry: From Monosaccharides to Complex Glycoconjugates, First Edition. Edited by Daniel B. Werz and Sébastien Vidal.

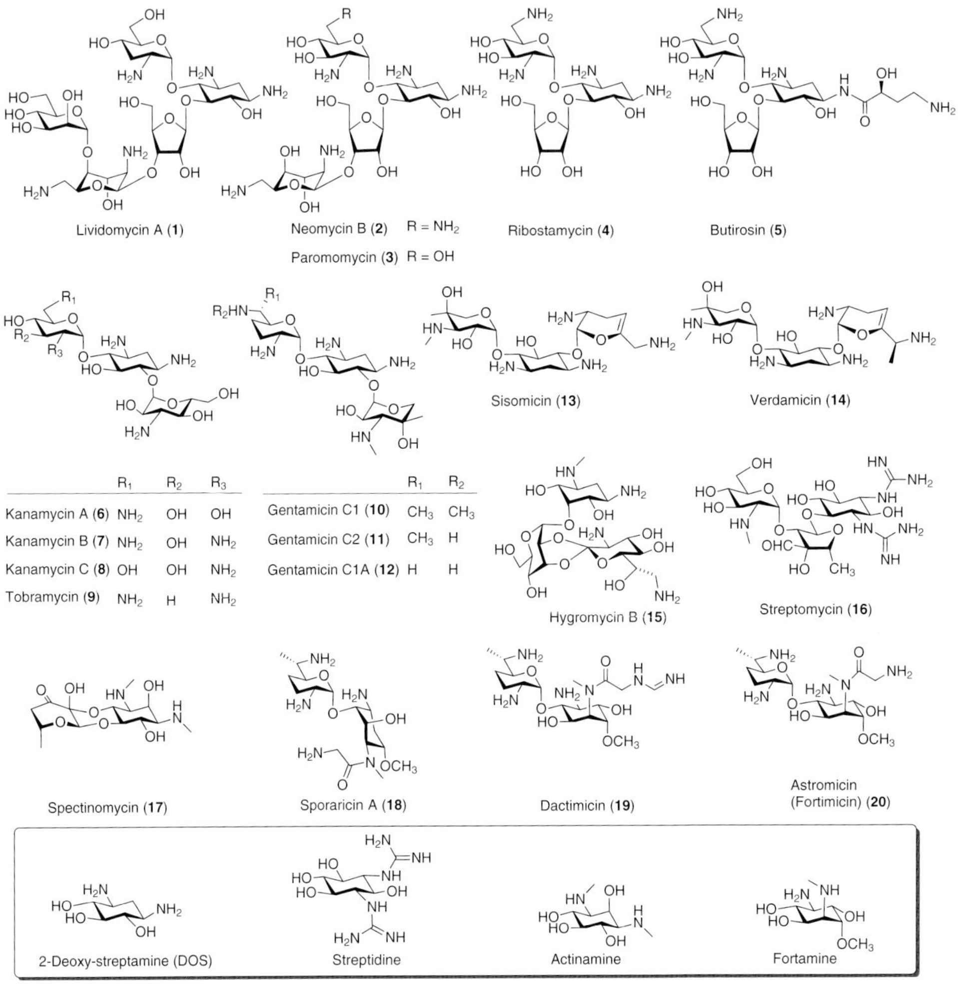

Figure 6.1 The structural diversity of natural AG antibiotics.

processes inhibited by AGs include the self-splicing of group I introns [15], the cleavage reactions of the hammerhead ribozyme[16] and of the hepatitis delta virus (HDV) ribozyme [17].

In recent years, AG derivatives have been preclinically evaluated as potential treatments for genetic disorders caused by nonsense mutations such as premature stop codons that lead to the formation of nonfunctional proteins [18, 19]. A significant effort has been made to develop small molecules that will induce selective read-through of pathogenic nonsense mutations, but not of normal termination codons, so that the expression of the full-length, functional protein is restored to some extent [20]. Semisynthetic AGs analogs demonstrated both *in vitro* and *in vivo* read-through activity and also a significant reduction in toxicity compared with their parent clinically used AGs [21, 22].

Finally, in an attempt to avoid the multiple intracellular resistance mechanisms that bacteria have evolved to inactivate AGs, several studies have demonstrated the potential of the positively charged AGs to be used as scaffolds for the development of membrane-targeting cationic amphiphilic antimicrobial agents. In these derivatives, hydrophobic residues are attached to one or more positions on the AG [23–25]. Several synthetic directions to develop AG-based cationic amphiphiles that will act as membrane-targeting antibiotics have resulted in AG derivatives with broad-spectrum activity against pathogenic bacteria with high levels of resistance to AGs and different levels of specificity for bacterial relative to eukaryotic membranes.

Drug resistance, toxic side effects, and the variety of biological activities of AGs and their synthetic analogs have spurred attempts to develop chemical strategies and methodologies for the preparation of novel AGs. This chapter describes selected practical strategies and methods for the preparation of AG derivatives.

6.2 Amine-Protecting Group Strategies

All AGs are water soluble and have limited solubility in organic solvents. The lack of solubility in organic solvents can be solved by partial or full protection of AG amine groups. The conditions for the full protection of the AG amines are demonstrated in Scheme 6.1 using the scaffold of tobramycin (**9**) as an example. The most commonly and widely used protecting groups are *tert*-butyloxycarbonyl carbamates (*t*-BOC, **21**) [26, 27], carboxybenzyl carbamates (Cbz, **22**) [28, 29], and azides [30, 31] (compound **23**).

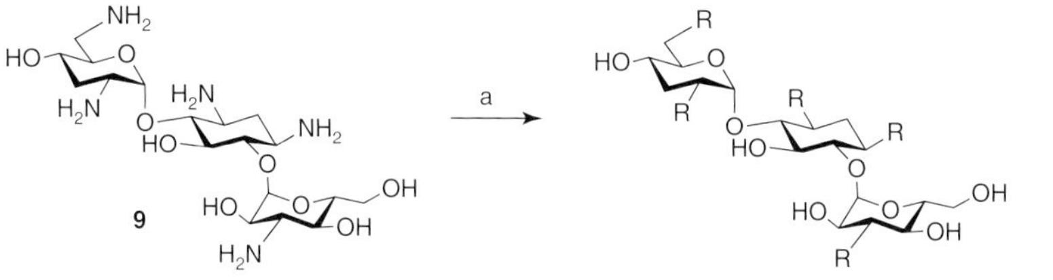

a.
R = NHBoc (**21**): Di-*tert*-butyl dicarbonate (6 eq.), DMSO/H_2O: 6/1, (94%)
R = NHCbz (**22**): Benzyl chloroformate (5.2 eq.), Na_2CO_3 (6 eq.), Acetone/H_2O: 1/1 (96%)
R = N_3 (**23**): TfN_3 (6.25 eq.), Et_3N (3 eq.), ZnCl2 (cat.), DCM/MeOH/H_2O: 3/10/3, $ZnCl_2$ (cat.), (95%)

Scheme 6.1 Common AG amine-protecting group strategies.

6.2.1 Chemoselective Amine Protecting Group Manipulations

Several efficient methodologies for chemoselective protection of specific amines on a variety of AG scaffolds have been reported in the literature. AG aminomethylene groups such as the 6′-position of **6**, **7**, and **9** and the 6′ and 6‴ amines of **2** (Figure 6.1) are in a more electron-rich environment and are less sterically hindered than the other amines on these AG scaffolds, making them better nucleophiles [32, 33].

This property has been used to enable selective protection of these functionalities by the use of *N*-hydroxysuccinimyl-ester-activated carbamylation and acylation reagents [34–36]. Good yields and selectivity were reported when the hindered, carbamylation reagent *N*-benzyloxycarbonyloxy-5-norbornene-2,3-dicarboximide (Cbz-NOR) was used for the selective 6′-amine protection of several AGs and, as described in Scheme 6.2, for the preparation of the 6′-NHCbz kanamycin A (**24**) [37, 38].

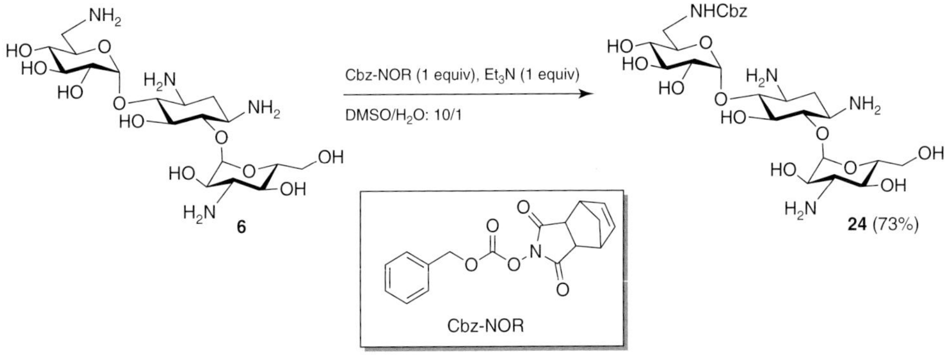

Scheme 6.2 Selective protection of AG aminomethylene groups.

To enable selective protection of amines other than the aminomethylene positions, other methods have been developed. The use of transition metal cations as a methodology for the chemoselective protection of amines on a variety of AGs has been developed and widely used [39–41]. The use of metal ions for the chemoselective N-carbamylation of the AG aparamycin (**25**, Scheme 6.3a) resulted in different products depending on the metal employed. When 4 equiv of $Ni(OAc)_2$ were used for the NH-carboxybenzylation reaction, the C-2′ NH-Cbz product (**26**) was obtained as the single product in 82% yield. When a similar reaction was performed using $Cu(OAc)_2$, the C-3 amine position of the 2-DOS was selectively N-carbamylated to yield the 3-NHCbz aparamycin derivative **27** in 87% yield. Finally, when the same reaction was catalyzed by $Zn(OAc)_2$, a single product, which was carbamylated on the C-1 amine position (compound **28**), was obtained in 31% yield. In the case of $Zn(OAc)_2$, 5% of both the C-2′ product **26** and the C-3 product **27** were obtained as well [42].

A complex of copper ions was used for the chemoselective preparation of the 6′,3-di-NHBoc derivative of kanamycin A (**29**) from the parent AG (**6**) in 72% yield (Scheme 6.3b) [43]. Another example for selective amine-protecting group manipulations on the scaffold of the pseudo-trisaccharide scaffold of the AG sisomicin (**13**) is demonstrated in Scheme 6.3c. Three amine groups of sisomicin (**13**) were selectively protected using the NH-trifluoroacetamide and NH-Boc groups. The 6′-aminomethylene of the free base form of **13** was selectively

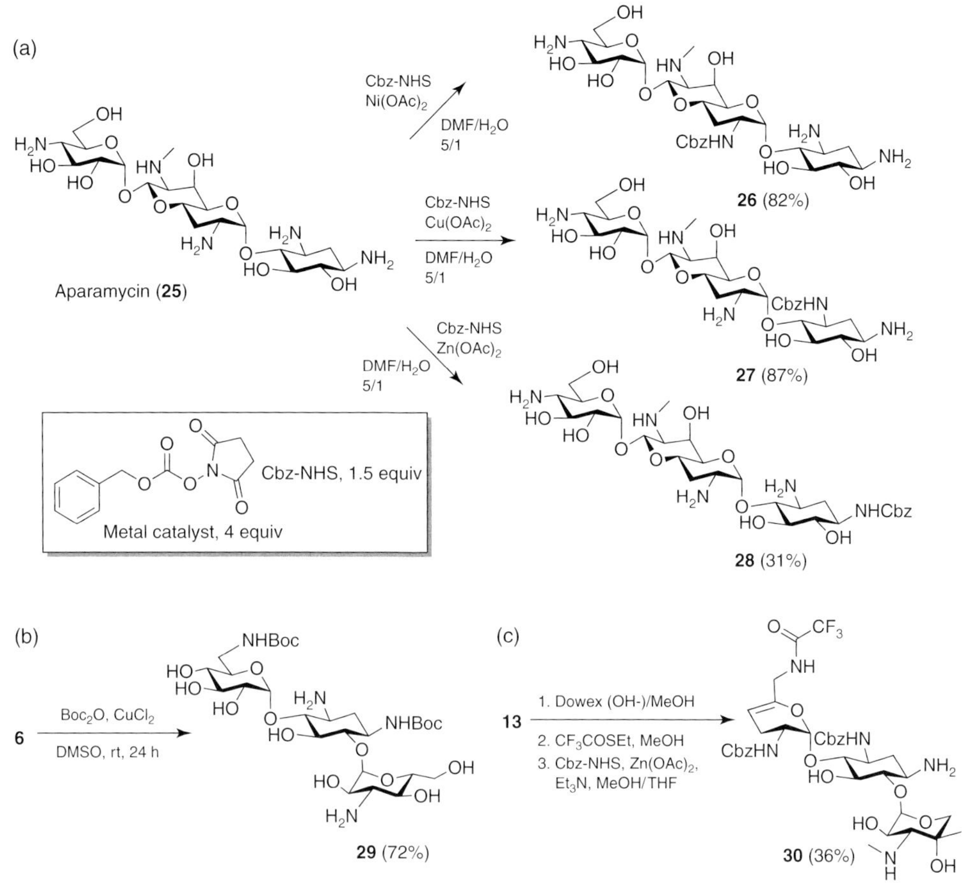

Scheme 6.3 Metal ion-mediated selective protection of AG amines.

converted to the corresponding NH-trifluoroacetamide derivative using *S*-ethyl trifluorothioacetate. Treatment of the 6′-trifluoroacetamide sisomicin analog with Zn(II) acetate and benzyloxycarbonyl-NHS (*N*-hydroxysuccinimide) ester resulted in selective protection of both the 2′ and 3 positions to form the tri-N-protected sisomicin derivative **30** in 36% yield from **13** [44].

6.3
Controlled Degradation of Aminoglycosides

Commercially available AGs have been used as a natural source for the generation of unusual aminosugar and aminocyclitol synthetic building blocks by controlled chemical degradation [22, 45, 46]. These semisynthetic AG fragments have been extensively used for the design and synthesis of novel AGs, thereby significantly reducing the synthetic workload required for the preparation of unnatural AG analogs through total synthesis [47–51]. Selective cleavages of glycosidic bonds on both protected and unprotected AGs can be performed through aqueous acidic hydrolysis or through Lewis-acid-mediated cleavage. Refluxing a mixture of gentamicins (**10–12**) in methanolic hydrogen chloride resulted in the methyl glycoside of the unusual monosaccharide building block garosamine (**31**) and the corresponding pseudo-disaccharide mixture (**32**) in high yields (Scheme 6.4a) [33]. Reflux of neomycin B (**2**) in a solution of methanol and 1N aqueous hydrogen chloride gave the methylglycoside of the disaccharide neobiose (**33**) and the pseudo-disaccharide neamine (**34**) (Scheme 6.4b) [52]. Similar reaction conditions were used for a similar glycosidic bond cleavage of the pseudo-tetrasaccharide paromomycin **3** [53].

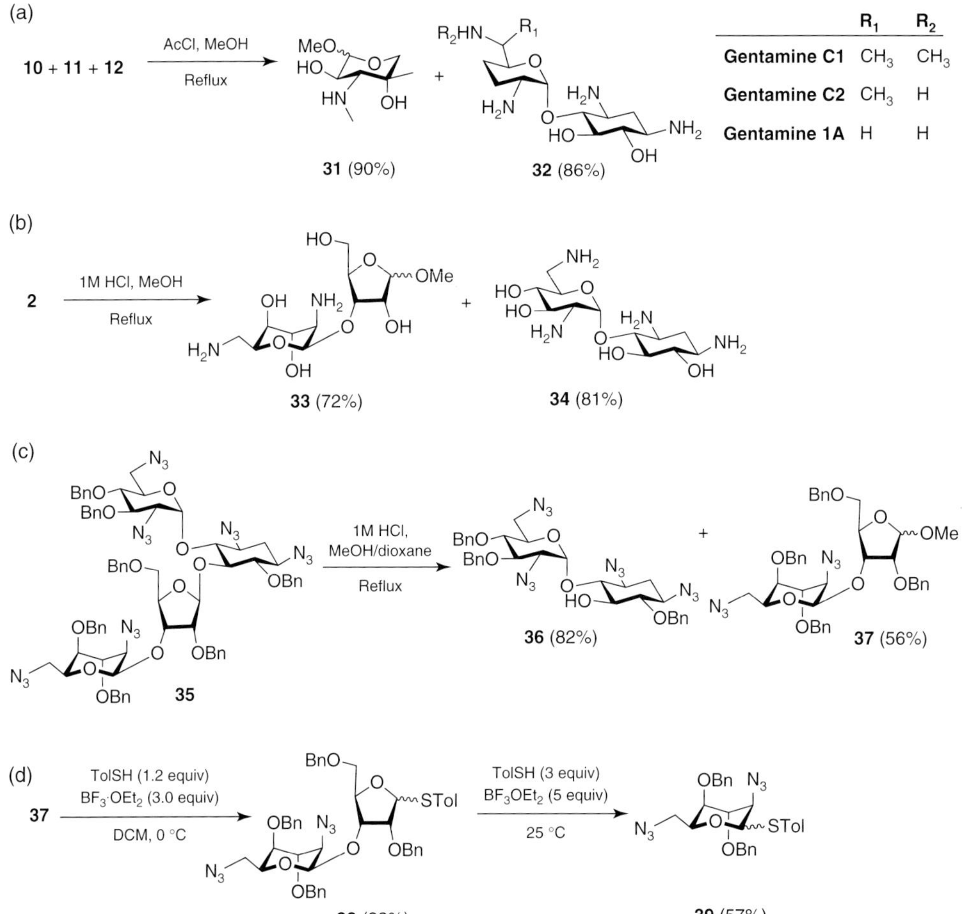

Scheme 6.4 Controlled cleavage of glycosidic bonds for the preparation of AG building blocks.

Hydrogen chloride catalyzed selective degradation of the per-azido, per-*O*-benzyl derivative of neomycin B (**35**) resulted in the neamine derived fragment **36**, with a single unprotected alcohol at the C-5 position of its 2-DOS ring, and the fully protected anomeric mixture neobiose derived methyl glycoside (**37**) (Scheme 6.4c) [54]. The methylglycoside **37** was directly converted to the corresponding thioglycoside donor **38** in 88% yield and was further degraded to the neosamine thioglycoside **39** using a similar strategy in 57% yield. The amounts of thiocresol, Lewis acid, and the temperature were optimized to control the degree of degradation and to prevent the formation of undesired di-thioacetal cleavage products.

Finally, acid-catalyzed methanolysis of the pseudo-pentasaccharide AG lividomycin A (**1**) resulted in the selective cleavage of the ribofuranoside glycosidic bond and the formation of the pseudo-disaccharide fragment (**40**, Scheme 6.5a) in 93% yield and the methyl glycoside of the trisaccharide fragment **41** in 50% yield [55]. The per-N-acetylated pseudo-disaccharide fragment **42** was further degraded under aqueous acidic conditions to afford the lividosamine **43** and the 2-DOS ring **44** in 77% and 90% yields, respectively (Scheme 6.5b) [56].

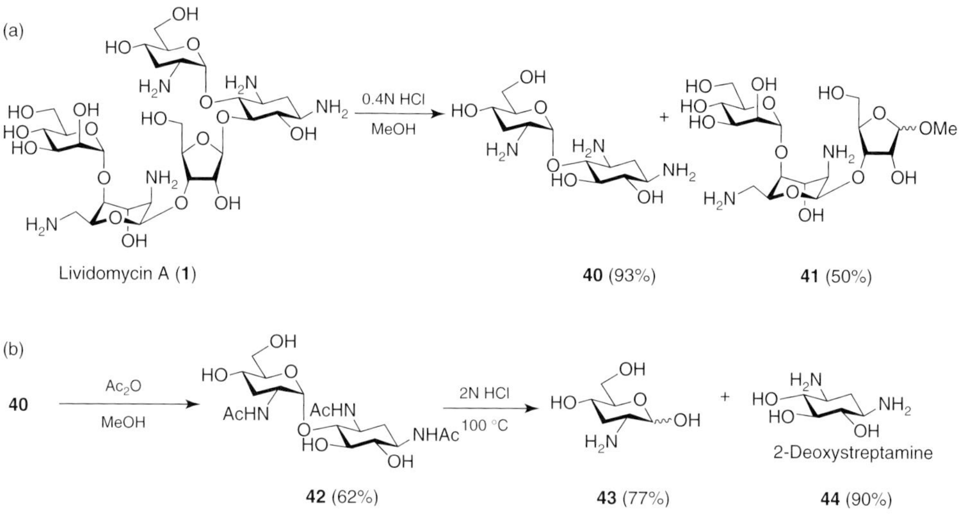

Scheme 6.5 Controlled degradation of lividomycin A (**1**).

6.4
Chemoselective Alcohol-Protecting Group Manipulations

The number of primary, secondary, and tertiary alcohols on AGs ranges from a single secondary alcohol in the pseudo-disaccharide AG sporaricin A (**18**, Figure 6.1) to 10 alcohols in the pseudo-pentasaccharide AG lividomycin A (**1**). Steric hindrance and chemical reactivity differences between AG alcohols were exploited to allow selective alcohol-protecting groups manipulations. Treatment of the tetra-azido neamine derivative **45** with the benzoylation reagent 1-*N*-benzoyloxybenzotriazole (BzOBT) afforded the tri-*O*-benzoyl derivative **46** as a major product in 55% yield (Scheme 6.6a). Under these conditions, the C-5

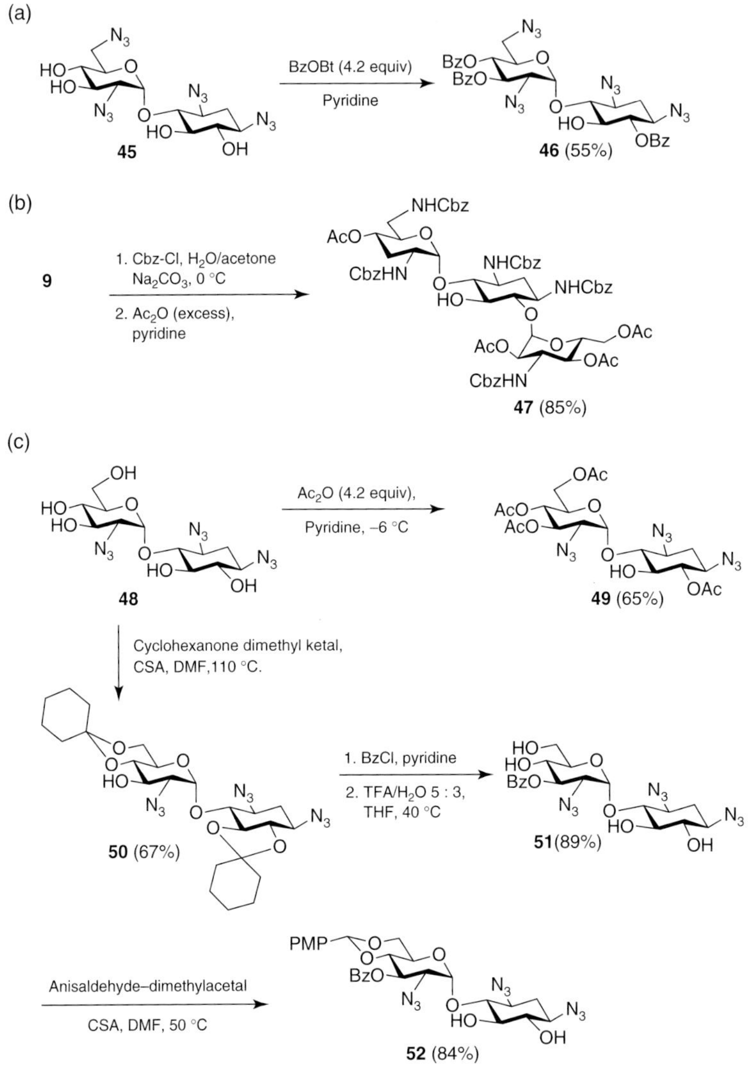

Scheme 6.6 Selected strategies for AG alcohol-protecting group manipulations.

hydroxyl group of the DOS ring remains unprotected [57]. A similar result was obtained when the intact AG structure of the pseudo-trisaccharide tobramycin **9** was used. When **9** was converted to the penta-NH-Cbz tobramycin and treated with an excess of acetic anhydride in pyridine, the partially acetylated derivative **47**, in which the C-5 hydroxyl group of the DOS unit was free, was obtained in 85% yield for the two steps (Scheme 6.6b) [58]. The tri-azido paromamine derivative **48** was selectively converted to the corresponding tetra-*O*-acetylated derivative (**49**, Scheme 6.6c) using standard acetylation conditions (4.2 equiv of acetic anhydride in pyridine) in 65% yield. When **48** was treated with cyclohexanone dimethyl ketal, the di-cyclohexylidene tetra-azido paromamine **50**, in which all functional groups except the sugar C-3′ secondary alcohol were protected, was obtained in 67% yield. Benzoylation of the C-3′ hydroxy group of **50** and removal of the two cyclohexylidene groups under mild acidic conditions gave the C-3′ *O*-benzoyl tri-azidoparomamine analog **51** in 89% yield for the two steps. The sugar of **51** was then selectively protected with the *p*-methoxybenzylidene to provide diol **52** in 84% yield (Scheme 6.6c) [59].

In an attempt to develop AG analogs with improved rRNA affinity, the structure of neamine **34**, prepared from neomycin B (Scheme 6.4b), was treated with benzyl chloroformate in the presence of sodium carbonate to afford the tetra-NH-Cbz-protected neamine derivative **53** (Scheme 6.7). Compound **53** was treated with cyclohexanone dimethyl ketal under acidic conditions and gave the mono-cyclohexylidene neamine analog **54**. Under these conditions, some of the

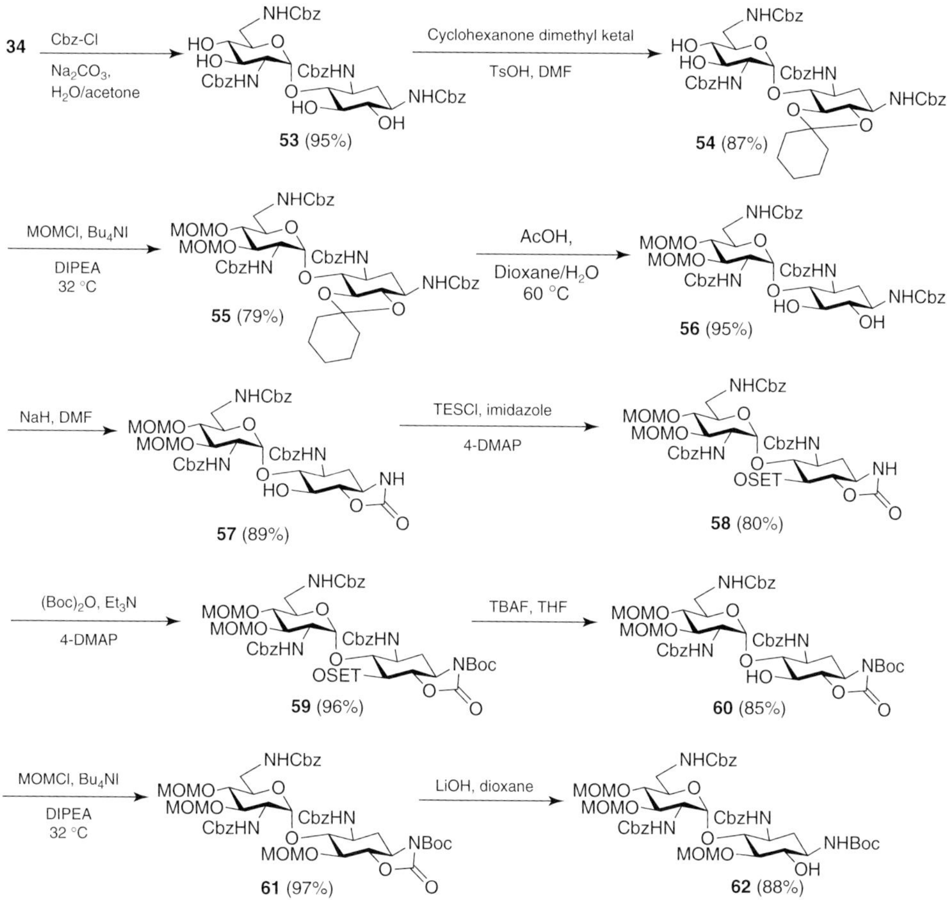

Scheme 6.7 Protecting group manipulations on the neamine building block (**34**).

di-cyclohexylidene product was formed; however, this byproduct was fully converted to compound **54** in the presence of *p*-toluenesulfonic acid (TsOH) and methanol in dimethylformamide (DMF).

Protection of the two remaining sugar alcohols at positions 3′ and 4′ of compound **54** was accomplished by treatment with chloromethyl methyl ether (MOMCl) in the presence of *N,N*-diisopropylethylamine (DIPEA) and tetrabutylammonium iodide (TBAI) to afford the di-*O*-methoxymethyl (MOM) protected **55**. Removal of the cyclohexylidene with acetic acid furnished compound **56**, which upon treatment with sodium hydride in DMF, gave the cyclic carbamate **57** in 89% yield. The C-5 alcohol of the 2-DOS of compound **57** was protected with a triethylsilyl (TES) group to afford neamine derivative **58** [60]. Treatment of **58** with di-*tert*-butyl dicarbonate in the presence of triethylamine and 4-dimethylaminopyridine (4-DMAP) resulted in compound **59**. The TES group of **59** was removed by tetrabutylammonium fluoride (TBAF) to yield compound **60**, which was treated with MOMCl to afford the *O*-MOM-protected **61**; this step made it possible to avoid the migration of the TES group that would otherwise occur under basic conditions. Treatment of compound **61** with 0.5N aqueous lithium hydroxide selectively removed the oxazolidinone ring-protecting group and gave compound **62** with a 2-DOS C-6 hydroxyl group free for chemical manipulation (Scheme 6.7) [60].

Remarkable chemoselectivity was demonstrated during the preparation of the paromomycin analog **66** (Scheme 6.8). The penta-NH-Cbz paromomycin **63** was

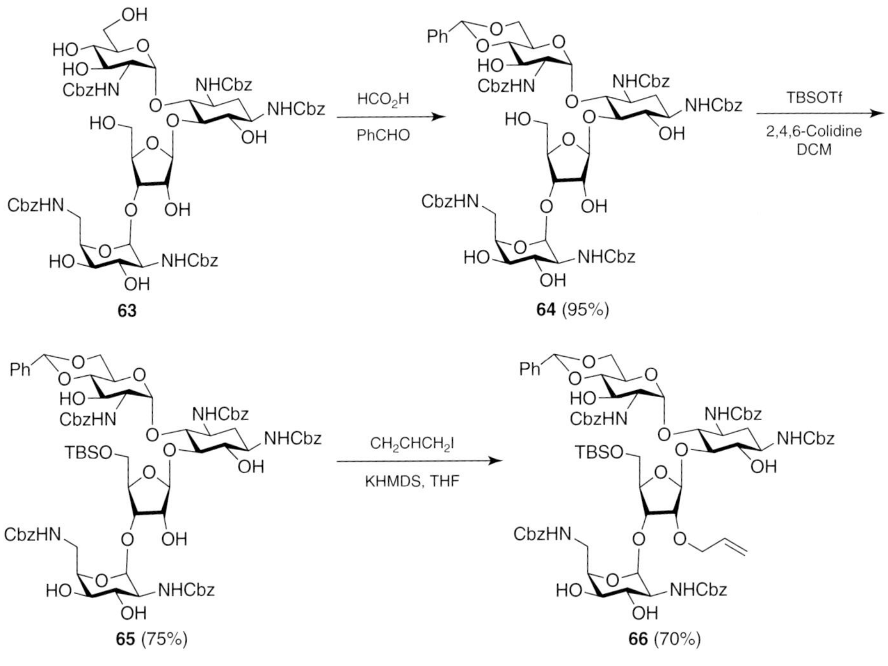

Scheme 6.8 Selective *O*-allylation of paromomycin.

converted to the 4′,6′-benzylidene-protected analog **64** in 95% yield by performing the reaction in benzaldehyde as the solvent and formic acid as catalyst [61]. Selective silylation of the 5″-primary alcohol of **64** with *tert*-butyldimethylsilyl trifluoromethansulfonate (TBSOTf) catalyzed by 2,4,6-collidine gave compound **65** in 75% yield. When compound **65** with five secondary alcohols was treated with allyl iodide and potassium bis(trimethylsilyl)amide (KHMDS, potassium hexamethyldisilazane) in tetrahydrofuran (THF), the single product **66**, selectively allylated on the 2″-alcohol of the ribofuranose ring, was obtained and isolated in 70% yield [62].

6.5
Strategies for Glycosylation of Aminoglycoside Scaffolds

In addition to a single aminocyclitol ring, aminosugars are the major components of all AGs. The specific combinations of alcohols and amines on these aminosugars are essential for the binding interactions with the target rRNA and contribute to cell permeability and high water solubility of this vast family of antibiotics. Hence, numerous AG analogs have been prepared through glycosylation reactions. Both intact AG structures and AG degradation carbohydrate products have been used as glycosyl acceptors as well as glycosyl donors for the preparation of novel AG derivatives [63–67].

In an attempt to optimize the carbohydrate structure of the AG tobramycin **9**, the tobramycin-derived pseudo-disaccharide and suitably protected nebramine analog **67** were obtained by the degradation of **9** as detailed in Scheme 6.9a. Treatment of **9** with trifluoromethanesulfonyl azide (TfN_3) followed by benzylation of all of the alcohols and chemoselective glycosidic bond acidic hydrolysis resulted in the nebramine-based pseudo-disaccharide glycosyl acceptor **67** with a free 6-OH group. Glycosyl acceptor **67** was glycosylated with a series of mono- and disaccharide *p*-tolylthioglycosyl donors with alcohols protected with benzyl ether groups and with amines protected with azides. These thioglycoside donors were activated by *N*-iodosuccinimide–trifluoromethanesulfonic acid (NIS/TfOH) to afford pseudo-trisaccharides in yields ranging from 42 to 92%. Deprotection of the pseudo-trisaccharides was carried out in two steps – first, a Staudinger reaction for the removal of the azido groups and subsequently debenzylation by hydrogenolysis, which was catalyzed by $Pd(OH)_2$/carbon – to obtain the final compounds (general structure **68**, Scheme 6.9a) [68].

In a search for AGs with improved antimicrobial properties, "glycodiversification" of neamine-based scaffolds resulted in libraries of pseudo-trisaccharide-based AG analogs named pyranmycins (Scheme 6.9b). Pyranmycin libraries were generated by the glycosylation of the C-5 alcohol of the neamine-based glycosyl acceptor **36** [47]. These libraries were generated from glycosyl trichloroacetimidate donors having a C-2 *O*-acetyl protecting group to favor the formation of *trans*-glycosidic bond in the resultant pyranmycins. The acetyl groups of the protected pyranmycins were removed under mild basic conditions followed by Staudinger reaction to

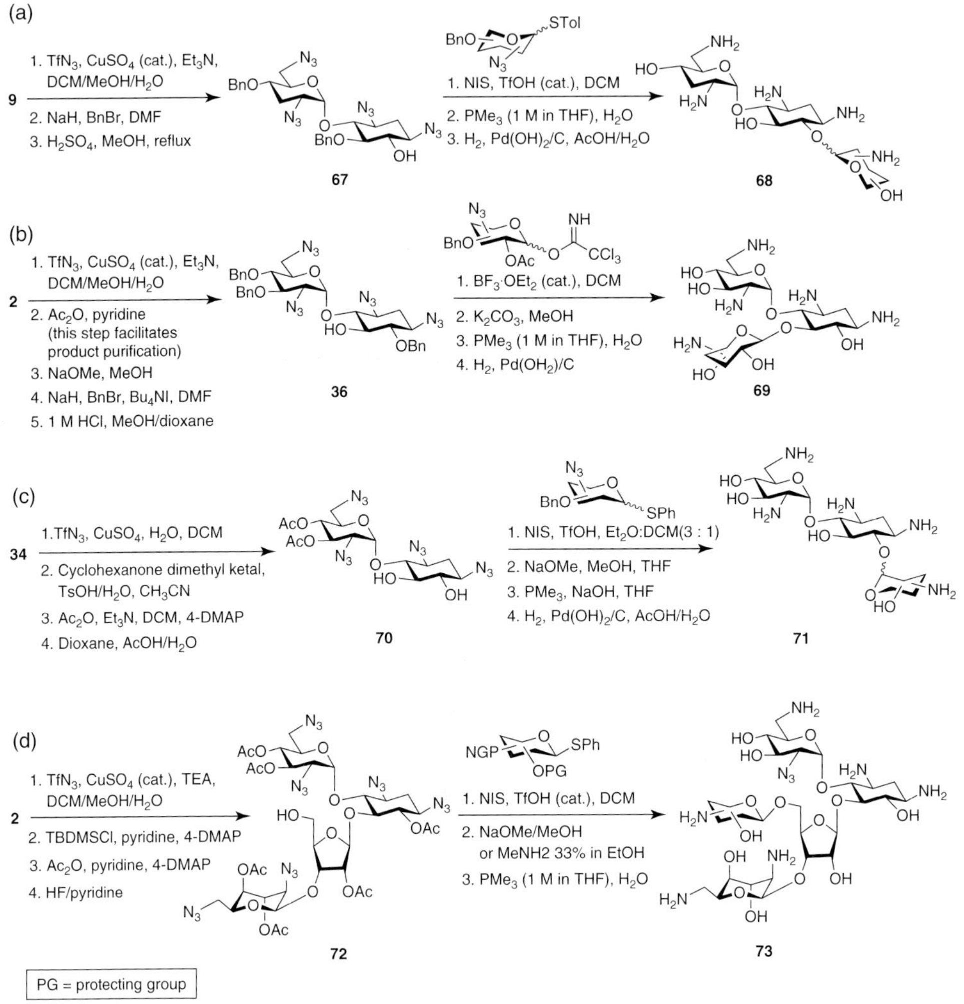

Scheme 6.9 Synthesis of glycosylated AG derivatives.

remove the azido-protecting groups, and finally the benzyl groups were reduced under catalytic hydrogenation to give the desired fully deprotected pyranmycins (general structures **69**, Scheme 6.9b) [69, 70]. A library of kanamycin B (**7**, Figure 6.1) based analogs was generated by the chemoselective glycosylation of the C-6 alcohol of the neamine-based pseudo-disaccharide acceptor **70**, which was generated in four protecting group manipulation steps from neamine **34** (Scheme 6.9c) [71]. Compound **70** readily underwent chemoselective glycosylation at the C-6 alcohol position using a library of thioglycoside donors, and after a three-step deprotection sequence resulted in a library of kanamycin B analogs (general structure **71**, Scheme 6.9c). [71].

In a study aimed at developing novel AG derivatives with improved A-site rRNA affinity and reduced susceptibility to AMEs, additional recognition and binding elements were added to neomycin B (**2**). Hence, the intact AG scaffold **2** was modified by attaching additional sugars to the C-5″ primary alcohol of its ribofuranose ring (Scheme 6.9d) [72]. In these neomycin B analogs, sugar rings with diamine and hydroxylamine functionalities designed to recognize the phosphodiester bond of the target RNA were added to improve target affinity. For that purpose, neomycin acceptor **72** was prepared in four chemical steps from the commercial neomycin B in an overall yield of 57% (Scheme 6.9d). NIS/TfOH-promoted glycosylation of **72** with a set of thioglycoside donors furnished protected pseudo-pentasaccharides in yields ranging from 58 to 89%. All ester and phthalimido groups of the protected AG derivatives were removed under basic conditions and azido groups were reduced to the corresponding free amines through the Staudinger reaction conditions to furnish the AG derivatives with the general structure **73** (Scheme 6.9d).

6.6 Synthesis of Amphiphilic Aminoglycosides

In recent years, several groups have focused on the utilization of the highly positively charged AGs for the design of bacterial membrane-disrupting antibiotics [23–25, 73, 74]. Such antimicrobial agents offer several advantages over antibiotics that target intracellular bacterial targets: activity does not depend on the bacterial cell cycle stage, the potency of these antimicrobials is not diminished due to intracellular resistance mechanisms, and no cell permeability considerations are required in their design. The idea of using positively charged compounds such as AGs as scaffolds for the development of membrane-disrupting antibiotics is based on the fact that relative to most eukaryotic cell membranes, both gram-positive and gram-negative bacteria membranes are significantly more negatively charged due to a high content of anionic lipids such as cardiolipins and phosphatidylglycerol [75, 76]. In addition, the outer leaflet of the outer membrane of Gram-negative bacteria membranes is composed of negatively charged lipopolysaccharide (LPS), whereas negatively charged teichoic acids are major constituents of gram-positive bacteria cell walls [77, 78]. Hence, both gram-positive and gram-negative bacteria effectively attract positively charged compounds.

A series of neomycin B-based lipid conjugates were designed and synthesized as described in Scheme 6.10. These derivatives have potent antimicrobial activity against several bacterial strains [79]. The NH-Boc-protected neomycin B derivative **74** was selectively tosylated on its C-5″ primary alcohol by treatment with 4-toluenesulfonyl chloride (TosCl) in pyridine. Nucleophilic displacement of the tosyl group by sodium azide in DMF at 60 °C resulted in the C-5″ azide derivative, which was converted to the corresponding C-5″ amine neomycin B analog **75** by catalytic hydrogenation. Compound **75** served as a precursor for conjugation to various lipophilic carboxylic acids using 2-(1*H*-benzotriazole-1-yl)-1,1,3,3-tetramethyluronium tetrafluoroborate (TBTU) as the coupling reagent of choice to

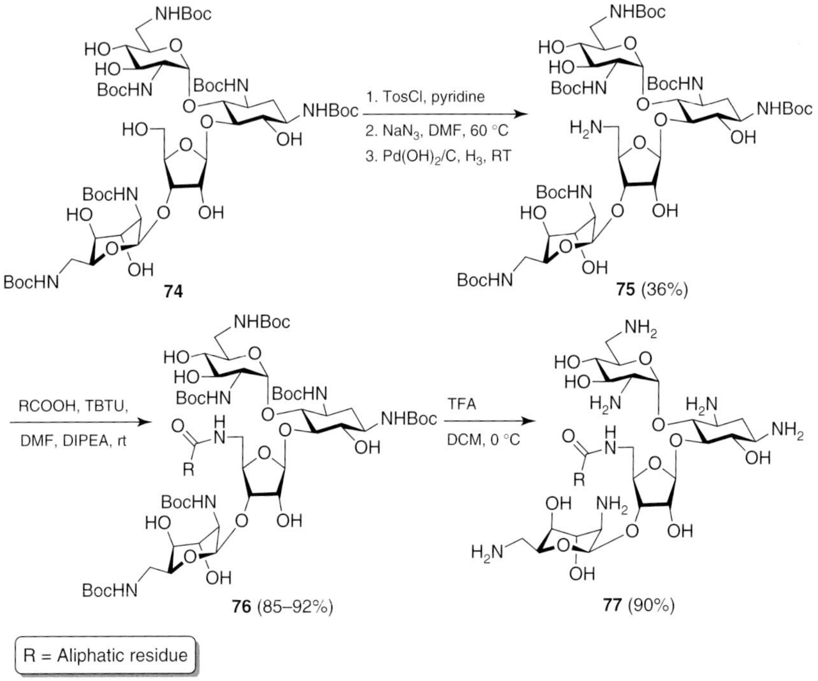

Scheme 6.10 Synthesis of amphiphilic neomycin B-based AGs.

yield the NH-Boc-protected amphiphilic neomycin B analogs (general structure **76**, Scheme 6.10). The final deprotected compounds (general structure **77**, Scheme 6.10) were obtained in their trifluoroacetate (TFA) salt forms after short treatment with TFA [79].

Several types of 6″-aliphatic chain tobramycin analogs differing in the chemical linkage between the AG and the hydrophobic chain (thioether, sulfone, triazole, and amide bonds) and in the length of their hydrophobic linear aliphatic chains were synthesized and evaluated as membrane-targeting antibiotics (Scheme 6.11a) [80]. Of all tested tobramycin-based cationic amphiphiles, those containing were the most potent antimicrobial agents, and the C_{12}, C_{14}, and C_{16} aliphatic chain analogs demonstrated the most potent antimicrobial activity.

Both the aliphatic chain length and the type of chemical linkage between the hydrophilic and hydrophobic parts of these tobramycin-based cationic amphiphiles affected their antimicrobial activity and levels of damage to red blood cell membranes [80].

The thioether tobramycin analogs (general structure **79**, Scheme 6.11a) were prepared from the penta-NH-Boc-6″-*O*-trisyl tobramycin **78** by reaction with alkylthiols followed by the removal of the NH-Boc protecting groups in neat TFA [26, 80]. Oxidation of the protected thioether analogs **79** using *meta*-chloroperoxybenzoic acid (*m*CPBA) followed by deprotection gave the sulfone analogs (general structure **80**). In addition, the 6″-*O*-trisyl group of **78** was replaced by an azide to yield a

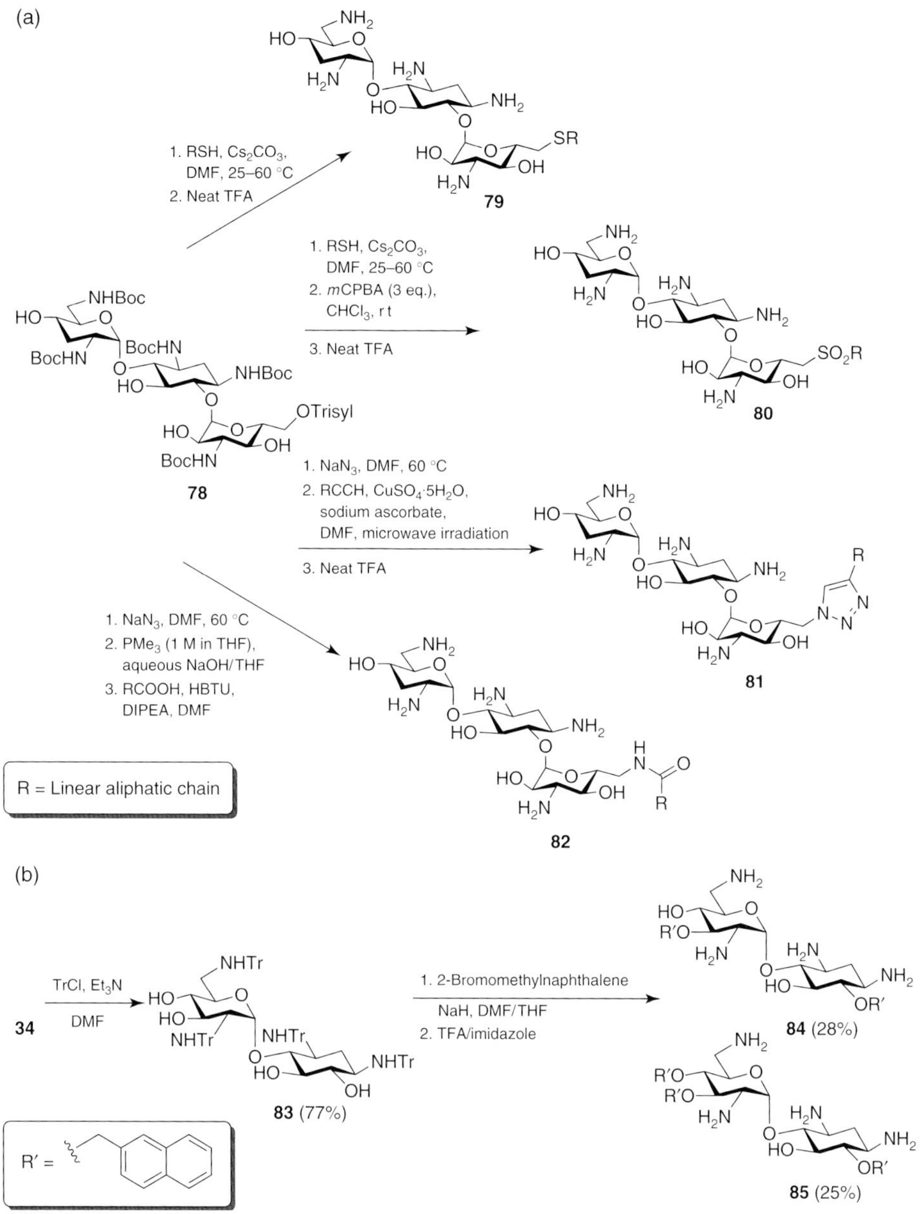

Scheme 6.11 Synthesis of tobramycin- and neamine-based amphiphilic AGs.

compound that served as a precursor for the preparation of the triazole analogs (general structure **81**, Scheme 6.11a). Reduction of the 6″-azido group using the Staudinger reaction conditions resulted in the formation of 6″-amino tobramycin in 80% yield, which served as the precursor for the preparation of the amide analogs (general structure **82**, Scheme 6.11a). This 6″-amino tobramycin was coupled to linear aliphatic carboxylic acids using 2-(1*H*-benzotriazol-1-yl)-1,1,3,3-tetramethyluronium hexafluorophosphate (HBTU) in 71–86% yield, followed by the acidic removal of the NH-Boc groups to yield the amide-linked analogs **82** [81].

A small library of amphiphilic AGs based on the pseudo-disaccharide neamine (**34**) were synthesized by the conversion of two or three of its alcohols to a corresponding aryl ether as illustrated in Scheme 6.11b. Some of the amphiphilic neamine analogs, such as the 2-naphthylmethylene derivative **84**, were active against gram-positive and gram-negative bacterial strains, including strains with different AG resistance mechanisms [23]. The synthesis of the di- or tri-*O*-naphtyl neamine analogs **84** and **85** is illustrated in Scheme 6.11b. Neamine **34** was treated with trityl chloride (TrCl) in triethylamine and DMF to yield the *tetra-N*-trityl protected neamine derivatives **83** [82]. Etherification of compound **83** with bromomethylnaphthalene followed by detritylation with TFA/anisole at 0 °C afforded the compounds **84** and **85** in (28% and 25% yields for the two steps, respectively) [23].

6.7 Chemoenzymatic Strategies for the Preparation of Aminoglycoside Analogs

Although a wide repertoire of synthetic methodologies have enabled the preparation of numerous synthetic AG analogs, chemical synthesis of AGs involves multiple protecting group manipulations and purification steps and as in all multistep syntheses, results in low overall synthetic yields. Hence, there was impetus to develop more efficient and rapid methods for the generation of AG analogs. Numerous enzymes are involved in the biosynthesis of AGs and in catalysis of chemical modifications that deactivate AGs as part of drug resistance mechanisms. A number of these enzymes have been isolated biochemically characterized [83], and X-ray crystallographic structures of some have been determined [7, 84–86]. Several studies have explored the substrate specificity of these enzymes, and the use of these enzymes for the generation of unnatural AGs in a chemoenzymatic approach has been demonstrated.

The synthetic methods for chemoselective N-acylation of AGs discussed previously in this chapter offer access to a variety of semisynthetic AG analogs; however, the diversity of AG structures makes it challenging to chemoselectively modify each of the multiple amine positions on the various AGs using the existing synthetic methodologies [87–91]. An additional approach to overcome this problem exploits the chemoselectivity of aminoglycoside acetyl transferases (AACs) for the *in vitro* chemoenzymatic generation of novel chemoselectively N-acylated

AGs. AACs evolved in bacteria to confer resistance to AGs by the chemoselective conversion of amines to the corresponding acetamides, and may therefore be used for chemoselective N-acylation of AGs and circumvent the need for multistep syntheses. Although in most cases limited amounts of products can be generated using a chemoenzymatic approach, the method is extremely valuable as it allows for an initial activity screen that can guide further decisions as to which compounds justify large-scale and multistep chemical syntheses.

A library of N-acylated AGs was generated using two AACs, AAC(3)-IV and AAC(6′)-APH(2″) (aminoglycoside phosphotransferase) [92]. The substrate and cosubstrate specificity of both these enzymes was studied using a variety of AGs and natural as well as semisynthetic acyl coenzyme A (CoA) derivatives as demonstrated for sisomicin **13** and glycinyl CoA **86** (Scheme 6.12a). In the case of AAC(3)-IV, of the 80 combinations of AGs and acyl CoAs evaluated, 25 had successful transfers. One such example is the synthesis of 3-NH-glycinyl-sisomicin **88** (Scheme 6.12a). AAC(6′)-APH(2″) was found to have relatively relaxed substrate specificity and readily transferred a diverse set of acyl-CoAs to a diverse set of AG scaffolds (e.g., to form product 6″-NH-glycinyl-sisomicin **89**). Out of the 96 combinations tested, 29 resulted in appreciable formation of the desired N-acylated product after incubation for 30 min. Moreover, sequential double acylation with AAC(3)-IV and AAC(6′)-APH(2″) yielded double-hetero- as well as homo-N-acylated AGs.

Using a similar approach, AAC(3)-IV, which readily accepted the semisynthetic azido acetyl CoA **87** as a substrate, was used for the generation of 3-NH-azidoacetyl AGs that were further diversified via click chemistry [93]. For example, aparamycin **25** was converted to the 3-NH-azidoacetyl aparamycin, which was further chemically diversified to the corresponding triazole derivatives (as shown in general structure **90**, Scheme 6.12b). This approach gave access to modified compounds with diversity beyond that possible using AAC(3)-IV and modified CoA thioesters.

The biosynthesis of butirosin (**5**, Figure 6.1) from ribostamycin (**4**, Figure 6.1) involves two sequential enzymatic steps in which the (*S*)-4-amino-2-hydroxy-butanoyl (AHB) group is transferred to the 2-DOS *N*-1-amine of **4**. The two enzymes BtrH and BtrG were used for the development of a chemoenzymatic generation of a variety of novel *N*-1-AHB-bearing AGs using N-terminal His6-tagged BtrH and BtrG [94]. To make it worthwhile to use this process, the structurally complex natural acyl donor was very efficiently replaced by the synthetic *N*-acetylcysteamine thioester γ-l-Glu-AHB-SNHAc (**91**, Scheme 6.12c) [95].

Using this method, several pseudo di- and tri-saccharides of 2-deoxystreptamine-containing AGs were acylated at the N-1 position with an AHB as described in Scheme 6.12c for the preparation of compound **94**. Compound **92** was converted to the corresponding γ-l-Glu-AHB analog **93** catalyzed by BtrH, followed by an intramolecular cleavage catalyzed by BtrG to yield the N-1-AHB AG analog **94**. The BtrH/BtrG catalytic system with the synthetic acyl donor reagent **91** provides an efficient chemoenzymatic method for the preparation of a wide variety of 2-DOS-containing AGs with an *N*-1-AHB group. This method significantly shortens and simplifies the preparation of AHB N-acylated AG analogs as compared to

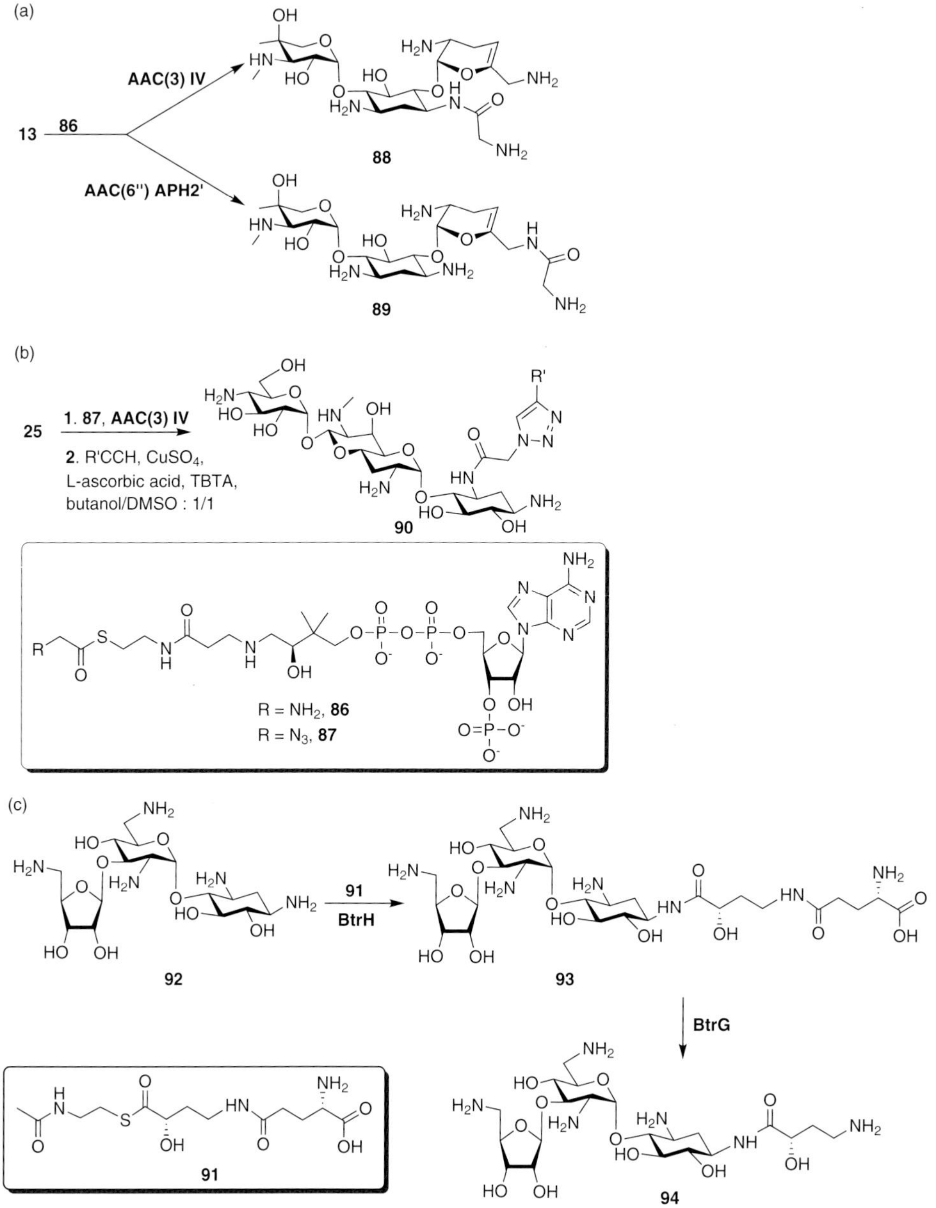

Scheme 6.12 Chemoenzymatic synthesis of AG derivatives.

a chemical synthesis approach. The diversity of substrates accepted by BtrH and availability of the recombinant BtrH and BtrG enzymes make this method attractive for high-throughput synthesis of libraries of 2-DOS-containing AGs [96].

Although the advantage of chemoenzymatic synthesis is the rapid access to multiple AG analogs that would otherwise require multistep synthetesis for their preparation, in the vast majority of the cases, this type of approach cannot yet be used for large-scale production. With the need to synthesize unnatural enzyme substrates and or cosubstrates, and with the need to obtain large quantities of the enzymes, it is still difficult and expensive to scale up the production even to single-digit milligram scales of the AG products. Hence, there is still a great need to further improve and develop chemoenzymatic methodologies for the preparation of AGs.

6.8 Novel Synthetic Strategies to Overcome Resistance to Aminoglycosides

Bacteria have evolved several types of resistance mechanisms to AGs (Section 6.1). AMEs are undoubtedly the most prevalent AG resistance mechanism; this mechanism is found in almost all AG-resistant bacteria. Hence, through the years, efforts have been made to develop AG analogs that are not susceptible to deactivation by one or several AMEs [29, 38, 97–99]. Some of the synthetic strategies that have been developed to prepare these AG derivatives have been described in this chapter. In this section, two recently reported and unique approaches to tackle with AME-based AG deactivation are described.

In recent years, a novel approach to inhibit the catalytic activity of the widespread AME *N*-6′-acetyltransferase (AAC-(6′)) that acetylates the 6′-amine of several AGs using acetyl coenzyme A (AcCoA) as the acetyl transfer cosubstrate was developed. A set of competitive inhibitors of the enzyme were developed with the rational that these compounds may be administered with the AG to regain efficacy of the antibiotic by inhibiting the AME-based AG deactivation. A selection of AG-CoA bisubstrate-based inhibitors of AAC(6′) were designed and synthesized. Some of the compounds did not show significant *in vitro* inhibition of the isolated enzyme and yet were potent nanomolar-scale inhibitors *in vivo* (compound **95**, Scheme 6.13a). Compounds such as **96** (Scheme 6.13a) were potent *in vitro* inhibitors but, because of their negatively charged phosphates, did not permeate cellular membranes. Compounds such as **95** are uncharged and able to penetrate the membrane. Inside the bacterial cells, these compounds are activated by enzymes of the CoA biosynthetic pathway [100]. Hence, chemically prepared compound **96** [101] was a potent *in vitro* inhibitor of AAC(6′) yet could not permeate the bacterial cell membrane; however, the same compound was formed by the AcCoA biosynthetic pathway inside the cell from the truncated synthetic precursor **95** where it inhibited the enzyme.

One recently proven concept is the introduction of not one but several chemical modifications to the AG scaffold to block the action of a broad spectrum of AMEs.

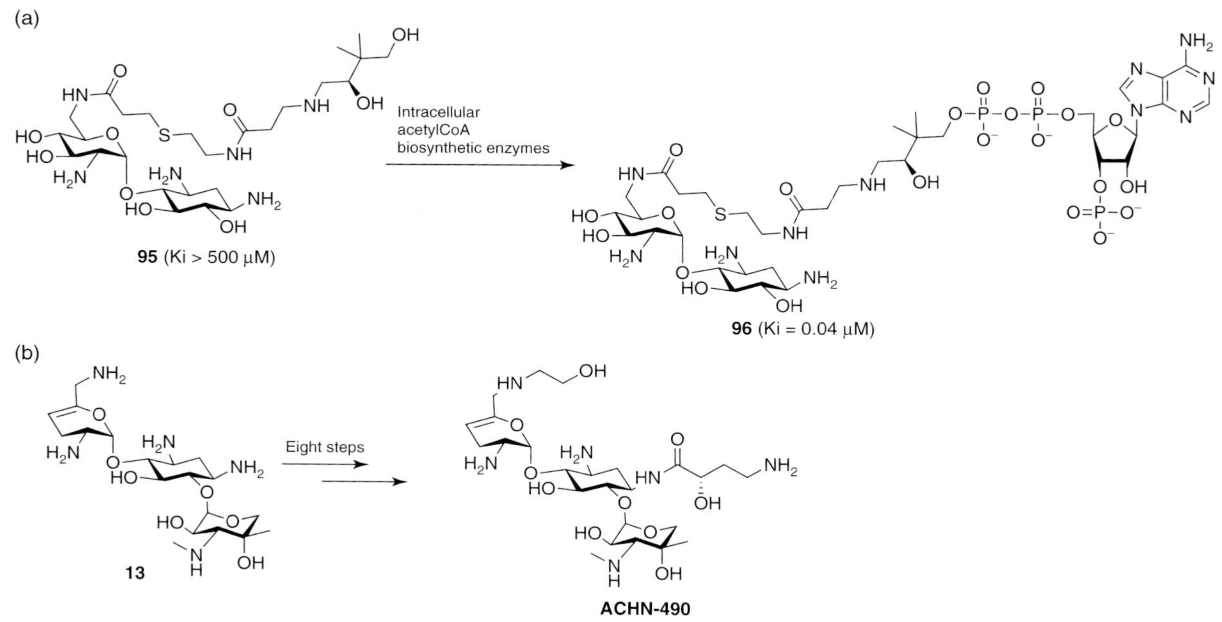

Scheme 6.13 Bisubstrate inhibitors of AAC(6′) and the broad-spectrum ACHN-490.

This strategy was very successfully employed in the recent development of the eight-step synthesis of the sisomicin derivative ACHN-490 (Scheme 6.13b). ACHN-490 that was selected from a library of over 400 semisynthetic sisomicin analogs, was found not to be modified by the AMEs that target its parent antibiotics. This sisomicin analog demonstrated highly potent and broad-spectrum antimicrobial activities against multiple AG-resistant bacterial strains, and was recently advanced into clinical development [44, 102, 103]. The discovery of ACHN-490 demonstrates that even after the synthesis and testing of thousands of synthetic AG analogs, new synthetic directions still yield novel AGs with improved antimicrobial activities and reduced toxicity [103].

6.9
Conclusions and Future Perspectives

Seven decades after the discovery of the first of the AG antibiotics, these bacterial ribosome-targeting antimicrobial agents remain clinically important. Their broad spectrum of biological activities maintains their status as one of the most widely studied families of biologically active molecules. In attempts to overcome the major problems of resistance and toxic side effects, synthetic chemistry efforts continue to focus on both rational design and combinatorial strategies. So far, the effort to develop AGs with resistance to bacterial enzyme-based deactivation has resulted in several clinically useful next-generation AGs; however, these next-generation antibiotics do not offer an answer to all of the combinations of AMEs that can be found in various AG-resistant bacteria. Furthermore, development of AG derivatives that overcome other AG resistance mechanisms such as structural alterations in the ribosomal A-site and the action of protein efflux pumps should become the focus of research effort as these resistance mechanisms have been poorly addressed. As bacteria with resistance to AGs usually evolve more than one resistance mechanism, efforts to develop new and more general strategies to cope with resistance and to revive the clinical efficacy of AGs are required in the near future.

In recent years, features of AGs such as their high water solubility and positively charged structures have raised interest in using these antimicrobial agents for the development of novel antibiotics that no longer target the bacterial ribosome. These studies have led to the development of AG-based bacterial membrane-targeting antibiotics with some selectivity for bacterial over mammalian membranes; these agents may have potential for the development of antibiotics for internal clinical use but require additional optimization to reach that stage. AGs also induce selective read-through of pathogenic nonsense mutations and thus may have value in treatment of certain genetic disorders. Given the abundance of potential applications of AGs in medicine, further development of synthetic methodologies for preparation of novel AG derivatives and libraries to be used in biological screening is likely to remain the focus of AG-based studies in the future.

6.10
Selected Synthetic Procedures

The syntheses of the pseudo-pentasaccharide derivative of neomycin B (Compound **99**, Scheme 6.14a) and of the pseudo-trisaccharide AG analog **107** (Scheme 6.14b) were chosen as examples of common and useful synthetic manipulations that are performed on AG scaffolds. The detailed synthetic procedures for the preparation and purification of these AG derivatives are described in this section.

Neomycin acceptor (**72**, Scheme 6.14a) [72]. Hexaazido-neomycin was prepared from the commercial neomycin B (tri-sulfate salt, 5 g, 5.5 mmol) [47] and was used for the next step as crude without purification. The crude hexaazido-neomycin in pyridine (40 ml) was added 4-DMAP (cat.) and stirred at 70 °C for 15 min after which *tert*-butyldimethylsilylchloride (1.66 g, 11 mmol) was added. The reaction was monitored by thin layer chromatography (TLC; 100% EtOAc,) which indicated completion of the reaction after 30 min. To the mixture, pyridine (20 ml), 4-DMAP (cat.), and Ac_2O (7.8 ml, 82.5 mmol) were then added. The reaction was monitored by TLC (EtOAc/hexane, 3 : 7), which indicated completion after 3 h. The mixture was then diluted with EtOAc and washed with brine, HCl (2%), $NaHCO_3$ (sat.), and once more with brine. The combined organic layer was dried over $MgSO_4$, evaporated, and purified by flash column chromatography (silica, EtOAc/hexane, 1 : 1), which resulted in the corresponding silyl ether as a white powder (3.5 g, 62% yield over three steps). The silyl ether (1.06 g, 0.93 mmol) was dissolved in pyridine (8 ml) and stirred in a polyethylene vessel at 0 °C for 10 min followed by the addition of hydrofluoride (HF)/pyridine (4 ml). The reaction was monitored by TLC (EtOAc/hexane, 1 : 4), which indicated completion after 5 min. The mixture was then diluted with EtOAc and neutralized with $NaHCO_3$ (sat.). The combined organic layer was dried over $MgSO_4$, evaporated, and purified by flash column chromatography (silica, EtOAc/hexane, 1 : 1) to yield acceptor **72** as a white powder (884 mg, 93%).

Compound 98 (Scheme 6.14a) [72]: To powdered, flame-dried 4 Å molecular sieves (0.4 g) dichloromethane, DCM (4 ml), the neomycin acceptor **72** (100 mg, 0.098 mmol), and the thioglycoside donor **97** (65 mg, 0.143 mmol) were added. After 10 min of stirring at room temperature, the mixture was treated with NIS (64.3 mg, 0.286 mmol) and stirred for additional 5 min at room temperature. The reaction mixture was then activated by the addition of a catalytic amount of TfOH. The reaction was monitored by TLC (EtOAc/hexane, 1 : 1), which indicated completion after 10 min. The solution was then diluted with EtOAc and filtered through celite. The organic layer was washed with 10% $Na_2S_2O_3$, saturated (aq.) $NaHCO_3$, brine, dried over $MgSO_4$, and concentrated. The crude product was purified by flash column chromatography (silica, EtOAc/hexane1 : 1) to yield compound **98** (112 mg, 81%).

Compound 99 (Scheme 6.14a) [72]: Compound **98** (0.11 g, 0.078 mmol) was dissolved in 33% solution of $MeNH_2$ in EtOH (40 ml), and the mixture was stirred at room temperature for 30 h. The reagent and the solvent were removed by evaporation, and the crude residue was dissolved in THF (10 ml), 0.1 M

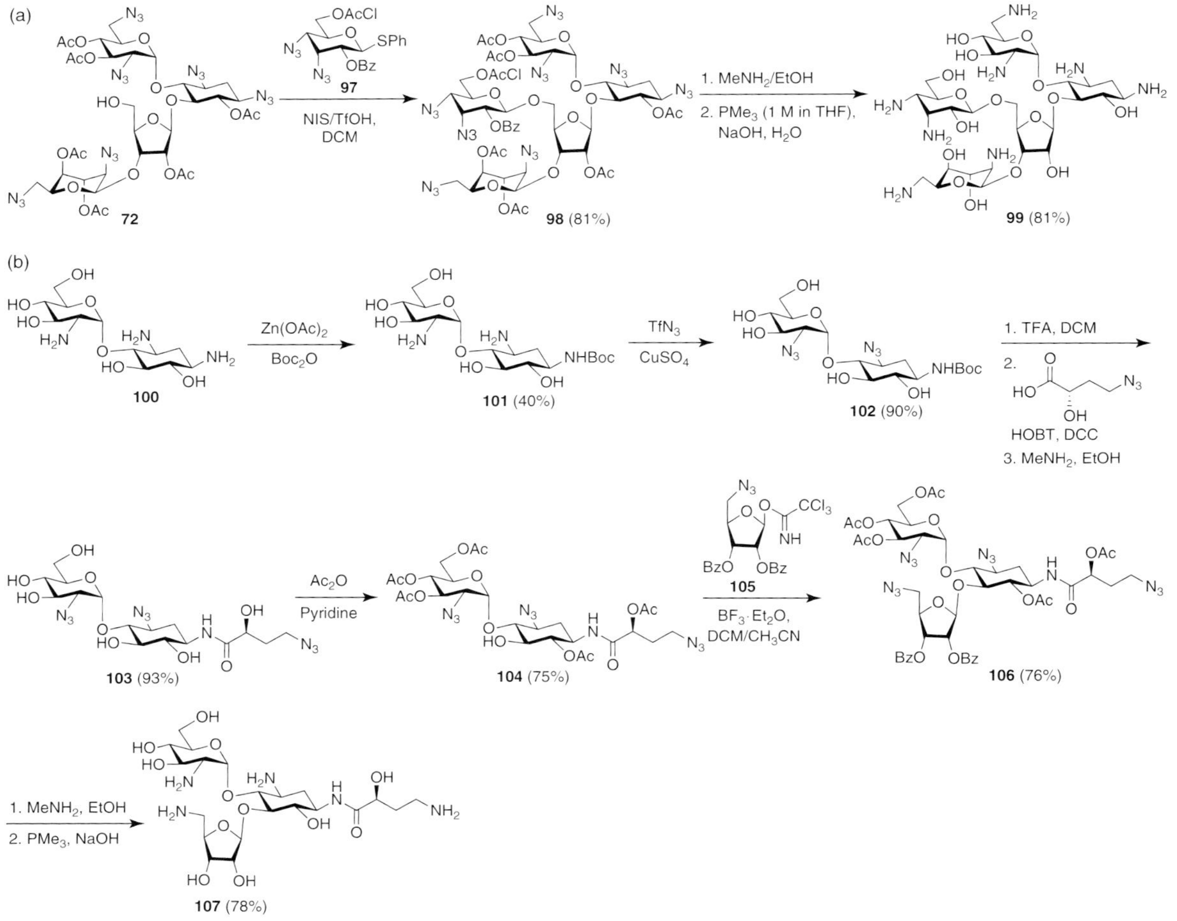

Scheme 6.14 Syntheses of the pseudo-pentasaccharide derivative of neomycin B (**99**) and of the pseudo-trisaccharide AG analog (**107**).

NaOH (2 ml) and stirred at 60 °C for 10 min after which PMe_3 (1 M solution in THF, 3.73 ml, 3.73 mmol) was added. The reaction was monitored by TLC (DCM/MeOH/H_2O/$MeNH_2$ (33% solution in EtOH), 10 : 15: 6: 15 in EtOH, product $R_f = 0.33$), which indicated completion after 3.5 h. The reaction mixture was then purified by flash chromatography on a short silica column that was eluted as follows: THF, EtOAc, MeOH/EtOAc (1 : 1), MeOH, and finally with $MeNH_2$ (33% solution in EtOH). The fractions containing the product were evaporated under vacuum, redissolved in water, and evaporated again to afford the product in its free amine base form (48.7 mg, 81%). This product was then dissolved in water, the pH was adjusted to 7.5 with 0.01 M H_2SO_4, and lyophilized to give the sulfate salt of **99** (88.5 mg) as a white foamy solid.

Compound 101 (Scheme 6.14b) [104]: $Zn(OAc)_2$ (14.75 g, 66.0 mmol) was added to a stirred solution of paromamine [46] (**100**), in its free base form (9.69 g, 30 mmol) in H_2O (30 ml) and DMF (150 ml), and the mixture was stirred for 12 h at room temperature. A solution of di-*tert*-butyldicarbonate (9.81 g, 45 mmol) in DMF (20 ml) was then added to the reaction mixture over 30 min and the mixture was stirred for an additional 24 h. Reaction progress was monitored by TLC [DCM/MeOH/H_2O/$MeNH_2$ (33% solution in EtOH), 10 : 15 : 6 : 15, in EtOH)]. Upon completion, the mixture was diluted with MeOH (250 ml) and loaded onto 50 × 300 mm ion-exchange column (Amberlite CG50, H+ form). The column was first washed extensively (about 10 column volumes) with a mixture MeOH/H_2O (60 : 40), followed by elution with the mixture of 25% MeOH/H_2O/NH_4OH (80 : 15 : 5) in water to afford the desired *N*-1-NHBoc derivative of paromamine (**101**) in 40% yield (5.03 g).

Compound 102 (Scheme 6.14b) [104]: Compound **101** (44.3 g, 0.1 mol) was converted to the corresponding 3,2′-diazido derivative, using Tf_2O (110 ml, 0.66 mol) and NaN_3 (100 g, 1.53 mol) [104]. The reaction progress was monitored by TLC (EtOAc/MeOH: 95/5), which indicated completion after 8 h. The crude product was purified by flash column chromatography (EtOAc/MeOH 95 : 5) to yield the title compound **102** (42.5 g, 90% yield).

Compound 103 (Scheme 6.14b) [104]: Compound **102** (3.0 g, 6.31 mmol) was dissolved in a mixture of TFA (12 ml) and DCM (30 ml) and stirred at ambient temperature. Reaction progress was monitored by TLC (DCM/MeOH 4 : 1), and indicated completion after 1 h. The reaction mixture was then evaporated under reduced pressure and the crude was dissolved in a mixture of Et_3N (10 ml) and DMF (10 ml) and cooled to −20 °C. In a separate flask, (*S*)-2-hyroxy-4-azidobutyric acid (3.94 g, 31.55 mmol) was dissolved in anhydrous DMF (30 ml) and cooled to 0 °C. To the cold solution, *N*,*N*′-dicyclohexylcarbodiimide (DCC; 7.10 g, 34.46 mmol) and HOBt (4.72 g, 34.96 mmol) were added, and the mixture was stirred at 0 °C for about 1 h. This mixture was carefully added by a syringe to the cold solution of the amine at −20 °C. The reaction was stirred at −20 °C for 1 h and then allowed to warm to room temperature for an additional hour. The mixture was then treated with a solution of $MeNH_2$ (33% solution in EtOH, 30 ml) and its progress was monitored by TLC (DCM/MeOH, 7 : 3). Upon completion (approximately 8 h), the reaction mixture was concentrated and purified by flash chromatography (MeOH/DCM, 1 : 9) to provide compound **103** (3.0 g, 93% yield).

Compound 104 (Scheme 6.14b) [104]: Compound **103** (3.0 g, 5.85 mmol) was dissolved in dry pyridine (10 ml), cooled at −12 °C, and then acetic anhydride (5.4 equiv, 3.0 ml, 31.80 mmol) was added. The reaction temperature was kept at −12 °C, and the reaction progress was monitored by TLC (EtOAc/hexane, 7 : 3), which indicated completion after 8 h. The reaction mixture was diluted with EtOAc (100 ml) and extracted with aqueous solution of HCl (2%), saturated aqueous $NaHCO_3$, and brine. The combined organic layer was dried over $MgSO_4$ and concentrated. The crude product was purified by flash column chromatography (EtOAc/hexane, 2 : 3) to afford **104** (3.15 g, 75% yield).

Compound 106 [104]: Anhydrous DCM (10 ml) was added to powdered, flame-dried 4 Å molecular sieves (3.0 g), followed by the addition of acceptor **104** (1.75 g, 2.46 mmol) and the trichoroacetimidate glycosyl donor **105** [59] (3.3 g, 6.27 mmol) dissolved in CH_3CN (10 ml). The mixture was stirred for 10 min at room temperature and was then cooled to −20 °C. A catalytic amount of $BF_3{\cdot}OEt_2$ (100 µl) was added and the mixture was stirred at −15 °C. The reaction progress was monitored by TLC (EtOAc/hexane, 3 : 2), which indicated the completion after 30 min. The reaction was diluted with DCM and filtered through celite. After thorough washing of the celite with DCM, the washes were combined and extracted with saturated aqueous $NaHCO_3$ and then brine, dried over $MgSO_4$, and concentrated. The crude product was purified by flash chromatography (EtOAc/hexane, 2 : 3) to afford compound **106** (2.01 g, 76% yield).

Compound 107 [104]: Compound **106** (1.55 g, 1.44 mmol) was treated with a solution of $MeNH_2$ (33% solution in EtOH, 50 ml) and the reaction progress was monitored by TLC (EtOAc/MeOH, 85 : 15), which indicated completion after 8 h. The reaction mixture was evaporated to dryness and the residue was dissolved in a mixture of THF (5 ml) and aqueous NaOH (1 mM, 5.0 ml). The mixture was stirred at room temperature for 10 min, after which PMe_3 (1 M solution in THF, 11.52 ml, 11.52 mmol) was added. The reaction progress was monitored by TLC ($DCM/MeOH/H_2O/MeNH_2$ (33% solution in EtOH), 10 : 15 : 6 : 15, in EtOH) and was complete after 1 h. The product was purified by flash chromatography on a short column of silica gel. The column was washed with the following solvents: THF (800 ml), DCM (800 ml), EtOH (200 ml), and MeOH (400 ml). The product was then eluted with $MeNH_2$ (33% solution in EtOH)/MeOH, 1 : 4. Fractions containing the product were combined and evaporated to dryness. The residue was dissolved in a small volume of water and evaporated again (two to three repeats) to afford the free amine form of **107**. The analytically pure product was obtained by passing the above-mentioned product through a short column of Amberlite CG50 (NH_4+ form). The column was first washed with $MeOH/H_2O$ (3 : 2), then the product was eluted with $MeOH/H_2O/NH_4OH$ (8 : 1 : 1) to afford compound **107** (628 mg, 78%). For the storage and biological tests, compound **107** was converted to its sulfate salt form: the free base was dissolved in water, the pH was adjusted to 6.5 with 0.1N H_2SO_4, and the compound was lyophilized to afford the sulfate salt as a white powder.

Acknowledgments

We thank the US–Israel Binational Science Foundation (grant no. 2006/301) and the FP7-PEOPLE-2009-RG Marie Curie Action: Reintegration Grants (Grant 246673) for their generous support that has made it possible for us to perform our studies in the field of AG antibiotics.

List of Abbreviations

AACs	Aminoglycoside acetyl transferases
AcCoA	Acetyl coenzyme A
AGs	Aminoglycosides
AHB	(*S*)-4-Amino-2-hydroxybutanoyl
AMEs	Aminoglycoside-modifying enzymes
APH	Aminoglycoside phosphotransferase
Bn	Benzyl
Bz	Benzoyl
BzOBT	Benzoyloxybenzotriazole
Cbz	Carboxybenzyl carbamates
Cbz-NOR	*N*-Benzyloxycarbonyloxy-5-norbornene-2,3-dicarboximide
CoA	Coenzyme A
CSA	Camphorsulfonic acid
DCC	*N*,*N'*-Dicyclohexylcarbodiimide
DCM	Dichloromethane
DIPEA	*N*,*N*-Diisopropylethylamine
4-DMAP	4-Dimethylaminopyridine
DOS	Deoxystreptamine
HBTU	2-(1*H*-Benzotriazol-1-yl)-1,1,3,3-tetramethyluronium hexafluorophosphate
HDV	Hepatitis delta virus
HIV	Human immunodeficiency virus
HOBt	Hydroxybenzotriazole
KHMDS	Potassium hexamethyldisilazane
LPS	Lipopolysaccharide
*m*CPBA	*meta*-Chloroperoxybenzoic acid
MOMCl	Chloromethyl methyl ether
MS	Molecular sieves
NHS	*N*-Hydroxysuccinimide
NIS	*N*-Iodosuccinimide
Rev	Regulator of virion expression
RRE	Rev response element
rt	Room temperature
TBAF	Tetrabutylammonium fluoride
TBAI	Tetrabutylammonium iodide

TBDMS	*tert*-butyldimethylsilyl
t-Boc	*tert*-butyloxycarbonyl
TBSOTf	*tert*-butyldimethylsilyl trifluoromethansulfonate
TBTU	2-(1*H*-benzotriazole-1-yl)-1,1,3,3-tetramethyluronium tetrafluoroborate
TES	Triethylsilyl
TfOH	Trifluoromethanesulfonic acid
TLC	Thin layer chromatography
TosCl	4-Toluenesulfonyl chloride
TsOH	*p*-Toluenesulfonic acid
TrCl	Trityl chloride (triphenyl methylchloride)

References

1. Davis, B.D. (1987) *Microbiol. Rev.*, **51**, 341–350.
2. Magnet, S. and Blanchard, J.S. (2005) *Chem. Rev.*, **105**, 477–498.
3. Ogle, J.M., Brodersen, D.E., Clemons, W.M., Tarry, M.J., Carter, A.P., and Ramakrishnan, V. (2001) *Science*, **292**, 897–902.
4. Vicens, Q. and Westhof, E. (2003) *Biopolymers*, **70**, 42–57.
5. Vakulenko, S.B. and Mobashery, S. (2003) *Clin. Microbiol. Rev.*, **16**, 430–450.
6. Schatz, A., Bugie, E., and Waksman, S.A. (1944) *Proc. Soc. Exp. Biol. Med.*, **55**, 66–69.
7. Davies, J. and Wright, G.D. (1997) *Trends Microbiol.*, **5**, 234–240.
8. Jana, S. and Deb, J.K. (2006) *Appl. Microbiol. Biotechnol.*, **70**, 140–150.
9. Lambert, P.A. (2005) *Adv. Drug Delivery Rev.*, **57**, 1471–1485.
10. Marquez, B. (2005) *Biochimie*, **87**, 1137–1147.
11. Forge, A. and Schacht, J. (2000) *Audiol. Neuro-Otol.*, **5**, 3–22.
12. Sciences, B. (2004) *Drug Metab. Pharmacokin.*, **19**, 159–170.
13. Moazed, D. and Noller, H.F. (1987) *Nature*, **327**, 389–394.
14. Zapp, M.L., Stern, S., and Green, M.R. (1993) *Cell*, **74**, 969–978.
15. von Ahsen, U., Davies, J., and Schroeder, R. (1992) *J. Mol. Biol.*, **226**, 935–941.
16. Clouet-d'Orval, B., Stage, T.K., and Uhlenbeck, O.C. (1995) *Biochemistry*, **34**, 11186–11190.
17. Rogers, J., Chang, A.H., von Ahsen, U., Schroeder, R., and Davies, J. (1996) *J. Mol. Biol.*, **259**, 916–925.
18. Kaufman, R.J. (1999) *J. Clin. Invest.*, **104**, 367–368.
19. Manuvakhova, M., Keeling, K.I.M., and Bedwell, D.M. (2000) *RNA*, **6**, 1044–1055.
20. Kellermayer, R. (2006) *Eur. J. Med. Genet.*, **49**, 445–450.
21. Rowe, S.M., Sloane, P., Tang, L.P., Backer, K., Mazur, M., Buckley-Lanier, J., Nudelman, I., Belakhov, V., Bebok, Z., Schwiebert, E., Baasov, T., and Bedwell, D.M. (2011) *J. Mol. Med.*, **89**, 1149–1161.
22. Nudelman, I., Glikin, D., Smolkin, B., Hainrichson, M., Belakhov, V., and Baasov, T. (2010) *Bioorg. Med. Chem.*, **18**, 3735–3746.
23. Baussanne, I., Bussière, A., Halder, S., Ganem-Elbaz, C., Ouberai, M., Riou, M., Paris, J.-M., Ennifar, E., Mingeot-Leclercq, M.-P., and Décout, J.-L. (2010) *J. Med. Chem.*, **53**, 119–127.
24. Bera, S., Zhanel, G.G., and Schweizer, F. (2010) *J. Med. Chem.*, **53**, 3626–3631.
25. Ouberai, M., El Garch, F., Bussiere, A., Riou, M., Alsteens, D., Lins, L., Baussanne, I., Dufrêne, Y.F., Brasseur, R., Decout, J.-L., and Mingeot-Leclercq, M.-P. (1808) *Biochim. Biophys. Acta*, **2011**, 1716–1727.

26. Michael, K., Wang, H., and Tor, Y. (1999) *Bioorg. Med. Chem.*, **7**, 1361–1371.
27. Van Schepdael, A., Delcourt, J., Mulier, M., Busson, R., Verbist, L., Vanderhaeghe, H.J., Mingeot-Leclercq, M.P., Tulkens, P.M., and Claes, P.J. (1991) *J. Med. Chem.*, **34**, 1468–1475.
28. Shitara, T., Kobayashi, Y., Tsuchiya, T., and Umezawa, S. (1992) *Carbohydr. Res.*, **232**, 273–290.
29. Bastida, A., Hidalgo, A., Chiara, J.L., Torrado, M., Corzana, F., Pérez-Cañadillas, J.M., Groves, P., Garcia-Junceda, E., Gonzalez, C., and Jimenez-Barbero, J. (2006) *J. Am. Chem. Soc.*, **128**, 100–116.
30. Nyffeler, P.T., Liang, C.-H., Koeller, K.M., and Wong, C.-H. (2002) *J. Am. Chem. Soc.*, **124**, 10773–10778.
31. Alper, P.B., Huang, S.-C., Wong, C.-H. (1996) *Tet. Lett.*, **37**, 6029–6032.
32. Shaul, P., Green, K.D., Rutenberg, R., Kramer, M., Berkov-Zrihen, Y., Breiner-Goldstein, E., Garneau-Tsodikova, S., and Fridman, M. (2011) *Org. Biomol. Chem.*, **9**, 4057–4063.
33. Chen, L., Hainrichson, M., Bourdetsky, D., Mor, A., Yaron, S., and Baasov, T. (2008) *Bioorg. Med. Chem.*, **16**, 8940–8951.
34. Kawaguchi, H. and Naito, T. (1972) *J. Antibiot.*, **25**, 695–708.
35. Yan, X., Gao, F., Yotphan, S., Bakirtzian, P., and Auclair, K. (2007) *Bioorg. Med. Chem.*, **15**, 2944–2951.
36. Nam, G., Kim, S.H., Kim, J., Shin, J., and Jang, E. (2002) *Org. Process Res. Dev.*, **6**, 206–208.
37. Sainlos, M., Belmont, P., Vigneron, J.-P., Lehn, P., and Lehn, J.-M. (2003) *Eur. J. Org. Chem.*, **15**, 2764–2774.
38. Roestamadji, J., Grapsas, I., and Mobashery, S. (1995) *J. Am. Chem. Soc.*, **117**, 11060–11069.
39. Nagabhushan, T.L., Cooper, A.B., Turner, W.N., Tsai, H., McCombie, S., Mallams, A.K., Rane, D., Wright, J.J., Reichert, P., Boxler, D.L., and Weinstein, J. (1978) *J. Am. Chem. Soc.*, **100**, 5253–5254.
40. Tsuchiya, T., Takagi, Y., and Umezawa, S. (1979) *Tetrahedron Lett.*, **51**, 4951–4954.
41. Hanessian, S. and Patil, G. (1978) *Tetrahedron Lett.*, **12**, 1031–1034.
42. Kirst, H.A., Truedell, B.A., and Toth, J.E. (1981) *Tetrahedron Lett.*, **22**, 295–298.
43. Kotretsou, S., Mingeot-Leclercq, M.P., Constantinou-Kokotou, V., Brasseur, R., Georgiadis, M.P., and Tulkens, P.M. (1995) *J. Med. Chem.*, **38**, 4710–4719.
44. Aggen, J.B., Armstrong, E.S., Goldblum, A.A., Dozzo, P., Linsell, M.S., Gliedt, M.J., Hildebrandt, D.J., Feeney, L.A., Kubo, A., Matias, R.D., Lopez, S., Gomez, M., Wlasichuk, K.B., Diokno, R., Miller, G.H., and Moser, H.E. (2010) *Antimicrob. Agents Chemother.*, **54**, 4636–4642.
45. Reimann, H. and Jaret, R. (1976) *Infection*, **4**, S289–S291.
46. Cohen, S., Lacher, J.R., and Park, J.D. (1959) *J. Am. Chem. Soc.*, **81**, 3480.
47. Greenberg, W.A., Priestley, E.S., Sears, P.S., Alper, P.B., Rosenbohm, C., Hendrix, M., Hung, S., and Wong, C. (1999) *J. Am. Chem. Soc.*, 6527–6541.
48. Kling, D., Hesek, D., Shi, Q., and Mobashery, S. (2007) *J. Org. Chem.*, **72**, 5450–5453.
49. Gernigon, N., Bordeau, V., Berrée, F., Felden, B., and Carboni, B. (2012) *Org. Biomol. Chem.*, **10**, 4720–4730.
50. Hamasaki, K., Woo, M.-C., and Ueno, A. (2000) *Tetrahedron Lett.*, **41**, 8327–8332.
51. Liang, F.-S., Wang, S.-K., Nakatani, T., and Wong, C.-H. (2004) *Angew. Chem. Int. Ed.*, **43**, 6496–6500.
52. Park, W.K.C., Auer, M., Jaksche, H., and Wong, C.-H. (1996) *J. Am. Chem. Soc.*, **118**, 10150–10155.
53. Ding, Y., Swayze, E.E., Hofstadler, S.A., and Griffey, R.H. (2000) *Tetrahedron Lett.*, **41**, 4049–4052.
54. Wu, B., Yang, J., He, Y., and Swayze, E.E. (2002) *Org. Lett.*, **4**, 3455–3458.
55. Oda, T., Mori, T., and Kyotani, Y. (1971) *J. Antibiot.*, **24**, 503–510.
56. Oda, T., Mori, T., Kyotani, Y., and Nakayama, M. (1971) *J. Antibiot. (Tokyo)*, **24**, 511–518.

57. Chou, C.-H., Wu, C.-S., Chen, C.-H., Lu, L.-D., Kulkarni, S.S., Wong, C.-H., and Hung, S.-C. (2004) *Org. Lett.*, **6**, 585–588.
58. Hanessian, S., Tremblay, M., and Swayze, E.E. (2003) *Tetrahedron*, **59**, 983–993.
59. Nudelman, I., Rebibo-Sabbah, A., Shallom-Shezifi, D., Hainrichson, M., Stahl, I., Ben-Yosef, T., and Baasov, T. (2006) *Bioorg. Med. Chem.*, **16**, 6310–6315.
60. Haddad, J., Kotra, L.P., Llano-Sotelo, B., Kim, C., Azucena, E.F., Liu, M., Vakulenko, S.B., Chow, C.S., and Mobashery, S. (2002) *J. Am. Chem. Soc.*, **124**, 3229–3237.
61. Hanessian, S., Takamoto, T., Massé, R., and Patil, G. (1977) *Can. J. Chem.*, **56**, 1482–1491.
62. François, B., Szychowski, J., Adhikari, S.S., Pachamuthu, K., Swayze, E.E., Griffey, R.H., Migawa, M.T., Westhof, E., and Hanessian, S. (2004) *Angew. Chem. Int. Ed.*, **43**, 6735–6738.
63. Rai, R., Mcalexander, I., and Chang, C.W. (2005) *Org. Prep. Proced. Int.*, **37**, 337–375.
64. Wang, J., Li, J., Chen, H.-N., Chang, H., Tanifum, C.T., Liu, H.-H., Czyryca, P.G., and Chang, C.-W.T. (2005) *J. Med. Chem.*, **48**, 6271–6285.
65. Suami, T., Nishiyama, S., Ishikawa, Y., and Katsura, S. (1977) *Carbohydr. Res.*, **56**, 415–418.
66. Suami, T., Nishiyama, S., Ishikawa, Y., and Katsura, S. (1976) *Carbohydr. Res.*, **52**, 187–196.
67. Suami, T., Nishiyama, S., Ishikawa, Y., and Katsura, S. (1977) *Carbohydr. Res.*, **53**, 239–246.
68. Yao, S., Sgarbi, P.W.M., Marby, K.A., Rabuka, D., O'Hare, S.M., Cheng, M.L., Bairi, M., Hu, C., Hwang, S.-B., Hwang, C.-K., Ichikawa, Y., Searsc, P., and Sucheck, S.J. (2004) *Bioorg. Med. Chem. Lett.*, **14**, 3733–3738.
69. Chang, C.-W.T., Hui, Y., Elchert, B., Wang, J., Li, J., and Rai, R. (2002) *Org. Lett.*, **4**, 4603–4606.
70. Elchert, B., Li, J., Wang, J., Hui, Y., Rai, R., Ptak, R., Ward, P., Takemoto, J.Y., Bensaci, M., and Chang, C.-W.T. (2004) *J. Org. Chem.*, **69**, 1513–1523.
71. Li, J., Wang, J., Czyryca, P.G., Chang, H., Orsak, T.W., Evanson, R., and Chang, C.-W.T. (2004) *Org. Lett.*, **6**, 1381–1384.
72. Fridman, M., Belakhov, V., Yaron, S., and Baasov, T. (2003) *Org. Lett.*, **5**, 3575–3578.
73. Hanessian, S., Pachamuthu, K., Szychowski, J., Giguère, A., Swayze, E.E., Migawa, M.T., François, B., Kondo, J., and Westhof, E. (2010) *Bioorg. Med. Chem. Lett.*, **20**, 7097–7101.
74. Bera, S., Dhondikubeer, R., Findlay, B., Zhanel, G.G., and Schweizer, F. (2012) *Molecules*, **17**, 9129–9141.
75. Weghuber, J., Aichinger, M.C., Brameshuber, M., Wieser, S., Ruprecht, V., Plochberger, B., Madl, J., Horner, A., Reipert, S., Lohner, K., Henics, T., and Schütz, G.J. (2011) *Biochim. Biophys. Acta*, **1808**, 2581–2590.
76. Epand, R.F., Savage, P.B., and Epand, R.M. (2007) *Biochim. Biophys. Acta*, **1768**, 2500–2509.
77. Silhavy, T.J., Kahne, D., and Walker, S. (2010) *Cold Spring Harbor Perspect. Biol.*, **2**, a000414.
78. Swoboda, J.G., Campbell, J., Meredith, T.C., and Walker, S. (2010) *ChemBioChem*, **11**, 35–45.
79. Bera, S., Zhanel, G.G., and Schweizer, F. (2008) *J. Med. Chem.*, **51**, 6160–6164.
80. Herzog, I.M., Green, K.D., Berkov-Zrihen, Y., Feldman, M., Vidavski, R.R., Eldar-Boock, A., Satchi-Fainaro, R., Eldar, A., Garneau-Tsodikova, S., and Fridman, M. (2012) *Angew. Chem. Int. Ed.*, **124**, 5750–5754.
81. Herzog, I.M., Feldman, M., Eldar-Boock, A., Satchi-Fainaro, R., and Fridman, M. (2012) *Med. Chem. Commun.* doi: 10.1039/c2md20162c.
82. Riguet, E., Désiré, J., Bailly, C., and Décout, J.-L. (2004) *Tetrahedron*, **60**, 8053–8064.
83. Jana, S., Karan, G., and Deb, J.K. (2005) *Protein Expres. Purif.*, **40**, 86–90.
84. Pedersen, L.C., Benning, M.M., and Holden, H.M. (1995) *Biochemistry*, **34**, 13305–133011.

85. Burk, D.L., Xiong, B., Breitbach, C., and Berghuis, A.M. (2005) *Acta Crystallogr.*, **D61**, 1273–1279.
86. Burk, D.L., Ghuman, N., Wybenga-groot, L.E., and Berghuis, A.M. (2003) *Prot. Sci.*, **12**, 426–437.
87. Rai, R., Chen, H.-N., Czyryca, P.G., Li, J., and Chang, C.-W.T. (2006) *Org. Lett.*, **8**, 887–889.
88. Kondo, S., Iinuma, K., Yamamotoy, H., Ikeda, Y., Maeda, K., and Umezawa, H. (1973) *J. Antibiot.*, **26**, 705–707.
89. Li, J., Chiang, F.-I., Chen, H.-N., and Chang, C.-W.T. (2007) *J. Org. Chem.*, **72**, 4055–4066.
90. Ikeda, D., Tsuchiya, T., Umezawa, S., and Umezawa, H. (1972) *J. Antibiot.*, **25**, 741–742.
91. Kondo, S., Iinuma, K., Yamamotoy, H., Ikeda, Y., Maeda, K., and Umezawa, H. (1973) *J. Antibiot.*, **26**, 412–415.
92. Green, K.D., Chen, W., Houghton, J.L., Fridman, M., and Garneau-Tsodikova, S. (2010) *ChemBioChem*, **11**, 119–126.
93. Tsitovich, P.B., Pushechnikov, A., French, J.M., and Disney, M.D. (2010) *ChemBioChem*, **11**, 1656–1660.
94. Llewellyn, N.M., Li, Y., and Spencer, J.B. (2007) *Chem. Biol.*, **14**, 379–386.
95. Llewellyn, N.M. and Spencer, J.B. (2008) *Chem. Commun.*, 3786–37888.
96. Nudelman, I., Chen, L., Llewellyn, N.M., Sahraoui, E.-H., Cherniavsky, M., Spencer, J.B., and Baasov, T. (2008) *Adv. Synth. Catal.*, **350**, 1682–1688.
97. Kondo, S. and Hotta, K. (1999) *J. Infect. Chemother.*, **5**, 1–9.
98. Haddad, J., Vakulenko, S., and Mobashery, S. (1999) *J. Am. Chem. Soc.*, **121**, 11922–11923.
99. McKay, G.A., Roestamadji, J., Mobashery, S., and Wright, G.D. (1996) *Antimicrob. Agents Chemother.*, **40**, 2648–2650.
100. Vong, K., Tam, I.S., Yan, X., and Auclair, K. (2012) *ACS Chem. Biol.*, **7**, 470–475.
101. Gao, F., Yan, X., Baettig, O.M., Berghuis, A.M., and Auclair, K. (2005) *Angew. Chem. Int. Ed.*, **44**, 6859–6862.
102. Endimiani, A., Hujer, K.M., Hujer, A.M., Armstrong, E.S., Choudhary, Y., Aggen, J.B., and Bonomo, R.A. (2009) *Antimicrob. Agents Chemother.*, **53**, 4504–4507.
103. Cass, R.T., Brooks, C.D., Havrilla, N.A., Tack, K.J., Borin, M.T., Young, D., and Bruss, J.B. (2011) *Antimicrob. Agents Chemother.*, **55**, 5874–5880.
104. Nudelman, I., Rebibo-Sabbah, A., Cherniavsky, M., Belakhov, V., Hainrichson, M., Chen, F., Schacht, J., Pilch, D.S., Ben-Yosef, T., and Baasov, T. (2009) *J. Med. Chem.*, **52**, 2836–2845.

7
Synthesis of Natural and Nonnatural Heparin Fragments: Optimizations and Applications toward Modulation of FGF2-Mediated FGFR Signaling

Pierre-Alexandre Driguez

7.1
Introduction

Heparin (HPN) and its structurally close analog heparan sulfate (HS) are natural, linear anionic polysaccharides that belong to the family of glycosaminoglycans (GAGs). They are both composed of repeated disaccharidic sequences of a uronic acid (D-glucuronic or L-iduronic) and D-glucosamine (GlcN) linked through (1 → 4)-glycosidic bonds (Figure 7.1).

Considerable heterogeneity with regard to their structure, given that hydroxyls may be *O*-sulfonated at the 2-position of uronic acids as well as at the 3- and 6-positions of glucosamine, and that amino groups can be *N*-acetylated (NA), *N*-sulfated (NS), or nonsubstituted, means that a large number of combinations in nature are possible, leading to GAG-specific properties. In particular, sulfate groups, like carboxylates, give high negative charges both locally and through patterns that impact GAG conformation and ability to interact with key biological targets [1]. Such interactions are involved in many essential biological processes like blood coagulation, cell–cell communication, cell growth, proliferation, migration, and differentiation and may also result in several pathological conditions including inflammation, cancer, tumoral angiogenesis, neurodegenerative diseases, atherosclerosis, rheumatoid arthritis, and parasite infections. Some of these have been well documented in the literature, as is the case with serpins, proteases, fibroblast growth factors (FGFs), chemokines, adhesion proteins, and viral pathogens [2–5]. A good illustration is the identification and total synthesis of the highly functionalized HPN pentasaccharide that targets the anticoagulation factor Xa through activation of the serine protease inhibitor antithrombin with high selectivity [6]. However, apart from their extensively studied anticoagulation properties, and despite major efforts devoted to study their biological properties, understanding the physiological roles and functions of GAGs is still a major goal in glycobiology. As part of our research program devoted to the synthesis of well-defined HPN fragments, we were particularly interested in their interactions with fibroblast growth factor 2 (FGF-2) and their receptors in relation with angiogenetic processes [5].

Modern Synthetic Methods in Carbohydrate Chemistry: From Monosaccharides to Complex Glycoconjugates, First Edition. Edited by Daniel B. Werz and Sébastien Vidal.

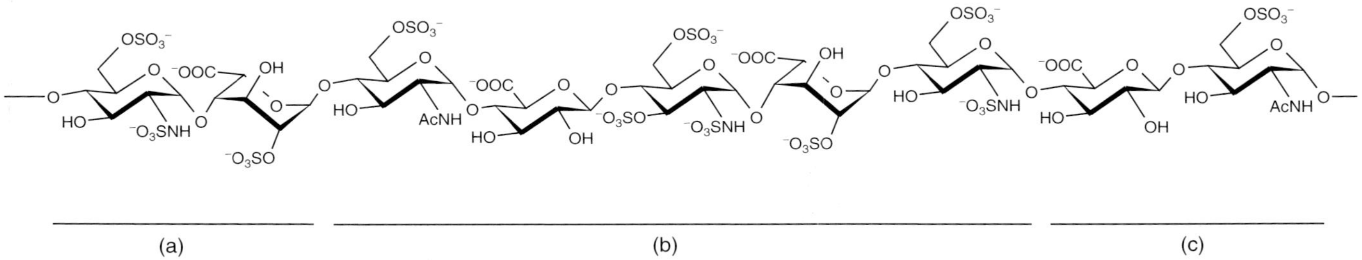

Figure 7.1 Schematic representation of HPN/HS polysaccharide. (a) Main HPN sequence, (b) HPN anticoagulant pentasaccharide, and (c) main HS sequence.

FGFs belong to a 23-member family of proteins involved in multiple roles in development and metabolism of numerous organs in animals [7]. The formation of a ternary complex involving HPN/HS, FGFs, and their tyrosine kinase receptor is a prerequisite for activation of the signal transduction cascade [8], through a symmetrical topology in a 2 : 2 : 2 ratio [9–11] for FGFR1 (fibroblast growth factor receptor) or R4-FGF2, or through a 1 : 2 : 2 stoichiometry model for FGFR2c-FGF1 [12].

Even if several studies have been performed with modified HPN/HS fragments obtained by chemo-enzymatic methods [13], or using chemistry for the preparation of the oligosaccharidic core followed by final enzymatic transformations [14], only synthetic aspects yielding highly pure fragments of repetitive HPN or HS sequences (Figure 7.1) of the so-called regular region, devoid of anticoagulant properties, are considered in this chapter.

Several preparations of fully synthetic oligosaccharides derived from the main HPN sequence have been described in the literature [15, 16]. The general strategies rely on careful selection of protecting groups on glucosamine and rare iduronic acid precursors, whose preparation have been recently reviewed and will therefore not be detailed here [17, 18]. In many cases, acetates or benzoates are chosen to protect hydroxyls to be sulfonated, whereas benzyl ethers allow temporary protection of hydroxyls that will be nonsubstituted in the final compounds. Owing to their convenient nonparticipating character in coupling reactions for the generation of α-D-glycosides, azides are often precursors of amines, but carbamates may also be found, as illustrated in this chapter. Because of the nature of the backbone, disaccharide building blocks are usually oligomerized in an iterative manner from the reducing end to the nonreducing end. Trichloroacetimidates or thioglycosides are the most common glycosyl donors that react with the help of an acidic catalyst on a free hydroxyl at position 4 of a glycosyl acceptor, which, in numerous cases, has been orthogonally deprotected from a levulinoyl ester or a *para*-methoxy aryl ether, except in one-pot synthesis that requires either a well-defined reactivity [19], or a preactivation of selected glycosyl donors [20]. Position 6 of idose may be oxidized into an iduronic acid either at the monosaccharidic [21, 22] or at the disaccharidic building block level [23], or once the elongation process is terminated [19, 20, 24]. The anomeric position of the reducing unit is permanently blocked through an acetal with methyl [21], or longer alkyl chains that may possibly be functionalized by an amine to allow engraftment of small amounts (<1 nmol) of saccharides onto solid plates for use in assays that require only few picomoles of protein for biological screening [16, 20, 25–27].

7.2
Total Synthesis of Standard HPN Fragments

A typical synthetic strategy is depicted in Scheme 7.1 for the preparation of tetrasaccharide **1** and hexasaccharide **2** [28].

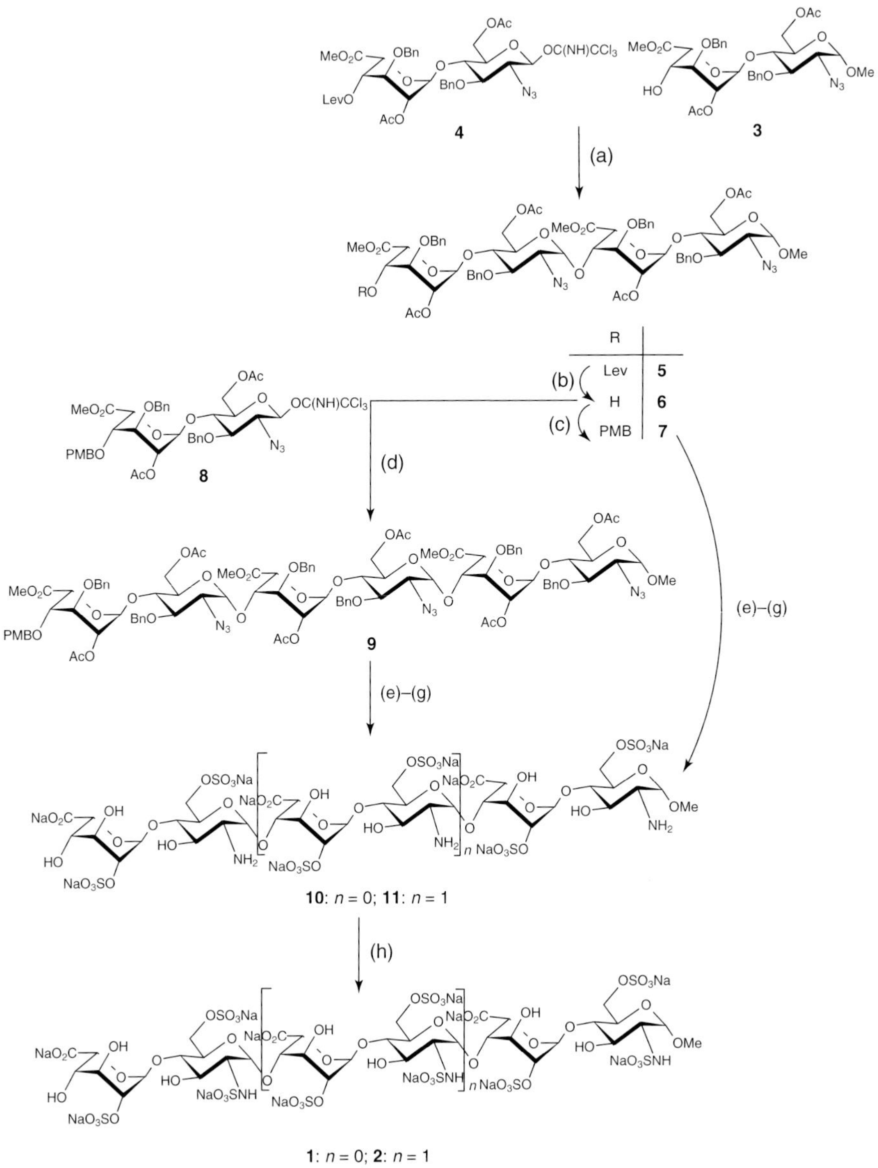

Scheme 7.1 Synthesis of tetrasaccharide **1** and hexasaccharide **2** according to [28]. *Reagents*: (a) TBSOTf (cat), 4 Å MS, CH_2Cl_2, −10 °C, 20 min, 82%; (b) NH_2NH_2, HOAc, pyridine, rt, 100%; (c) PMB trichloroacetimidate, TfOH (triflic acid) (cat), Et_2O/CH_2Cl_2, rt, 2 h, 86%; (d) TBSOTf (cat), 4 Å MS, toluene, −20 °C, 3 h, 80%; (e) aq. LiOH (0.7 M), H_2O_2, THF, −5 °C 1 h, 0 °C 2 h, then rt 16 h, followed by addition of MeOH and 4N NaOH, 0 °C then rt, 12 h; (f) $Et_3N{\cdot}SO_3$, DMF, 50 °C, 16 h; (g) H_2 (60 bar), 10% Pd/C (cat), *tert*-butanol/H_2O 2/3, 4 h; and (h) pyridine·SO_3, H_2O, 3 h, 46% over the four steps (**2**).

Elongations were performed starting from acceptor disaccharide **3**, with the use of levulinoyl glycosyl trichloroacetimidate **4** [29] in CH_2Cl_2 at $-10\,^{\circ}C$ in the presence of TBSOTf to furnish tetrasaccharide **5** (82%). No β-coupled product was detected by ^1H-NMR spectroscopy. Tetrasaccharide **5** was quantitatively delevulinoylated into **6** with hydrazine in a mixture of pyridine and acetic acid [30], which was reacted in similar conditions either with *para*-methoxybenzyl (PMB) trichloroacetimidate to furnish tetrasaccharide **7** (86%), or with PMB glycosyl trichloroacetimidate **8** to give hexasaccharide **9** (80%). Both tetrasaccharide **7** and hexasaccharide **9** were further deprotected and functionalized under the same conditions. Saponification of acetates with lithium hydroxide in H_2O_2/THF (tetrahydrofuran) was followed by sulfonatation of free hydroxyls with triethylamine–SO_3 complex in DMF (*N,N*-dimethylformamide) at $55\,^{\circ}C$. Finally, catalytic reduction of benzyl and PMB ethers, as well as azides, with H_2 at 60 bar in *tert*-butanol/water, yielded diamine **10** or triamine **11**, which were subjected to pyridine·SO_3 complex in water to generate **1** and **2** (46% over the four steps).

The same general strategy has been recently applied with slight modifications (Scheme 7.2) for the synthesis of octasaccharide **12** [31]. We decided to focus on the size of eight saccharidic units that our data suggest as being the optimal size to enhance FGFR activation and subsequent signaling, which is in agreement with literature data [32], although different results have been reported by others [33].

Elongations were performed from acceptor disaccharide **14** [34], with the use of levulinoyl glycosyl trichloroacetimidate **4** to furnish successively tetra- and hexasaccharides, while 4′-*O*-benzyl glycosyl trichloroacetimidate **13** was used to terminate the elongation and give octasaccharide **15**. The yield of each successive elongation step, catalyzed by TBSOTf, was about 60–80%. The absence of β-anomer-coupling products detected in the reaction mixtures may be explained by the axially oriented hydroxyl at C-4 because of the probable $^1C_4$ conformation of the iduronic acid acceptor [35], which enhances an α-selective formation of the new glycosidic bond [36, 37]. Hydrazine acetate was used for delevulinoylations of tetra- and hexasaccharides (87–94% yield). Octasaccharide **15** was finally deprotected and functionalized as above to generate **12**.

Within the series of unmodified synthetic HPN fragments, the preparation of hexasaccharide **16** and octasaccharide **17** has been reported (Scheme 7.3) [32].

It should be noted that both oligosaccharides differ from **1** and **12** in the carbohydrate unit sequence, which is not available either by standard enzymatic or chemical degradation of HPN. Tetrasaccharide **21** was prepared by elongating nonreducing disaccharide **18** with glycosyl donor **19** in the presence of TMSOTf (79% yield, together with 19% of recovered glycosyl acceptor), followed by benzylidene hydrolysis (75%) and selective benzoylation of position 6 (93%). Coupling of trichloroacetimidate **20** with acceptor **21** (58% yield, together with 36% of recovered glycosyl acceptor) and subsequent deprotection/functionalization reactions finally yielded hexasaccharide **16** (46% overall yield). Octasaccharide **17** was synthesized according to the same strategy, and the effects of both compounds on the mitogenic activity of FGF1 have been assessed.

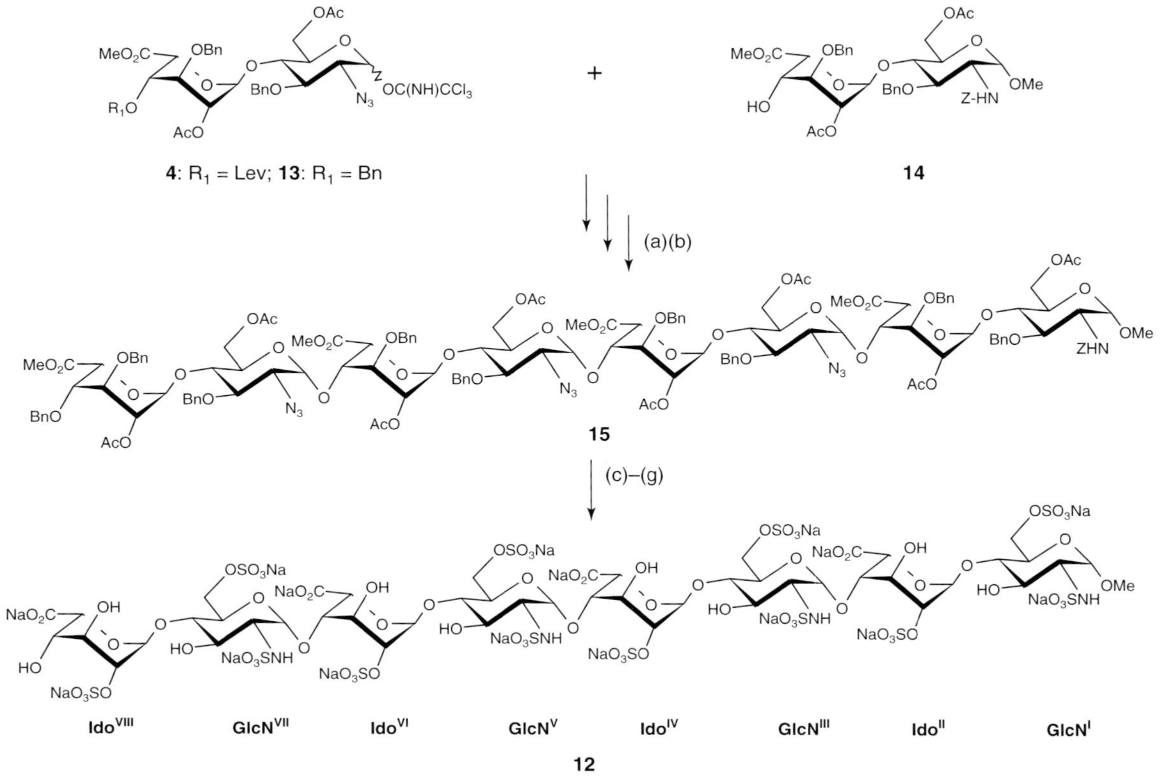

Scheme 7.2 Synthesis of octasaccharide **12**. *General conditions*: (a) Coupling reactions for the preparation of tetra-, hexa-, and octasaccharides: TBSOTf (cat), 4 Å MS, CH_2Cl_2, −20 °C, 1 h, 60–80%; (b) delevulinoylation of tetra- and hexasaccharides: NH_2NH_2·HOAc, toluene/ethanol 1/2, 30 min, 87–94%; (c) transesterification: NaOMe, MeOH; (d) $Et_3N{\cdot}SO_3$, DMF, 55 °C, 16 h; (e) aq. LiOH (0.7 M), THF/MeOH 1/1, 0 °C, 1 h, then rt, 16 h; (f) ammonium formate, 10% Pd/C (cat), *tert*-butanol/H_2O 1/1, 4 h; and (g) $NaHCO_3$, pyridine·SO_3, aq. $NaHCO_3$, 0 °C then rt, 16 h.

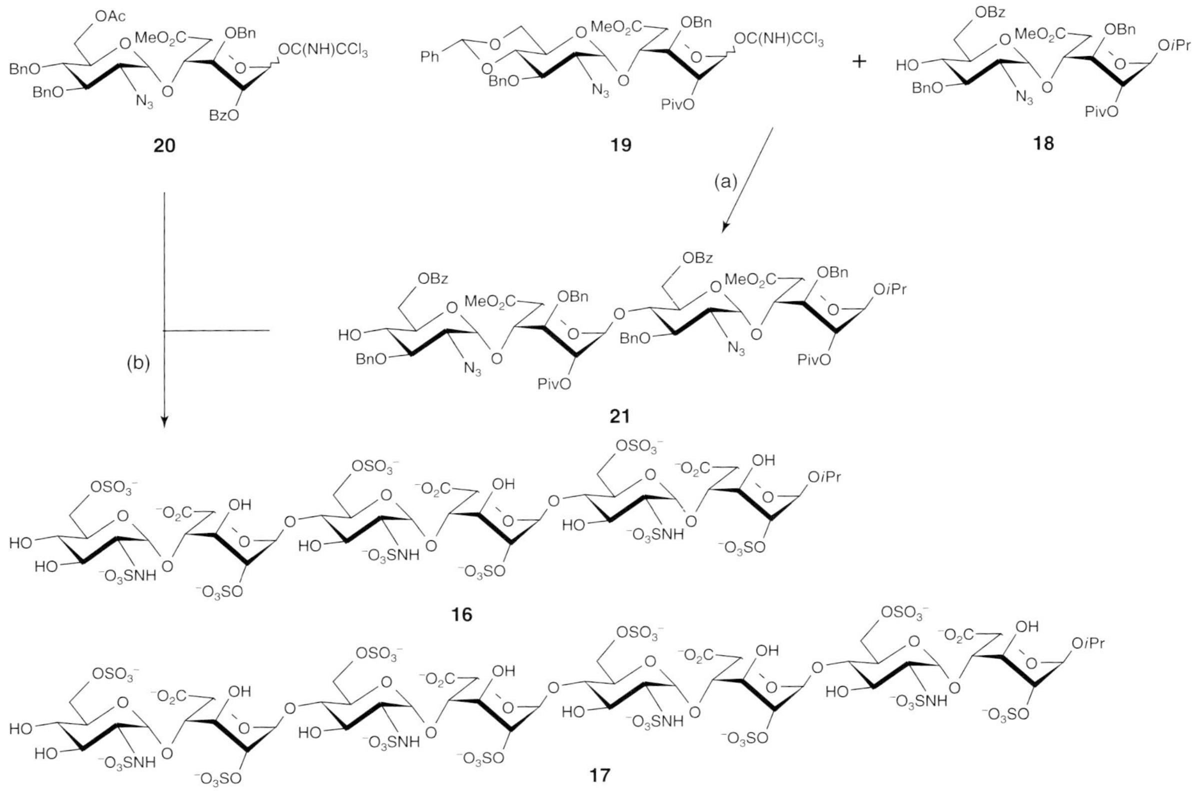

Scheme 7.3 Synthesis of hexasaccharide **16** and octasaccharide **17** according to [32]. *Reagents and conditions*: (a) (i) 3% TMSOTf, CH_2Cl_2, 79% (19% of recovered **18**) and (ii) EtSH, PTSA (*para*-toluenesulfonic acid) (cat), 75%; (iii) BzCN, Et_3N (cat), MeCN, −40 °C, 93%. (b) (i) 3% TMSOTf, 58% (36% of recovered **21**); (ii) KOH, 74%; (iii) $Me_3N{\cdot}SO_3$; Dowex 50WX4 (Na^+), 71%; and (iv) 10% Pd/C, H_2; pyridine·SO_3, 87%.

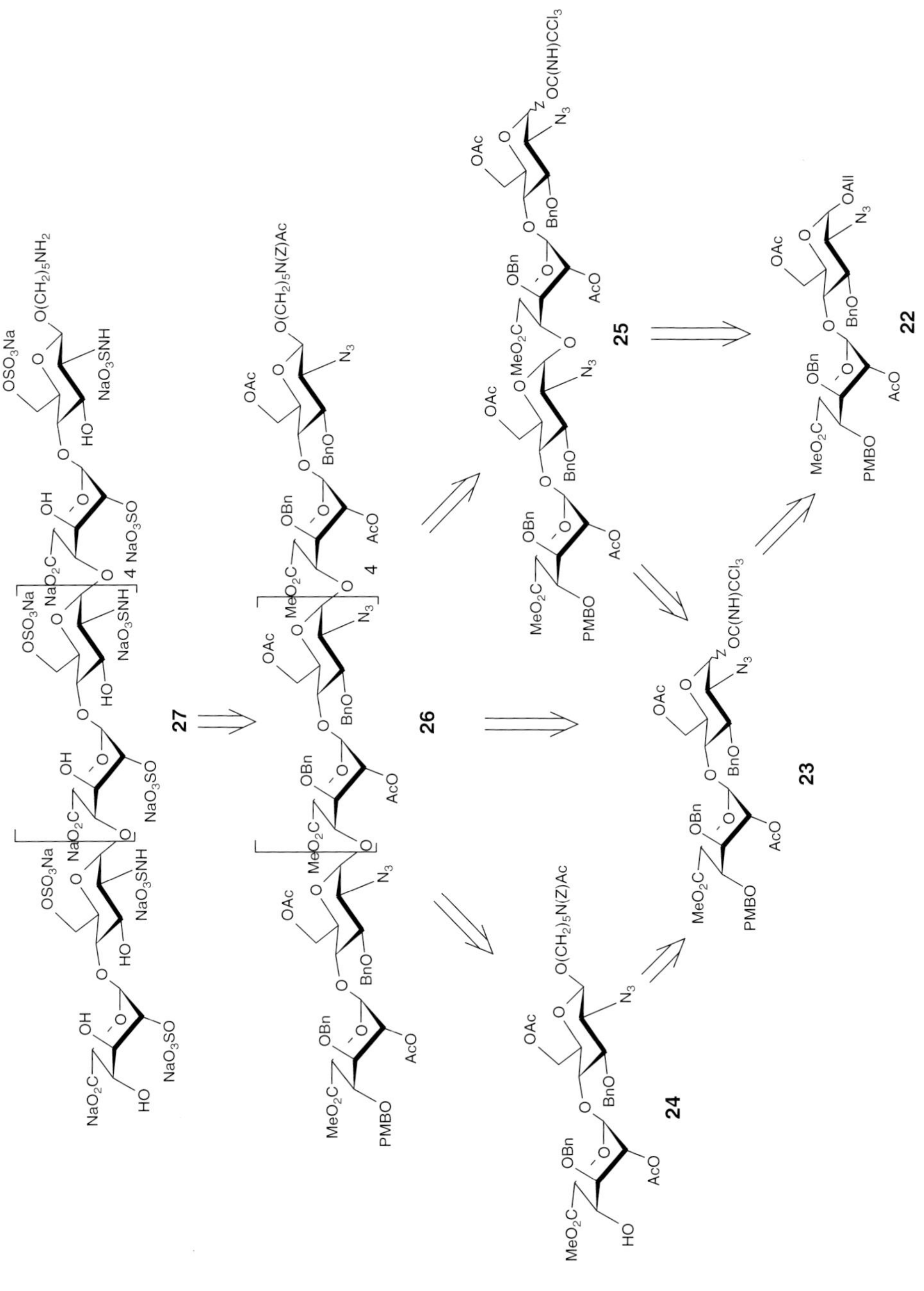

Scheme 7.4 Retrosynthesis of dodecasaccharide **27** from key building block **22** according to [38].

Furthermore, a convergent synthesis based on the ability to oligomerize the suitably protected disaccharidic building block **22** has been developed and allowed the preparation of dodecamer **27** with high yield and stereoselectivity (Scheme 7.4) [38].

Thus, **22** was transformed into trichloroacetimidate **23**, which was glycosylated with 5-(acetamido-*N*-benzyloxycarbonyl)-pentanol using TMSOTf as catalyst. Subsequent trifluoroacetic acid (TFA) hydrolysis of the PMB ether gave glycosyl acceptor **24**, which was elongated into the corresponding tetramer with the help of glycosyl donor **23**. Similarly, the latter has also been used for the elaboration of tetrasaccharidic donor **25** for straightforward synthesis of fully protected dodecasaccharide **26**. As for the above-described synthesis, full α-selectivity has been observed for each glycosylation reaction, and the overall yield of the five-step oligomerization process was 45%. Deprotection and functionalization gave the target **27** in a highly pure form.

7.3 Total Synthesis of Modified HPN Fragments: Some Synthetic Clues

Synthetic compounds are probably the most powerful tool to perform structure–activity relationship (SAR) studies of a ligand to its receptor [25, 39]. However, a range of 65–70 steps are necessary to synthesize octasaccharide **12**. Accurate modifications in its complex structure require tailored-made building blocks that need to be prepared at early stages of the synthesis and, in consequence, the number of steps may increase dramatically, depending on the chemical strategy. One solution is to generate diversification at late stages whenever chemistry allows it, as illustrated in the following examples.

7.3.1 Modifications on the Aglycon Moiety

Preliminary pharmacokinetic studies performed on rats with octasaccharide **12** showed rapid elimination of the compound associated with a short half-life [31]. We thought that the incorporation of lipophilic groups on specific allowed positions, associated with a decrease in the number of nonessential sulfates (Section 7.3.3) would probably generate compounds with more favorable pharmacokinetic parameters, hopefully associated with improved biological activities. However, fully protected octasaccharide **15** as well as all the synthetic intermediates up to the final target **12** (Scheme 7.2) are highly functionalized structures that need to be handled with appropriate care to avoid their partial destruction.

X-ray crystallographic analysis and molecular modeling [31] indicated that the reducing end does not seem to be involved in the FGF2–FGFR1 ternary complex, and therefore was identified as a potential starting point of diversification, in particular for the incorporation of lipophilic groups.

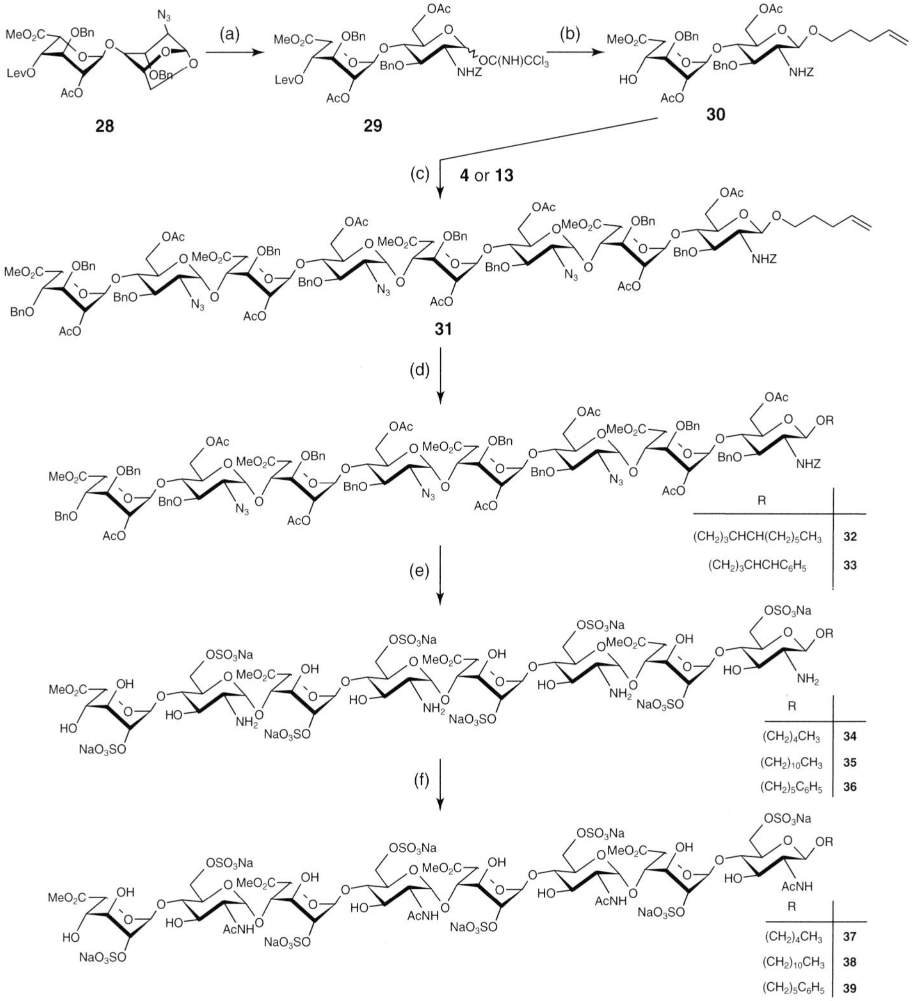

Scheme 7.5 Synthesis of octasaccharides **37–39**. *Reagents*: (a) (i) 1,3-propanedithiol, Et_3N, DMF, 17 h, 76%; (ii) BnOCOCl, $NaHCO_3$, CH_2Cl_2, 0 °C, 15 min, crude; (iii) TFA, Ac_2O, 0 °C then rt, 18 h, 80% (two steps); (iv) $BnNH_2$, Et_2O, 0 °C, 1 h, then rt, 4.5 h, 78%; and (v) Cs_2CO_3, CCl_3CN, rt, CH_2Cl_2, 80%. (b) (i) $CH_2CH(CH_2)_3OH$, TBSOTf (cat), CH_2Cl_2, −20 °C, 20 min, 85% and (ii) $NH_2NH_2 \cdot HOAc$, toluene/ethanol 1/2, 2.5 h, 74%. (c) (i) *General conditions for the preparation of tetra-, hexa-, and octasaccharides*: TBSOTf (cat), 4 Å MS, CH_2Cl_2, −20 °C, 1 h, 58–69% and (ii) *general conditions for delevulinoylation of tetra- and hexasaccharides*: $NH_2NH_2 \cdot HOAc$, toluene/ethanol 1/2, 30 min, 85–96%. (d) for **32**: 1-octene, Grubbs I ($C_{44}H_{74}Cl_2P_2Ru$), CH_2Cl_2, reflux, 6 h, then rt, 16 h; for **33**: styrene, Grubbs II ($C_{46}H_{65}Cl_2N_2PRu$), CH_2Cl_2, reflux, 8 h, then 35 °C, 16 h, 40%. (e) (i) Transesterification: NaOMe, MeOH, 61–69%; (ii) $Et_3N \cdot SO_3$, DMF, 55 °C, 16 h, 91–99%; (iii) aq. LiOH (0.7 M), THF/MeOH 1/1, 0 °C, 1 h, then rt, 16 h, crude and (iv) ammonium formate, 10% Pd/C (cat), *tert*-butanol/H_2O 1/1, 4 h, 49–83% (two steps). (f) Ac_2O, $NaHCO_3$, satd aq. $NaHCO_3$, 0 °C then rt, 16 h, 73–91%.

It is well known from the literature that the olefin metathesis reaction with Grubbs ruthenium catalyst allows smooth organic moieties' engraftment in the presence of a large variety of functional groups [40]. We decided to synthesize octasaccharide **31** as a key intermediate, corresponding to the β-pentenyl analog of α-methyl **15**, and using the conditions described earlier, to assess the chemical feasibility of hydrocarbon group introduction at this stage. Five steps only would then be needed afterward for each of the octasaccharidic glycosides **31–33** to terminate the sequence and obtain the chemical targets **37–39** (Scheme 7.5).

Although a straightforward and efficient method has been described in the literature for its 2-azido α-D-allyl glycoside analog [41], the key pentenyl building block **30** was synthesized from disaccharide **28** [42]. First, the nonparticipating 2-azido group was replaced by a carbamate to capitalize on the anchimeric assistance and avoid an anomeric α/β-mixture during the formation of the pentenyl glycoside. Thus, selective reduction of the azide with 1,3-propanedithiol in the presence of triethylamine [43] yielded the corresponding amine (76%) that was protected using benzyloxycarbonyl chloride under basic conditions. The 1,6-anhydro bridge was acetolyzed with the help of TFA in acetic anhydride (80% yield, two steps); the resulting anomeric acetate reacted selectively with benzylamine, thus giving rise to a hemiacetal (78%) that was transformed into glycosyl donor **29** with trichloroacetonitrile in the presence of cesium carbonate (80% yield). As expected, activation of **29** with Lewis acid yielded pure β-D-pentenyl glucoside, which was subjected to delevulinoylation using hydrazine acetate to give acceptor **30** in 74% yield. The latter was elongated as described for the preparation of **15** with the successive use of glycosyl donors **4** to furnish the tetra- (69%) and hexasaccharides (61%), then **13** to generate the pentenyl octasaccharide **31** (58% yield). At this stage, olefin metathesis reaction was carried out with octene in the presence of Grubbs first-generation catalyst to give **32**, whereas second-generation catalyst gave better results when reacted with styrene to yield **33** (40%). Deprotection and functionalization of **31–33** were performed as previously described for the synthesis of **12**, and yielded glycosides **37–39**, respectively.

7.3.2
Modifications at Position 2 of Glucosamines

Despite previous reports suggesting that *N*-sulfate groups are essential for binding to FGF2 [16, 20, 44], we hypothesized, based on X-ray crystallographic analysis and molecular modeling [31], that such groups on glucosamines would not be critical for ternary complex formation with FGF2/FGFR1. Therefore, another alternative for modulating pharmacokinetic parameters together with biological activity may be found in *N*-sulfate replacement by a nonionic moiety such as an acetamido group, and possibly bulkier amides. As the latter may be introduced at the very last step of the synthesis, such modifications seem attractive from the synthetic point of view. It should be mentioned that the preparation of hexasaccharide **40** and octasaccharide **41**, two synthetic oligosaccharides bearing both *N*-sulfates and acetamides on glucosamines, has been described in the literature [45] (Figure 7.2).

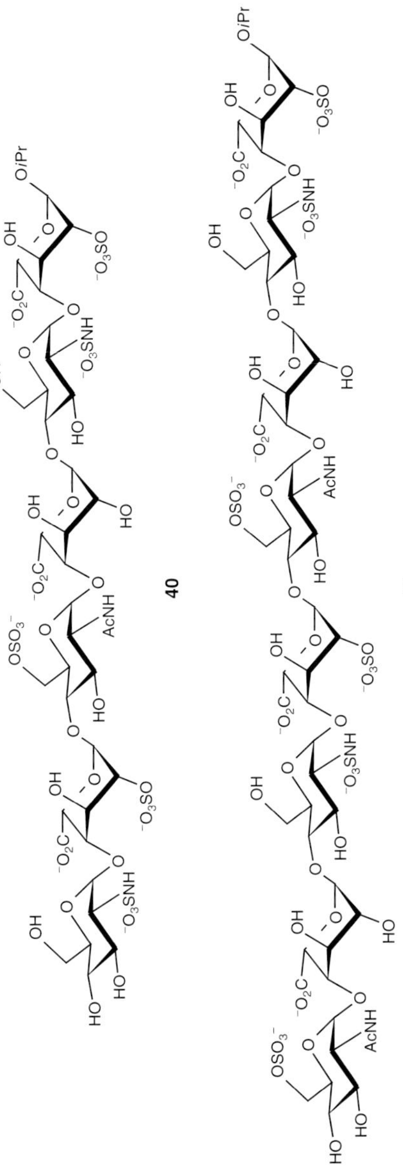

Figure 7.2 Schematic representation of synthetic hexasaccharide **40** and octasaccharide **41** [45].

We decided to use pentyl glycoside **34** as a backbone, which was reacted directly with *N*-hydroxysuccinimide-activated acids in the presence of an organic base in DMF to furnish 1-oxobutylamino derivative **42** (89%), 1-oxohexylamino derivative **43** (46%), and cyclopentylcarbonylamino derivative **44** (55%, Scheme 7.6).

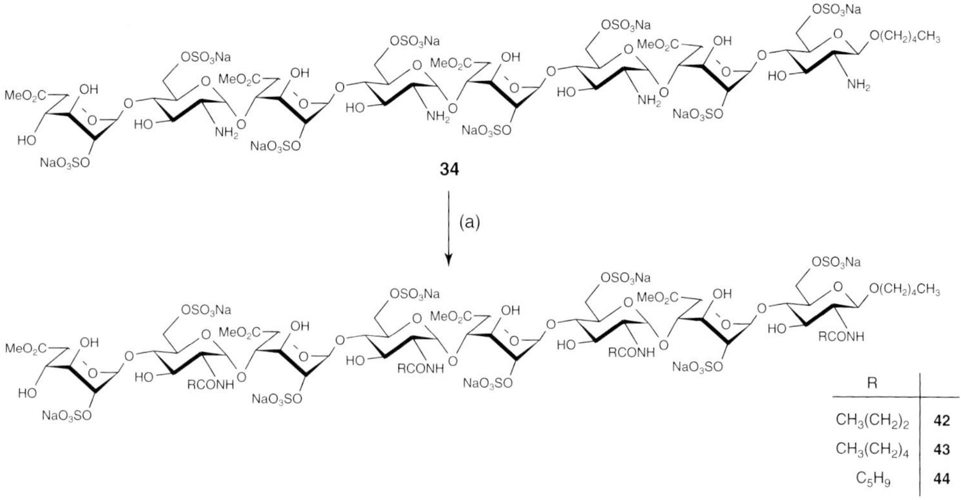

Scheme 7.6 Synthesis of pentyl octasaccharides **42–44**. *Reagents*: (a) RCONHS, diisopropylethylamine, DMF (+H_2O in the case of **42**), 0 °C then rt, 18 h, 46–89%.

7.3.3
Modifications of the *O*-Sulfonatation Pattern

Apart from introducing some lipophilic groups on the saccharidic structure, another obvious option to modulate the pharmacokinetic parameters and also possibly the specificity toward FGF2 and/or its receptors is to lower the overall negative charges on octasaccharide **12** by decreasing the number of *O*-sulfonates, which would also allow identification of critical positions involved in ternary complex formation, if any. In spite of the controversies regarding the beneficial effect of subtle sulfonatation pattern changes [20, 46, 47] versus a cluster of negative charges [5, 48] on FGF-binding specificities, our studies were motivated by some results in the literature showing that a precise arrangement of *O*-sulfonates on hexasaccharide **40** (Figure 7.2) is able to enhance FGF1-mediated mitogenic activity more efficiently than hexasaccharide **16** (Scheme 7.3), which corresponds to the standard repetitive HPN disaccharide [49].

The first *O*-sulfonatation pattern modification on the HPN sequence was performed at position 6 of glucosamine on a disaccharide (compounds **45–46**) [23], on tetrasaccharides (compounds **47–49** [50]), and on trisaccharides (**50–51**, Figure 7.3) [51].

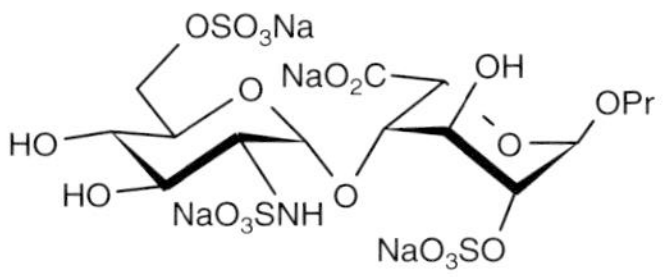

45

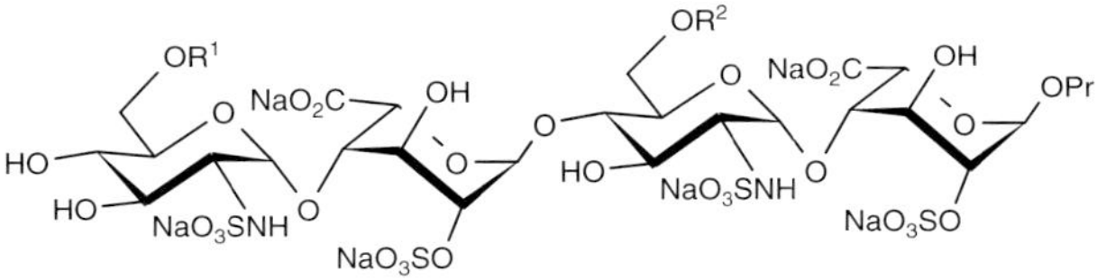

R^1	R^2	
OSO_3Na	OSO_3Na	**47**
H	OSO_3Na	**48**
H	H	**49**

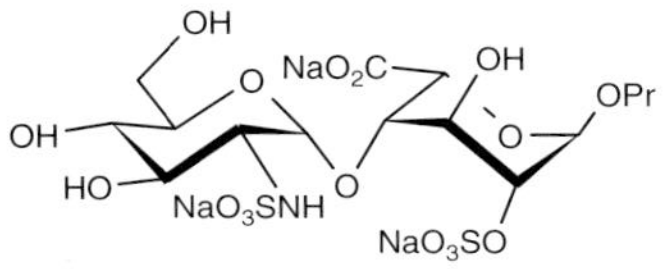

46

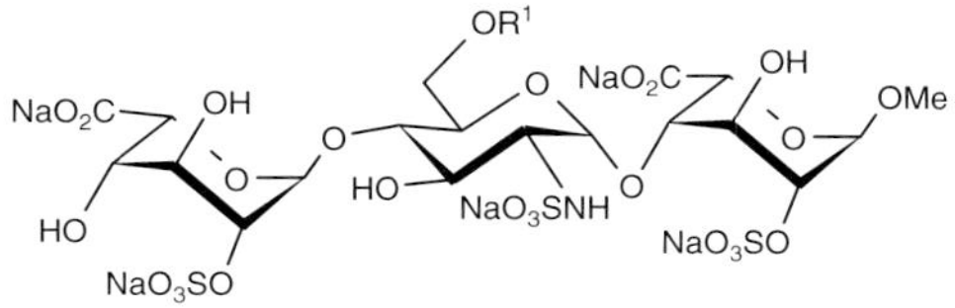

R^1	
OSO_3Na	**50**
H	**51**

Figure 7.3 Schematic representation of synthetic disaccharides **45–46** [23], tetrasaccharides **47–49** [50], and trisaccharides **50–51** [51].

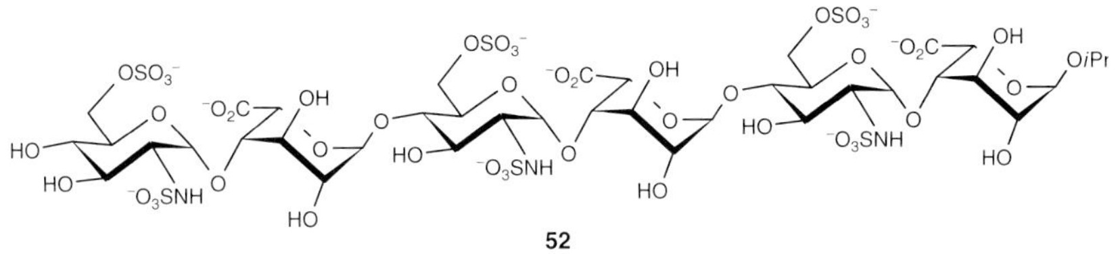

52

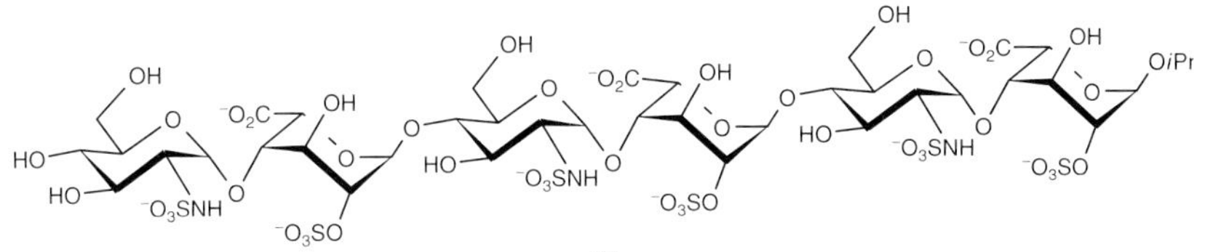

53

Figure 7.4 Schematic representation of synthetic hexasaccharides **52** and **53** [23].

Moreover, the synthesis of hexasaccharides **52–53** and octasaccharide **41**, partly *O*-desulfonated at positions 2 and/or 6 of glucosamine as well as position 2 of iduronic acid have been reported (Figures 7.2 and 7.4) [32, 45, 52].

In order to go a step further in the optimization of the FGF–FGFR–oligosaccharide complex formation and subsequent signaling, we decided to focus our attention on *O*-sulfonatation pattern modifications at the primary positions of the glucosamines. Desulfation of high-valued octasaccharide **12** was not considered as specific sulfate hydrolysis on all different positions is highly improbable using chemical or enzymatic methods. Furthermore, the synthesis of the theoretically 16 final octasaccharides using conventional strategies would be an overwhelming task. We therefore decided to assess the generation of several 6-*O*-desulfonated analogs of **12** from one single orthogonally protected octasaccharidic precursor bearing one group on the four primary positions that display a unique reactivity. To this end, we believed that *tert*-butyldiphenylsilyl (TBDPS) ether groups would be appropriate, capitalizing on the optimization of statistical desilylation conditions for the generation of partially 6-*O*-sulfonated targets after deprotection and functionalization (Scheme 7.7). Therefore, we embarked on the preparation of octasaccharide **59** bearing secondary hydroxyls that were either free, ready for *O*-sulfonatation, or temporarily protected as benzyl ethers that could be regenerated after catalytic reduction.

The strategy selected was based on the preparation of three disaccharide building blocks designed for each end (**56**, **58**) and one for the central part of the molecule (**57**). Both disaccharide donors **56** and **57** were prepared from the corresponding 1,6,2′-triacetates **54** and **55** that had been used for the synthesis of **13** and **4** (Scheme 7.2). Thus, **54** and **55** were selectively deacetylated on the primary positions using neutral organotin catalyst [*tert*-$Bu_2SnOH(Cl)]_2$ to give the corresponding primary alcohols [53, 54], which were substituted without purification into their TBDPS ether using the corresponding chloride under basic conditions (51–58% yield, two steps). Subsequent treatment with benzylamine for anomeric deacetylations (90%),

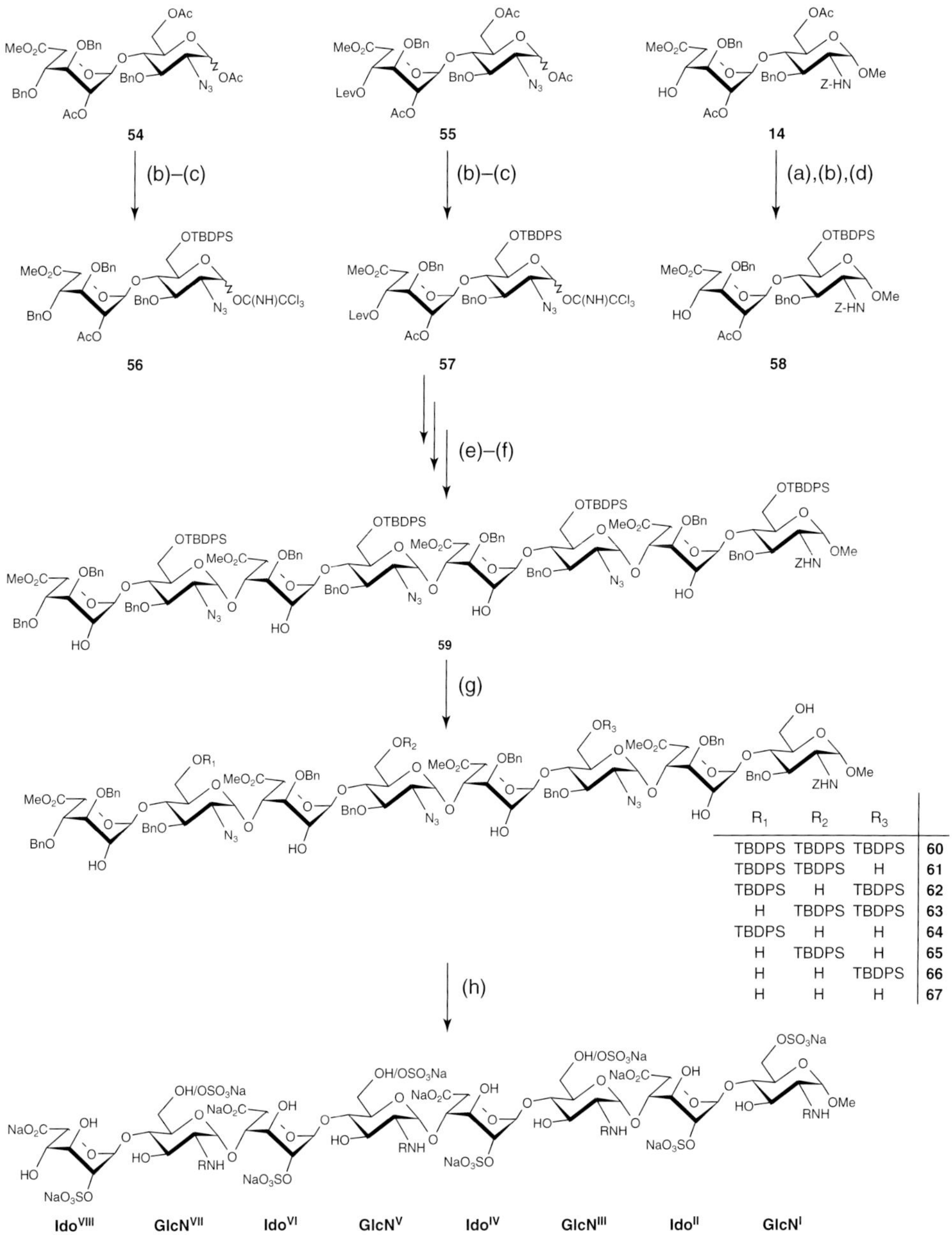

R_1	R_2	R_3	
TBDPS	TBDPS	TBDPS	**60**
TBDPS	TBDPS	H	**61**
TBDPS	H	TBDPS	**62**
H	TBDPS	TBDPS	**63**
TBDPS	H	H	**64**
H	TBDPS	H	**65**
H	H	TBDPS	**66**
H	H	H	**67**

68–78: R = $-SO_3Na$ and/or CH_3CO-

Scheme 7.7 Synthesis of partially 6-*O*-desulfonated octasaccharides **68–78**. *Reagents*: (a) Levulinic acid, DMAP (4-dimethylamino-pyridine), 1-(3-dimetylaminopropyl)-3-ethylcarbodiimide hydrochloride, CH_2Cl_2, 0 °C then rt, 88%. (b) (i) [*tert*-$Bu_2SnCl(OH)]_2$, THF/MeOH 1/1, 38 h, 35 °C, crude; (ii) TBDPSCl, DMAP (cat), Et_3N, CH_2Cl_2, 16 h, 51–94% (two steps); (c) $BnNH_2$, Et_2O, 0 °C, 1 h, then rt, 4.5 h, 90%; and (ii) Cs_2CO_3, CCl_3CN, RT, 85%. (d) NH_2NH_2·HOAc, toluene/ethanol 1/2, 30 min, 90%. (e) *General conditions for the generation of octasaccharides from disaccharides*: (i) coupling reaction: TBSOTf (cat), 4 Å MS, CH_2Cl_2, −20 °C, 1 h, 63–80%; (ii) delevulinoylation: NH_2NH_2·HOAc, toluene/ethanol 1/2, 30 min, 77–95%. (f) NaOMe, MeOH, 94%. (g) NH_4F, MeOH, rt, 20 h. (h) (i) $Et_3N·SO_3$, DMF, 55 °C, 16 h; (ii) NH_4F, MeOH, 55 °C, 48 h; (iii) aq. LiOH (0.7 M), THF/MeOH 1/1, 0 °C, 1 h, then rt, 16 h; (iv) ammonium formate, 10% Pd/C (cat), *tert*-butanol/H_2O 1/1, 4 h; and (v) $NaHCO_3$, pyridine·SO_3, aq. $NaHCO_3$, 0 °C then rt, 16 h, or Ac_2O, $NaHCO_3$, satd aq. $NaHCO_3$, 0 °C, 3 h, then rt, 15 h.

and then with trichloroacetonitrile in the presence of cesium carbonate afforded glycosyl donors **56** and **57** in 85% yield.

Temporary protection of disaccharide **14** [34] at position 4′ with levulinic acid under standard conditions (88% yield) and subsequent selective deacetylation of the primary position using [*tert*-$Bu_2SnOH(Cl)]_2$ afforded a free alcohol at position 6, which was further protected as a TBDPS ether (94% yield over the two steps) and finally treated with hydrazine acetate in toluene/ethanol [55] to yield glycosyl acceptor **58** (90%).

The construction of the octasaccharide scaffold was performed as previously described in this chapter, and the yields were consistent with those given for the above-mentioned syntheses, that is, 63–80% for the coupling reactions and 77–95% for delevulinoylations. The glycosyl acceptor **58** was coupled with trichloroacetimidate **57** in dichloromethane at −20 °C, using *tert*-butyldimethylsilyl (TBDMS) triflate as a catalyst. Hydrazine acetate was used for delevulinoylation to give a tetrasaccharide acceptor that was coupled again with **57** to yield a hexasaccharide. The latter was delevulinoylated, then reacted with capping trichloroacetimidate **56** as explained previously to generate an octasaccharide that was transesterified using Zemplén conditions to finally give compound **59** (94% yield).

The next task was devoted to the statistical desilylation study. The aim of this key reaction was to generate the highest number of octasaccharides bearing free primary hydroxyls prior to *O*-sulfonatation. Several reagents and conditions were tested, and the most satisfactory results were obtained with the use of a large excess of ammonium fluoride (80 M equiv) in methanol. Normal-phase thin layer chromatography (TLC) analysis of the reaction after stirring for 20 h at room temperature showed total disappearance of starting material and highlighted six distinct spots of similar intensities that could easily be separated on silica gel column chromatography and characterized by NMR spectroscopy. Surprisingly, all the analyzed compounds lacked the silyl group at the reducing-end glucosamine unit, showing that this position is the first to be hydrolyzed, but also meaning that only 8 compounds out of the 16 theoretically possible could be obtained using this strategy in a straightforward manner. Luckily, all eight combinations were identified within the six spots. Compounds **60–62** and **67** could be obtained pure after flash column chromatography on silica gel in a 3 : 2 : 2 : 1 ratio. Compounds **63** and **64** were isolated as a 1 : 2 ratio mixture but could be separated after performing a second silica gel chromatography. A C-18 reverse-phase column chromatography was required to separate the 1 : 1 mixture of compounds **65** and **66**. Although the eight compounds **60–67** were obtained in sufficient amounts, the termination of the synthetic sequence was performed for compounds **61–67** only. Therefore, after *O*-sulfonatation, total desilylation was achieved with the use of ammonium fluoride at 55 °C for 48 h. Finally, methyl ester saponification, catalytic reduction, and *N*-sulfatation or *N*-acetylation were performed as described earlier in this chapter, leading to targets **68–78** that display two or more primary *O*-sulfonates.

The method described yielded a series of pure octasaccharides with various sulfonatation patterns on the primary position of glucosamines. However, as already mentioned, only 8 out of the 16 possible combinations were obtained because of the outcome of the statistical desilylation reaction. To overcome this drawback, we capitalized on the major reaction product, that is, the mono-desilylated octasaccharide **60**, that was selectively reprotected, with the use of 1-(benzoyloxy)benzotriazole, as a benzoate ester at position 6 of glucosamine into compound **79** with 68% yield (Scheme 7.8).

The switch from a silyl ether toward an ester should allow the preparation of the remaining half of the combinations, with the reducing unit being non-*O*-sulfonated on the primary position. This opportunistic approach is illustrated with the synthesis of octasaccharide **82**. Full desilylation of compound **79** into **80** was followed by *O*-sulfonatation to give **81** (45% yield over the two steps). Saponification of the benzoate together with the methyl esters (80%), reduction of the benzyl ethers, the azides, and the benzyl carbamate (55%), and finally *N*-sulfatation as previously described in this chapter yielded target derivative **82** (78% yield). The synthesis of the seven remaining products, although not undertaken, should theoretically be achieved using the statistical desilylation reaction on monobenzoyl octasaccharide **79**, as described for **59**.

7.4
Alternative Synthetic Methods: Means to Build Libraries

The synthetic tips derived from standard approaches described in the previous sections of this chapter to prepare HPN oligomers proved highly valuable for obtaining chemical diversity with modifications being either standard (regarding the sulfation pattern) or nonnatural (incorporation of hydrocarbon moieties), thus giving rise to small libraries. Other alternatives for this purpose exist, as the entire set of 48 disaccharidic building blocks suitable for the synthesis of all the theoretically possible combinations of HPN/HS oligomers has been prepared [56]. Furthermore, several chemical targets have been effectively synthesized by several groups. Some selected examples of alternative methods described in the literature are briefly listed in the following text.

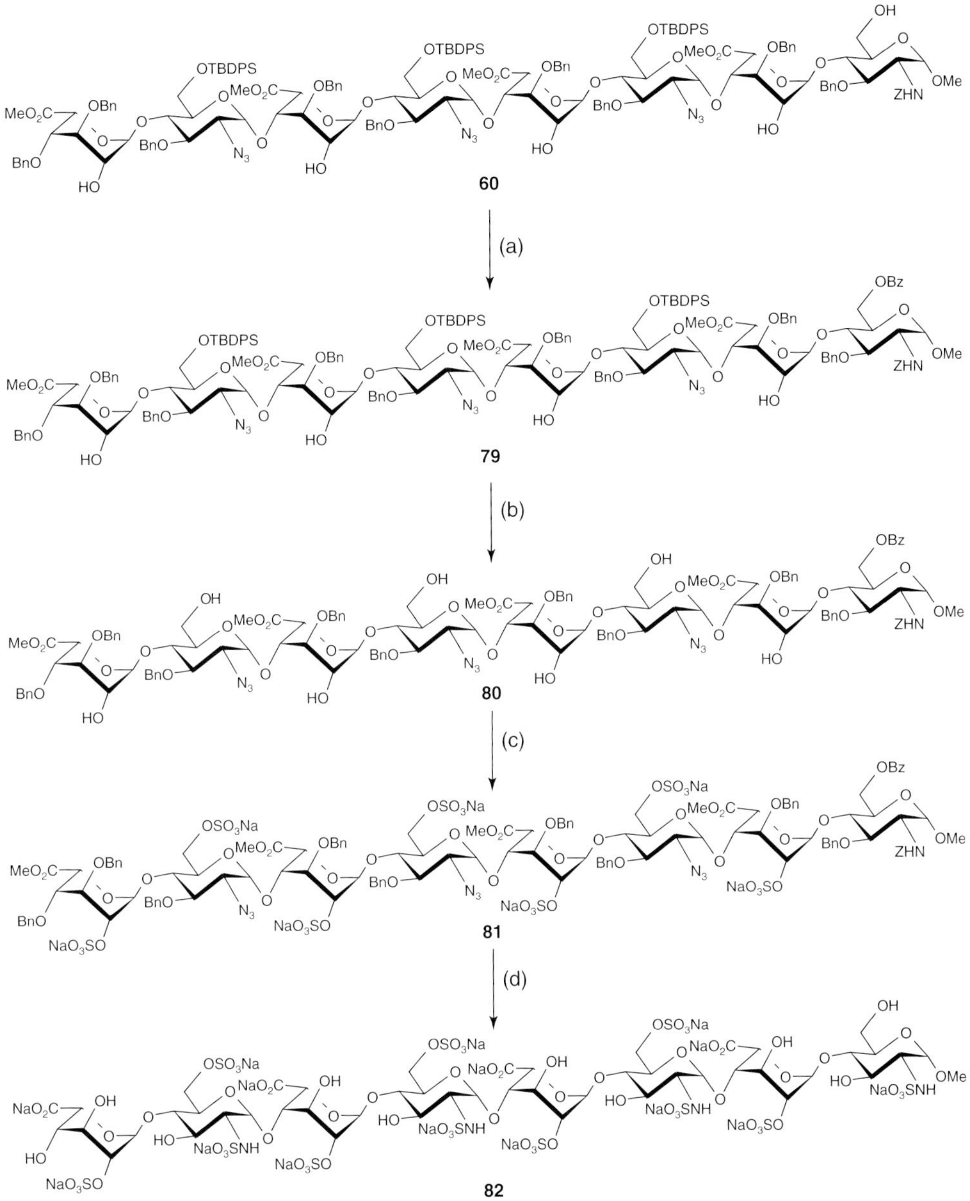

82

Scheme 7.8 Synthesis of octasaccharide **82**. *Reagents*: (a) BzOBT, Et_3N, 1,2-dichloroethane, rt, 24 h, 68%. (b) NH_4F, MeOH, 55 °C, 48 h, crude. (c) $Et_3N{\cdot}SO_3$, DMF, 55 °C, 16 h, 45% (two steps). (d) (i) Aq. LiOH (0.7 M), THF/MeOH 1/1, 0 °C, 1 h, then rt, 16 h, 80%; (ii) ammonium formate, 10% Pd/C (cat), *tert*-butanol/H_2O 1/1, 4 h, 55%; and (iii) $NaHCO_3$, pyridine$\cdot SO_3$, aq. $NaHCO_3$, 0 °C then rt, 16 h, 78%.

7.4.1
Synthesis of Tetrasaccharide Mixtures Followed by Purification

An efficient strategy for the preparation of a HS library of tetrasaccharides starting from suitably protected disaccharide building blocks has been developed (Scheme 7.9) [57].

Equimolar amounts of disaccharidic glycosyl acceptors **83–85** were mixed and glycosylated at once with 1.3 equiv of disaccharidic glycosyl donor **23** in the presence of a Lewis acid. Once the reaction was complete, a Sephadex LH-20 size-exclusion chromatography allowed separation of the pool of the three tetrasaccharides **86–88** from unwanted disaccharide starting material or their degradation products. The two iduronic acid-containing acceptors gave the corresponding tetrasaccharides in quantitative yields and full α-stereoselectivities. However, the coupling reaction involving the glucuronic acid derivative gave lower yields (12–18%) and both anomers (90/10 to 86/14 α/β ratios) were obtained (Section 7.2). The mixture of tetrasaccharides **86–88** was then subjected to acetate transesterification (K_2CO_3/MeOH), azide-/allyl-selective reduction (H_2 and Pd/$BaSO_4$), concomitant *O*/*N*-sulfation (pyridine·SO_3 complex), and finally methyl ester saponification (lithium, then potassium hydroperoxide) to yield a mixture of tetrasaccharides **89–91**. Owing to the different sulfate/benzyl group ratios, every compound was separated from its counterparts first by RP-C18 flash chromatography and then by semipreparative high-performance liquid chromatography (HPLC). The tetrasaccharides containing the two iduronic acids were obtained in 20% and 22% yields over the five mixture synthesis steps, whereas a lower yield (4%) was observed for the third tetrasaccharide with the glucuronic acid residue, for the reasons explained earlier. Finally, the three pure targets **92–94** were obtained in 73% to quantitative yields after benzyl cleavage (H_2 and $Pd(OH)_2$/C).

7.4.2
Modular Synthesis of HPN/HS Oligosaccharides

Recently, some efforts have been directed toward the modular synthesis of HPN/HS oligosaccharides using a relatively small number of key mono- [22] or disaccharidic building blocks [20, 26, 37], that may be transformed into multiple glycosyl donors and acceptors to be used in a parallel combinatorial manner for the elaboration of libraries for SAR studies. To illustrate the relevance of this approach, 11 tetrasaccharides **100–110** differing in the nature of the uronic acid (IdoA/GlcA, L-iduronic acid/D-glucuronic acid), the presence (or not) of an *O*-sulfonate on $IdoU^{II}$, and the amino substitution (NA/NS) have been prepared starting from three glycosyl acceptors **97–99** and two glycosyl donors **95–96** only (Scheme 7.10) [26]. The hexasaccharide **111** has also been synthesized with the help of the same building blocks.

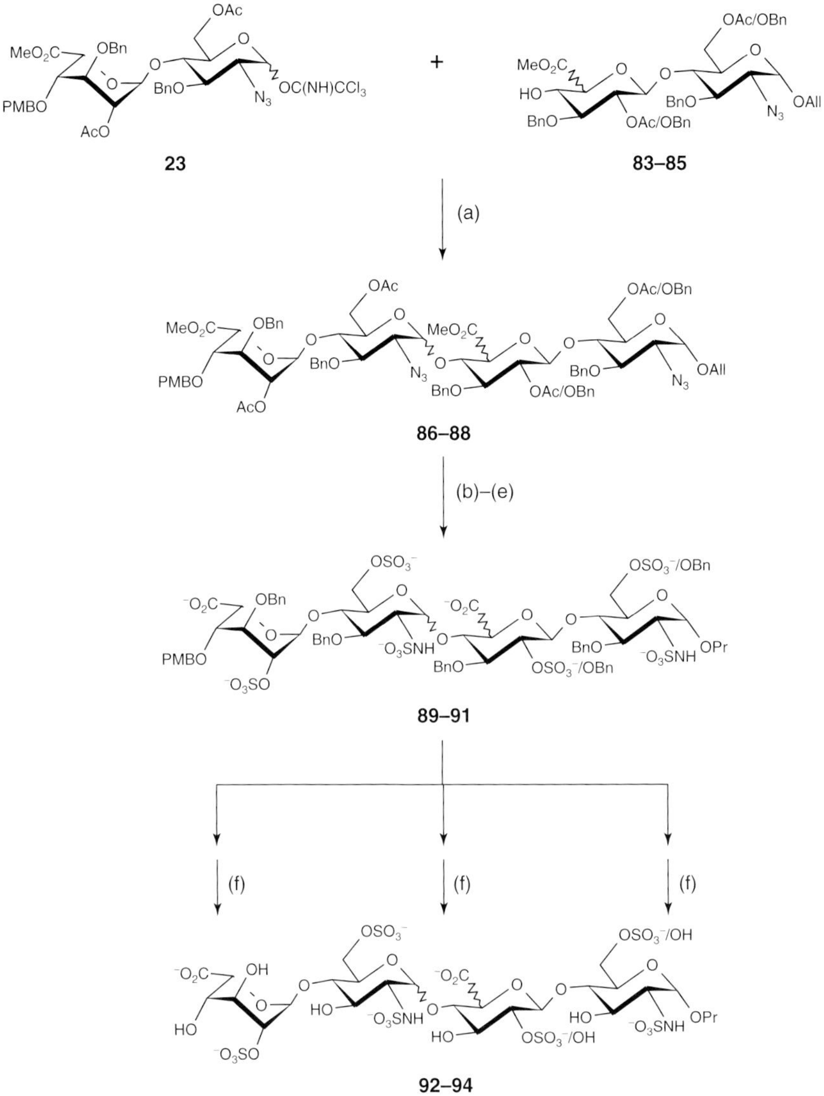

Scheme 7.9 Synthesis of tetrasaccharides **92–94**, according to [57]. *Reagents*: (a) TMSOTf, CH_2Cl_2, −30 °C, 3 h; (b) K_2CO_3, MeOH, rt, 18 h; (c) H_2, Pd/$BaSO_4$, 0.1 M pyridine, MeOH/THF 2/1, 24 h; (d) pyridine·SO_3, pyridine, rt, 24 h, then 50 °C, 16 h; (e) LiOH, H_2O_2, *n*BuOH/THF 1/1, 0 °C, 3 h, then KOH, rt, 57 h (20 and 22% yield over the five steps (4% for the glucuronic acid-containing tetrasaccharide)); and (f) H_2, $Pd(OH)_2$/C, 100 mM phosphate buffer pH 7.0/*t*BuOH 3/2 (73–100%).

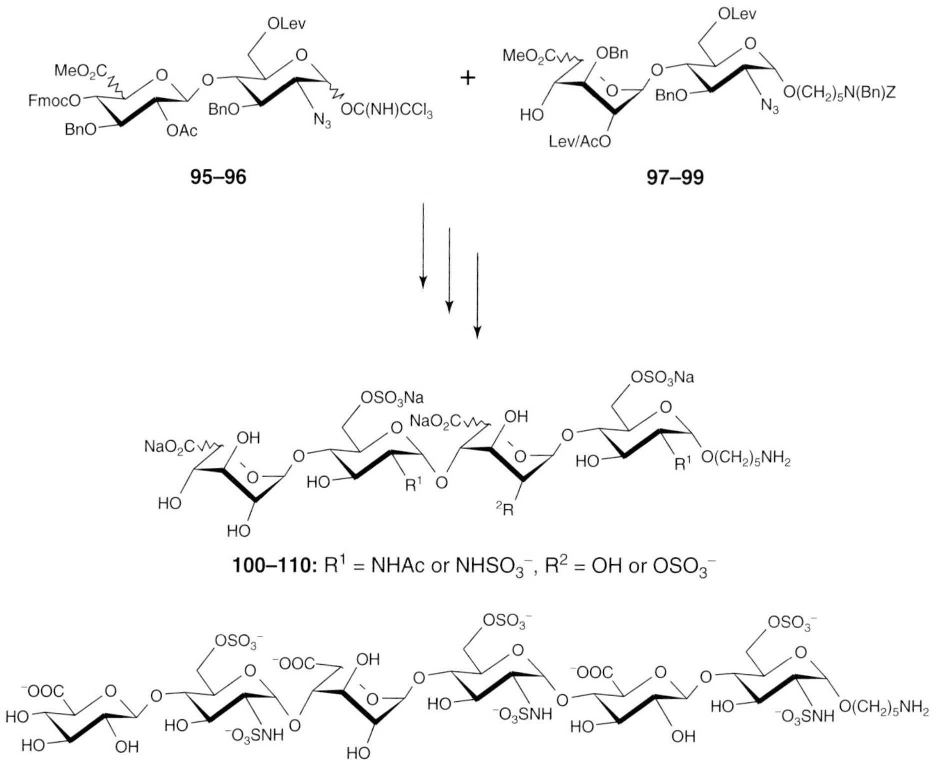

Scheme 7.10 Synthesis of tetrasaccharides **100–110** and hexasaccharide **111** according to [26].

7.5 Biological Evaluation

The abilities of compounds **37–39**, **42–44**, **68–78**, and **82** to enhance pseudo-tubule sprouting on endothelial cells are reported in Table 7.1, and have been compared to the parent octasaccharide **12**, used as the reference compound because of its standard sulfation pattern.

The first striking result is that no loss of activity is observed with the replacement of *N*-sulfates by acetamides (compound **68**, entry 1) as might be expected from literature data. Second, incorporation of pentyl and pentylphenyl groups as an aglycon (compounds **37**, **39**, entry 2) led to a significant increase in the pseudo-tubule formation, but an undecyl chain is not well tolerated at the anomeric position as compound **38** is less active. Further, introduction of butyramides (compound **42**, entry 3) results in an increase in activity, whereas the slightly bulkier amides **43–44** display the same biological activity as **12**. Finally, the exact location of 6-*O*-sulfonates is directly related to the biological response. Indeed, if compounds **78** and **82**, bearing three primary *O*-sulfonates on $GlcN^{I,V,VII}$ and on

Table 7.1 Pseudo-tubule formation assessment of *N*-sulfated and amido octasaccharides **12**, **37–39**, **42–44**, **68–78**, and **82** on endothelial cells.

Entry	Compound	Main features of GlcN residue		Tubule formation on endothelial cells
		Sulfonation pattern on 6-OH $GlcN^{VII}$-$GlcN^{V}$-$GlcN^{III}$-$GlcN^{I}$	Amino substitution/ aglycone	
1	12	OS-OS-OS-OS	NS/OMe	++
	68	OS-OS-OS-OS	NA/OMe	++
2	37	OS-OS-OS-OS	NA/OPentyl	+++
	38	OS-OS-OS-OS	NA/OUndecyl	−
	39	OS-OS-OS-OS	NA/OPentylphenyl	+++
3	42	OS-OS-OS-OS	1-Oxobutylamino/OPentyl	+++
	43	OS-OS-OS-OS	1-Oxohexylamino/OPentyl	++
	44	OS-OS-OS-OS	Cyclopentylcarbonylamino/ OPentyl	++
4	75	OS-OH-OS-OS	NS/OMe	+
	76	OH-OS-OS-OS	NA/OMe	+
	77	OS-OH-OS-OS	NA/OMe	+
	78	OS-OS-OH-OS	NA/OMe	++
	82	OS-OS-OS-OH	NS/OMe	++
5	69	OH-OH-OS-OS	NS/OMe	+
	70	OH-OS-OH-OS	NS/OMe	+
	71	OS-OH-OH-OS	NS/OMe	+
	72	OS-OH-OS-OH	NA/OMe	−
	73	OH-OS-OS-OH	NA/OMe	−
	74	OS-OS-OH-OH	NA/OMe	++

(+++) corresponds to a higher activity than the reference compound **12** in the model (which displays (++) activity), whereas a lower activity is expressed by (+) and (−).

$GlcN^{III,V,VII}$ retained the activity of the reference compound, their counterparts **75** (*O*-sulfonated on primary positions at $GlcN^{I,III,VII}$), **76** (*O*-sulfonated on primary positions at $GlcN^{I,III,V}$), and **77** (*O*-sulfonated on primary positions at $GlcN^{I,III,VII}$) were less efficient (entry 4). These results indicate that the primary *O*-sulfonates on $GlcN^{V}$ and $GlcN^{VII}$ are more critical than those on $GlcN^{I}$ and $GlcN^{III}$. This tendency is confirmed and even more evident with compounds displaying only two primary *O*-sulfonates (entry 5). Indeed, compounds **69–73** lacking one or two *O*-sulfonates on $GlcN^{V/VII}$ are less active than **12** whereas **74**, where those *O*-sulfonates are preserved, retains the same biological activity as the reference compound.

Taken together, these data clearly show that the number as well as the exact location of primary *O*-sulfonates plays a major role in the FGF2/FGFR activation system. The results presented here also indicate that the introduction of carefully selected alkyl/aryl moieties at the anomeric position and/or at position 2 of GlcN results in an increase of biological activity.

7.6 Conclusion and Outlook

The synthesis of well-defined complex carbohydrates such as GAGs is a major challenge for chemists. Although enzymatic methods are useful for the rapid preparation of such fragments to establish general structural requirements for selected targets, the specificities of substrates toward enzymes are a drawback with respect to diversity. Total synthesis provides access to the pure and well-defined molecules required to perform structure–activity relationship studies that are mandatory to improve biological response and/or absorption, distribution, metabolism, and excretion (ADME) parameters of potential leads. As polymer-supported coupling reactions still suffer from low yields that necessitate the use of a large excess of high-value glycosyl donors, to date, solution-phase synthesis remains the most efficient method to yield modified HPN/HS fragments. Several strategies have been reported in this chapter that dramatically reduce the number of synthetic steps. Those include the use of versatile building blocks, concomitant coupling reactions followed by demixing targets, or the late-stage introduction of chemical diversification on highly functionalized oligomers. The latter method was successfully applied for the optimization of an octasaccharidic structure involved in the FGF-2-mediated pseudo-tubule formation on endothelial cells, and may open the field for new drugs remodeling the vascular system in the context of peripheral arterial occlusive disease or for improvement of wound healing. Indeed, the withdrawal of unnecessary *O*-sulfonates together with the introduction of hydrocarbon moieties at some specific sites allowed significant enhancement of the "natural" HPN octasaccharide biological activity.

7.7 Experimental Section (General Procedures)

7.7.1 General Conditions for Coupling Reactions

A solution of glycosyl acceptor (1.0 equiv) and glycosyl donor (1.2 equiv) in anhydrous dichloromethane (35 l mol^{-1}/glycosyl donor) was stirred in the presence of preactivated (300 °C) finely grounded 4 Å molecular sieves (MS, 750 g mol^{-1}/glycosyl donor) under argon atmosphere at room temperature for 1 h. The temperature was then cooled to −20 °C, and a 1 M solution of TBSOTf (0.15 equiv of glycosyl donor) in dichloromethane was added to the preceding mixture. After total disappearance (1–18 h) of either glycosyl donor or glycosyl acceptor (TLC), the Lewis acid was neutralized by the addition of sodium hydrogencarbonate, and the reaction mixture was filtered (*Celite*®). The organic layer was successively washed with 2% aqueous sodium hydrogencarbonate solution and then water and finally dried (Na_2SO_4). Concentration under reduced pressure gave a residue that was purified using silica gel column chromatography to give the coupling reaction product.

7.7.2 General Conditions for Delevulinoylations

To a solution of the saccharide in toluene/ethanol (1 : 2, 100 l mol^{-1}), hydrazine acetate (5 equiv) was added. After about 1 h of magnetic stirring (TLC), the solvent was concentrated under reduced pressure and the residue was purified using silica gel column chromatography to furnish a glycosyl acceptor.

7.7.3 General Conditions for Olefin Cross Metathesis Reactions

A solution of pentenyl glycoside, alkene (2 equiv), and generation 1 or 2 Grubbs catalyst (0.1 equiv) in dichloromethane (50 l mol^{-1}) was refluxed for 6 h. After concentration under reduced pressure, the residue was purified using silica gel column chromatography to yield the desired glycoside.

7.7.4 General Conditions for Transesterifications

To a solution of glycoside in MeOH/dichloromethane (3 : 2, 300 l mol^{-1}) finely grounded 3 Å MS and then, at 0 °C, a 1 M solution of sodium methoxide (3 equiv) were added. After 1 h at 0 °C, the reaction mixture was stirred for 15 h at room temperature and then neutralized with H^+ Dowex 50WX4 resin. After filtration and partial concentration, the oligosaccharide was first purified over a size-exclusion LH-20 Sephadex® chromatography column using MeOH/dichloromethane (1 : 1) as eluant, and then silica gel column chromatography afforded the desired polyol.

7.7.5 General Conditions for Desilylations

To a solution of silylated glycoside in MeOH (130 l mol^{-1}), ammonium fluoride (20 equiv/TBDPS ether) was added. After 20 h of magnetic stirring at room temperature (for statistical desilylation), or 48 h at 55 °C (for total desilylation), the reaction mixture was purified first with a size-exclusion LH-20 Sephadex chromatography column using EtOH/dichloromethane (1 : 1) as eluant, then by flash column chromatography on silica gel, and finally, if required, by silica gel HPLC C18 (Waters®) using acetonitrile/H_2O (9 : 1 + 0.1% TFA).

7.7.6 General Conditions for *O*-Sulfonatations

To a solution in DMF (90 l mol^{-1}) of the saccharide, Et_3N·sulfur trioxide complex (5 equiv) was added. After 17 h of magnetic stirring at 55 °C with protection from light, the reaction mixture was cooled to 0 °C and an excess of MeOH was added. After 30 min of stirring at 0 °C and then 2 h at room temperature, the oligosaccharide was purified over a size-exclusion LH-20 Sephadex chromatography column using MeOH/DMF (9 : 1) as eluant to furnish the expected *O*-sulfonated saccharide.

7.7.7 General Conditions for Saponifications

To a solution in MeOH/THF (1 : 1, 160 l mol^{-1}) of the glycoside, a 1N solution of lithium hydroxide (30 equiv) was added at 0 °C. After 1 h of stirring at 0 °C and then 16 h at room temperature, the saccharide was purified over a size-exclusion LH-20 Sephadex chromatography column using MeOH/DMF (9 : 1) as eluant to give the desired acid.

7.7.8 General Conditions for the Catalytic Reductions

To a *tert*-butanol/water solution (1 : 1, 230 l mol^{-1}) of the glycoside, ammonium formate (160 equiv) and then 10% Pd/C (26 g $mmol^{-1}$) were added. After 4 h of vigorous stirring, the reaction mixture was filtered (*Celite*), partially concentrated under reduced pressure, and then loaded onto a size-exclusion Sephadex G-25 chromatography column using 0.2N sodium chloride as eluant. The fractions containing the compound were pooled and partially concentrated under vacuum and then desalted over the same gel filtration column equilibrated with water to finally give the expected polyol.

7.7.9 General Conditions for *N*-Sulfations

To a saturated aqueous sodium hydrogencarbonate solution (100 l mol^{-1}) of the polyamine, pyridine·sulfur trioxide complex (180 equiv) was added at 0 °C. After 16 h of magnetic stirring at room temperature, the reaction mixture was purified as described in Section 7.7.8.

7.7.10 General Conditions for *N*-Acylations

Acetamido derivatives were prepared as described in Section 7.7.8 with the pyridine·sulfur trioxide complex replaced by acetic anhydride. The following procedure was used for the other amides.

To a solution of the glycoside in water (70 l mol^{-1}), first a DMF solution of *N*,*N*-diisopropylethylamine (28 equiv) and then *N*-hydroxysuccinimide (NHS) ester (20 equiv) were added at 0 °C. After stirring at 0 °C for 15 min and then at room temperature overnight, the target compound was purified as described in Section 7.7.8.

Acknowledgments

This work has been performed under the supervision of Françoise Bono and Jean-Marc Herbert. The author wishes to thank them for their support, as well as Aline Barbier, Alexandre Froidbise, Jérome Gabriel, Corine Garnier, Jean-Marc Strassel, and Gilbert Lassalle for the preparation of the octasaccharides and technical discussions. The author also acknowledges Pierre Fons and his team for the biological evaluation of the compounds in the tubule formation model on endothelial cells.

List of Abbreviations

Ac	Acetyl
ADME	Absorption, distribution, metabolism, and excretion
All	Allyl
aq.	aqueous
Bn	Benzyl
BT	Benzotriazole
Bu	Butyl
Bz	Benzoyl
cat	Catalytic
DMAP	4-Dimethylaminopyridine
DMF	*N,N*-Dimethylformamide
FGF	Fibroblast growth factor
FGFR	Fibroblast growth factor receptor
GAG	Glycosaminoglycan
GlcA	D-Glucuronic acid
GlcN	D-Glucosamine
HPN	Heparin
HS	Heparan sulfate
*i*Pr	Isopropyl
IdoA	L-Iduronic acid
Lev	Levulinoyl
Me	Methyl
MS	Molecular sieves
NA	*N*-Acetylated
NHS	*N*-Hydroxysuccinimide
NS	*N*-Sulfated
Piv	Pivaloyl
PMB	*para*-Methoxybenzyl
Pr	*n*-Propyl
PTSA	*para*-Toluenesulfonic acid
rt	Room temperature
satd	saturated

TBDMS	*tert*-Butyldimethylsilyl
TBDPS	*tert*-Butyldiphenylsilyl
TFA	Trifluoroacetic acid
TfOH	Triflic acid
THF	Tetrahydrofuran
TLC	Thin layer chromatography
TMS	Trimethylsilyl
Z	Benzyloxycarbonyl

References

1. (a) Lindahl, U., Kusche-Gullberg, M., and Kjellen, L. (1998) *J. Biol. Chem.*, **273**, 24979–24982. (b) Gallagher, J.T. (2001) *J. Clin. Invest.*, **108**, 357–361.
2. Rosenberg, R.D., Shworak, N.W., Liu, J., Schwartz, J.J., and Zhang, L.J. (1997) *Clin. Invest.*, **99**, 2062–2070.
3. Lindahl, U. and Li, J.-P. (2009) *Int. Rev. Cell Mol. Biol.*, **276**, 105–157.
4. Gandhi, N.S. and Mancera, R.L. (2008) *Chem. Biol. Drug Des.*, **72**, 455–482.
5. Kreuger, J., Spillmann, D., Li, J.P., and Lindahl, U. (2006) *J. Cell Biol.*, **174**, 323–327.
6. (a) Sinaÿ, P. and Jacquinet, J.-C. (1984) *Carbohydr. Res.*, **132**, C5–C9. (b) Jaquinet, J.-C., Petitou, M., Duchaussoy, P., Lederman, I., Choay, J., Torri, G., and Sinaÿ, P. (1984) *Carbohydr. Res.*, **130**, 221–241. (c) van Boeckel, C.A.A., Beetz, T., Vos, J.N., De Jong, A.J.M., Van Aelst, S.F., Van den Bosch, R.H., Mertens, J.M.R., and van der Vlugt, F.A. (1985) *Carbohydr. Res.*, **4**, 293–321. (d) van Boeckel, C.A.A. and Petitou, M. (1993) *Angew. Chem., Int. Ed. Engl*, **32**, 1671–1818. (e) Petitou, M. and van Boeckel, C.A.A. (2004) *Angew. Chem. Int. Ed.*, **43**, 3118–3133.
7. Sleeman, M., Fraser, J., McDonald, M., Yuan, S., White, D., Grandison, P., Kumble, K., Watson, J.D., and Murison, J.G. (2001) *Gene (Amst.)*, **271**, 171–182.
8. (a) Yarden, Y. and Ullrich, A. (1988) *Annu. Rev. Biochem.*, **57**, 443–478. (b) Esko, J.D. and Selleck, S.B. (2002) *Annu. Rev. Biochem.*, **71**, 435–471.
9. Schlessinger, J., Plotnikov, A.N., Ibrahimi, O.A., Eliseenkova, A.V., Yeh, B.K., Yayon, A., Linhardt, R.J., and Mohammadi, M. (2000) *Mol. Cell*, **6**, 743–750.
10. Naimy, H., Buczek-Thomas, J.A., Nugent, M.A., Leymarie, N., and Zaia, J. (2011) *J. Biol. Chem.*, **286**, 19311–19319.
11. Saxena, K., Schieborr, U., Anderka, O., Duchardt-Ferner, E., Elshorst, B., Lakshmi Gande, S., Janzon, J., Kudlinzki, D., Sreeramulu, S., Dreyer, M.K., Wendt, K.U., Herbert, C., Duchaussoy, P., Bianciotto, M., Driguez, P.-A., Lassalle, G., Savi, P., Mohammadi, M., Bono, F., and Schwalbe, H. (2010) *J. Biol. Chem*, **285**, 26628–26640.
12. Pellegrini, L., Burke, D.F., von Delft, F., Mulloy, B., and Blundell, T.L. (2000) *Nature*, **407**, 1029–1034.
13. (a) Bhaskar, U., Sterner, E., Hickey, A.M., Onishi, A., Zhang, F., Dordick, J.S., and Linhardt, R.J. (2012) *Appl. Microbiol. Biotechnol.*, **93**, 1–16. (b) Kreuger, J., Jemth, P., Sanders-Lindberg, E., Eliahu, L., Ron, D., Basilico, C., Salmivirta, M., and Lindahl, U. (2005) *Biochem. J.*, **389**, 145–150. (c) Jemth, P., Kreuger, J., Kusche-Gullberg, M., Sturiale, L., Gimenez-Gallego, G., and Lindahl, U. (2002) *J. Biol. Chem.*, **277**, 30567–30573. (d) Chen, J., Avci, F.Y., Munoz, E.M., McDowell, L.M., Chen, M., Pedersen, L.C., Zhang, L., Linhardt, R.J., and Liu, J. (2005) *J. Biol. Chem.*, **280**, 42817–42825. (e) Peterson, S., Frick, A., and Liu, J. (2009) *Nat. Prod. Rep.*, **26**, 610–627. (f) Powell, A.K., Yates, E.A., Fernig, D.G., and Turnbull, J.E. (2004) *Glycobiology*, **14**, 17R–30R.
14. Xu, Y., Wang, Z., Liu, R., Bridges, A.S., Huang, X., and Liu, J. (2012) *Glycobiology*, **22**, 96–106.

15. (a) Poletti, L. and Lay, L. (2003) *Eur. J. Org. Chem.*, 2999–3024. (b) Noti, C. and Seeberger, P.H. (2005) *Chem. Biol.*, **12**, 731–756. (c) Codée, J.D.C., Overkleeft, H.S., van der Marel, G.A., and van Boeckel, C.A.A. (2004) *Drug Discov. Today: Technol.*, **1**, 317–326.
16. Noti, C., de Paz, J.L., Polito, L., and Seeberger, P.H. (2006) *Chem. Eur. J.*, **12**, 8664–8686.
17. Dulaney, S.B. and Huang, X. (2012) *Adv. Carbohydr. Chem. Biochem.*, **67**, 95–136.
18. Hsu, C.-H., Hung, S.-C., Wu, C.-Y., and Wong, C.-H. (2011) *Angew. Chem. Int. Ed.*, **50**, 11872–11923.
19. Polat, T. and Wong, C.-H. (2007) *J. Am. Chem. Soc.*, **129**, 12795–12800.
20. Wang, Z., Xu, Y., Yang, B., Tiruchinapally, G., Sun, B., Liu, R., Dulaney, S., Liu, J., and Huang, X. (2010) *Chem. Eur. J.*, **16**, 8365–8375.
21. Kovensky, J., Duchaussoy, P., Petitou, M., and Sinaÿ, P. (1996) *Tetrahedron: Asymmetry*, **7**, 3119–3128.
22. Codée, J.D.C., Stubba, B., Schiattarella, M., Overkleeft, H.S., van Boeckel, C.A.A., van Boom, J.H., and van der Marel, G.A. (2005) *J. Am. Chem. Soc.*, **127**, 3767–3773.
23. La Ferla, B., Lay, L., Guerrini, M., Poletti, L., Panza, L., and Russo, G. (1999) *Tetrahedron*, **55**, 9867–9880.
24. Lee, J.-C., Lu, X.-A., Kulkarni, S.S., Wen, Y.-S., and Hung, S.-C. (2004) *J. Am. Chem. Soc.*, **126**, 476–477.
25. Maza, S., Macchione, G., Ojeda, R., López-Prados, J., Angulo, J., de Paz, J.L., and Nieto, P.M. (2012) *Org. Biomol. Chem.*, **10**, 2146–2163.
26. Arungundram, S., Al-Mafraji, K., Asong, J., Leach, F.E. III,, Amster, I.J., Venot, A., Turnbull, J.E., and Boons, G.-J. (2009) *J. Am. Chem. Soc.*, **131**, 17394–17405.
27. de Paz, J.L., Noti, C., and Seeberger, P.H. (2006) *J. Am. Chem. Soc.*, **128**, 2766–2767.
28. Tabeur, C., Mallet, J.-M., Bono, F., Herbert, J.-M., Petitou, M., and Sinaÿ, P. (1999) *Bioorg. Med. Chem.*, **7**, 2003–2012.
29. Tabeur, C., Machetto, F., Mallet, J.-M., Petitou, M., and Sinaÿ, P. (1996) *Carbohydr. Res.*, **281**, 253–276.
30. Koeners, H.J., Verhoeven, J., and van Boom, J.H. (1981) *Recl. Trav. Chim. Pays-Bas*, **100**, 65.
31. Fons, P., Herbert, C., Driguez, P.-A., Gueguen, G., Desjobert, J., Michaux, C., Barbier, A., Gabriel, J., Garnier, C., Strassel, J.-M., Sibrac, D., Dol, F., Naimi, S., Lassalle, G., Augé, F., Daveu, C., Mohammadi, M., Hérault, J.-P., Bono, F., and Herbert, J.-M. *J. Biol. Chem.* submitted.
32. de Paz, J.-L., Angulo, J., Lassaletta, J.-M., Nieto, P.M., Redondo-Horcajo, M., Lozano, R.M., Giménez-Gallego, G., and Martín-Lomas, M. (2001) *ChemBioChem*, **2**, 673–685.
33. (a) Wu, Z.L., Zhang, L., Yabe, T., Kuberan, B., Beeler, D.L., Love, A., and Rosenberg, R.D. (2003) *J. Biol. Chem.*, **278**, 17121–17129. (b) Zhou, F.Y., Kan, M., Owens, R.T., McKeehan, W.L., Thompson, J.A., Linhardt, R.J., and Hook, M. (1997) *Eur. J. Cell Biol.*, **73**, 71–80. (c) Pye, D.A., Vives, R.R., Turnbull, J.E., Hyde, P., and Gallagher, J.T. (1998) *J. Biol. Chem.*, **273**, 22936–22942. (d) Delehedde, M., Lyon, M., Gallagher, J.T., Rudland, P.S., and Fernig, D.G. (2002) *Biochem. J.*, **366**, 235–244.
34. Petitou, M., Duchaussoy, P., Lederman, I., Choay, J., Jacquinet, J.C., Sinaÿ, P., and Torri, G. (1987) *Carbohydr. Res.*, **167**, 67–75.
35. (a) Duchaussoy, P., Jaurand, G., Driguez, P.-A., Lederman, I., Ceccato, M.-L., Gourvenec, F., Strassel, J.-M., Sizun, P., Petitou, M., and Herbert, J.-M. (1999) *Carbohydr. Res.*, **317**, 85–99. (b) Das, S.K., Mallet, J.-M., Esnault, J., Driguez, P.-A., Duchaussoy, P., Sizun, P., Hérault, J.-P., Herbert, J.-M., Petitou, M., and Sinaÿ, P. (2001) *Angew. Chem. Int. Ed.*, **40**, 1670–1673. (c) Das, S.K., Mallet, J.-M., Esnault, J., Driguez, P.-A., Duchaussoy, P., Sizun, P., Hérault, J.-P., Herbert, J.-M., Petitou, M., and Sinaÿ, P. (2001) *Chem. Eur. J.*, **7**, 4821–4834.
36. (a) Lucas, R., Hamza, D., Lubineau, A., and Bonnaffé, D. (2004) *Eur. J. Org. Chem.*, **2004**, 2107–2117. (b) Cid, M.B., Alfonso, F., and Martin-Lomas, M. (2005) *Chem. Eur. J.*, **11**, 928–938. (c) Hamza, D., Lucas, R., Feizi, T., Chai,

W., Bonnaffé, D., and Lubineau, A. (2006) *ChemBioChem*, **7**, 1856–1858.

37. Orgueira, H.A., Bartolozzi, A., Schell, P., Litjens, R.E.J.N., Palmacci, E.R., and Seeberger, P.H. (2003) *Chem. Eur. J.*, **9**, 140–169.
38. Baleux, F., Loureiro-Morais, L., Hersant, Y., Clayette, P., Arenzana-Seisdedos, F., Bonnaffé, D., and Lortat-Jacob, H. (2009) *Nat. Chem. Biol.*, **5**, 743–748.
39. Faham, S., Linhardt, R.J., and Rees, D.C. (1998) *Curr. Opin. Struct. Biol.*, **8**, 578–586.
40. (a) Trnka, T.M. and Grubbs, R.H. (2001) *Acc. Chem. Res.*, **34**, 18–29. (b) Chatterjee, A.K., Morgan, J.P., Scholl, M., and Grubbs, R.H. (2000) *J. Am. Chem. Soc.*, **122**, 3783–3784. (c) Grubbs, R.H. and Chang, S. (1998) *Tetrahedron*, **54**, 4413–4450. (d) Nguyen, S.T. and Grubbs, R.H. (1993) *J. Am. Chem. Soc.*, **115**, 9858–9859.
41. Gavard, O., Hersant, Y., Alais, J., Duverger, V., Dilhas, A., Bascou, A., and Bonnaffé, D. (2003) *Eur. J. Org. Chem.*, 3603–3620.
42. van Boeckel, C.A.A., Beetz, T., Vos, J.N., De Jong, A.J.M., van Aelst, S.F., van den Bosch, R.H., Mertens, J.M.R., and van der Vlugt, F.A. (1985) *J. Carbohydr. Chem.*, **4**, 293–321.
43. Bayley, H., Standring, D.N., and Knowles, J.R. (1978) *Tetrahedron Lett.*, **39**, 3633–3634.
44. Lundin, L., Larsson, H., Kreuger, J., Kanda, S., Lindahl, U., Salmivirta, M., and Claesson-Welsh, L. (2000) *J. Biol. Chem.*, **275**, 24653–24660.
45. Ojeda, R., Angulo, J., Nieto, P.M., and Martín-Lomas, M. (2002) *Can. J. Chem.*, **80**, 917–936.
46. (a) Ashikari-Hada, S., Habuchi, H., Kariya, Y., Itoh, N., Reddi, A.H., and Kimata, K. (2004) *J. Biol. Chem.*, **279**, 12346–12354. (b) Kreuger, J., Salmivirta, M., Sturiale, L., Gimenez-Gallego, G., and Lindahl, U. (2001) *J. Biol. Chem.*, **276**, 30744–30752.
47. Gama, C.I., Tully, S.E., Sotogaku, N., Clark, P.M., Rawat, M., Vaidehi, N., Goddard, W.A., Nishi, A., and Hsieh-Wilson, L.C. (2006) *Nat. Chem. Biol.*, **2**, 467–473.
48. Jastrebova, N., Vanwildemeersch, M., Rapraeger, A.C., Gimenez-Gallego, G., Lindahl, U., and Spillmann, D. (2006) *J. Biol. Chem.*, **281**, 26884–26892.
49. Angulo, J., Ojeda, R., de Paz, J.L., Lucas, R., Nieto, P.M., Lozano, R.M., Redondo-Horcajo, M., Gimenez-Gallego, G., and Martin-Lomas, M. (2004) *ChemBioChem*, **5**, 55–61.
50. Poletti, L., Fleischer, M., Vogel, C., Guerrini, M., Torri, G., and Lay, L. (2001) *Eur. J. Org. Chem.*, 2727–2734.
51. Tatai, J. and Fügedi, P. (2008) *Tetrahedron*, **64**, 9865–9873.
52. Lucas, R., Angulo, J., Nieto, P.M., and Martín-Lomas, M. (2003) *Org. Biomol. Chem.*, **1**, 2253–2256.
53. Driguez, P.-A. (2012) *Encyclopedia of Reagents for Organic Synthesis*, John Wiley & Sons, Ltd, Chichester. doi: 10.1002/047084289X.RN01513
54. Orita, A., Hamada, Y., Nakano, T., Toyoshima, S., and Otera, J. (2001) *Chem. Eur. J.*, **7**, 3321–3327.
55. Slaghek, T.M., Hyppönen, T.K., Ogawa, T., Kamerling, J.P., and Vliegenthart, J.F.G. (1994) *Tetrahedron: Asymmetry*, **5**, 2291–2301.
56. Lu, L.-D., Shie, C.-R., Kulkarni, S.S., Pan, G.-R., Lu, X.-A., and Hung, S.-C. (2006) *Org. Lett.*, **8**, 5995–5998.
57. Dilhas, A., Lucas, R., Loureiro-Morais, L., Hersant, Y., and Bonnaffé, D. (2008) *J. Comb. Chem.*, **10**, 166–169.

8
Light Fluorous-Tag-Assisted Synthesis of Oligosaccharides

Rajarshi Roychoudhury and Nicola L. B. Pohl

8.1
Introduction

Oligosaccharides are ubiquitous in biological systems, but largely exist in such microheterogenous mixtures that the chemical synthesis of structurally well-defined carbohydrates is very attractive compared to their isolation from natural sources. The use of solid-phase methods has been highly successful for the automation of peptide and nucleic acid syntheses [1, 2]. By contrast, the synthesis of oligosaccharides, including their production on solid phases [3–6] has been complicated by the presence of multiple hydroxyl groups and the need to control the stereochemistry of the glycoside linkage – a process that has not approached the >95% yields as usually required for good peptide and nucleic acid synthetic processes. The use of a soluble light fluorous tag (those that by definition contain fluorine content ≤40% by molecular weight [7] for the synthesis of such oligosaccharides from smaller carbohydrate building blocks has several unique advantages over more traditional polymeric supports. Automated solid-phase methods suffer from the inherent limitation of biphasic kinetics and therefore require large excesses of building blocks [8–11] at each coupling cycle. Additionally, solid phases are not amenable to convergent synthetic strategies – ones in which two oligosaccharides are built separately and then are linked together – and are more complex to monitor than solution-phase reactions. In solution-phase glycosylation reactions, both coupling partners are in solution; therefore, the need for large excesses of building blocks to improve biphasic kinetics is eliminated. The small tag required, unlike polymeric supports, can also be readily obtained as a homogeneous compound with no risk of retaining reagents between coupling cycles to contaminate subsequent steps. Reaction intermediates can even be readily characterized using standard mass spectrometry, nuclear magnetic resonance (NMR), and chromatographic techniques. At the same time, the strong carbon–fluorine bond renders the requisite fluorocarbon tag inert to a wide variety of reaction conditions to maintain flexibility in the synthetic strategy. However, like a solid phase, the fluorous tag can still be used to simplify the purification of the growing oligosaccharide chain. Heavy fluorous tags (those that contain fluorine content by molecular weight of >60%) can

Modern Synthetic Methods in Carbohydrate Chemistry: From Monosaccharides to Complex Glycoconjugates,
First Edition. Edited by Daniel B. Werz and Sébastien Vidal.

aid small molecule syntheses through purification by liquid–liquid extraction protocols using fluorous solvents; their use in the synthesis of oligosaccharides has been summarized elsewhere [12]. Unlike these heavy tags, the so-called light fluorous tags can be used to mark carbohydrate molecules for isolation from nonfluorous-tagged side products and excess reagents as needed in the various stages of complex oligosaccharide synthesis by a simpler fluorous solid-phase extraction (FSPE) [13]. Given these advantages, much effort has gone into designing a range of protecting groups with minimal fluorous content for use in carbohydrate synthesis. The general use of various fluorous tags and protecting groups in organic synthesis was recently reviewed [14]. In this chapter, we put these fluorous tag design efforts into the context of carbohydrate synthesis and then focus on the construction of oligosaccharides using these tags along with their capacity for applications in automated protocols and in fluorous-based carbohydrate microarrays.

8.2 Fluorous-Protecting Groups and Tags Amenable to Fluorous Solid-Phase Extraction in Carbohydrate Synthesis

Since the initial development of the FSPE protocol by Curran and Luo [13] for light fluorous-tagged compounds, many new fluorous-protecting groups have been reported [15]. Not surprisingly, given the ubiquity of protecting groups in carbohydrate synthesis, several of these fluorous compounds have been employed to differentially protect functional groups in carbohydrate building blocks as outlined in the following text.

8.2.1 Mono- and Diol Protecting Groups

As carbohydrates are dominated by the hydroxyl functional group as a rule, protecting groups for alcohols figure prominently in synthetic strategies. Acetates and benzyl or allyl ethers are particularly common as carbohydrate-protecting groups, followed by various silyl ethers for single alcohols and acetals for neighboring diols. Examples of fluorous variants of these common protecting groups have now been reported.

A fluorous carboxylic acid (**1**, Scheme 8.1) to protect the alcohol functional group as an ester has been shown to be readily appended and later removed from a saccharide [16]. This fluorous acetate-protecting group, once cleaved, could also be recycled for further use [16]. However, to date, the properties of this group have not been modified to create other fluorous tags, such as a fluorous pivaloyl analog, with variable cleavage reaction rates for carbohydrate building block syntheses.

Fluorous silyl ethers have also been developed to serve as protecting groups for carbohydrates. A light fluorous silicon group (**2**, Scheme 8.1) was wielded for the protection of a donor building block, and a trisaccharide was synthesized

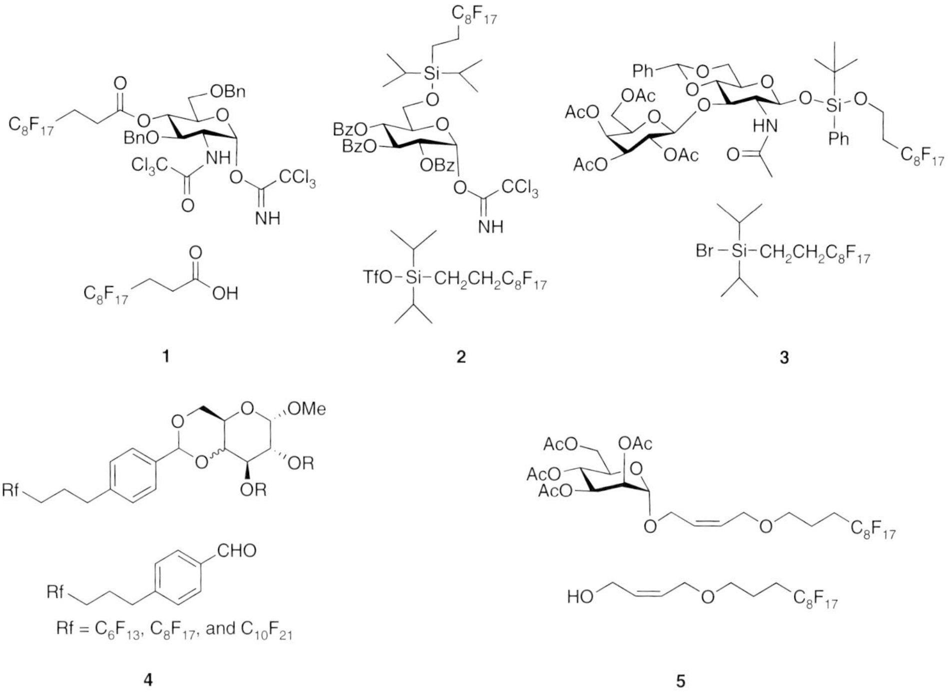

Scheme 8.1 Fluorous tags and protecting groups for alcohol and diol used in oligosaccharide synthesis.

using FSPE [17]. After the final deprotection, the product was collected in the nonfluorous fraction using FSPE – a nice example of purification of an extremely polar molecule by FSPE. Neither reverse-phase column chromatography nor expensive fluorous solvents were required for the purification [17]. A related fluorous silyl-protecting group (**3**, Scheme 8.1) has been used to block the anomeric hydroxyl group for the synthesis of a disaccharide [18]. In the case of protection of the relatively highly reactive anomeric hydroxyl group, a less reactive fluorous silyl bromide group was used, while for the protection of a less reactive 6-OH, a more reactive triflate version was used to generate the silyl ether linkage. As silyl ethers are not always stable under either acidic or basic conditions, care must be exercised in utilizing these protecting groups in oligosaccharide synthetic schemes.

An acid-labile fluorous acetal-protecting group (**4**, Scheme 8.1) was reported in 2007 and found to be stable in basic medium, similar to its nonfluorous counterpart benzylidene acetal [19]. The fluorous tag also did not interfere with protocols, allowing the regioselective opening of the acetal group to reveal either the 4-position or 6-position as a free hydroxyl group. Several disaccharides were successfully synthesized using this fluorous acetal, and FSPE could ease the purification process [19].

An alkenyl fluorous tag (**5**, Scheme 8.1) has also been developed for use in carbohydrate synthesis [20]. The alkene and the oxygen serve as a sufficient spacer between the anomeric carbon and the electron-withdrawing fluorous portion so as to rarely affect glycosylation reactions carried out by the modified carbohydrate building block. The presence of the oxygen atom helps in solubilizing the tag at low temperature, which was an issue with an analogous fluorous tag synthesized later without this oxygen atom [21]. In addition to removal by standard techniques involving allylic isomerization [22], the fluorous allyl group could easily be removed by ozonolysis of the double bond followed by oxidation to provide a carboxylic acid-terminated oligosaccharide. These oligosaccharides can then conveniently be used for biological studies after attachment via amide coupling to other supports [23–25]. Alternatively, the double bond can be reduced by hydrogenation and the fluorous-tagged carbohydrates can be used, as discussed in the following text, directly for noncovalent fluorous microarrays [20].

8.2.2
Amine Protection

After hydroxyl groups, amines are the next most prevalent functional group found in glycans that requires masking. However, relatively little work has been reported yet using fluorous variants for nitrogen protection in the realm of carbohydrates. The *N*-sulfates and *N*-acetyl groups found in natural products are too polar or basic to undergo many other common protection or other reaction steps in a typical oligosaccharide synthetic sequence and therefore require blocking. To this end, the synthesis and use of a fluorous alkyl carbamate-type nitrogen-protecting group (**6**, Scheme 8.2) were described for the protection and purification of amino sugars. The protecting group is stable in basic medium and can be cleaved in acidic media as well as with zinc in acetic anhydride to provide an *N*-acetylated sugar [26]. Very recently, a fluorous Cbz-type amine-protecting tag (**7**, Scheme 8.2) was also used to protect the amine of a linker used in the modular synthesis of heparan sulfate oligosaccharides [27]. However, unlike compound **6**, fluorous tag was not used in the direct protection of an amino sugar.

8.2.3
Phosphate Protection

In addition to hydroxyls and amines, sulfates and phosphates are also often found in biologically active oligosaccharides. Normally, phosphates and sulfates are introduced selectively toward the end of a synthesis, and their introduction makes the resulting compound extremely polar. This excessive polarity makes the compounds challenging to purify – an issue potentially mitigated by masking the charge of phosphates and sulfates. Unfortunately, to date, no fluorous sulfate-protecting groups have been developed; however, recently two options for phosphate protection with a fluorous tag have been published. The synthesis and application of a fluorous phosphate-protecting group were first reported in 2012 (**8**, Scheme 8.2)

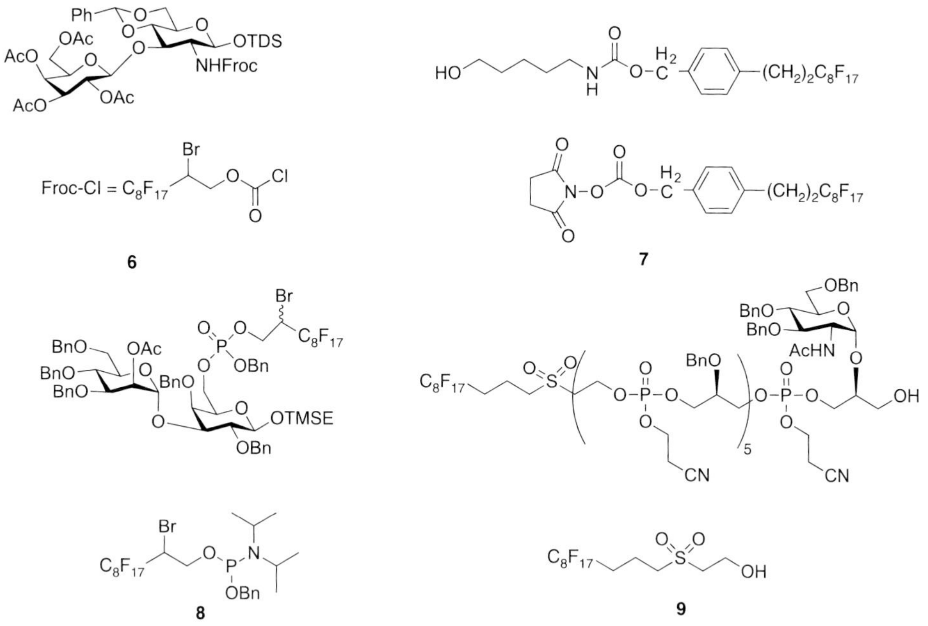

Scheme 8.2 Fluorous-protecting groups for amine and phosphate used in oligosaccharide synthesis.

[28]. This fluorous moiety was employed for the synthesis of various biomolecules such as a disaccharide present in the lipophosphoglycan of the *Leishmania* parasite [28]. Interestingly, this work sprang out of failed attempts to develop an analogous protecting group for sulfated sugars. Initially, the presence of a stereocenter at the bromoalkyl part coupled with generation of another stereocenter from the phosphorous center raised concerns about getting mixtures of diastereomers. Eventually, from ^{1}H NMR studies, it was found that the remote presence of the haloalkyl stereocenter from the carbohydrate center did not overly complicate the proton spectrum.

This protecting group was found to be stable under both acidic and basic conditions. Shortly after this initial report, another new fluorous phosphate-protecting group (**9**, Scheme 8.2) was published with application to the production of teichoic acid fragments [29]. Both the protecting groups are stable to regular acidic conditions required in carbohydrate synthesis. Protecting group **9** requires 25% ammonium hydroxide solution for deprotection, which may cleave some base-labile protecting groups already present in the molecule, whereas **8** requires relatively milder conditions using zinc and ammonium formate. The same protecting group was then used in the synthesis of a series of methyl maltosides (Scheme 8.3) to test the utility of the fluorous tag in both early and late stage introductions. When the protecting group was introduced during the synthesis of several maltose-type derivatives at an early stage of the synthesis, mixture of diastereomers arose that were challenging to separate [30].

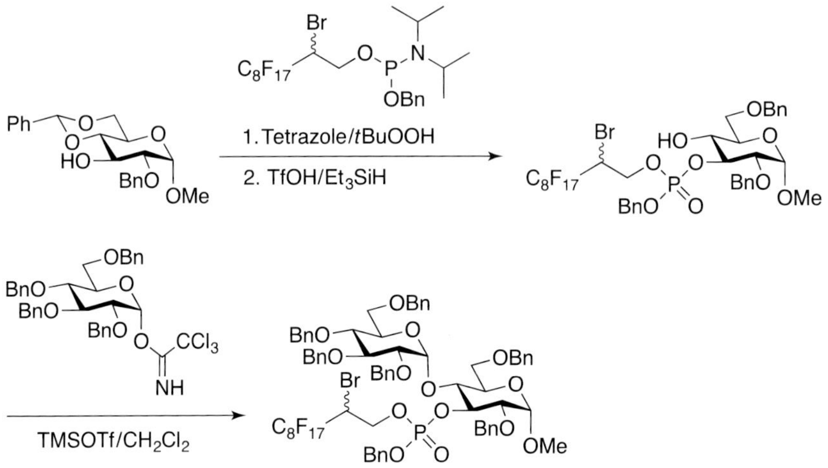

Scheme 8.3 Early stage introduction of a fluorous phosphate-protecting group in the synthesis of methyl maltotriose phosphates. (Reproduced with permission from [30]. Copyright 2013, Elsevier.)

8.3
Light Fluorous-Protecting Groups with Potential Use in Oligosaccharide Synthesis

In addition to the various fluorous-containing protecting groups designed to ease the purification of carbohydrate intermediates, a range of fluorous groups have been reported with the potential for applications in the construction of oligosaccharides.

8.3.1
Alcohol Protection

As the main challenge in sugar synthesis is to control the relative reactivities of various hydroxyl groups, a large arsenal of alcohol-protecting groups that allow orthogonal deprotection strategies are needed. Several fluorous alcohol-protecting groups have been reported, which could have some utility in carbohydrate synthesis. Benzyl groups are a common protecting group for alcohols. Normally a benzyl group is introduced under basic conditions, for example, using sodium hydride. However, if the molecule already contains a base-labile protecting group such as an acetate, then basic conditions cannot be used. For this purpose, fluorous benzyl trichloroacetimidate (**10**, Scheme 8.4) was synthesized [31]. This protecting group could certainly be used in masking alcohol functional groups in sugars in the presence of base-labile protecting groups using trimethylsilyl trifluoromethanesulfonate or any other mild Lewis acid. In a related work, a fluorous version of the *p*-methoxybenzyl group (**11**, Scheme 8.4) was designed, which can be introduced under sodium hydride conditions, but can be cleaved

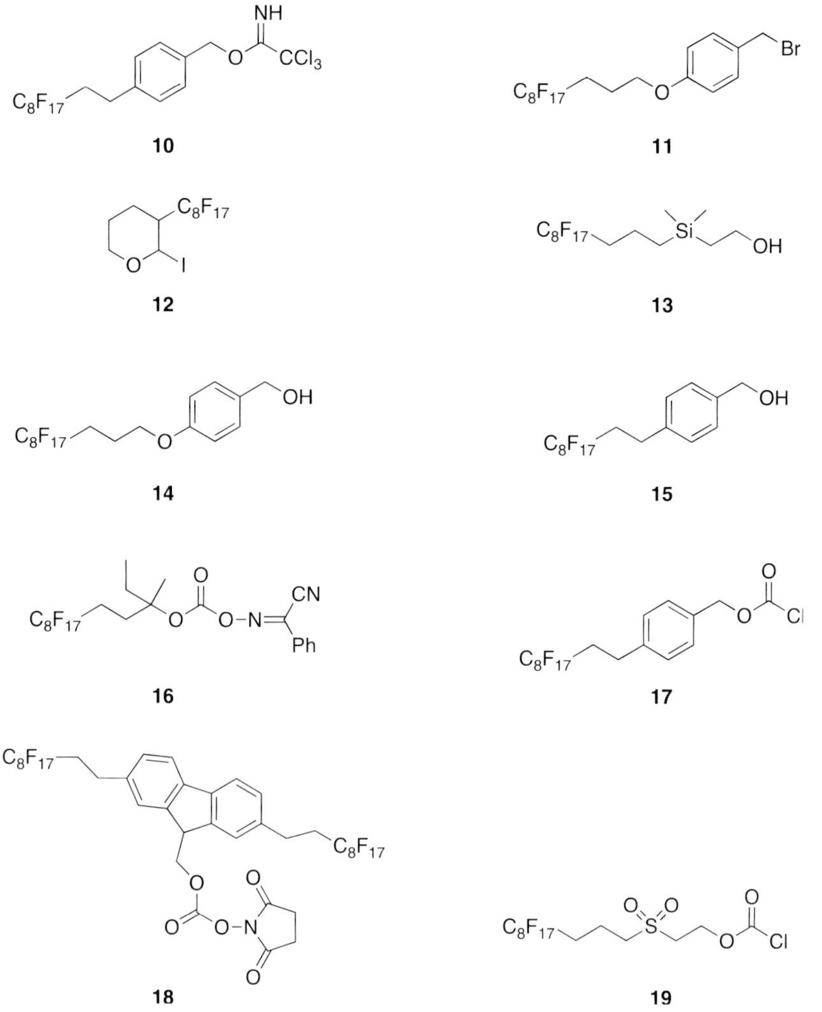

Scheme 8.4 Several fluorous-protecting groups have been reported, which could potentially be readily applied to carbohydrate synthesis.

under mild oxidative conditions using 2,3-dichloro-5,6-dicyano-1,4-benzoquinone (DDQ) [32].

A recyclable fluorous version of tetrahydropyranyl (THP) (**12**, Scheme 8.4) has been used to protect and purify small molecules [33]. The use of THP as a protecting group is generally problematic as a stereocenter is generated at the halogen-bearing carbon that becomes an acetal. The combination of carbohydrates with this type of protecting group can generate inseparable mixtures of diastereomers. In rare cases when the generation of diastereomers is not an issue, the fluorous THP could find use as a temporary protecting group in oligosaccharide synthesis with the advantage that the THP by-product can be recycled after cleavage of the protecting group.

8.3.2
Carboxylic Acid Protection

Uronic acid-type sugars are key components in hyaluronic acid, and glycosaminoglycan-type structures and their use often require masking groups for the reactive carboxylic acids. Although to date no published work describes the use of any fluorous-protecting groups for such uronic acids, several fluorous carboxylic-acid-protecting groups have been reported in other contexts that could find use in carbohydrate chemistry. A 2-trimethylsilylethyl-type fluorous linker (**13**, Scheme 8.4) was developed to synthesize a tripeptide. This group could easily be cleaved using tetrabutylammonium fluoride (TBAF) [34]. TBAF conditions can be extremely basic, and some common base-labile protecting groups (acetate or fluorenylmethoxycarbonyl (Fmoc), for example) can sometimes be inadequately cleaved. However, use of TBAF buffered with acetic acid or hydrofluoric acid [35] can solve this problem and provide an avenue for use of this protecting group in carbohydrate chemistry. A fluorous *p*-methoxybenzyl (**14**, Scheme 8.4) and a fluorous benzyl (**15**, Scheme 8.4) protecting group have been reported for the protection of carboxylic acids during peptide synthesis [36, 37]. Both of these benzyl derivatives likely could be effectively used for the protection of uronic acids in oligosaccharide synthesis.

8.3.3
Amine Protection

Several fluorous amine-protecting groups were developed by various groups over the years to facilitate peptide and small molecule synthesis, and these protecting groups could in the future expedite oligosaccharide syntheses containing amino sugars.

The first fluorous versions of (*tert*-butoxycarbonyl) Boc-anhydride were reported in 2001 [38]. In its initial design, the protecting group had two fluorous chains and an ethylene spacer between the carbamate and the fluorous chain. However, this carbamate was too stable to be removed. Even after treatment with trifluoroacetic acid (TFA) for 63 h, only a 69% deprotection was achieved. The proximity of the highly electron-withdrawing fluorous content contributed to the stability of the carbamate linkage under these conditions. A longer spacer between the fluorous chain and the carbamate linkage was hypothesized to facilitate the desired hydrolysis reaction. Indeed, protecting groups with two fluorous chains or one with a propylene spacer were found to be more suitable for temporary protection purposes. Complete deprotection was achieved in 40 min. From these observations, it was also concluded that the number of fluorous chains had little or no effect in the hydrolysis rate. The fluorous alcohol portion of the protecting group (**16**, Scheme 8.4) could easily be recycled by evaporation in the form of a trifluoroacetate ester, which was a by-product in the TFA-mediated hydrolysis reaction. Hydrolysis of that ester using lithium hydroxide regenerated the fluorous alcohol, which could be reused.

A fluorous benzyloxycarbonyl-type of carbamate linker [39] has also been used for solid-phase peptide synthesis, in this case involving a 22-amino acid sequence (**17**, Scheme 8.4). Upon cleavage from the solid support, the mixtures were purified using fluorous high-performance (high-pressure) liquid chromatography (HPLC). Compound **17** can be stored for a long time and this observation likely makes **17** a good choice as a protecting group in many cases despite the fact that toxic and explosive chemicals such as phosgene or triphosgene are needed for its synthesis. Also, a fluorous 9-Fmoc protecting group was employed (**18**, Scheme 8.4) for peptide synthesis. As the Fmoc group with its carbonyl group and high steric size can influence the stereoselectivity of glycosylation reactions, this protecting group could readily be applied to oligosaccharide synthesis as well [40].

As fluorous Cbz-type linkers are known to be partially deprotected in acidic medium, a base-labile amine-protecting group (**19**, Scheme 8.4) was developed as an orthogonal alternative. The peptide products were purified by both fluorous HPLC and FSPE [41]. The deprotection of the fluorous tag was carried out using 2% aqueous ammonia, a reagent that would raise concerns in the presence of common carbohydrate-protecting groups such as acetates.

8.4
"Cap-Tag" Strategies or Temporary Fluorous-Protecting Group Additions

Although most common as a protecting group integrated into the overall oligosaccharide synthesis scheme, fluorous tags have also served as brief temporary tags in the solid-phase synthesis of oligosaccharides. Solid-phase synthesis approaches suffer from the inability to remove unreacted resin-bound materials until the final product cleavage from the solid support; therefore, a common solution is to cap the failed sequence to mark it for easier removal in the final step [42]. The unique properties of the fluorous tag can serve nicely for this purpose. For example, a trisaccharide was synthesized sequentially on a resin [43]. At the final step, just before the deprotection from the solid support, a fluorous silyl group (**20**, Scheme 8.5) was introduced. Then the trisaccharide was cleaved from the solid support and was purified using FSPE. Tag **20** was used again successfully for the same "cap-tag strategy" at the final stage of the synthesis of another oligosaccharide where it was used to remove failed reaction sequences [44].

one-pot oligosaccharide(This "cap-tag" strategy was also applied to a one-pot oligosaccharide synthesis in a case in which a fluorous hydrazide (**21**, Scheme 8.5) was synthesized in order to "catch" ketone-terminated tri- and tetrasaccharides. The two oligosaccharides were synthesized in a one-pot fashion and were then purified by FSPE after the attachment of the fluorous hydrazide (Scheme 8.6) [45]. one-pot oligosaccharide)Surprisingly, additional reports of "cap-tag" strategies have not followed. In most of the work published with fluorous tags and carbohydrates, a fluorous tag/protecting group has been introduced in the early stages of a synthesis and used repeatedly for purification.

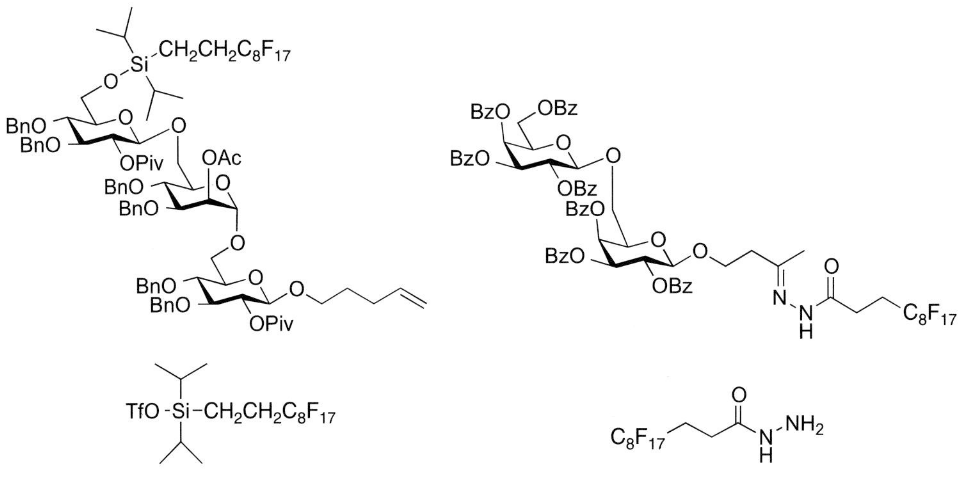

20

21

Scheme 8.5 Temporary fluorous-protecting groups used in "cap-tag" strategies for oligosaccharide synthesis.

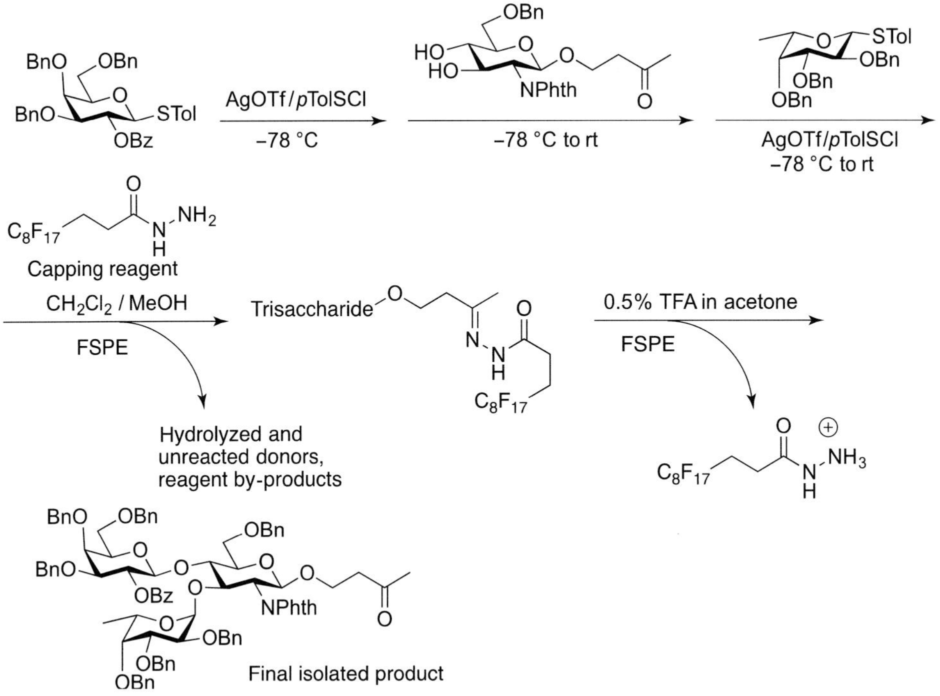

Scheme 8.6 A "cap-tag" strategy employed in a one-pot oligosaccharide synthesis. (Reproduced with permission from [45]. Copyright 2010, John Wiley & Sons.)

8.5
Double-Tagging Carbohydrates with Fluorous-Protecting Groups

One major drawback of the FSPE strategy using a mono-tagged acceptor to build oligosaccharides is the incomplete conversion of the fluorous-tagged acceptor during glycosylation. In that case, FSPE does not adequately purify the resulting mixture, as both the fluorous-tagged acceptor and the fluorous-tagged product will be eluted together in a fluorophilic fraction. One way to counter this problem is to introduce a light fluorous tag in the donor as well as the acceptor, so that the resulting glycosylation product would contain two fluorous tags rather than one [16]. Then, using an FSPE cartridge, the di-tagged product should easily be separated from the mono-tagged acceptor as the molecule with the higher fluorous content should be retained on the column for a longer period of time [13]. Such a strategy was shown indeed to be practical in the context of oligosaccharide synthesis in the separation of a growing di-tagged glucosamine chain from a mono-tagged glucosamine building block (Scheme 8.7).

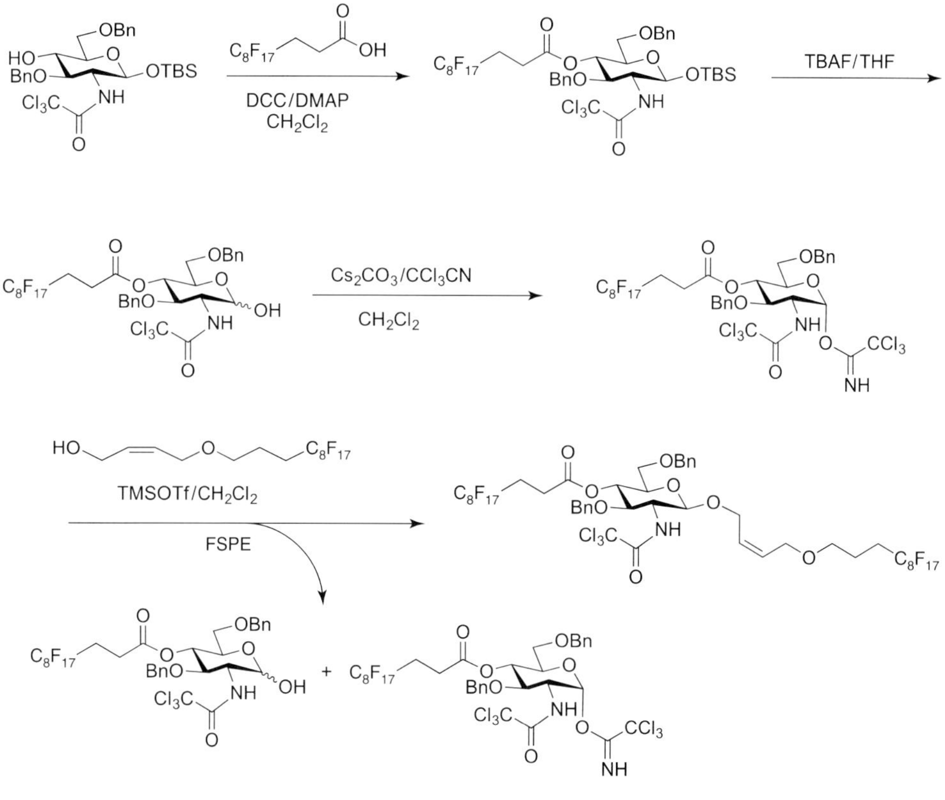

Scheme 8.7 Double-tagging strategy for oligosaccharide synthesis.

8.6
Other Advantages to Fluorous-Assisted Oligosaccharide Synthesis

8.6.1
Automated Oligosaccharide Synthesis Using Fluorous Tags

In addition to serving as an excellent handle for the purification of products with little interference in the desired reaction chemistry, fluorous tags have proven to have additional advantages. The automation of oligosaccharide synthesis is more challenging than that of peptide synthesis for a variety of reasons including the more complex protecting group patterns required for carbohydrates, and therefore building block syntheses, and the issue of controlling stereochemistry when linking building blocks together in a glycosylation reaction.

Automated solution-phase synthesis using soluble light fluorous tags provides an intriguing alternative to these traditional solid-phase methods (Scheme 8.8) [46]. This solution-phase method requires only 1.5–3 equiv of the relatively costly glycosyl donors for glycosylation and allows standard solution-phase reaction monitoring techniques such as thin layer chromatography and mass spectrometry. The acceptor contains a fluorous-protecting group such as a fluorous allyl (**5**, Scheme 8.1), usually at the anomeric position, which enables ready purification of the growing oligosaccharide chain through FSPE. Although work in this area is still in its infancy, to date noncovalent fluorous interactions have proven robust

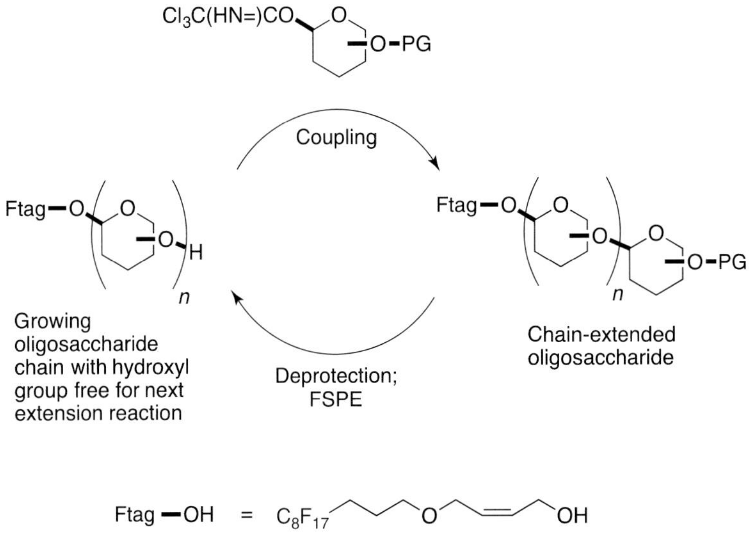

Scheme 8.8 Automated iterative solution-phase synthesis of oligosaccharides using fluorous solid-phase extraction (FSPE) to purify the growing oligosaccharide chain; *n* grows by one after each coupling cycle.

enough to reliably separate growing oligosaccharide chains with a variety of protecting groups and tags using standardized automated FSPE routines. In this context, a branched pentamannoside was synthesized using conditions amenable to automation (Scheme 8.9). Various protecting groups were used to synthesize this branched oligomannose and provide the desired regio- and stereoselectivities in the final product [47].

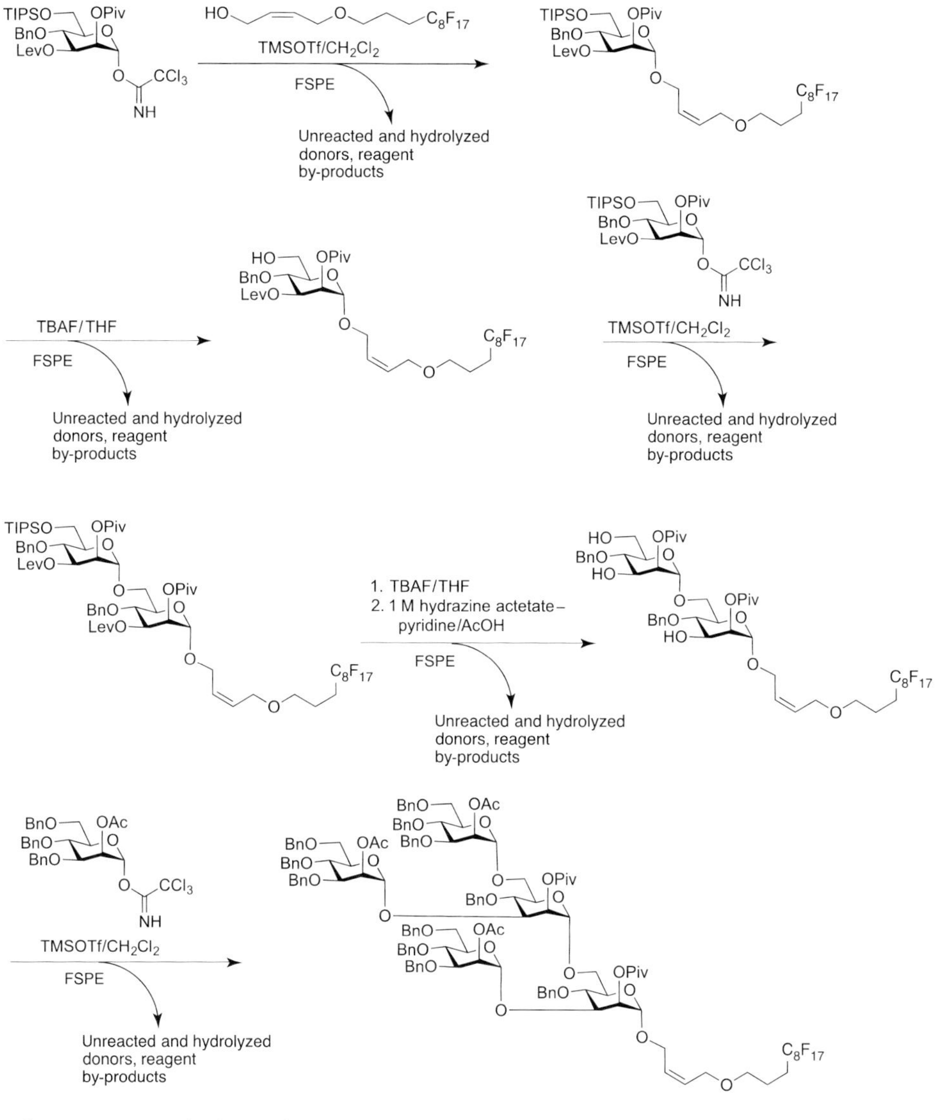

Scheme 8.9 Toward solution-phase-automated iterative synthesis: fluorous-tag-assisted solution-phase synthesis of branched mannose oligomers.

8.6.2
Fluorous-Based Carbohydrate Microarrays

Introduction of a light fluorous tag not only facilitates oligosaccharide purification by methods amenable to automated routines, but can also facilitate the study of carbohydrate–protein interactions. Given the challenges and limited availability of most carbohydrate structures, the use of microarrays, which require less than milligram quantities of material for the production of numerous spots for interrogation by possible binding partners, is appealing. Interestingly, the same noncovalent fluorous–fluorous interactions that permit the use of FSPE for the separation of fluorous-tagged compounds also allow the spatial patterning of such compounds [46, 48–51].

While most of microarray strategies involve covalent immobilization of the analyte on the slide, a fluorous microarray approach involves noncovalent hydrophobic interactions between the fluorous-coated glass slide and the fluorous compound. This method can thereby save synthetic steps required for sugar functionalization and conjugation (Figure 8.1), especially in cases when a fluorous-protecting group is already used to aid the synthesis as in cases discussed earlier. Normally, a fluorous-tagged sugar is immobilized in a multivalent display on a fluorous silica-coated glass slide. Excess carbohydrate is then washed off before the glass slide is incubated with a fluorescently labeled protein of interest. Once the excess unbound protein is rinsed from the microarray slide, sugar–protein binding events are observed under a standard fluorescence scanner [48–52].

8.7
Conclusions and Outlook

With the advent of automation methods amenable to the use of fluorous-protecting group tags, explorations of the scope and limitations of fluorous-protecting groups

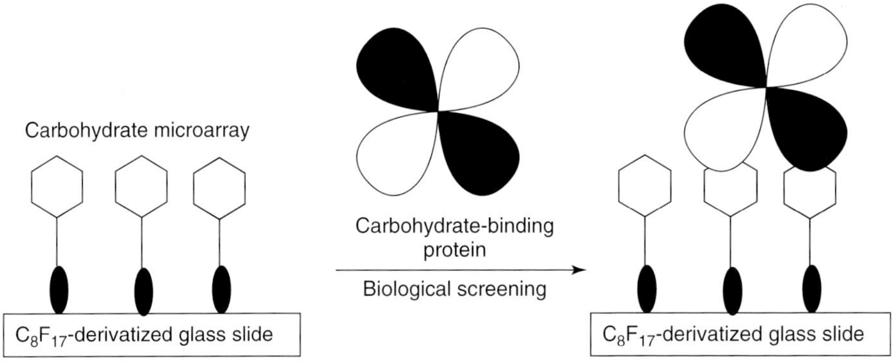

Figure 8.1 Fluorous-based carbohydrate microarrays allow the patterning of fluorous-tagged carbohydrates based on noncovalent fluorous–fluorous interactions.

as well as the development of new fluorous tags become even more important. Tuning of the reactivity of fluorous-tagged compounds has barely been examined. So far, in almost all the protecting group designs, the fluorous chain is placed away from the reaction center, leading the resulting compound to behave largely like the nonfluorous versions of the protecting groups. However, if spacing between the fluorous chain and the active reaction center is reduced, entirely different reactivity patterns could be accessed. The electronic properties of suitably placed protecting groups are already known to affect the stereoselectivities of glycosylation reactions, and fluorous tags could possibly be enlisted to set the anomeric ratios of products. Also, fluorous-tagged oligosaccharides could form micelle-like structures and access different conformations depending on their concentration and the nature of the solvents. These properties could also be used to alter the reaction patterns of the carbohydrate moieties. Clearly, the strong electron-withdrawing properties and the solvent-dependent reaction kinetics of long perfluorinated chains have only begun to be fully exploited in the realm of carbohydrate chemistry.

8.8 Experimental Section

8.8.1 Synthesis of 6-(Benzyl 2-bromo-3,3,4,4,5,5,6,6,7,7,8,8,9,9,10,10,10-heptadecafluorodecyl phosphate)-1,2,3,4-di-*O*-isopropylidene-α-D-galactopyranose

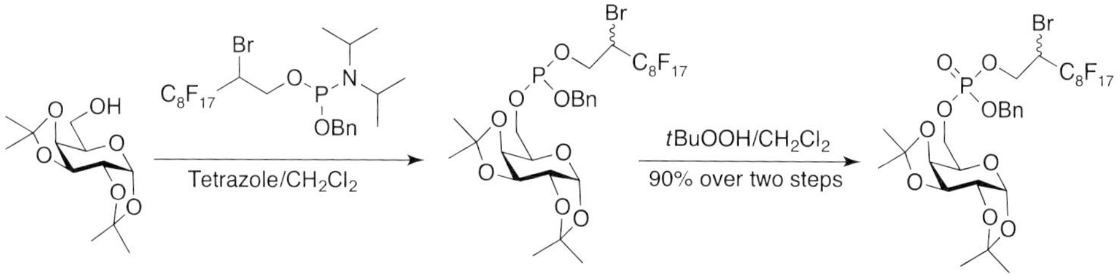

Benzyl 2-bromo-3,3,4,4,5,5,6,6,7,7,8,8,9,9,10,10,10-heptadecafluorodecyldiisopropylphosphoramidite (0.17 g, 0.22 mmol) and di-acetone-D-galactose (0.17 g, 0.66 mmol) were dissolved in CH_2Cl_2 (10 ml) at ambient temperature [15]. A solution (1 ml, 0.45 mmol) of tetrazole in CH_3CN (0.45 M) was added dropwise and the mixture was stirred overnight. A solution (200 μl) of *t*BuOOH (5–6 M) in nonane was added, and the reaction was stirred for another 4 h. The reaction was diluted with CH_2Cl_2, washed with saturated $NaHCO_3$ (aq), and dried over anhydrous Na_2SO_4. Solvents were removed under reduced pressure and the mixture was purified using FSPE. The crude reaction mixture was dissolved in anhydrous dimethylformamide (DMF) (0.4 ml) and was loaded onto the fluorous cartridge. The nonfluorous fractions were eluted using 80% methanol/water (8 ml). The final fluorous compound was eluted using methanol (8 ml). Methanol was removed under reduced pressure to afford the product as a white solid (0.2 g, 0.2 mmol, 90%).

8.8.2
Synthesis of 3-(Perfluorooctyl)propanyloxybutenyl-3,4,6-tri-*O*-acetyl-2-deoxy-2-(*p*-nitrobenzyloxycarbonylamino)-β-D-glucopyranoside

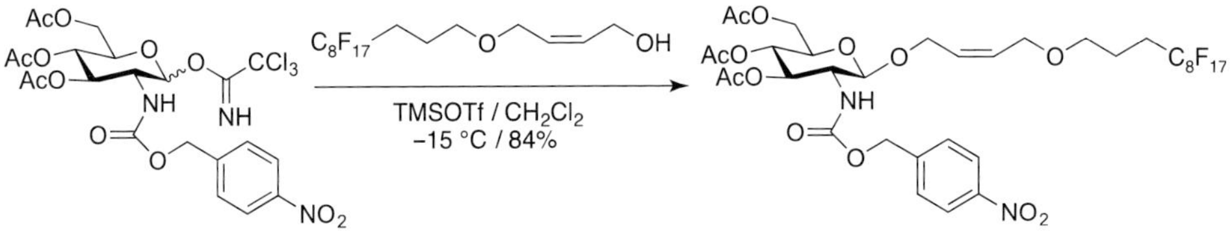

To a solution of 3,4,6-tri-*O*-acetyl-2-deoxy-2-(*p*-nitrobenzyloxycarbonylamino)-α/β-glucopyranosyltrichloroacetimidate (1.70 g, 2.70 mmol) and 3-(perfluorooctyl)propanyloxybutenyl alcohol (1.00 g, 1.80 mmol) in dichloromethane (20 ml), TMSOTf (0.20 ml, 0.90 mmol) was added at −15 °C [9]. The reaction mixture was stirred at −15 °C for 30 min. The reaction mixture was quenched with triethylamine (0.50 ml) and then concentrated under reduced pressure. The crude product was purified by solid-phase extraction using a FSPE cartridge. Nonfluorous compounds were eluted with 80% MeOH/water and the desired product was eluted by 100% MeOH. The solvent was removed under reduced pressure to provide the product (1.53 g, 1.51 mmol, 84%) as a solid.

8.8.3
Synthesis of 3-(Perfluorooctyl)propanyloxybutenyl-4-*O*-benzyl-3,6-di-*O*-(2-*O*-acetyl-3,4,6-*O*-tribenzyl-α-D-mannopyranoside)-2-*O*-pivaloyl-α-D-mannopyranoside

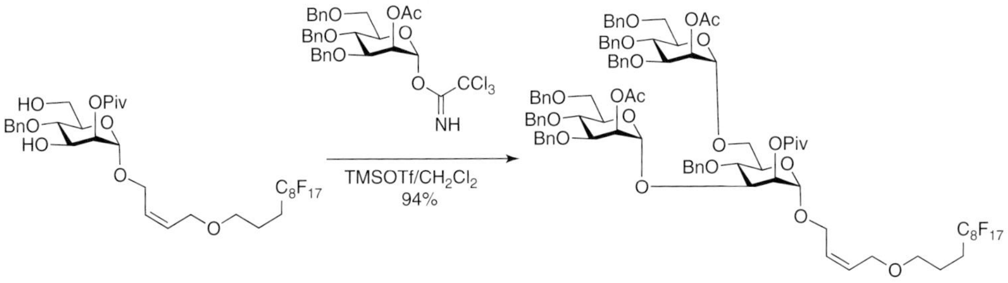

A solution of 3-(perfluorooctyl)propanyloxybutenyl-4-*O*-benzyl-3,6-dihydroxy-2-*O*-pivaloyl-α-D-mannopyranoside (0.030 g, 0.03 mmol) and 2-*O*-acetyl-3,4,6-*O*-tribenzyl-α/β-D-mannopyranosyl trichloroacetimidate (0.07 mg, 0.10 mmol) in dichloromethane (3 ml) was cooled to 5 °C and TMSOTf (15 μl, 0.80 mmol) was added [47]. The reaction mixture was stirred for 30 min. The reaction mixture was quenched with triethylamine (30 μl) and then concentrated under reduced pressure. The crude product was purified by solid-phase extraction using a FSPE cartridge. Nonfluorous compounds were eluted with 80% MeOH/water (8 ml) and the desired product was eluted by 100% MeOH (8 ml). The solvent was removed under reduced pressure to obtain mannose trisaccharide (0.06 mg, 0.03 mmol, 94%) as a yellow gel.

Acknowledgments

This work was supported in part by the National Institutes of Health (1R01GM090280, U19AI091031) and by the National Science Foundation (CHE-0911123/CHE-1261046).

List of Abbreviations

Ac	Acetate
Boc	*tert*-Butoxycarbonyl
Bn	Benzyl
Bz	Benzoyl
Cbz	Carboxybenzyl
DDQ	2,3-Dichloro-5,6-dicyano-1,4-benzoquinone
DMF	*N,N*-dimethylformamide
Fmoc	Fluorenylmethoxycarbonyl
Froc	Fluorous version of the trichlorethoxycarbonyl group
FSPE	Fluorous solid-phase extraction
Ftag	Tag with fluorous content
HPLC	High-performance (high-pressure) liquid chromatography
Me	Methyl
NMR	Nuclear magnetic resonance
PG	Protecting group
Ph	Phenyl
TBAB	Tetrabutylammonium bromide
TBAF	Tetrabutylammonium fluoride
TFA	Trifluoroacetic acid
THP	Tetrahydropyranyl
TMSE	2-(Trimethylsilyl)ethyl
TMSOTf	Trimethylsilyl trifluoromethanesulfonate

References

1. Merrifield, R.B. and Stewart, J.M. (1965) *Nature*, **207**, 522–523.
2. Fields, C.G., Lloyd, D.H., Macdonald, R.L., Otteson, K.M., and Noble, R.L. (1991) *Pept. Res.*, **4**, 95–101.
3. Schuerch, C. and Frechet, J.M. (1971) *J. Am. Chem. Soc.*, **93**, 492–496.
4. Nicolaou, K.C., Winssinger, N., Pastor, J., and DeRoose, F. (1997) *J. Am. Chem. Soc.*, **119**, 449–450.
5. Nicolaou, K.C., Watanabe, N., Li, J., Pastor, J., and Winssinger, N. (1998) *Angew. Chem., Int. Ed. Engl.*, **37**, 1559–1561.

6. Plante, J.O., Palmacci, E.R., and Seeberger, P.H. (2001) *Science*, **291**, 1523–1527.
7. Zhang, W. (2003) *Tetrahedron*, **59**, 4475–4489.
8. Ratner, D.M., Swanson, E.R., and Seeberger, P.H. (2003) *Org. Lett.*, **5**, 4717–4720.
9. Love, K.R. and Seeberger, P.H. (2004) *Angew. Chem. Int. Ed.*, **43**, 602–605.
10. Werz, D.B., Castagner, B., and Seeberger, P.H. (2007) *J. Am. Chem. Soc.*, **129**, 2770–2771.
11. Liu, X., Wada, R., Boonyarattanakalin, S., Castagner, B., and Seeberger, P.H. (2008) *Chem. Commun.*, **30**, 3510–3512.
12. Pohl, N.L. (2007) in *Current Fluoroorganic Chemistry: New Synthetic Directions, Technologies, Materials, and Biological Applications*, ACS Symposium Series, Vol. 949 (eds V.A. Soloshonok, K. Mikami, T. Yamazaki, J.T. Welch, and J.F. Honek), American Chemical Society, Washington, DC, pp. 261–270.
13. Curran, D.P. and Luo, Z. (1999) *J. Am. Chem. Soc.*, **121**, 9069–9072.
14. Zhang, W. (2009) *Chem. Rev.*, **109**, 749–795.
15. Zhang, W. (2004) *Curr. Opin. Drug Discovery Dev.*, **7**, 784–797.
16. Park, G., Ko, K.-S., Zakharova, A., and Pohl, N.L. (2008) *J. Fluorine Chem.*, **129**, 978–982.
17. Zhang, F., Zhang, W., Zhang, Y., Curran, D.P., and Liu, G. (2009) *J. Org. Chem.*, **74**, 2594–2597.
18. Manzoni, L. (2003) *Chem. Commun.*, **23**, 2930–2931.
19. Kojima, M., Nakamura, Y., and Takeuchi, S. (2007) *Tetrahedron Lett.*, **48**, 4431–4436.
20. Mamidyala, S.K., Ko, K.-S., Jaipuri, F.A., Park, G., and Pohl, N.L. (2006) *J. Fluorine Chem.*, **127**, 571–579.
21. Carrel, F.R., Geyer, K., Codée, J.D.C., and Seeberger, P.H. (2007) *Org. Lett.*, **9**, 2285–2288.
22. Mantilli, L. and Mazet, C. (2009) *Tetrahedron Lett.*, **50**, 4141–4144.
23. Song, E.-H., Osanya, A.O., Petersen, C.A., and Pohl, N.L.B. (2010) *J. Am. Chem. Soc.*, **132**, 11428–11430.
24. Carrillo-Conde, B., Song, E.-H., Chavez-Santoscoy, A., Phanse, Y., Ramer-Tait, A.E., Pohl, N.L.B., Wannemuehler, M.J., Bellaire, B.H., and Narasimhan, B. (2011) *Mol. Pharm.*, **8**, 1877–1886.
25. Chavez-Santoscoy, A., Roychoudhury, R., Pohl, N.L.B., Wannemuehler, M.J., and Narasimhan, B. (2012) *Biomaterials*, **33**, 4762–4772.
26. Manzoni, L. and Castelli, R. (2006) *Org. Lett.*, **8**, 955–957.
27. Zong, C., Venot, A., Dhamale, O., and Boons, G.-J. (2013) *Org. Lett.*, **15**, 342–345.
28. Liu, L. and Pohl, N.L.B. (2011) *Org. Lett.*, **13**, 1824–1827.
29. Hogendorf, W.F.J., Lameijer, L.N., Beenakker, T.J.M., Overkleeft, H.S., Filippov, D.V., Codée, J.D.C., and van der Marel, G.A. (2012) *Org. Lett.*, **14**, 848–851.
30. Liu, L. and Pohl, N.L.B. (2013) *Carbohydr. Res.*, **369**, 14–24.
31. Dakas, P.Y., Barluenga, S., Totzke, F., Zirrgiebel, U., and Winssinger, N. (2007) *Angew. Chem. Int. Ed.*, **46**, 6899–6902.
32. Zhang, Q.S., Lu, H.J., Richard, C., and Curran, D.P. (2004) *J. Am. Chem. Soc.*, **126**, 36–37.
33. Wipf, P. and Reeves, J.T. (1999) *Tetrahedron Lett.*, **40**, 4649–4652.
34. Fustero, S., Sancho, A.G., Chiva, G., Sanz-Cervera, J.F., del Pozo, C., and Aceña, J.L. (2006) *J. Org. Chem.*, **71**, 3299–3302.
35. Schwarz, J.B., Kuduk, S.D., Chen, X.-T., Sames, D., Glunz, P.W., and Danishefsky, S.J. (1999) *J. Am. Chem. Soc.*, **121**, 2662–2673.
36. Wang, X., Nelson, S.G., and Curran, D.P. (2007) *Tetrahedron*, **63**, 6141–6145.
37. Flögel, O., Codée, J.D.C., Seebach, D., and Seeberger, P.H. (2006) *Angew. Chem. Int. Ed.*, **45**, 7000–7003.
38. Luo, Z., Williams, J., Read, R.W., and Curran, D.P. (2001) *J. Org. Chem.*, **66**, 4261–4266.
39. Filippov, D.V., van Zoelen, D.J., Oldfield, S.P., van der Marel, G.A., Overkleeft, H.S., Drijfhout, J.W., and van Boom, J.H. (2002) *Tetrahedron Lett.*, **43**, 7809–7812.
40. Matsugi, M., Yamanaka, K., Inomata, I., Takekoshi, N., Hasegawa, M., and

Curran, D.P. (2006) *QSAR Comb. Sci.*, **25**, 713–715.

41. de Visser, P.C., van Helden, M., Filippov, D.V., van der Marel, G.A., Drijfhout, J.W., van Boom, J.H., Noort, D., and Overkleeft, H.S. (2003) *Tetrahedron Lett.*, **44**, 9013–9016.
42. Seeberger, P.H. (2003) *Chem. Commun.*, **10**, 1115–1121.
43. Carrel, F.R. and Seeberger, P.H. (2008) *J. Org. Chem.*, **73**, 2058–2065.
44. Palmacci, E.R., Hewitt, M.C., and Seeberger, P.H. (2001) *Angew. Chem. Int. Ed.*, **40**, 4433–4437.
45. Yang, B., Jing, Y., and Huang, X. (2010) *Eur. J. Org. Chem.*, **7**, 1290–1298.
46. Pohl, N.L. (2008) in *Automated Solution-Phase Oligosaccharide Synthesis and Carbohydrate Microarrays: Development of Fluorous-Based Tools for Glycomics*, ACS Symposium Series, Vol. 990 (eds X. Chen, R. Halcomb, and G.P. Wang), American Chemical Society, Washington, DC, pp. 272–287.
47. Jaipuri, F.A. and Pohl, N.L. (2008) *Org. Biomol. Chem.*, **6**, 2686–2691.
48. Jaipuri, F.A., Collet, B.Y.M., and Pohl, N.L. (2008) *Angew. Chem. Int. Ed.*, **47**, 1707–1710.
49. Pohl, N.L. (2008) *Angew. Chem. Int. Ed.*, **47**, 3868–3870.
50. Ko, K.S., Jaipuri, F.A., and Pohl, N.L. (2005) *J. Am. Chem. Soc.*, **127**, 12162–12163.
51. Chen, G.S. and Pohl, N.L. (2008) *Org. Lett.*, **10**, 785–788.
52. Edwards, H.D., Nagappayya, S.K., and Pohl, N.L.B. (2012) *Chem. Commun.*, **48**, 510–512.

9
Advances in Cyclodextrin Chemistry

Samuel Guieu and Matthieu Sollogoub

9.1
Introduction

Cyclodextrins are a class of cyclic oligomers composed of glucopyranosidic units in the $^4C_1$ conformation, linked in an α-1,4 manner (Figure 9.1). They were isolated by Villiers [1] in 1891, but their structure was elucidated only in 1942 [2]. They are obtained by enzymatic degradation of starch, using cyclodextrin glucosyltransferase (CGTase), as a mixture of α-cyclodextrin **1**, β-cyclodextrin **2**, and γ-cyclodextrin **3**, possessing six, seven, or eight glucosides, respectively, and other oligomers [3]. Industrially, they are separated by selective precipitation in yields ranging from 40 to 60%.

α-, β-, and γ-Cyclodextrins adopt a conformation in the shape of a truncated cone with C_n symmetry, n being the number of glucose units. This cone possesses a hydrophobic internal cavity and an external hydrophilic surface. As a consequence, cyclodextrins are moderately soluble in polar solvents. Their cavity can accept apolar guests to form inclusion complexes in polar solvents [4].

The height of all cyclodextrins is the same (0.79 nm), dictated by the size of the glucose units. The diameter of their cavity increases with the number of glucose units contained in the structure (Figure 9.2): 0.49 nm for α-cyclodextrin, 0.62 nm for β-cyclodextrin, and 0.79 nm for γ-cyclodextrin [4].

In native cyclodextrins, hydroxyls in position 2 are pointing inward, and hydroxyls in position 3 are pointing outward (Figure 9.1). The distance between the hydroxyls in positions 2 and 3 of two consecutive glucose units varies from 2.82 (α-cyclodextrin) to 2.98 Å (γ-cyclodextrin). Hydrogen bonds can easily form through these distances (Figure 9.3) and restrain molecular motion. Native cyclodextrins have a relatively rigid structure.

These intramolecular hydrogen bonds modulate the solubility of cyclodextrins in water (145 $g\,l^{-1}$ for α-cyclodextrin **1**, 18.5 $g\,l^{-1}$ for β-cyclodextrin **2**, and 232 $g\,l^{-1}$ for γ-cyclodextrin **3**) [3]. The structure of β-cyclodextrin **2** is particularly favorable to the intramolecular hydrogen bonds, which explains its particularly poor solubility in water. α-Cyclodextrin **1** is partially distorted, and only four hydrogen bonds can

Modern Synthetic Methods in Carbohydrate Chemistry: From Monosaccharides to Complex Glycoconjugates, First Edition. Edited by Daniel B. Werz and Sébastien Vidal.

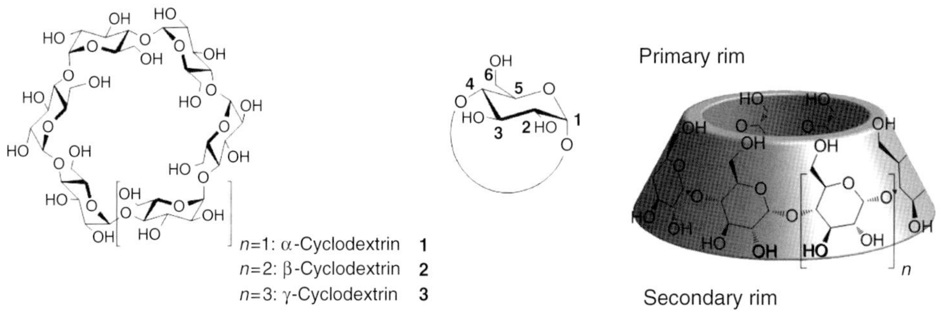

Figure 9.1 Structure of cyclodextrins.

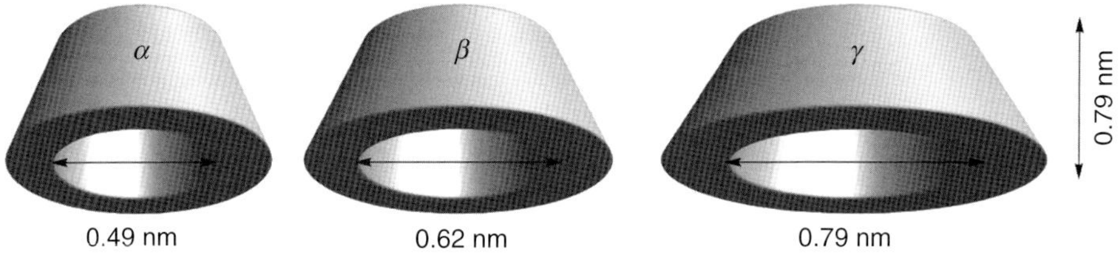

Figure 9.2 Dimensions of cyclodextrins.

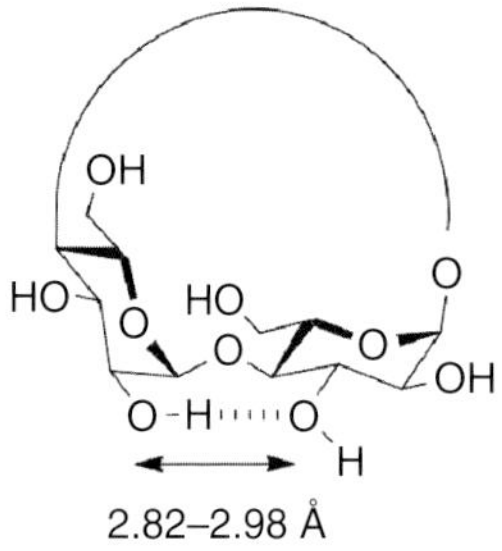

Figure 9.3 Intramolecular hydrogen bonds on the secondary rim.

form. The bigger size of γ-cyclodextrin **3** allows more flexibility, so that hydrogen bonds are less favored. All cyclodextrins are soluble in polar aprotic solvents.

Most industrial uses of cyclodextrins are based on their ability to form supramolecular inclusion complexes with diverse guests [4]. They are used as solubilizers or drug transporters for pharmaceutical applications [5], as stabilizers and taste protectors in food and cosmetic industries [6], and as static phase in chromatography [7] for the separation of chiral entities. The required properties of cyclodextrin derivatives vary depending on the desired application, such as enantioselectivity for the separation of enantiomers or stability of the inclusion complex for encapsulation of drugs. Native cyclodextrins rarely match perfectly to the application; so their modification is necessary to fine-tune their properties. In that context, efficient and selective methodologies have to be developed to access new derivatives that could be adapted to the desired application. Cyclodextrins are a particular class of complex carbohydrate derivatives, as they are easily produced on an industrial scale. Their synthesis is not a problem; so most of the research in this area is focused on the chemical modifications of native cyclodextrins.

9.1.1 Nomenclature of Modified Cyclodextrins

Following the IUPAC (International Union of Pure and Applied Chemistry) rules, glucose units are designated by roman numerals I, II, III, and so on, but the use of majuscule letters A, B, C, and so on proposed by Breslow [8] is the usual nomenclature in the literature. The glycosidic link α-1,4 imposes the numbering direction, and Tabushi proposed a view of the cyclodextrin from the primary rim: unit B follows unit A in the trigonometric direction (Figure 9.4). If more than one glucose unit is modified, numbering should follow the shortest way.

In this chapter, we use the following definitions:

To functionalize a position on a cyclodextrin means that this position has been modified. More than one position can be modified in the same manner. The cyclodextrin is *monofunctionalized* if a single position is modified, *difunctionalized* if two positions are modified, *trifunctionalized* if three positions are modified, and *perfunctionalized* if all hydroxyls have been modified in the same way.

To differentiate the positions means that different modifications have been made on these positions. The cyclodextrin is *differentiated* if two different groups are present, *tridifferentiated* if three different groups are present, *tetradifferentiated* if four different groups are present, and so on.

Different strategies have been developed to access functionalized and differentiated cyclodextrins. Perfunctionalization of the different positions relies on small differences of reactivity of the hydroxyls. Monofunctionalization methods use supramolecular complexes or precise control of the quantity of reagent. Difunctionalization is more difficult, as many regioisomers can be obtained, and the methodology is different for each case [9] (Figure 9.5). Capping reagents have been mostly developed with β-cyclodextrin, on both the primary and secondary

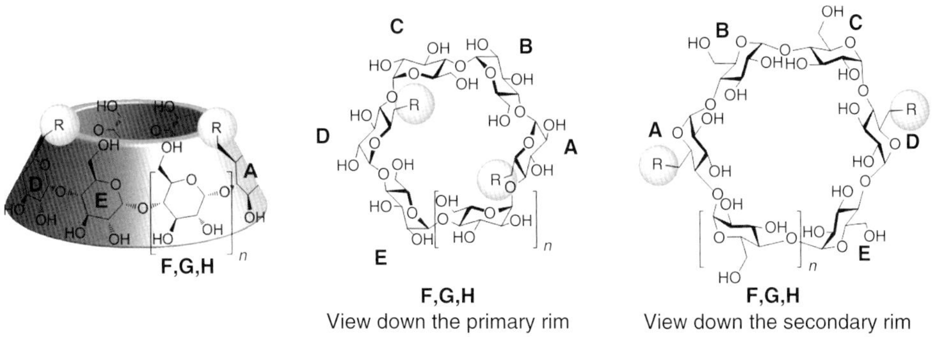

Figure 9.4 Nomenclature of modified cyclodextrins.

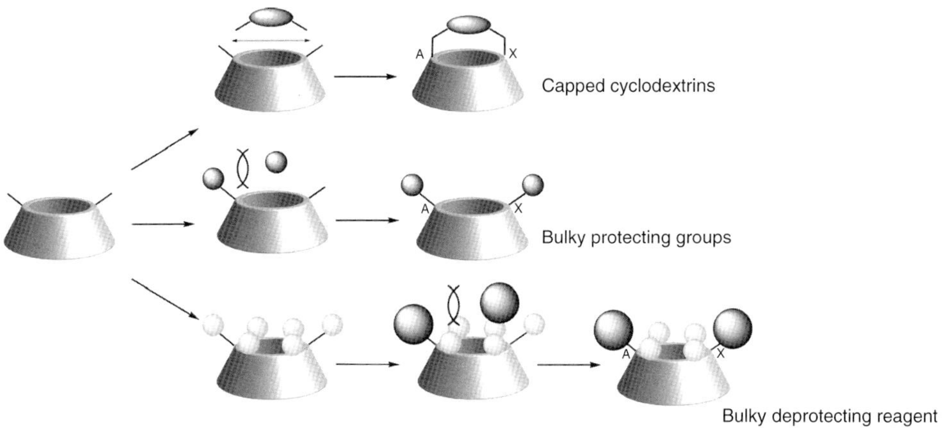

Figure 9.5 Different approaches for the selective multifunctionalization of cyclodextrins.

rims. The direct difunctionalization of the primary rim taking advantage of the steric hindrance created by bulky protecting groups has proven successful with α-cyclodextrin. Selective deprotections taking advantage of the steric hindrance created during the reaction process works on both α- and β-cyclodextrins. Owing to the flexibility of γ-cyclodextrin, very few methods give good results for its selective modification.

In this chapter, we present methodologies that have proven successful to access different patterns of functionalization and differentiation based on the approach developed toward selectivity: difference of reactivity between the different positions, control of the number of equivalents of reagents, use of supramolecular inclusion complexes, capping reagents, bulky protecting groups, and steric hindrance during deprotection.

9.2
General Reactivity, Per- and Monofunctionalization

9.2.1
General Reactivity of Cyclodextrins

Cyclodextrins possess three different hydroxyl groups (i.e., OH-2, OH-3, and OH-6) exhibiting different reactivities (Figure 9.6). Primary hydroxyls, which compose the primary rim, are less hindered, more basic, and more nucleophilic than the others. They can react with electrophiles, even if they are sterically hindered, using only a weak base. The hydrogen bond between OH-2 and OH-3 of two following glucose units reinforces the acidity of the hydroxyls OH-2, also exalted by the proximity with the electron-withdrawing anomeric acetal function, and they hence become more acidic than the others ($pK_a = 12.2$) [10]. They can be selectively deprotonated in anhydrous basic conditions. Hydroxyls OH-3 are less reactive, and can only be

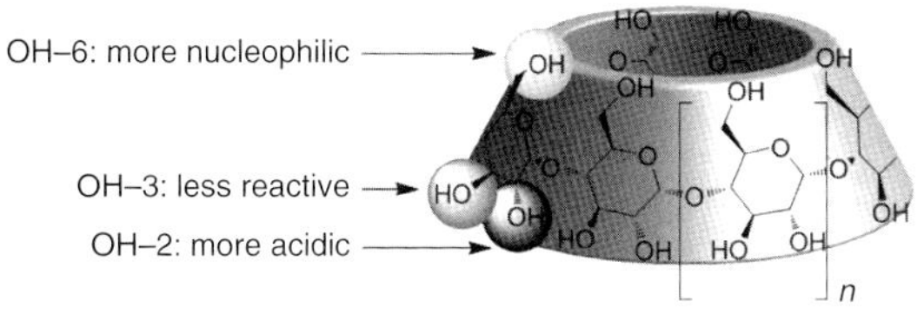

Figure 9.6 Reactivity of the different positions.

modified after protection of the other positions. Nevertheless, selectivity between the positions cannot be achieved if the conditions used are too basic or if the electrophile is too strong.

This difference of reactivity has been used to direct the perfunctionalization of each position, or the monofunctionalization if a default of reagent is employed. But this cannot direct the multifunctionalization toward a particular pattern.

9.2.2 Perfunctionalization of Each Position

The hydroxyls OH-2 are more acidic than the others, and can be selectively deprotonated using sodium hydride in anhydrous dimethylsulfoxide (DMSO). The alkoxide formed is more nucleophilic than the hydroxyls in position 3 or 6, and can be the sole to react to give cyclodextrins **4a–d** permethylated or pertosylated on their positions 2 (Scheme 9.1) [11, 12]. Nevertheless, this selectivity is altered as the number of functionalized positions increases, because the steric hindrance on the secondary rim makes the alkoxides less reactive. The electrophile can then react with the hydroxyls of the primary rim, and some side products are formed. To avoid this, the primary hydroxyls are usually protected beforehand.

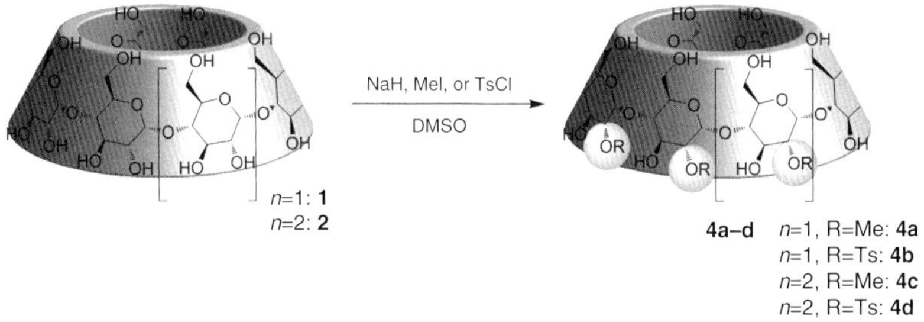

Scheme 9.1 Perfunctionalization of position 2 of α- and β-cyclodextrins.

The primary hydroxyls are better nucleophiles than the secondary ones; so they can be selectively modified using a weak base. Using *tert*-butyldimethylsilyl chloride (TBSCl) in dimethylformamide (DMF) with imidazole as a base, or in pyridine, gives access to cyclodextrins **5a–c** persilylated on their primary rim (Scheme 9.2) [13]. Once these positions are protected, the position 2 can be efficiently functionalized

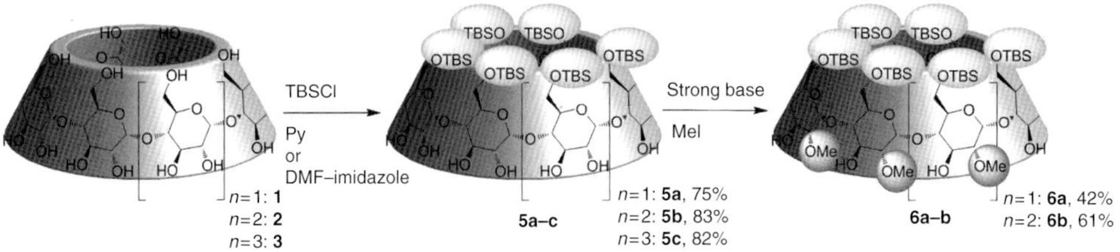

Scheme 9.2 Persilylation of the primary rim and peralkylation of the position 2 of cyclodextrins.

using a strong base, giving cyclodextrins **6a–b** with their positions 2, 3, and 6 differentiated [14].

Cyclodextrins can also be directly perhalogenated on the primary rim. Using triphenylphosphine and a halogen donor, such as iodine, bromine, or *N*-chlorosuccinimide (NCS), perhalogenated cyclodextrins **7a–e** were obtained in good yields (Scheme 9.3) [15].

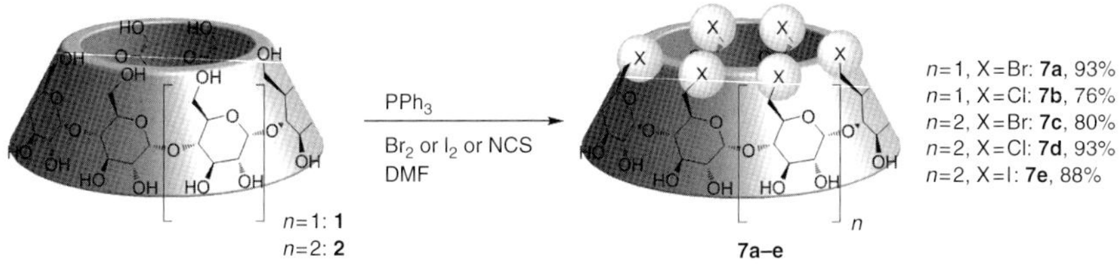

Scheme 9.3 Perhalogenation of the primary rim of α- and β-cyclodextrins.

The positions 2 and 6 can also be modified simultaneously. Using conditions similar to the selective silylation of the primary hydroxyls, but at higher temperature, positions 2 and 6 can be protected with silyl ethers to give cyclodextrins **8a–b** (Scheme 9.4) [13a,b, 16]. Using the inorganic base BaO/$Ba(OH)_2$ in a mixture of DMF and DMSO, positions 2 and 6 can be simultaneously protected as methyl ether to give cyclodextrins **9a–b** (Scheme 9.4) [17].

Another way to perfunctionalize cyclodextrins is to use the difference of reactivity of each position toward deprotection. Perbenzylated α-cyclodextrin **10a** is obtained quantitatively using benzyl chloride and sodium hydride in anhydrous DMSO [18].

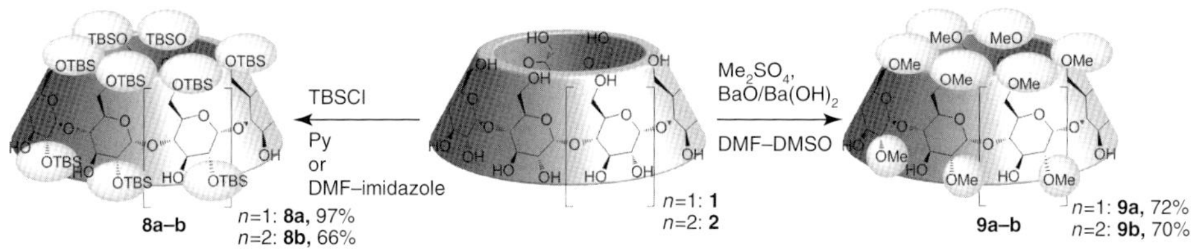

Scheme 9.4 Persilylation and permethylation of the positions 2 and 6 of α- and β-cyclodextrins.

When subjected to the action of trimethylsilyl triflate and acetic anhydride, careful control of the reaction conditions leads to cyclodextrin **11** by selective replacement of the benzyl groups from the primary rim by acetates (Scheme 9.5) [19]. The position 3 can also be selectively deprotected by the action of triethylsilane and iodine at low temperature, giving access to cyclodextrin **12** bearing a functionalization pattern similar to cyclodextrin **9a** [20]. These reactions illustrate another approach toward the selective modification of cyclodextrins: selective removal of already installed groups, which is the counterpart to the selective installation of functional groups.

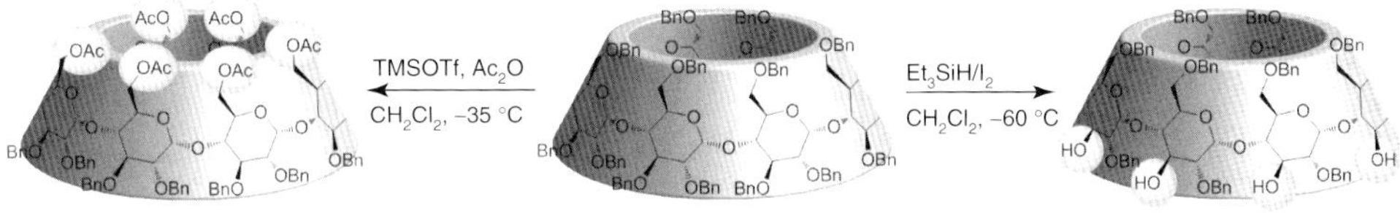

11, 95% **10a** **12**, 58%

Scheme 9.5 Perfunctionalization of position 6 or 3 by selective de-*O*-benzylations.

It is hence possible to access cyclodextrins selectively perfunctionalized in good yields without long purification processes. Potential applications of these derivatives include stationary phase for chiral separations, model for the design of artificial enzymes, and building blocks for supramolecular architectures [4, 21].

9.2.3 Monofunctionalization

Monofunctionalization of cyclodextrins has been for a long time the almost exclusive reaction used to functionalize those molecules in a practical manner. It mostly relies on the use of a reagent in default. A few examples have been developed in which a sterically hindered intermediate is formed, preventing the approach of a second molecule of reagent.

9.2.3.1 Use of Reagent in Default

Monotosylation of the primary rim is the most widely used method to access monofunctionalized cyclodextrins. Tosyl chloride reacts in pyridine to give the monosubstituted cyclodextrins **13a–b** in about 30% yield (Scheme 9.6) [22]. This low yield is partially due to the secondary elimination of the tosylate forming an alkene [11]. Many nucleophiles can react with this tosylate, giving access to a wide range of cyclodextrins monofunctionalized on their primary rim. Sodium or lithium azide reacts in DMF to give the cyclodextrins monoazide **14a–b** with yields as high as 95%. Oxidation of the tosylate using DMSO and a non-nucleophilic base gives the cyclodextrins monoaldehyde **15a–b** [23].

Cyclodextrins monoazide **14a–b** can also be synthesized by direct substitution reaction, using triphenylphosphine, carbon tetrabromide, and lithium azide in DMF, but the yields are modest [24].

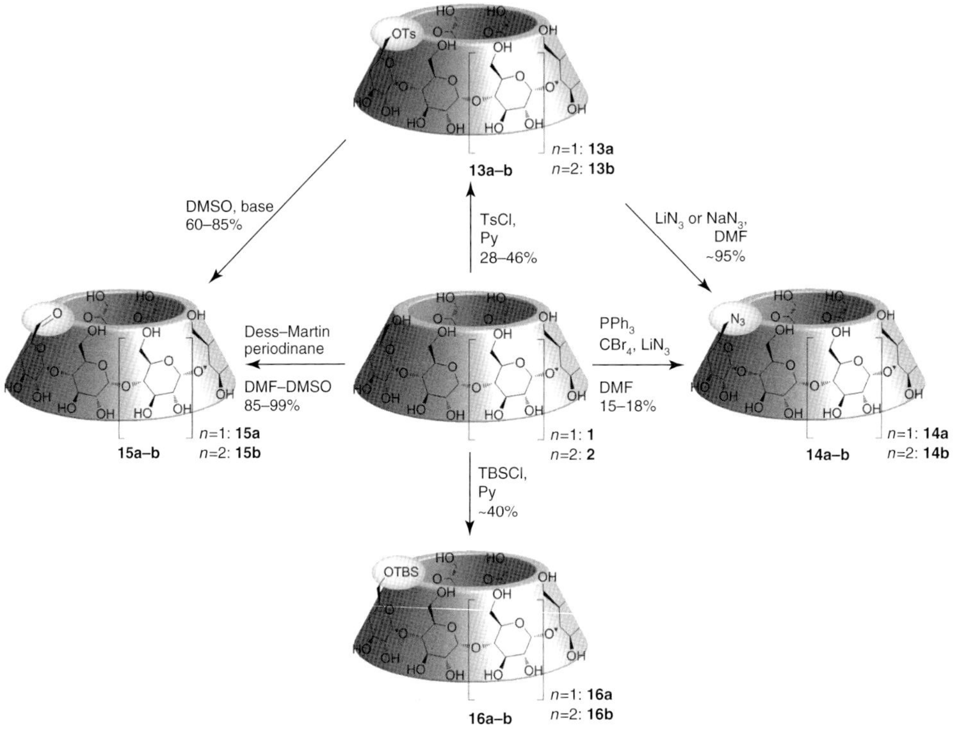

Scheme 9.6 Monofunctionalization of the primary rim of α- and β-cyclodextrins.

Sterically hindered silyl chlorides, such as *tert*-butyldimethylsilyl chloride, can also be used to protect a single hydroxyl on the primary rim, giving cyclodextrins **16a–b** in about 40% yield. After alkylation of all the other positions and deprotection of the silyl ether, cyclodextrins with a unique primary hydroxyl can be obtained [25].

The most efficient reaction to date is the monooxidation of a primary hydroxyl using Dess–Martin periodinane (Scheme 9.6) [26]. A highly hindered intermediate is formed, preventing the approach of a second periodinane on the primary rim, and this explains that cyclodextrins **15a–b** can be obtained in yields as high as 85%.

Direct monofunctionalization of cyclodextrins is possible using a reagent in default, but has its limitations: yields are usually modest, and the products are obtained after careful and sometimes difficult purification.

9.2.3.2 Use of Supramolecular Inclusion Complex

There are few examples using a supramolecular inclusion complex as intermediate for the monofunctionalization of cyclodextrins, when this approach should give good results.

Mono-2-tosyl-β-cyclodextrin **17** was prepared using a group transfer strategy [27]. *meta*-Nitrophenyl tosylate reacts with β-cyclodextrin **2** in a DMF/water buffer

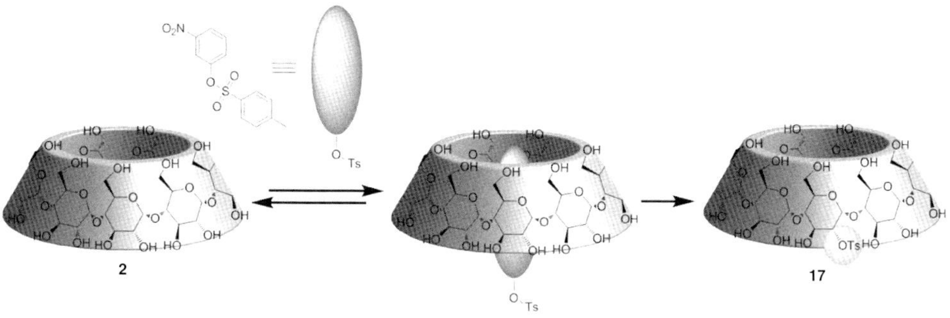

Scheme 9.7 Monofunctionalization of the position 2 of β-cyclodextrin.

to form an inclusion complex (Scheme 9.7). The orientation of the reagent into the cavity is such that the tosyl group is transferred selectively to one hydroxyl in position 2. After purification, the product is obtained in about 10% yield. Even if the yield is low, this methodology avoids a random functionalization of different positions, which would imply a long separation process to get the pure compound.

9.2.4 Random Multifunctionalization and Multidifferentiation

The first strategy to access multifunctionalized cyclodextrins relied on a controlled number of equivalents of reagent. But these reactions were not selective, a mixture of compounds being usually obtained needing difficult purification processes to isolate the different regioisomers [28]. Even using bulky groups, such as naphthylsulfonyl [29], tosyl [30], mesitylsulfonyl [31], trityl [32], or *tert*-butyldimethylsilyl [33], did not allow to increase the yields.

However, a defined structure is not always needed. For example, random methylation can produce a β-cyclodextrin derivative in which an average of 1.5 hydroxyl functions per glucose unit is methylated. This pattern enhances the solubility of the cyclodextrin in water, without affecting its ability to form supramolecular inclusion complexes. This has been proven useful in applications such as the extraction of cholesterol or the detoxification of polluted waters [3].

Tridifferentiation of the primary rim of β-cyclodextrin has been initially achieved in a random manner (Scheme 9.8). After a first monofunctionalization, a second substituent was introduced to give a mixture of starting material **18a** and all six possible regioisomers **18b–g**, which were then separated by chromatography [29]. All patterns of differentiation can be obtained by this method, but the yields are so low, and the separation so difficult, that it is not possible to use it on appreciable quantities.

Clearly, a better strategy is needed in order to access multidifferentiated cyclodextrins in an efficient way.

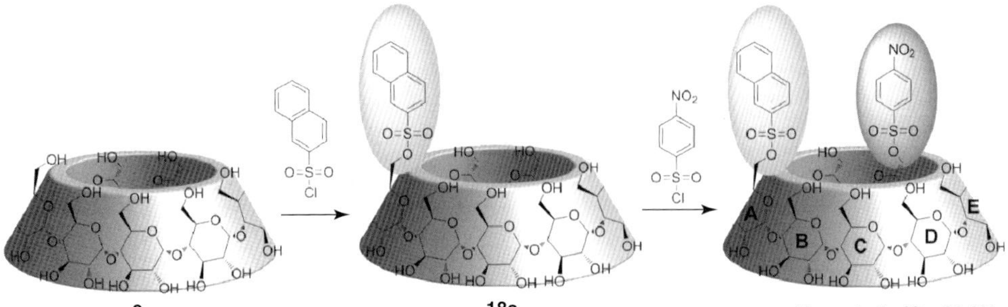

Scheme 9.8 Random tridifferentiation of the primary rim of β-cyclodextrin.

9.3 Capping Reagents for Direct Modification

Capping the primary or secondary rim to obtain difunctionalized cyclodextrins has been mostly developed on β-cyclodextrin. The rigidity of its backbone imposes a precise distance between the different hydroxyls (Section 9.1), making the design of bridging reagents easier and their reaction more selective. Nevertheless, a few capping reagents have given good results on α- and γ-cyclodextrins.

9.3.1 Difunctionalization: Capping the Cyclodextrin

9.3.1.1 Single Cap

Many capping reagents have been developed to selectively difunctionalize the primary or secondary rims of β-cyclodextrin.

Difunctionalization of the primary positions 6A and 6B was initially obtained using a default quantity of *meta*-benzenedisulfonyl dichloride in 40% yield [34], but the product was unstable. Introduction of stabilizing methoxy groups on the phenyl ring increased the yield in cyclodextrin **19** up to 32% [35] (Scheme 9.9, cap 1). Another capping reagent, 1,3-bis[bis(4-*tert*-butylphenyl)chloromethyl]benzene, possesses a similar geometry and bulkier substituents, and can selectively cap the positions 6A and 6B to give cyclodextrin **20** in 35% yield [36] (Scheme 9.9, cap 2).

Positions 6A and 6C can be functionalized selectively using 3,3′-disulfonylbenzophenone dichloride, giving cyclodextrin **21** in 20% yield [37] (Scheme 9.9, cap 3). Longer caps allow the selection of farther hydroxyls: bis(phenylsulfonyl) dichloride (Scheme 9.9, cap 4) and 4,4′-disulfonyl-*trans*-stilbene dichloride (Scheme 9.9, cap 5) selectively bridge the positions 6A and 6D, giving cyclodextrins **22** and **23** in 17% and 40% yields, respectively [37, 38]. Cyclodextrins capped with

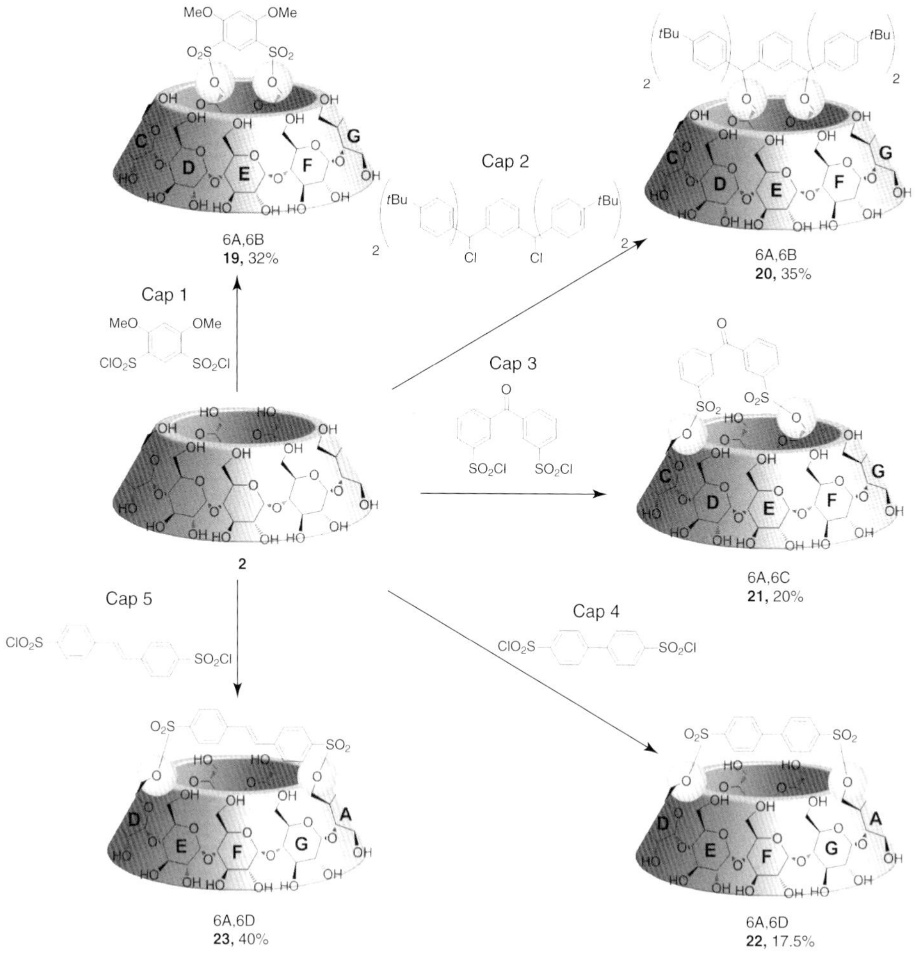

Scheme 9.9 Capping reagents give access to different patterns of functionalization on the primary rim of β-cyclodextrin.

bis-sulfonyl derivatives can then react with diverse nucleophiles to give β-cyclodextrins difunctionalized on their primary rim.

Even if the capping of α-cyclodextrin has been less studied, some interesting results have been obtained for the selective difunctionalization of its primary rim. A capping reagent successfully used on β-cyclodextrin, 3-bis[bis(4-*tert*-butylphenyl)chloromethyl]benzene, can selectively bridge the positions 6A and 6B to give cyclodextrin **24** in 55% yield [36] (Scheme 9.10, cap 1). Difunctionalization of the positions 6A and 6C can be achieved using a dibenzofuran derivative, giving cyclodextrin **25** in 18% yield [39] (Scheme 9.10, cap 2).

Selective difunctionalization of the secondary rim has been described on α-, β-, and γ-cyclodextrins, and is based on the use of imidazolyl leaving groups instead

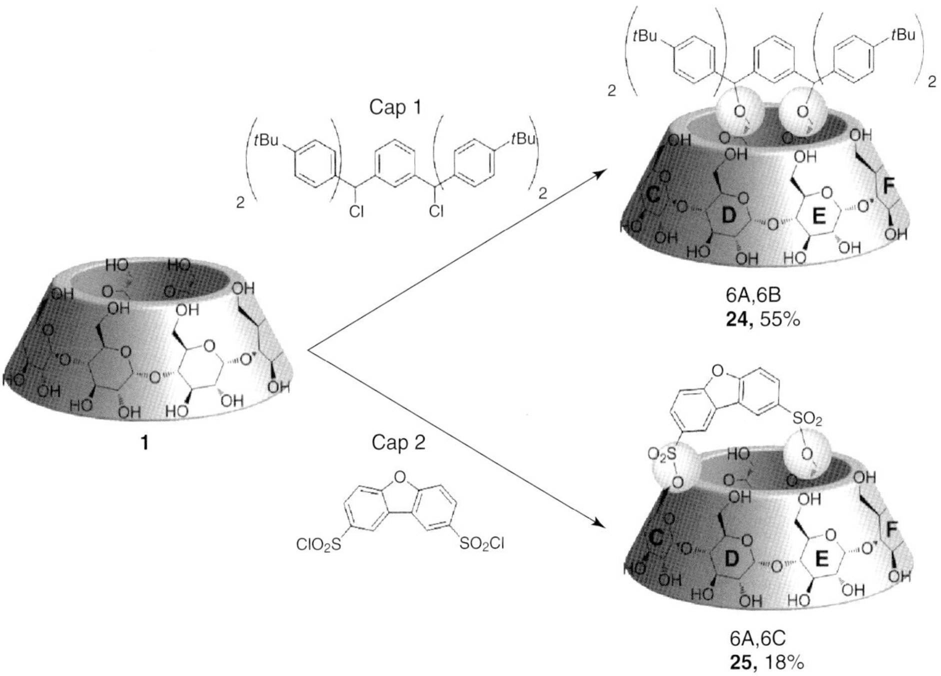

Scheme 9.10 Capping reagents give access to different patterns of functionalization on the primary rim of α-cyclodextrin.

of chlorides [40]. Reaction of native cyclodextrins **1–3** with benzophenone-3,3′-di(sulfonylimidazole) afforded 2A,2B-capped derivatives **26a–c** in about 30% yield [41] (Scheme 9.11, cap 1). Even if the positions 2 have not been activated, they are the only ones to react, and no substitution on the positions 6 happens. Using longer capping reagents, 2A,2C-difunctionalized cyclodextrins **27a–c** can be obtained in 25–50% yields [42] (Scheme 9.11, cap 2) and 2A,2D-difunctionalized cyclodextrins **28a–c** in 15–40% yields [43] (Scheme 9.11, cap 3).

Bridging different positions has rarely been described, but is an interesting way to differentiate the hydroxyls of the cyclodextrins. An example of mixed bridge was obtained by the reaction of benzaldehyde dimethylacetal with α-, β-, or γ-cyclodextrins **29a–c** protected on their primary rim with pivaloyl esters. The cyclodextrin derivatives **30a–c**, bridged between positions 2A and 3B, are obtained in 37–54% yields [44], depending on the cyclodextrin used (Scheme 9.12). After deprotection of the primary rim and perbenzylation, the bridging benzylidene can be cleaved under acidic conditions, giving a 2A,3B-dihydroxyl-perbenzylated cyclodextrin. The benzylidene can also undergo a selective reductive opening on position 2, leading to cyclodextrins monofunctionalized at position 2. An alternative has been proposed for β-cyclodextrin, which has been permethylated instead of perbenzylated, and the tridifferentiated 2A-hydroxyl,3B-benzyl-permethylated β-cyclodextrin **31** was obtained [45].

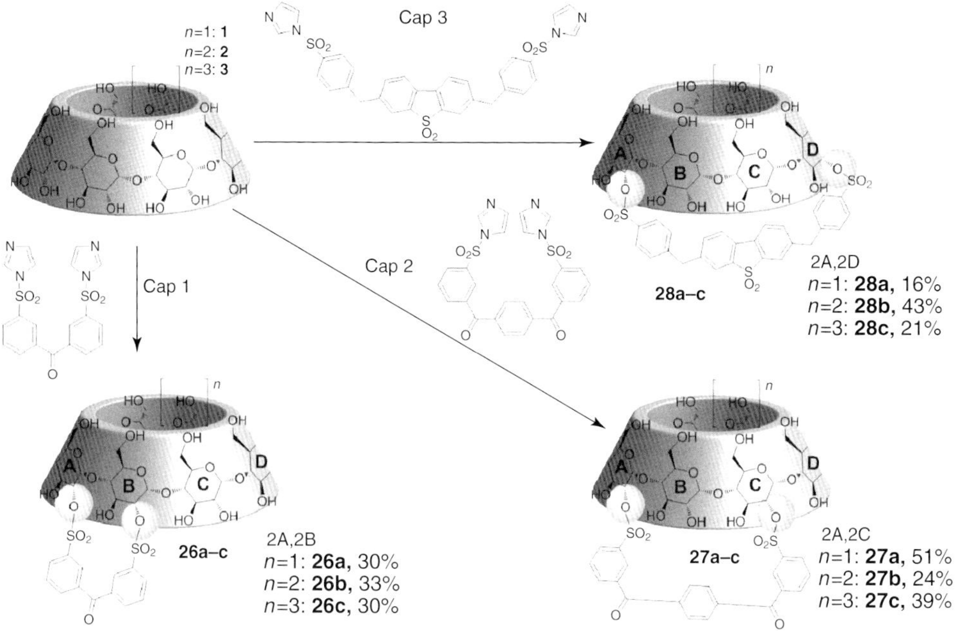

Scheme 9.11 Capping reagents give access to different patterns of functionalization on the secondary rim of α-, β-, and γ-cyclodextrins.

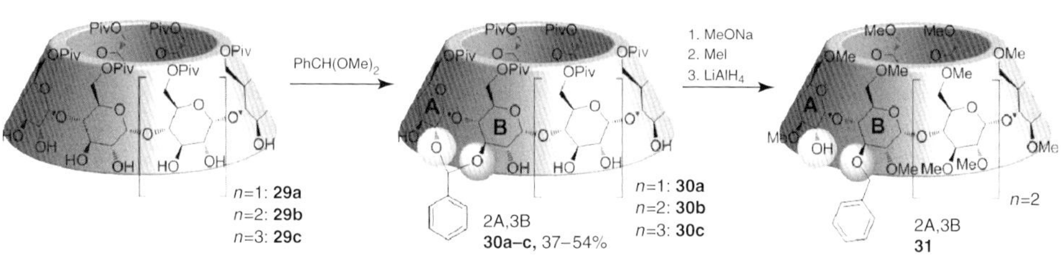

Scheme 9.12 Capping acetal and successive regioselective opening gives access to an original pattern of differentiation on the secondary rim of β-cyclodextrin.

9.3.1.2 **Double Capping**

As cyclodextrins possess more than two hydroxyls on their primary rim, a second capping reagent can be attached. This would lead to tetrafunctionalized cyclodextrins. This second reaction is more difficult, as the primary rim becomes more crowded after the introduction of the first cap, but as a consequence it is also more selective. Some of the few tetrafunctionalizations of β-cyclodextrin have been achieved through installation of two capping groups on its primary rim.

Reacting β-cyclodextrin **2** with 2 equiv of 1,3-bis[bis(4-*tert*-butylphenyl)chloromethyl]benzene, followed by permethylation of all remaining hydroxyls and purification, gave β-cyclodextrin **32** tetrafunctionalized on positions 6A,6B and 6D,6E in 50% overall yield [46] (Scheme 9.13, cap 1). Applying the same strategy to

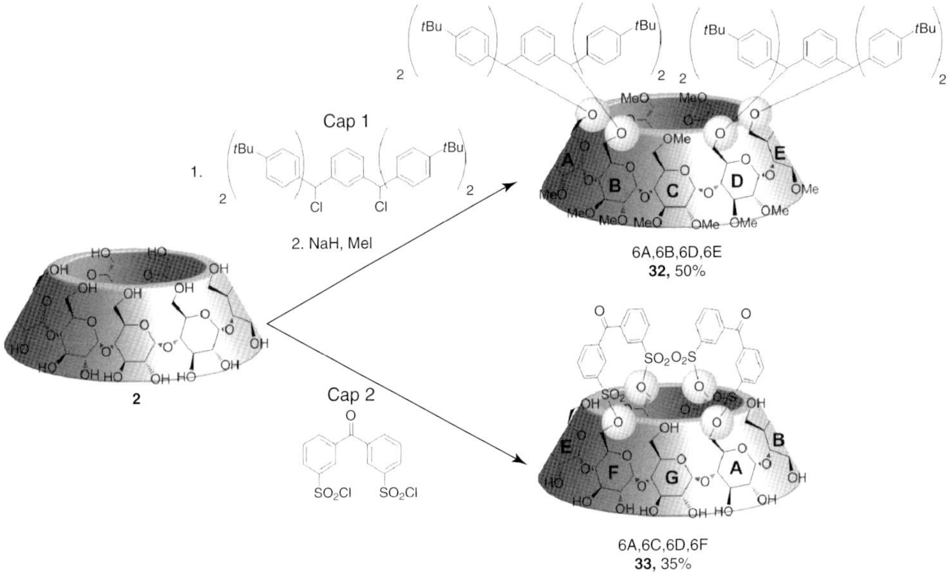

Scheme 9.13 Double capping gives access to original patterns of functionalization on the primary rim of β-cyclodextrin.

α-cyclodextrin did not give any appreciable results, probably because the primary rim is too small to accommodate two such bulky bridging groups.

Increasing the quantity of 3,3′-disulfonylbenzophenone dichloride used for the difunctionalization of positions 6A and 6C (Scheme 9.9, cap 3) leads to β-cyclodextrin **33** bearing two caps on positions 6A,6C and 6D,6F in 35% yield [47] (Scheme 9.13, cap 2).

9.3.2 Unsymmetrical Caps

Introduction of unsymmetrical caps in a single step on the primary rim of β-cyclodextrin would directly give a tridifferentiated pattern. This has been done with the so-called "flamingo cap," giving capped products **34a–d** in 17% yield [48] (Scheme 9.14). Only four regioisomers out of the six possible were obtained, but unfortunately their separation was not possible.

To be selective, heterocapping of cyclodextrins has to control both the distance and the direction between the grafting hydroxyls. This was achieved in a two-step manner on the primary rim of β-cyclodextrin between the positions 6A and 6G. The thiol group of L- or D-cysteine reacted with the 6A-tosyl group of cyclodextrin **13b** to give monofunctionalized β-cyclodextrins **35a–b** (Scheme 9.15). The amino group was functionalized with a dansyl moiety, and the free carboxylic acid was subjected to esterification conditions. The esterification only happened on the position 6G, leading to cyclodextrins **36a–b** in about 80% yield [49]. It has to be noted that this

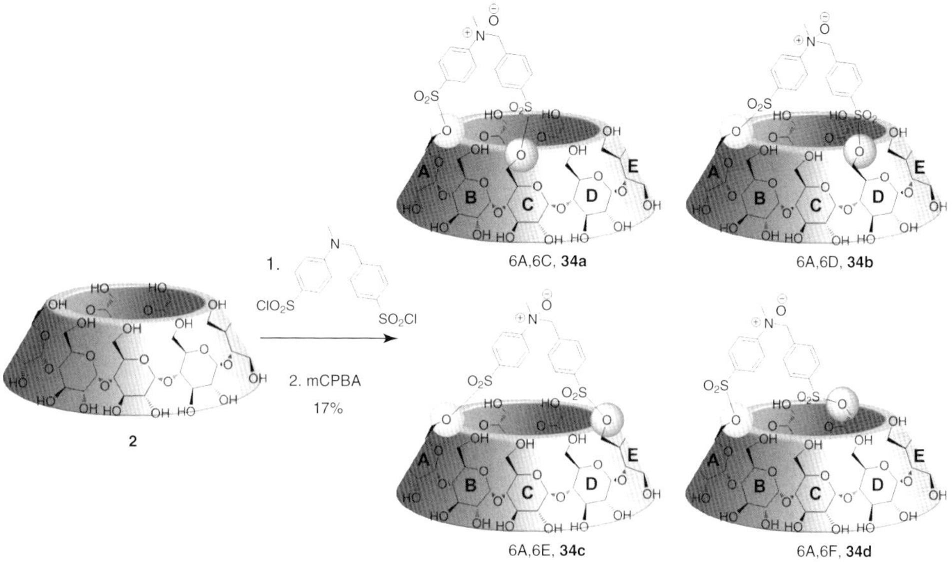

Scheme 9.14 Flamingo cap on β-cyclodextrin.

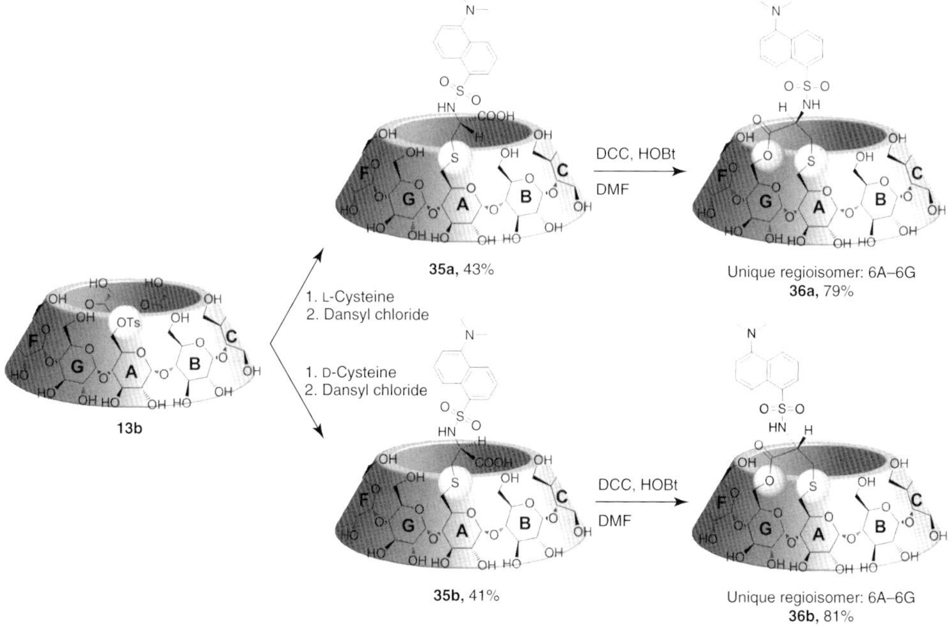

Scheme 9.15 Regioselective heterocapping of the primary rim of β-cyclodextrin.

selectivity is the same with L- and D-cysteines, which means that the regioselectivity is ruled by the chirality of the cyclodextrin and neither by the one of the cysteine nor by the preferential obtention of an *endo* or *exo* cap. A similar selectivity has been observed on γ-cyclodextrin, using L-cysteine as capping reagent [50].

9.3.3 Modification of Capped Cyclodextrins

9.3.3.1 Addition of Another Functionality

Only one example of selectively trifunctionalized β-cyclodextrin exists, and it is also the only utilization of a capping reagent on a cyclodextrin already substituted. Monotosylated β-cyclodextrin **13b** was treated with a dibenzofuran derivative bearing two sulfonyl chloride functions (see Scheme 9.10 for a difunctionalization using this reagent), leading to the bridging of two positions opposite the tosyl group (Scheme 9.16). After purification by high-performance liquid chromatography (HPLC), a mixture of two regioisomers **37a–b** is obtained in 23% yield. After substitution of the bridging sulfonyls and of the tosyl group by thiophenol, these two isomers lead to a unique regioisomer **38** trifunctionalized on the positions 6A, 6C, and 6E [51].

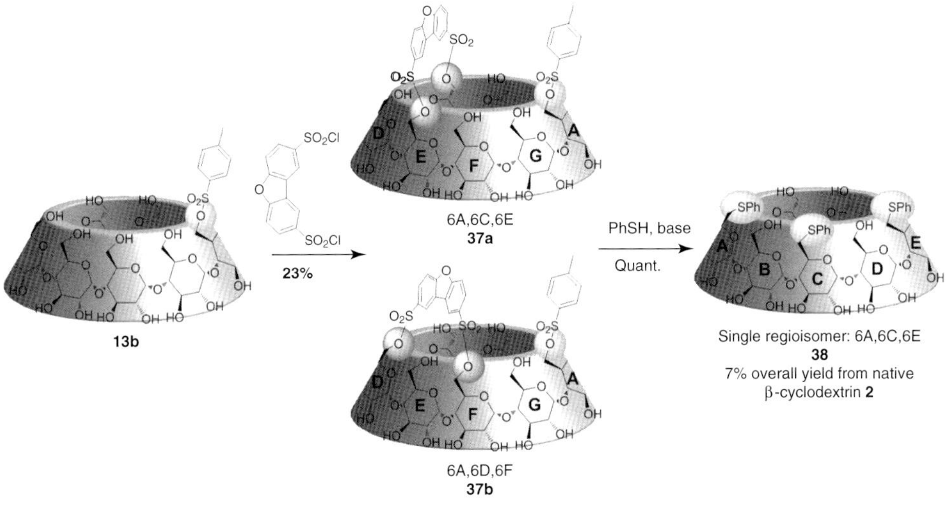

Scheme 9.16 Capping a monofunctionalized cyclodextrin followed by substitution of the three functions gives access to trifunctionalization on the primary rim of β-cyclodextrin.

9.3.3.2 Opening the Caps

Another way to tridifferentiate the primary rim of cyclodextrins is the selective opening of a symmetrical cap.

Mesitylene disulfonyl dichloride reacted with native β-cyclodextrin **2** to give selectively cyclodextrin **39** capped on the positions 6A and 6B in 19% yield. One of the sulfonyl groups can be selectively displaced by imidazole, preferentially the one on the position 6A, giving cyclodextrin **40a** in 42% yield [52] (Scheme 9.17).

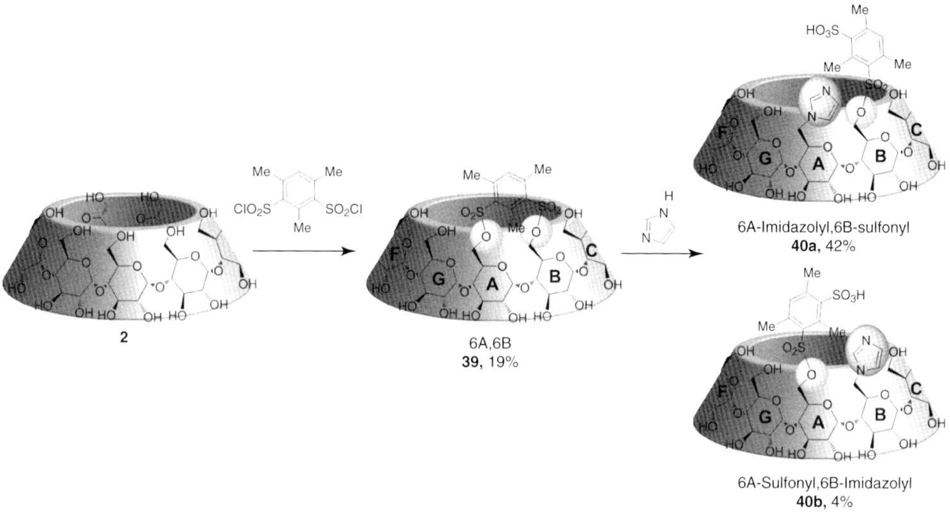

Scheme 9.17 Regioselective opening of a symmetrical cap on β-cyclodextrin.

The difference of reactivity between the positions 6A and 6B can explain the observed regioselectivity (Figure 9.7). During the opening of the bridge, the approach of imidazole on the position 6B (which would lead to cyclodextrin **40b**) is more difficult because of the steric hindrance created by the methyl group between the sulfonyls. The position 6A is more available, and substitution of this position is preferred, giving cyclodextrin **40a** as the major product.

Positions 6A and 6D can also be differentiated using the selective opening of a symmetrical sulfonate bridge. This bridge is introduced on α- or β-cyclodextrins **41a–b** bearing two free hydroxyls on the positions 6A and 6D, all other positions

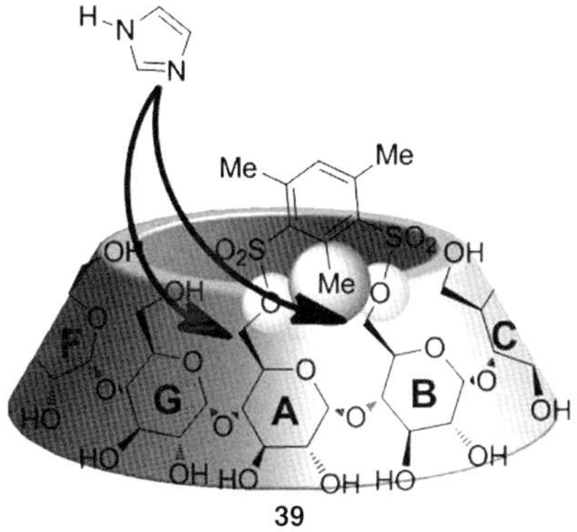

39

Figure 9.7 Rationalization of the regioselectivity of the cap opening.

being protected with benzyl ethers obtained through deprotection methodology developed in Section 9.5. First, the two hydroxyls react with thionyl chloride to form a sulfite bridge, which is then oxidized to the sulfate bridge to give cyclodextrins **42a–b**. Sodium azide reacts with one side of the bridge, the other one being left as a sulfate, which is removed in acidic conditions, selectively giving the trifunctionalized 6A-hydroxyl,6D-azido α- or β-cyclodextrins **43a–b** in about 90% yield [53] (Scheme 9.18).

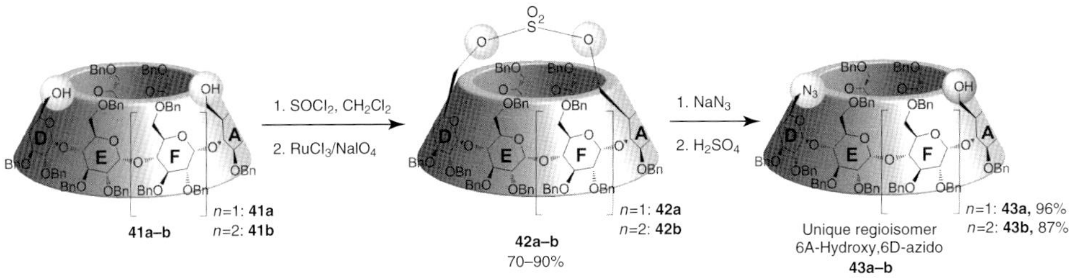

Scheme 9.18 Selective opening of a sulfato bridge.

For α-cyclodextrin **42a**, the positions 6A and 6D of the bridge are equivalent, and substitution of one or the other leads to a unique regiosiomer 6A-hydroxyl,6D-azido **43a**. The regiochemical outcome is more complex with β-cyclodextrin **42b**, as the two sides of the bridge are different, and substitution of one or the other would lead to the different regioisomers 6A-azido,6D-hydroxyl or 6A-hydroxyl,6D-azido. The experiment proved that the 6A-hydroxyl,6D-azido isomer **43b** is selectively obtained. Molecular model showed that the cone conformation is distorted in the bridged β-cyclodextrin **42b**, obstructing the approach of the nucleophile on the position 6A. The position 6D, being more available, is the sole to react, leading to the observed regioisomer. Other nucleophiles have been used, such as ammonia, piperidine, or sodium allylate, and gave the same regioselectivity, even if the yields are sometimes lower.

This methodology has been extended to the opening of sulfate bridging contiguous positions. Proximal capping of permethylated α-cyclodextrins can be achieved in 96% yield using the same reaction conditions as described earlier, starting from the corresponding diol **44** [46] (Scheme 9.19). Reaction of the sulfato group of cyclodextrin **45** with lithium diphenylphosphide selectively happened on the

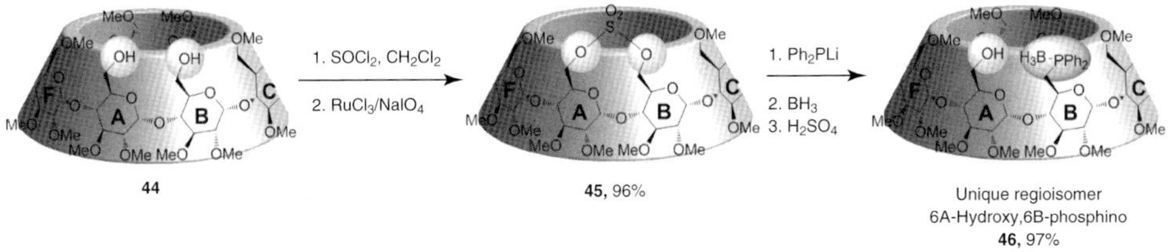

Scheme 9.19 Regioselective opening of 6A,6B-capped α-cyclodextrin.

position 6B, and after protection of the phosphine with borane and hydrolysis of the sulfate anion, a single regioisomer **46** was isolated in 97% overall yield [54]. A rationalization of the regioselectivity is given by the crystal structure of the α-cyclodextrin bearing a sulfate bridge. The carbon on position 6B is more available for nucleophilic attack than the one on position 6A, which is partially hindered by the glucose unit F. The selectivity is achieved under kinetic control, and for that reason, a better selectivity is obtained with bulky nucleophiles, strong enough to react at low temperature. Smaller nucleophiles gave moderate selectivity: for example, reaction with phthalimide ion gave 93% of attack on the position 6B, and reaction with azide gave 84% of attack on the same position. The same methodology has been applied to β-cyclodextrin, but the regioselectivity was much lower.

This procedure also gave good results when applied to doubly capped α-cyclodextrin **48**, obtained through the same strategy starting from the corresponding tetrahydroxy cyclodextrin **47**, even if the yields are lower (Scheme 9.20). The two sulfato bridges are opened by diphenylphosphide, leading to a single regioisomer **49** in good yields after protection of the phosphines and hydrolysis of the sulfate ion. When nucleophiles less bulky than diphenylphosphide are used, the selectivity is also lower: reaction with phthalimide ion gave 62% of attack on the positions 6B and 6E, the regiosiomer 6A,6E-dihydroxy,6B,6D-diphthalimido being the other product.

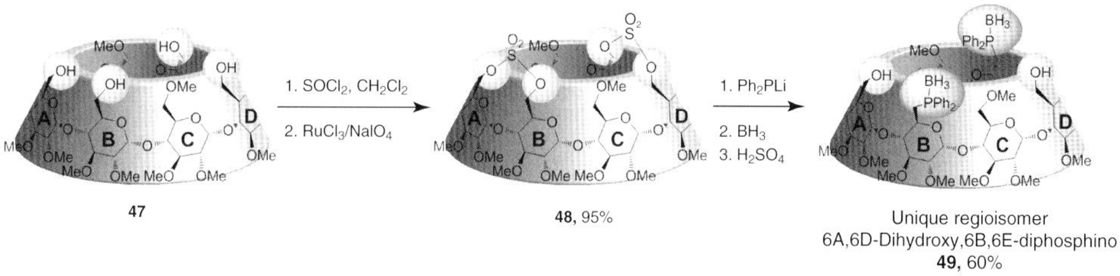

Scheme 9.20 Regioselective opening of doubly capped α-cyclodextrin.

Capping functionalization relies on the rigidity of the cyclodextrin and the geometry of the cap, another logical mean to access multifunctional cyclodextrin is to rely on the steric hindrance of the reagent that will be able to react only a certain number of times at certain positions.

9.4
Bulky Reagents for Direct Modifications

Bulky reagents have been mainly developed for direct modification of the primary rim of α-cyclodextrins. Being the smallest one, the effect of steric repulsion is more efficient. The reactions used are reversible, or at least the step installing the bulky group on the hydroxyl, to allow self-correction and to produce the most stable intermediate.

9.4.1
Trityl and Derivatives

Direct protection of some hydroxyls on the primary rim of native α-cyclodextrin **1** using trityl chloride is an interesting method to access multifunctionalized derivates. For example, α-cyclodextrin **50** selectively protected on the positions 6A,6C,6E can be obtained in 23% yield [55], using trityl chloride in pyridine at 55 °C (Scheme 9.21). All reaction conditions have to be precisely controlled, as for example, an increase in the temperature leads to the additional formation of regioisomers 6A,6B,6C, 6A,6B,6D, and 6A,6B,6E [56]. The free hydroxyls can then be methylated, and after deprotection of the trityl groups, 6A,6C,6E-trihydroxy α-cyclodextrin is obtained, these hydroxyls being available for further transformation.

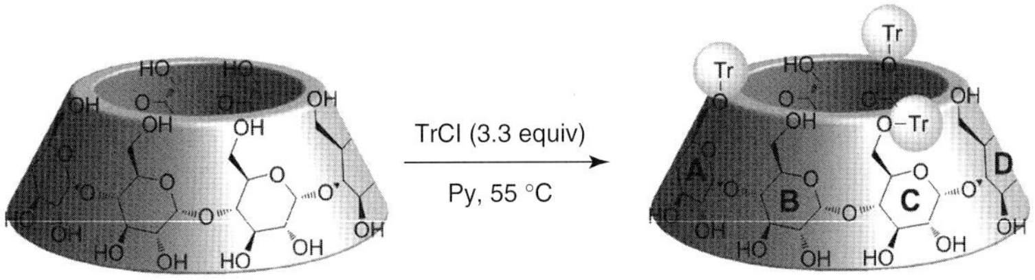

Scheme 9.21 Direct trifunctionalization of the primary rim of α-cyclodextrin.

Using trityl groups to protect the primary rim of α-cyclodextrin, followed by permethylation and separation by chromatography, regioisomers 6A,6B,6C,6D (15%), 6A,6B,6D,6E (25%), and 6A,6B,6C,6E (32%) can also be obtained [56].

In order to access α-cyclodextrins selectively difunctionalized on their primary rim, a bulkier protecting group is needed. Tris(4-*tert*-butylphenyl)methyl (supertrityl) proved successful in the synthesis of new patterns of functionalization. Reaction of 3 equiv of supertrityl chloride with α-cyclodextrin **1**, followed by permethylation of all hydroxyls remaining free gave a mixture of mono-, di-, and trifunctionalized cyclodextrins. Separation of the different products by chromatography over silica gel gave the 6A-monofunctionalized α-cyclodextrin **51a** (12%) and the three regioisomers 6A,6B **51b** (8%), 6A,6C **51c** (27%), and 6A,6D **51d** (30%) as pure compounds [57] (Scheme 9.22). This reaction has been done on quantities up to tens of grams. Even if the selectivity is not perfect, it is a useful methodology, as it gives access to all difunctionalized regiosomers in usable quantities.

Protection of the primary rim with supertrityl has proven particularly successful in the obtention of tetrafunctionalized α-cyclodextrins. Using 6 equiv of supertrityl chloride in pyridine and increasing the reaction temperature to 70 °C, four primary hydroxyls can be protected. After permethylation of all remaining hydroxyls and purification, the α-cyclodextrin **52** tetrafunctionalized on the positions 6A,6B,6D,6E is obtained in 70% overall yield [58] (Scheme 9.23). Supertrityl groups can then

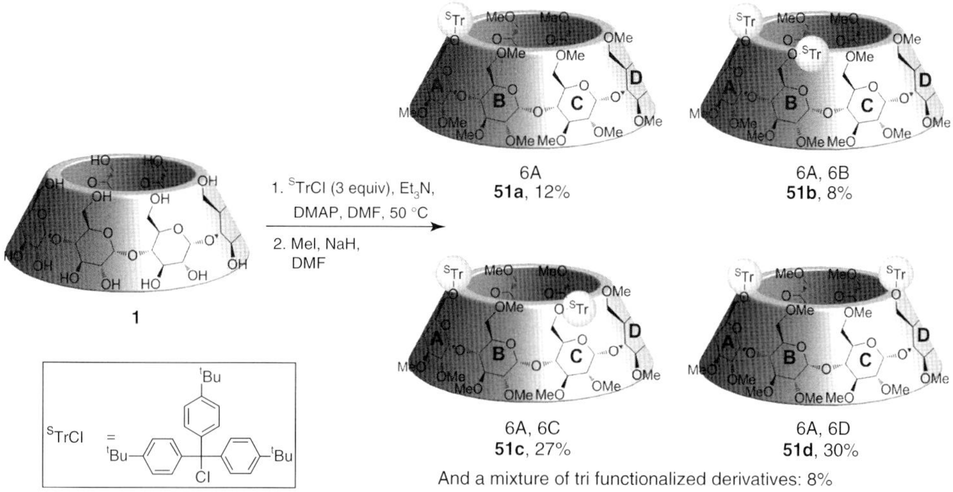

Scheme 9.22 Direct mono- and difunctionalizations of the primary rim of α-cyclodextrin.

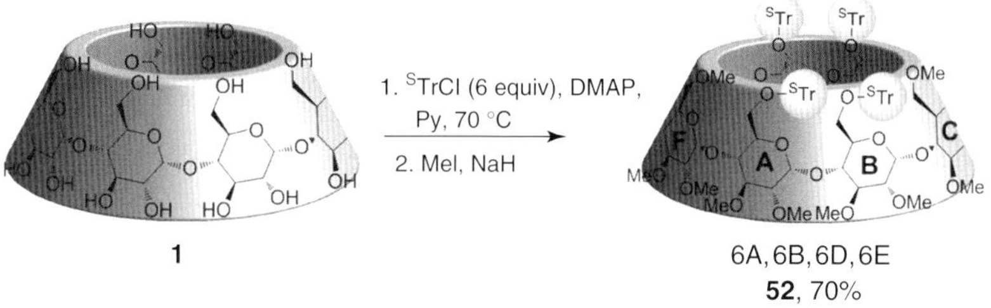

Scheme 9.23 Direct tetrafunctionalization of the primary rim of α-cyclodextrin.

be deprotected to get the tetrahydroxy cyclodextrin **47**, and these positions can be modified employing usual chemistry.

It can be noted that more than four supertrityl groups could not be introduced, even when using more equivalents of reagent and in harder conditions.

9.4.2 Triphenylphosphine

Triphenylphosphine has been used in a Mitsunobu reaction for the perhalogenation of the primary rim of α- and β-cyclodextrins (Scheme 9.3). Controlling the reaction conditions gave access to the α-cyclodextrin **53** trifunctionalized on the positions 6A,6C,6E (Scheme 9.24). Mitsunobu reaction of α-cyclodextrin **1** using triphenylphosphine and *N*-bromosuccinimide, followed by treatment with sodium azide and permethylation of all remaining hydroxyls, gave after simple crystallization the 6A,6C,6E-tri-azido permethylated α-cyclodextrin **53** in 52% overall yield [59]. This method is the only one providing this regioisomer selectively and in good yield.

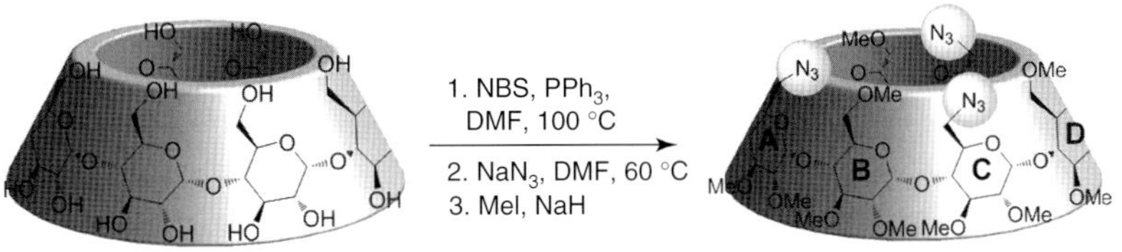

1

6A, 6C, 6E
53, 52%

Scheme 9.24 Direct trifunctionalization of the primary rim of α-cyclodextrin.

9.4.3 Selective Transfer

β-Cyclodextrin **54a** monofunctionalized on its primary rim with an imidazolyl group catalyzes the hydrolysis of *para*-nitrophenyl acetate [60]. This reaction proceeds by transfer of the acetate on a unique hydroxyl of the primary rim. The same reaction using mesitylenesulfonyl chloride has been studied (Scheme 9.25), and analysis of the product **55a** showed that the mesitylenesulfonyl group has been transferred onto the position 6D. The proposed mechanism takes place in two steps: first, the imidazolyl group reacts with the sulfonyl, and then transfers it to the hydroxyl on the position 6D, which happens to be at a suitable distance [61]. In the transition state, a bridge is formed between the positions 6A and 6D. The same reaction has been reported on α- and γ-cyclodextrins with the same regioselectivity. In order to access tetradifferentiated β-cyclodextrins, the reaction has been applied to β-cyclodextrins **54b–g** bearing an iodo and an imidazolyl on their primary rim, and the same regioselectivity has been observed (Scheme 9.25).

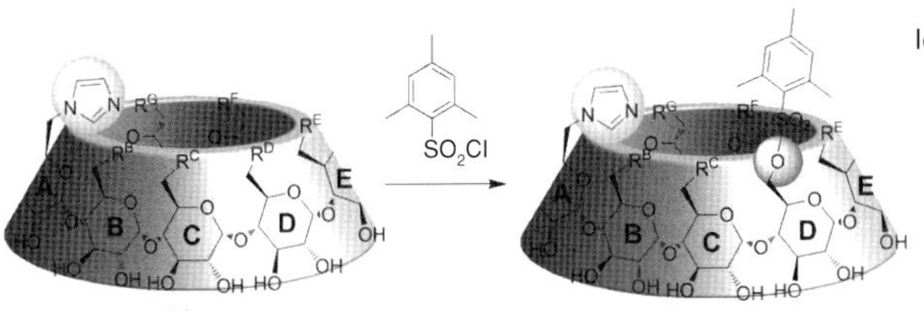

54a–g **55a–g**

Iodine position[a]	Yield (%)	Starting material
None	**55a**, 13	**54a**, 63%
R^B	**55b**, 42	**54b**, 46%
R^C	**55c**, 40	**54c**, 31%
R^D and R^E	**55e**, 28[b]	**54e**, 35%[c]
R^F	**55f**, 22	**54f**, 38%
R^G	**55g**, 57	**54g**, 0%

a All other positions are hydroxyl groups.
b Only the product 6A-imidazolyl, 6D-mesitylene, 6E-iodo is formed.
c Mainly the reagent 6A-imidazolyl, 6D-iodo.

Scheme 9.25 Tetradifferentiation using a selective delivery on the primary rim of β-cyclodextrin.

The tetradifferentiated β-cylodextrins **55b–g** have been obtained in low yields, about 1% from the native cyclodextrin **2**, after purification by HPLC for each step. Even if the method is selective, it is difficult to obtain appreciable quantities of products.

Direct functionalization of cyclodextrins appears to be efficient on α-cyclodextrin for the synthesis of 6A,6C,6E and 6A,6B,6D,6E patterns of functionality, but was not successful on β-cyclodextrin. A more general strategy is hence still needed to access more patterns.

9.5 Selective Deprotections

Recently, a third multifunctionalization strategy has emerged relying on selective deprotection, which now appears as not only a practical alternative to the previous methods but also as a means to go beyond toward polyheterofunctionalized cyclodextrins.

9.5.1 Diisobutylaluminum Hydride (DIBAL-H) as Deprotecting Agent

9.5.1.1 General Mechanism

Diisobutylaluminum hydride (DIBAL-H) does not usually react with alkyl ethers, such as **56**, and for this reason has not been used for their deprotection (Scheme 9.26). However, it has been shown that DIBAL-H can deprotect alkyl ethers in particular cases if a second oxygen is present on the molecular fragment and available to chelate the aluminum atom, as in **57** [62].

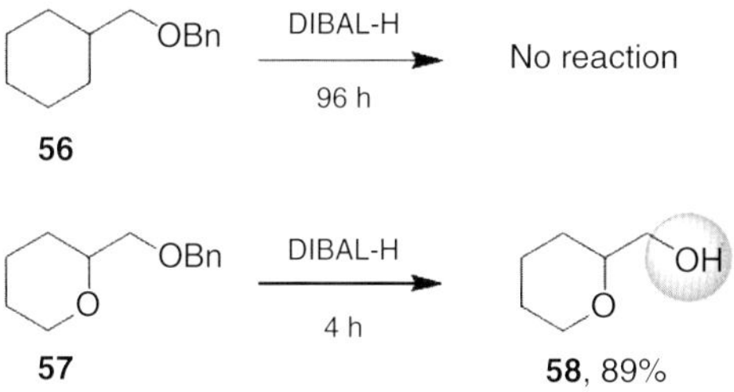

Scheme 9.26 Deprotection of alkyl ether by DIBAL-H.

Systematic study of Lewis acids derived from aluminum in the de-*O*-benzylation of compounds bearing multiple alkyl ethers gave an insight into the mechanism. The proposed mechanism rationalizes also the regioselectivity observed in the de-*O*-alkylation reaction.

A first aluminum derivative is complexed between the two oxygen atoms of **57**, which have to be in an appropriate configuration to allow this chelation. A second aluminum reagent is then complexed by the oxygen bearing the most available electron lone pair. This second complexation directs the reaction of de-*O*-alkylation, and rationalizes the observed regioselectivity leading to **58** as the sole product (Scheme 9.27). The mechanism involves two aluminum derivatives, one as a Lewis acid, which activates the alkyl ether, the other as a hydride donor, which reacts with the activated position [63].

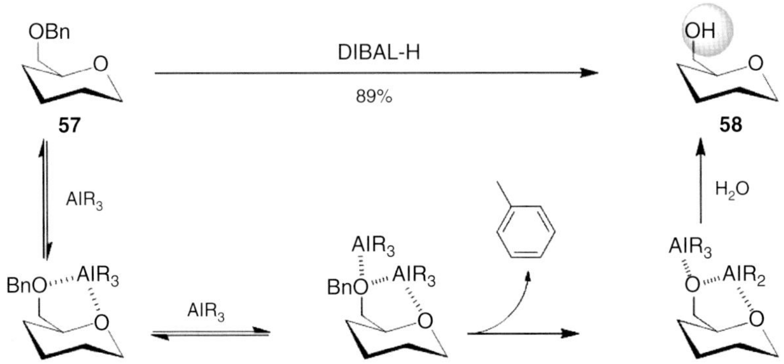

Scheme 9.27 Proposed mechanism for the deprotection of benzyl ethers by DIBAL-H.

In the particular case of carbohydrate derivatives bearing multiple alkyl ethers, such as **59**, de-*O*-benzylation of different positions can happen (Scheme 9.28). As an example, de-*O*-benzylation can be induced by the chelation of aluminum between oxygen atom at position 6 and the endocyclic oxygen, or by the chelation between oxygen atom at position 2 and the anomeric oxygen, resulting in a mixture of the two possible regioisomers **60a–b** [63].

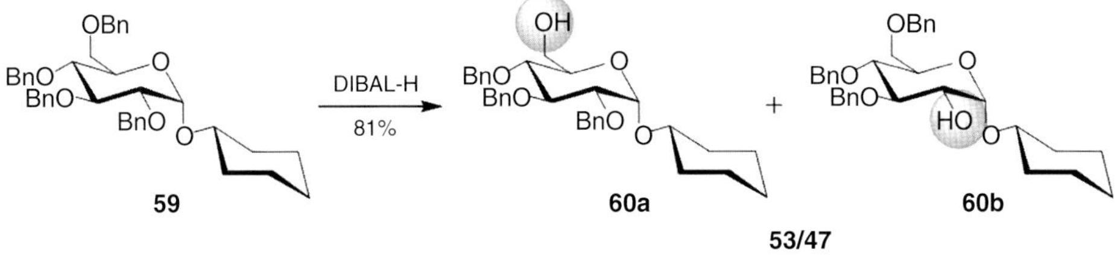

Scheme 9.28 Selective deprotection on sugar derivative.

If only one site is available for chelation of the aluminum derivative, then only one product of de-*O*-benzylation is selectively obtained. This happens when steric hindrance around some oxygen makes them unavailable, or on compounds in which an oxygen atom has been removed [63]. For example, the anomeric oxygen is missing in compound **61**, consequently de-*O*-benzylation reaction takes place only at position 6 (Scheme 9.29), leading to a unique product **62**.

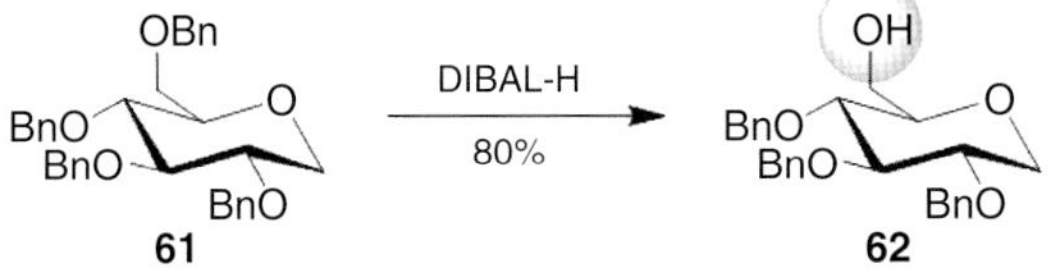

Scheme 9.29 Selective deprotection on sugar derivative.

Deprotection of primary silyl ethers can also be achieved using DIBAL-H [64]. The proposed mechanism is similar, with the necessary chelation of an aluminum derivative between two oxygen atoms. This deprotection happens chemoselectively on primary silyl ethers if secondary silyl ethers are present on the substrate.

9.5.1.2 Application to Cyclodextrins

In the particular case of cyclodextrins, the anomeric oxygen has its two lone electron pairs pointing toward the inside of the cavity, which means that they are not available to complex the aluminum derivative. Consequently, cyclodextrins will not be opened by action of DIBAL-H.

Perbenzylated Cyclodextrins Perbenzylated α- and β-cyclodextrins **10a–b** were obtained quantitatively by the action of sodium hydride and benzyl chloride in anhydrous DMSO [18]. On these compounds, the secondary rim is much more crowded than the primary rim. Therefore, when submitted to the action of DIBAL-H, the chelation happens between the oxygen at position 6 and the endocyclic one, as for the carbohydrate derivative **61** lacking the anomeric oxygen (Scheme 9.29). After complexation of a second aluminum derivative by the free lone electron pair of the oxygen at position 6, the de-*O*-benzylation occurs selectively on this position [63] (Scheme 9.30), leading to monohydroxy cyclodextrins **63a–b**. If the reaction conditions are stronger (more equivalents of DIBAL-H, higher concentration, and higher temperature), a second de-*O*-benzylation takes place on the diametrically opposed sugar unit, and a single regioisomer 6A,6D-dihydroxy **41a–b** is obtained [63, 65].

The mechanism of the de-*O*-benzylation rationalizes the observed regioselectivity (Scheme 9.31). After the first de-*O*-benzylation, two aluminum derivatives are linked to the hydroxylate formed, which creates a huge steric hindrance on one side of the primary rim of intermediates **64a–b**, also enhanced by the presence of benzyl groups. Two other aluminum derivatives can only approach and react on the farthest position, leading to the 6A,6D-dihydroxy perbenzylated cyclodextrins **41a–b** [63]. This rationalization is consistent with the reaction of monohydroxy α-cyclodextrin **63a** with DIBAL-H [63] (Scheme 9.30). DIBAL-H first reacts with the free hydroxyl to form an adduct **64a**, which creates a steric hindrance on the primary rim, and a second de-*O*-benzylation happens on the farthest position, leading to a single regioisomer **41a**. In the case of β-cyclodextrin, the positions 6D and 6E are equally distant from the position 6A, but reaction of one or the other leads to the same regioisomer 6A,6D-dihydroxy cyclodextrin **41b**.

This methodology has been applied to perbenzylated γ-cyclodextrin, but the selectivity is lower and two regioisomers 6A,6D and 6A,6E are obtained. This is due to the bigger size of the primary rim, making the steric hindrance of the aluminum derivatives too small to direct the second deprotection [65].

The selective de-*O*-benzylation of perbenzylated α-cyclodextrin can go further if different reaction conditions are used [66]. Using more equivalents of DIBAL-H, different solvents, and a longer reaction time, tri- and tetrafunctionalized α-cyclodextrins **65** and **66** can be obtained in reasonable yields (Scheme 9.32). To

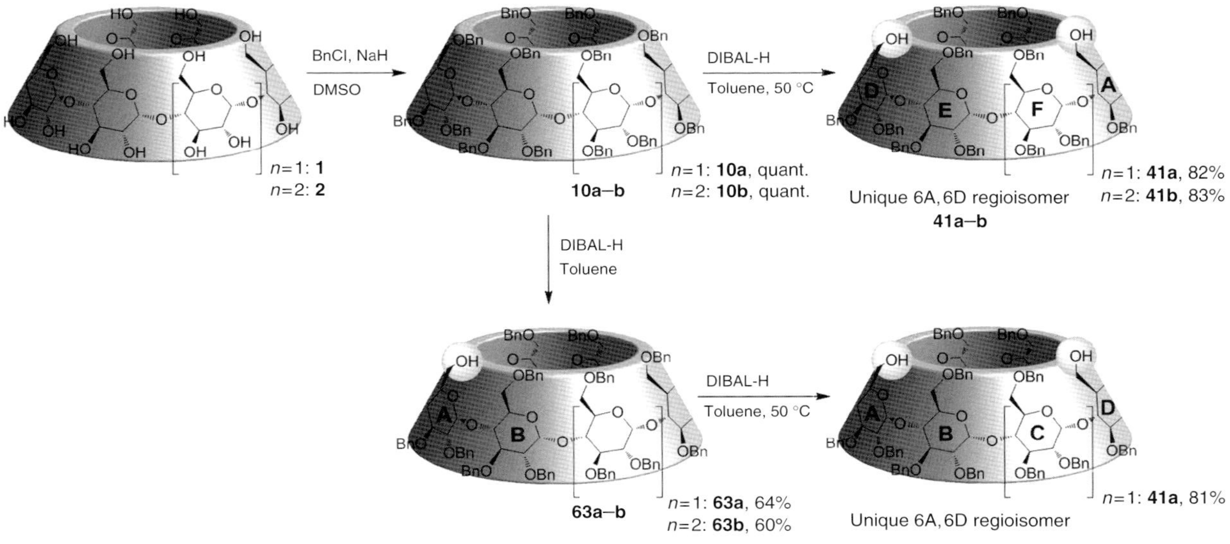

Scheme 9.30 Selective debenzylation on the primary rim of α- and β-cyclodextrins.

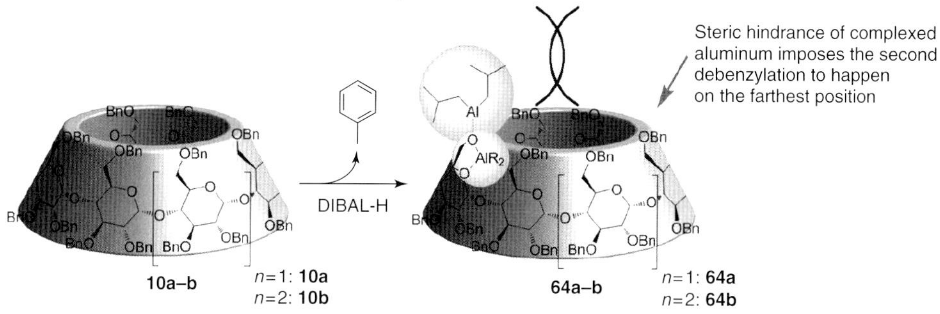

Scheme 9.31 Rationalization of the selectivity of the double deprotection.

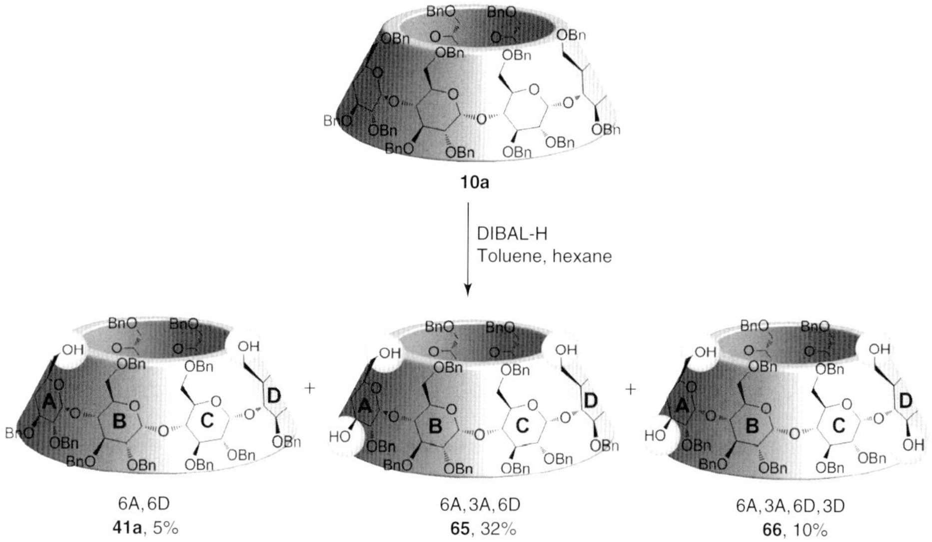

Scheme 9.32 Changing the reaction conditions with α-cyclodextrin.

rationalize the observed selectivity, the reaction is proposed to occur in a stepwise manner. The primary rim would be the first to react, and an aluminum derivative complexed on the hydroxylate on position 6 would be chelated with the oxygen on position 3 of the same glucose unit. This would direct further de-*O*-benzylations on positions 3A and/or 3D.

Applying the same conditions as mentioned earlier to perbenzylated β-cyclodextrin **10b** also resulted in further de-*O*-benzylations [67], but only on the primary rim (Scheme 9.33). This gave access to two trifunctionalized β-cyclodextrins **67** and **68** and a tetrafunctionalized one **69** in moderate yields. The observed regioselectivity is also justified by the stepwise successive de-*O*-benzylations. The positions 6A and 6D would be the first to react, and then an aluminum derivative would be chelated between the hydroxylate formed and the

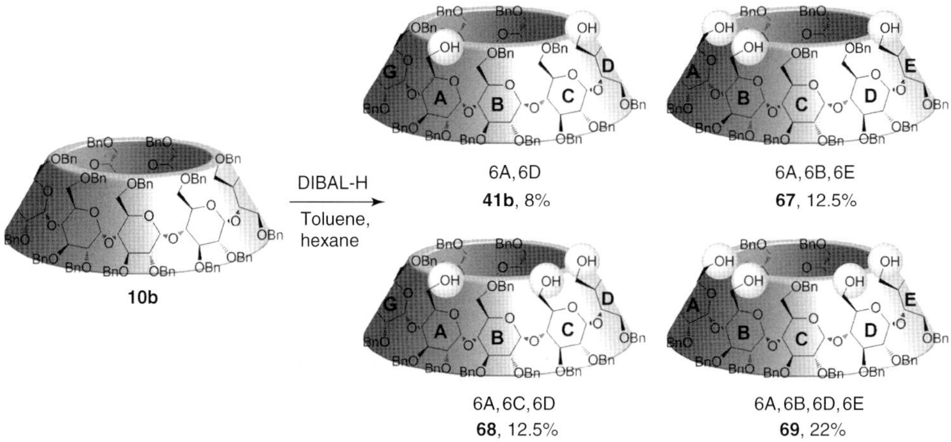

Scheme 9.33 Changing the reaction conditions with β-cyclodextrin.

oxygen on position 6 of the preceding glucose unit. This would direct the further de-*O*-benzylations on neighboring positions in a clockwise direction.

Cyclodextrins Persilylated on the Primary Rim When cyclodextrins bearing *tert*-butyldimethylsilyl ethers on their primary rim were subjected to the action of DIBAL-H, the same selective de-*O*-silylation happened on one or two positions [68]. The regioselectivity is the same as for the de-*O*-benzylation, and deprotections occurred selectively on the farthest positions, leading to 6A,6D-dihydroxy α- and β-cyclodextrins. Application of this methodology to γ-cyclodextrin also gave a mixture of regioisomers that could not be separated.

Permethylated Cyclodextrins Deprotection of methyl ethers has also been used in the selective functionalization of cyclodextrins. This method proved successful on α- and β-cyclodextrins, and gave surprising results.

When permethylated α- or β-cyclodextrins **70a–b** were subjected to the action of DIBAL-H in toluene, de-*O*-methylations happened at positions 2 and 3 of two contiguous glucose units. In contrast with perbenzylated cyclodextrins **10a–b**, the secondary rim is less crowded, and an aluminum derivative can be chelated between the oxygen atoms of the methyl ethers on positions 2A and 3B as in the case of the benzylidene bridge (Scheme 9.12). A second aluminum derivative can be complexed by the free electron lone pair of the oxygen on position 3B, which is the less hindered, and de-*O*-methylation occurs first on this position. The residual aluminum derivative is then chelated between the hydroxylate formed, the methyl ether on position 2A and the anomeric oxygen linking the two glucose units. The orientation resulting from this chelate leaves more space for a second DIBAL-H molecule to approach the oxygen on position 2A, which leads to the second de-*O*-methylation on this position [69]. This second deprotection is faster than the first one, which explains why the mono-deprotected cyclodextrin is not observed.

After deprotection, some aluminum derivatives stay linked to the hydroxylates, which creates a steric hindrance, and other de-*O*-methylation cannot happen on contiguous positions.

For α-cyclodextrin **70a**, reaction conditions can be adjusted to obtain 2A,3B-dihydroxy cyclodextrin **71** as the major product [70] (Scheme 9.34), or the 2A,3B,2D,3E-tetrahydroxy cyclodextrin **72** [71]. In the latter case, formation of some 2A,3B,2C,3D-tetrahydroxy cyclodextrin **73** as a side product could not be avoided, the steric hindrance on the secondary rim being probably too small for the reaction to be completely selective.

When permethylated β-cyclodextrin **70b** was subjected to the action of DIBAL-H, de-*O*-methylations also happened on the positions 2 and 3 of two contiguous glucose units. Reaction conditions can be adapted to the major formation of 2A,3B-dihydroxy cyclodextrin **74** [70], or to the formation of the 2A,3B,2D,3E-tetrahydroxy **75** [72] or 2A,3B,2C,3D,2E,3F-hexahydroxy cyclodextrin **76** [73] (Scheme 9.35).

9.5.2 Second Deprotection

From the initial deprotection, a few strategies have been delineated to further heterofunctionalize cyclodextrins through successive deprotections.

9.5.2.1 Monoazide Cyclodextrins

Monoazide α- and β-cyclodextrins **77a–b** were easily obtained from the corresponding monohydroxy cyclodextrins **63a–b**. When submitted to DIBAL-H, the azide is reduced first. Aluminum derivatives complexed on the formed amide create steric hindrance, which directs the following de-*O*-benzylation toward the farthest position [74] (Scheme 9.36) as in the case of monohydroxy cyclodextrin **63a** (Scheme 9.30 and Scheme 9.31). For α-cyclodextrin, the farthest position is 6D, which leads to the 6A-amino,6D-hydroxyl regioisomer **78a** in 74% yield. For β-cyclodextrin, a molecular model study revealed that the farthest position is 6E, leading to the formation of the 6A-amino,6E-hydroxyl regioisomer **78b** in 89% yield. It has to be noted that this pattern of differentiation is the same as for 6A-hydroxyl,6D-azido cyclodextrins **43a–b** (Scheme 9.18).

9.5.2.2 Deoxy and Bridged Cyclodextrins

De-*O*-benzylation of the primary rim of cyclodextrins using DIBAL-H occurs on diametrically opposed positions. If the two free hydroxyls are protected with groups that do not react with DIBAL-H, a second double de-*O*-benzylation can take place on two other diametrically opposed positions. Hence, two regioisomers 6C,6F and 6B,6E can be formed (Scheme 9.37), depending if the deprotection happens in the clockwise or counterclockwise direction, respectively (looking from the primary face).

Chelation of aluminum reagent between position 6 and the endocyclic oxygen on unit **A** is highly sensitive to the steric hindrance exerted by the groups on adjacent sugars, especially by the one on the primary position of the counterclockwise

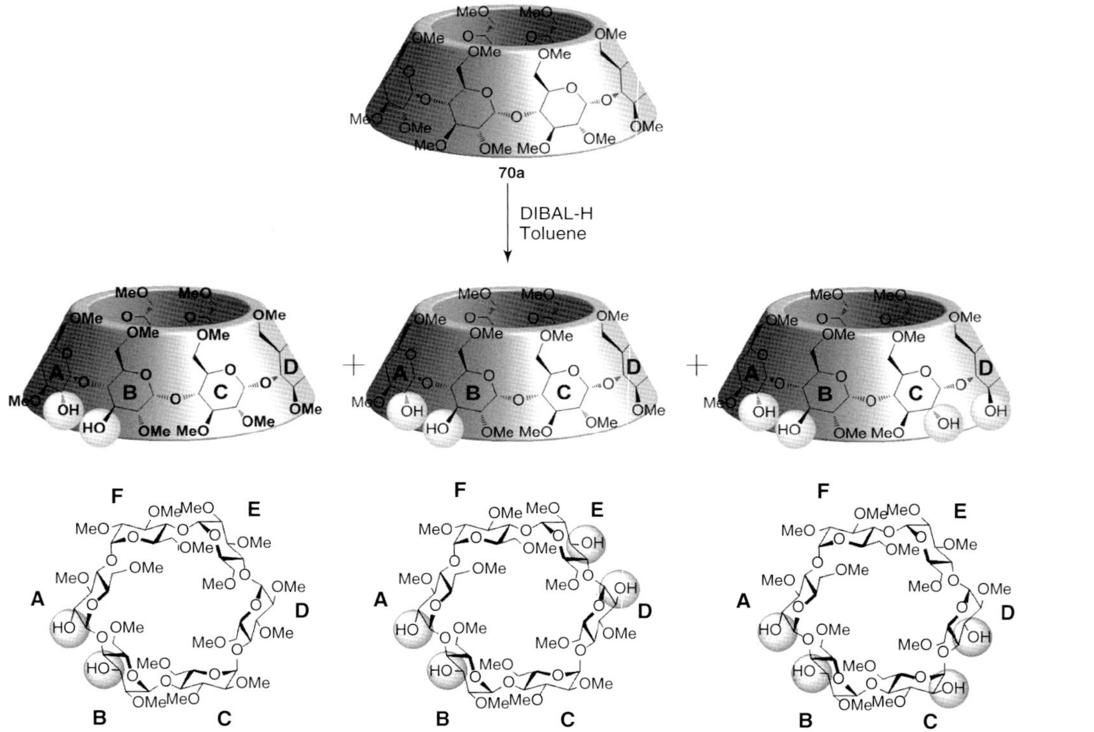

DIBAL-H	Temperature	Time (h)	2A, 3B **71**	2A, 3B, 2D, 3E **72**	2A, 3B, 2C, 3D **73**
6 equiv	50 °C	3	55%	0%	0%
40 equiv	RT	6	0%	45%	19%

Scheme 9.34 De-*O*-methylation of α-cyclodextrin.

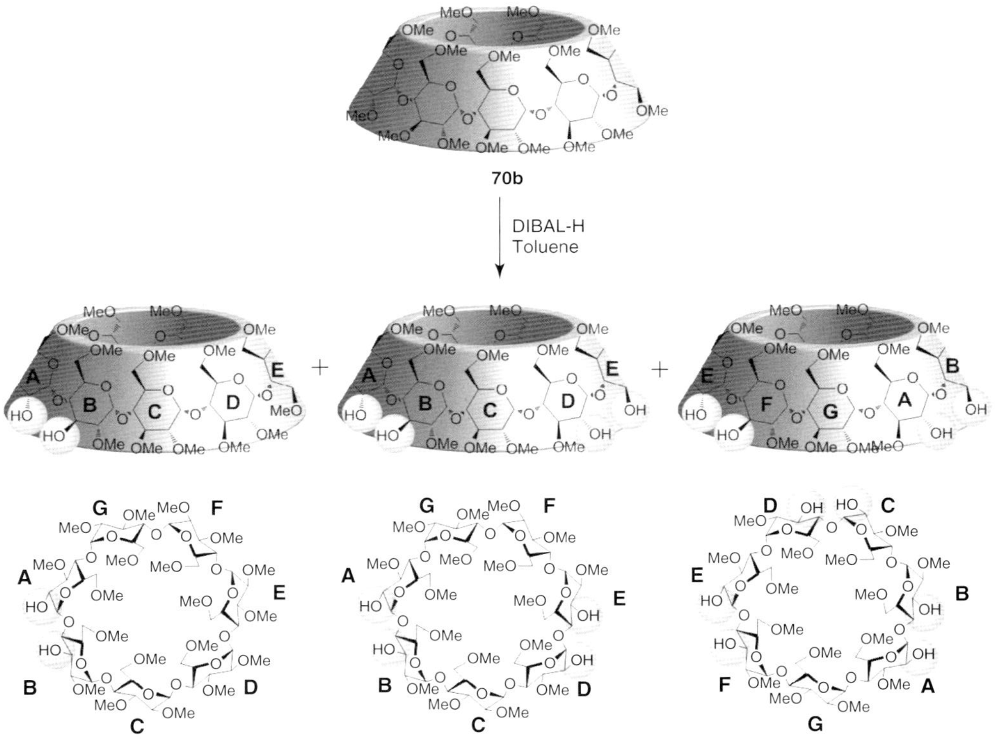

DIBAL-H	Temperature	Time (h)	2A,3B **74**	2A,3B, 2D,3E **75**	2A,3B, 2C,3D,2E,3F **76**
7 equiv	50 °C	3	55%	0%	0%
50 equiv	RT	3	8%	51%	0%
50 equiv	RT	28	0%	0%	45%

Scheme 9.35 Selective de-*O*-methylation of the secondary rim of β-cyclodextrin.

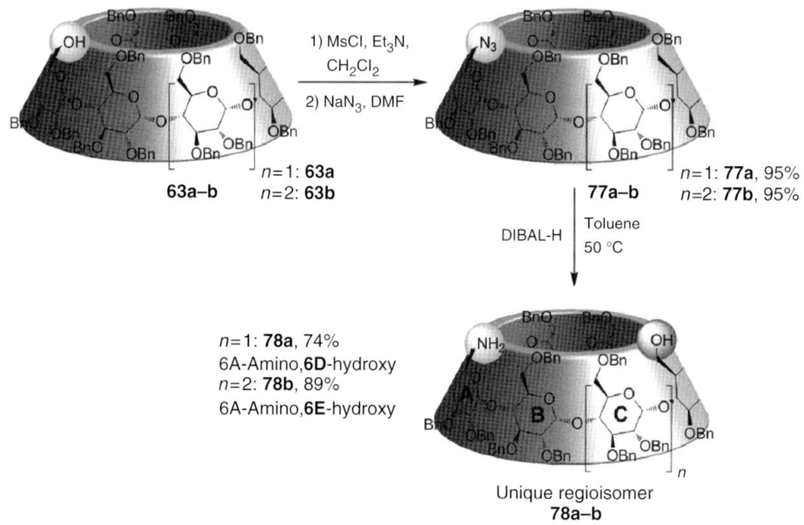

Scheme 9.36 Selective tridifferentiation of the primary rim using azide reduction.

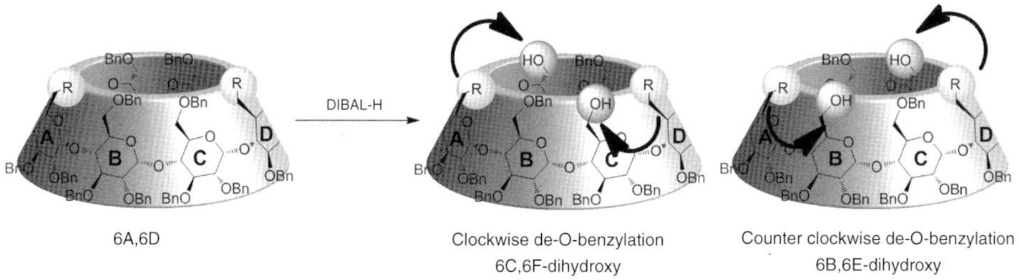

Scheme 9.37 Possible clockwise and counterclockwise second de-*O*-benzylation.

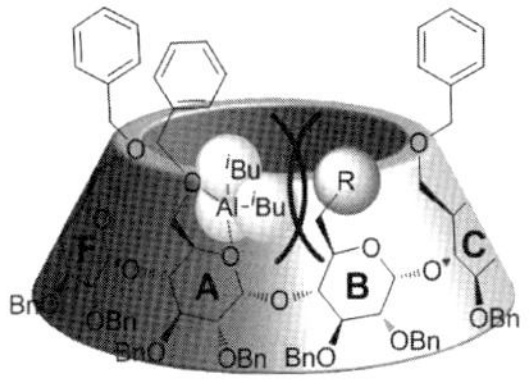

Figure 9.8 Steric hindrance during de-*O*-benzylation reaction on functionalized cyclodextrin.

sugar **B** (Figure 9.8). Owing to the cyclic directionality of the cyclodextrin, the group on the clockwise sugar **F** is too far away to interact with the aluminum derivative attached on **A** [75]. Consequently, introduction on the primary rim of a substituent R less bulky than a benzyl group, for example, on unit **B** should orientate the second de-*O*-benzylation toward the glucose unit in clockwise position: unit **A**.

Usual protecting groups do not resist the action of DIBAL-H in the conditions used for the de-*O*-benzylation; so, unconventional groups have to be used. Hydroxyls on 6A,6D-dihydroxy α-cyclodextrin **41a** can be converted into vinyl groups by oxidation followed by Wittig olefination. When divinyl cyclodextrin **79** was submitted to the action of DIBAL-H, the double de-*O*-benzylation occurred selectively on the positions preceding the vinyl groups (Scheme 9.38), leading to tridifferentiated cyclodextrin **80**, with no trace of the other regioisomer **81** [75]. This reaction also works with methyl groups instead of vinyl, albeit with no possibility of further functionalization [76].

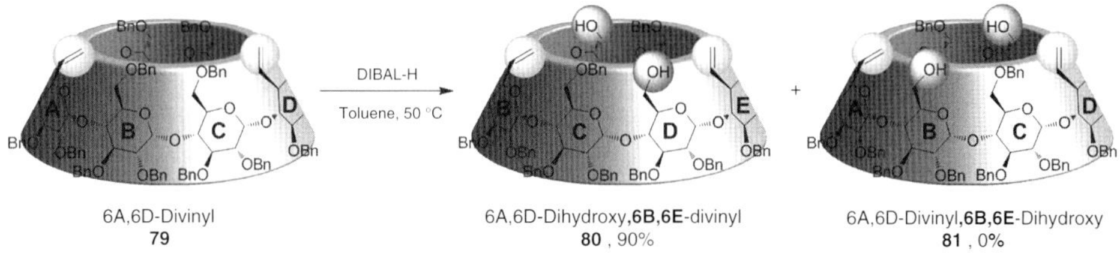

Scheme 9.38 Selective second deprotection on deoxy-α-cyclodextrin.

A bridging allylic derivative, such as the one on cyclodextrin **82**, can also create a similar steric decompression on the preceding glucose unit. This orientates the second de-*O*-benzylation toward the clockwise positions (Scheme 9.39) [77], leading to cyclodextrin **83** in good yields, with no traces of the other regioisomer **84**.

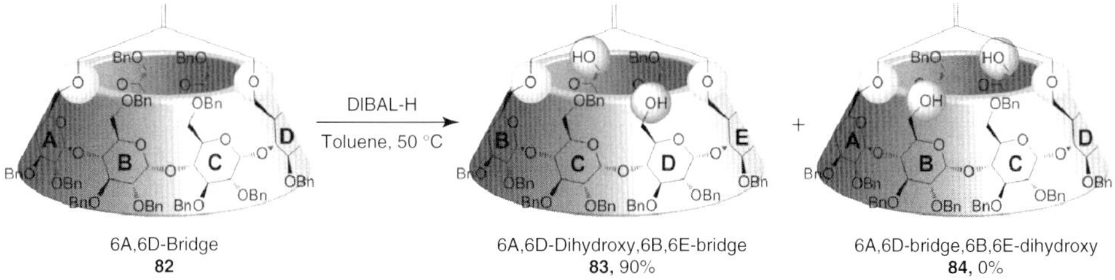

Scheme 9.39 Selective second deprotection on bridged-α-cyclodextrin.

In the case of β-cyclodextrin, the lack of symmetry makes the duplication of the double de-*O*-benzylation more challenging. Indeed, if we consider that the second deprotection will take place on opposite positions, de-*O*-benzylation can lead to four regioisomers.

Hence, 6A,6D-dihydroxy β-cyclodextrin **41b** was converted into the divinyl derivative **85** through Swern oxidation and a Wittig olefination. When subjected to the action of DIBAL-H, only two dihydroxy derivatives **86** and **87** were obtained out of the four possible (Scheme 9.40) [75]. As expected, the major regioisomer **86** is the result of the clockwise de-*O*-benzylation on positions 6C and 6G, the

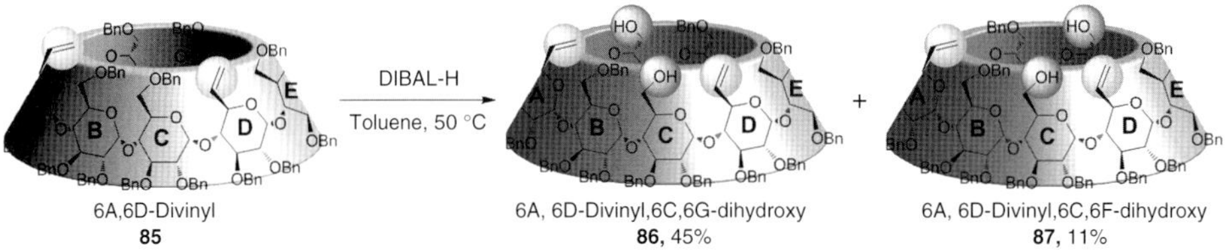

Scheme 9.40 Selective second deprotection on deoxy-β-cyclodextrin.

minor regioisomer **87** being obtained through clockwise de-*O*-benzylation of position 6C followed by deprotection of the alternative farthest position 6F. It has to be noticed that these two regioisomers could only be separated after protection of the hydroxyls with silyl ethers and conversion of the vinyl groups into hydroxyls.

6A,6D-dihydroxy β-cyclodextrin **41b** can be capped in two steps through bis-allylation and ring-closing metathesis. When capped cyclodextrin **88** was treated with DIBAL-H, two regioisomers out of the four possible were obtained (Scheme 9.41) [75]. They could be separated after protection of the hydroxyls with silyl ethers and removal of the bridge. The major compound is not the expected 6C,6G-dihydroxy **89** resulting from the clockwise de-*O*-benzylation, but the 6C,6F-dihydroxy **90**, obtained through a deprotection clockwise to the bridge on position 6C and one on position 6F, clockwise to a position bearing a benzyl group. Molecular modeling showed that the presence of the bridge drastically changes the conformation of the β-cyclodextrin, a consequence being that the sugar unit F is pointing outward and is hence more available.

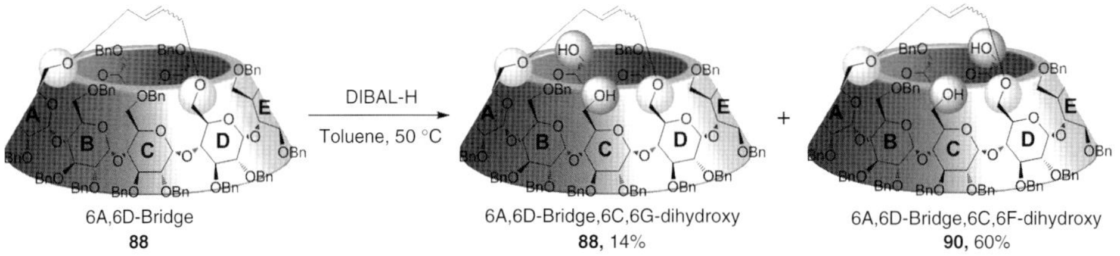

Scheme 9.41 Selective second deprotection on bridged-β-cyclodextrin.

These two approaches, the so-called bascule-bridge and the deoxy-sugars, are complementary and allow a regioselective access to two tridifferentiated β-cyclodextrins.

Reaction conditions can be adapted to favor the mono-de-*O*-benzylation. Lowering the concentration and the temperature of the reaction gave access to the 6A,6D-divinyl,6C-hydroxy α-cyclodextrin **91** as the major product (Scheme 9.42). The regioselectivity of the mono-deprotection is the same as for the double debenzylation, and the position clockwise to one of the two vinyl groups is selectively deprotected [78]. This has been proven by a second deprotection using DIBAL-H, converting monohydroxyl cyclodextrin **91** into dihydroxyl cyclodextrin **80**. The vinyl-protecting groups can be converted into hydroxyls by ozonolysis followed by

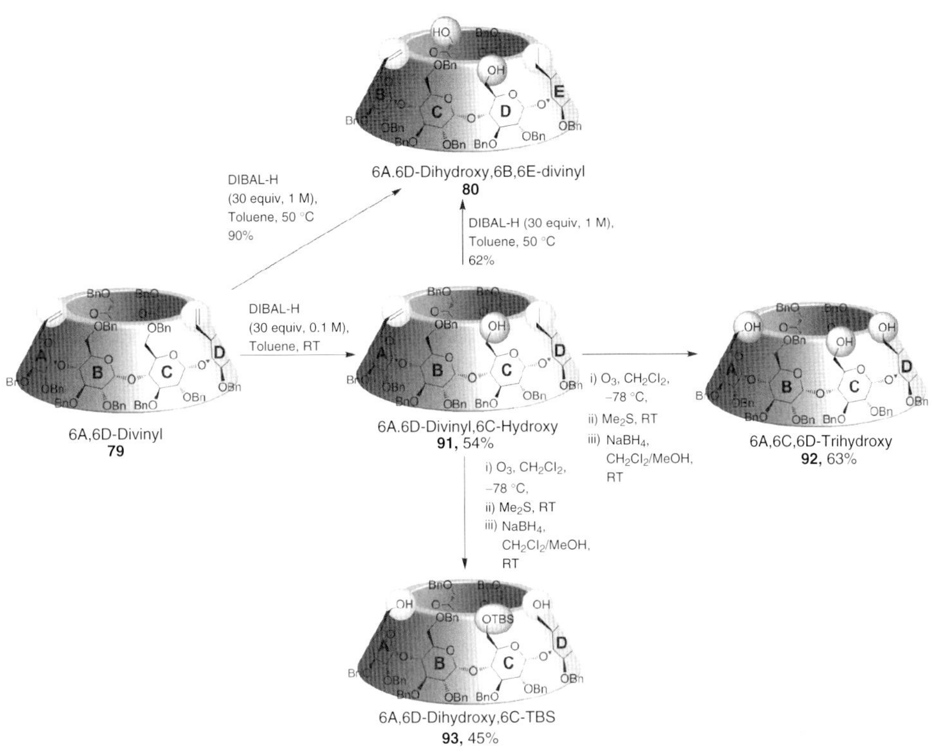

Scheme 9.42 Mono-de-*O*-benzylation on divinyl α-cyclodextrin.

reduction to give the 6A,6C,6D-trihydroxy α-cyclodextrin **92** in good yields. Alternatively, the hydroxyl can be protected with a silyl ether before the conversion of the vinyl groups into hydroxyls, leading to cyclodextrin **93** bearing usual protecting groups.

The same strategy can be applied to the bridged cyclodextrin **82**, giving the 6A,6D-bridge,6C-hydroxyl α-cyclodextrin **94** in 52% yield (Scheme 9.43) [78]. The bridge can be removed using a palladium-catalyzed double de-*O*-allylation to give the 6A,6C,6D-trihydroxy α-cyclodextrin **92**, or cyclodextrin **93** if the hydroxyl has been protected with a silyl ether before.

This second de-*O*-benzylation can also be performed on an α-cyclodextrin bearing a single substituent creating a steric decompression. The monovinyl α-cyclodextrin **95** can be obtained in good yield from the corresponding monohydroxyl **63a** by Swern oxidation followed by a Wittig olefination. When submitted to the action of DIBAL-H, a first de-*O*-benzylation occurred on the position preceding the vinyl group, which is the less hindered one (Scheme 9.44), and the 6A-hydroxyl,6B-vinyl α-cyclodextrin **96** was isolated. Two aluminum derivatives staying complexed on the hydroxylate create a steric hindrance that directs the second de-*O*-benzylation on the farthest position, leading after double deprotection to the 6A,6D-dihydroxyl,6B-vinyl α-cyclodextrin **97** [78]. The vinyl groups can then be converted into hydroxyls by

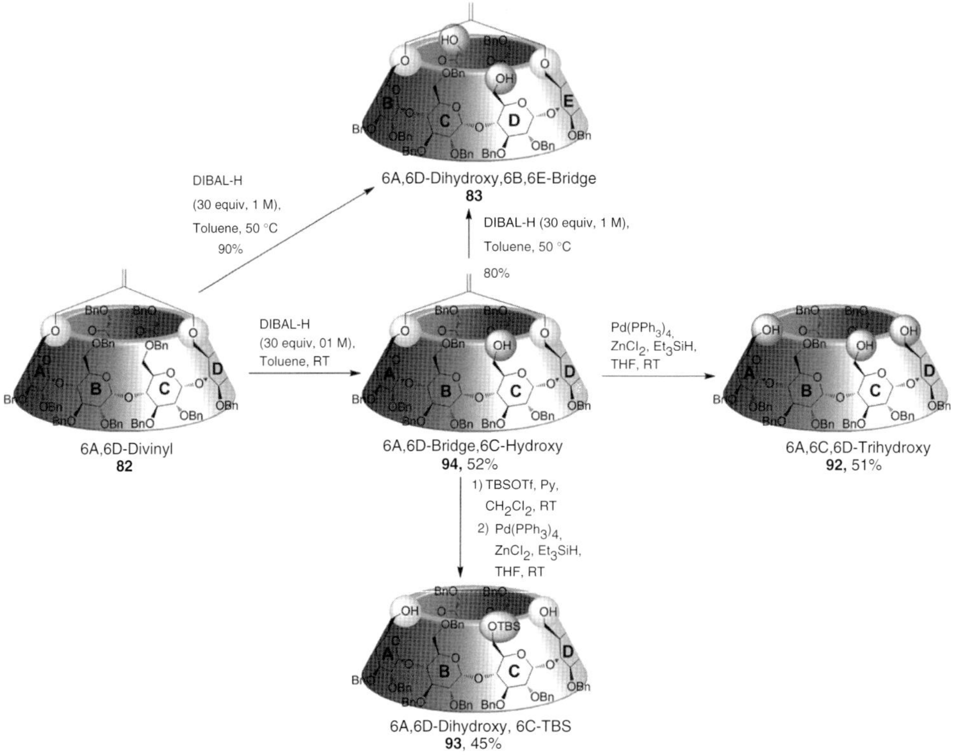

Scheme 9.43 Mono-de-*O*-benzylation on bridged α-cyclodextrin.

reductive ozonolysis followed by reduction, leading to 6A,6B-dihydroxyl cyclodextrin **98** and 6A,6B,6D-trihydroxyl cyclodextrin **99**.

It is worth noting that this strategy affords α-cyclodextrin triol **99** with mirror image arrangement of functions compared to cyclodextrin **92** obtained in Scheme 9.42 and Scheme 9.43.

9.5.3
Third Deprotection

Using a combination of the protecting groups that resist the action of DIBAL-H, a third de-*O*-benzylation is possible. The hydroxyls of the divinyl-dihydroxy α-cyclodextrin **80** were protected by the vinyl bridge, and cyclodextrin **100** was submitted another time to the action of DIBAL-H (Scheme 9.45). The double de-*O*-benzylation leads to α-cyclodextrin **101** bearing three different groups on its primary rim [75], and a different group on its secondary rim.

The substituents used to obtain α-cyclodextrin **101** tridifferentiated on its primary rim can be converted into usual protecting groups (Scheme 9.46). After protection of the hydroxyls as silyl ethers, the bridge can be removed using a palladium-catalyzed

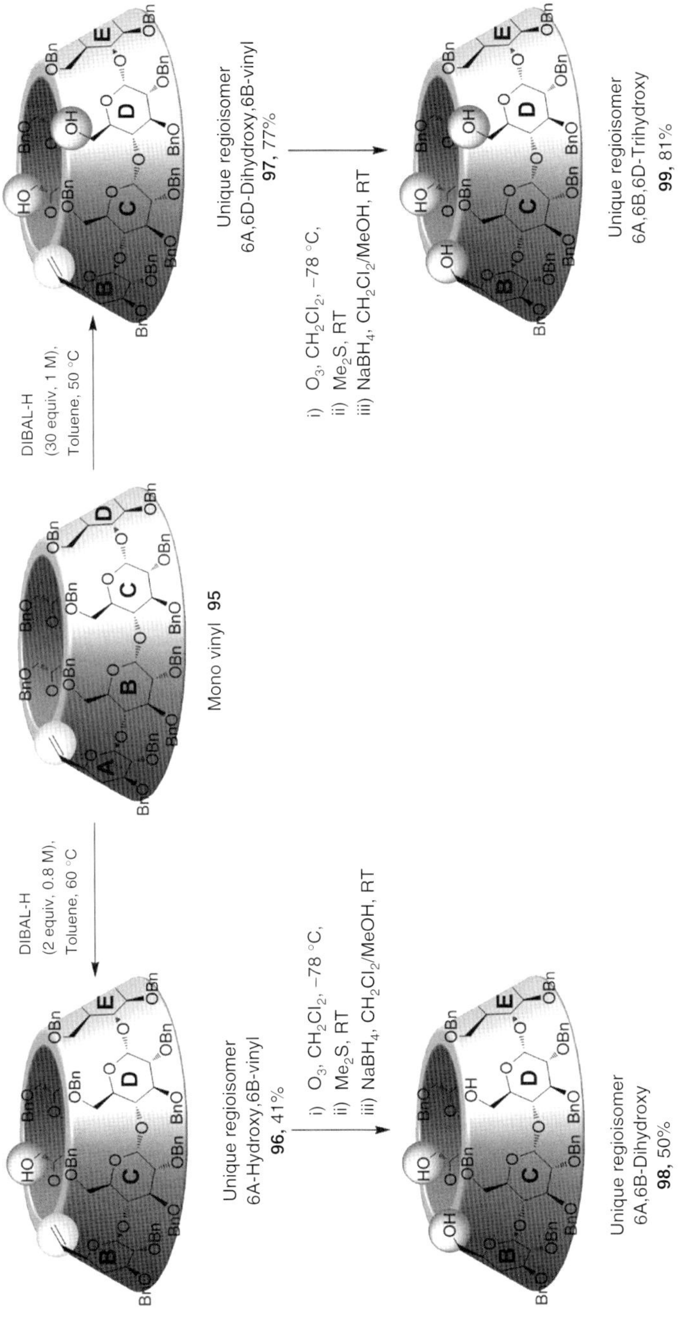

Scheme 9.44 Mono- and di-de-*O*-benzylation on monovinyl α-cyclodextrin.

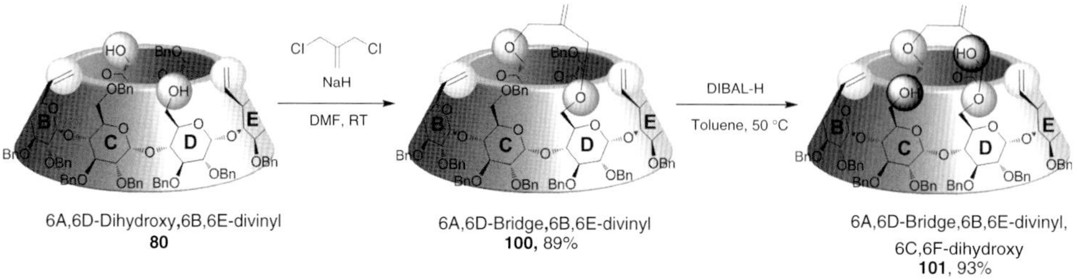

Scheme 9.45 Tridifferentiation of the primary rim of α-cyclodextrin.

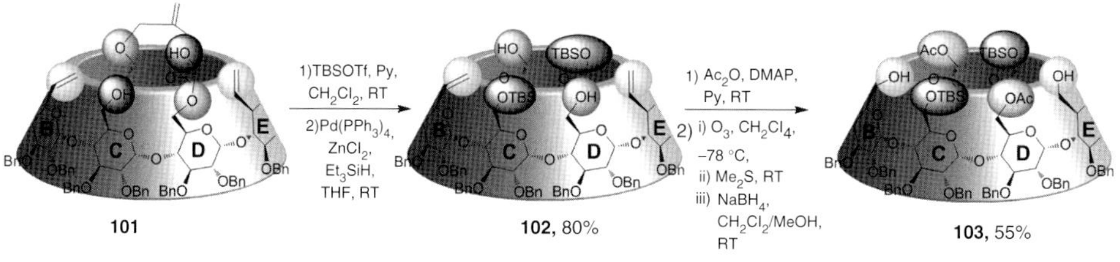

Scheme 9.46 Back to usual protection groups.

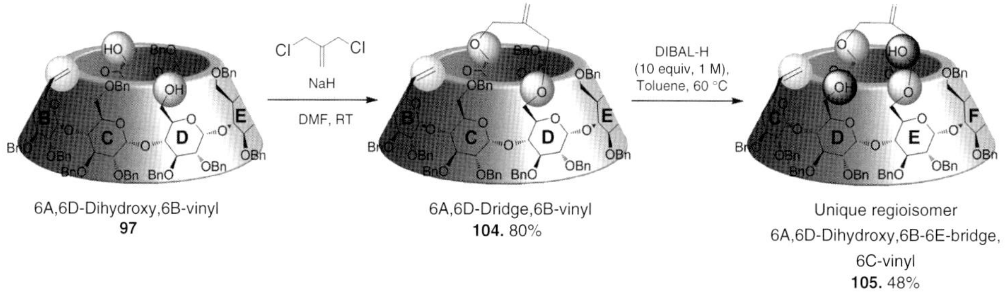

Scheme 9.47 Ultimate tetradifferentiation of the primary rim of α-cyclodextrin.

double de-*O*-allylation. The revealed hydroxyls of cyclodextrin **102** were protected as acetates, and the vinyl groups were cleaved by ozonolysis and reduction, leading to cyclodextrin **103** bearing usual protecting groups [75].

A cyclodextrin tetradifferentiated on its primary rim can be obtained through combination of mono- and double de-*O*-benzylations. The two hydroxyls of the 6A,6D-dihydroxy,6B-vinyl α-cyclodextrin **97** (Scheme 9.44) can be protected using the vinyl bridge (Scheme 9.47). Another action of DIBAL-H on cyclodextrin **104** leads to a double de-*O*-benzylation on the glucose units preceding the bridge, which are the less hindered, giving a unique regioisomer **105** in 48% yield [78].

9.6
Conclusion and Perspectives

Selective functionalization and differentiation of cyclodextrins is particularly challenging because of the complexity of their chemistry and the high number of isomers that can be obtained. Starting with random introduction of substituents, the methods became more selective. Three approaches have been successfully developed: bridging two positions for difunctionalization, steric repulsion of protecting groups for multifunctionalization, and selective deprotections for multidifferentiation. Even if many patterns of substitution can be selectively obtained, a long way remains before each hydroxyl can be individually modified.

Selective modification of cyclodextrin was motivated by the potential applications of these derivatives. Now, the new patterns of functionalization have to be included in the synthesis of new molecular architectures, supramolecular tectons, and materials.

9.7 Experimental Procedures

Native cyclodextrins **1–3** are very hydroscopic. They should be dried under reduced pressure until constant weight before use in reactions necessitating anhydrous conditions.

9.7.1 Tetrafunctionalization of the Primary Rim of α-Cyclodextrin Using Supertrityl

Tris-(*p*-*tert*-butylphenyl)methyl chloride (6 equiv) and 4-dimethylaminopyridine (0.6 equiv) were added to a solution of α-cyclodextrin **1** (1 equiv) in pyridine (30 $ml\,g^{-1}$ of cyclodextrin) [58]. The solution was stirred at 70 °C for 1 day, concentrated under reduced pressure, and poured into water. The precipitate was collected by filtration and dried under vacuum at 50 °C. Silica gel column chromatography of the residue ($MeOH/CH_2Cl_2$: 1/9) gave the pure compound in about 70% yield. Column chromatography was only possible in the presence of a small amount of pyridine to avoid the loss of supertrityl groups. The authors noted that this purification step is not compulsory for the following methylation step.

9.7.2 Double Deprotection of Perbenzylated α- or β-Cyclodextrins Using DIBAL-H

DIBAL-H (30 equiv, 1.5 M in toluene) was added to a solution of perbenzylated α- or β-cyclodextrin **10a–b** in anhydrous toluene (so that the concentration of DIBAL-H was 1 M) at room temperature [78]. The solution was then stirred at 50 °C for 2 h. Water was carefully added dropwise at 0 °C and the solution was stirred vigorously for 15 min. Extraction with EtOAc followed by evaporation of the solvent under reduced pressure afforded a white foam. Silica gel column chromatography of the residue (EtOAc/cyclohexane: 1/5) gave the pure compound **41a–b** in about 85% yield.

The same procedure can be used with α- or β-cyclodextrin monoazide **77a–b** for the tandem reaction: reduction of the azide and selective debenzylation on the opposite position [74].

List of Abbreviations

Ac	Acetyl
Bn	Benzyl
CGTase	Cyclodextrin glucosyltransferase
DIBAL-H	Diisobutylaluminum hydride
DMAP	4-Dimethylaminopyridine
DMF	Dimethylformamide
DMSO	Dimethylsulfoxide
Et	Ethyl
HPLC	High-performance liquid chromatography
IUPAC	International Union of Pure and Applied Chemistry
Me	Methyl
Ms	Mesyl, methanesulfonyl
NBS	*N*-Bromosuccinimide
NCS	*N*-Chlorosuccinimide
Py	Pyridine
RT	Room temperature
^{S}Tr	"Supertrityl," tris(4-*tert*-butylphenyl)methyl
TBS	*tert*-Butyldimethylsilyl
Tf	Triflate
THF	Tetrahydrofuran
TMS	Trimethylsilyl
Ts	Tosyl
Tr	Trityl

References

1. Villiers, A. (1891) *C.R. Hebd. Seances Acad. Sci.*, **112**, 536–538.
2. Martin Del Valle, E.M. (2004) *Process. Biochem.*, **39**, 1033–1046.
3. Szejtli, J. (1998) *Chem. Rev.*, **98**, 1743–1753.
4. Wenz, G. (1994) *Angew. Chem., Int. Ed. Engl.*, **33**, 803–822.
5. Uekama, K., Hrayama, F., and Irie, T. (1998) *Chem. Rev.*, **98**, 2045–2076.
6. (a) Munoz-Botalla, S., del Castillo, B., and Martin, M.A. (1995) *Ars. Pharm.*, **36**, 187–198. (b) Buschmann, H.J. and Schollmeyer, E. (2002) *J. Cosmet. Sci.*, **53**, 185–191.
7. Li, S. and Purdy, W.C. (1992) *Chem. Rev.*, **92**, 1457–1470.
8. Breslow, R., Doherty, J.B., Guillot, G., and Lipsey, C. (1978) *J. Am. Chem. Soc.*, **100**, 3227–3229.
9. Sollogoub, M. (2009) *Eur. J. Org. Chem.*, 1295–1303.
10. Gelb, R.I., Schwartz, L.M., Bradshaw, J.J., and Laufer, D.A. (1980) *Bioorg. Chem.*, **9**, 299–304.
11. (a) Croft, A.P. and Bartsch, R.A. (1983) *Tetrahedron*, **39**, 1417–1474. (b) Khan,

A.R., Forgo, P., Stine, K.J., and D'Souza, V.T. (1998) *Chem. Rev.*, **98**, 1977–1996.

12. (a) Rong, D. and D'Souza, V.T. (1990) *Tetrahedron Lett.*, **31**, 4275–4278. (b) Ward, S. and Ling, C.C. (2011) *Eur. J. Org. Chem.*, 4853–4861.
13. (a) Takeo, K., Uemura, K., and Mitoh, H. (1988) *J. Carbohydr. Chem.*, **7**, 293–308. (b) Fügedi, P. (1989) *Carbohydr. Res.*, **192**, 366–369.
14. Takeo, K., Mitoh, H., and Uemura, K. (1989) *Carbohydr. Res.*, **187**, 203–221.
15. (a) Gadelle, A. and Defaye, J. (1991) *Angew. Chem., Int. Ed. Engl.*, **30**, 78–80. (b) Chmurski, K. and Defaye, J. (2000) *Supramol. Chem.*, **12**, 221–224.
16. (a) Lai, C.S.I., Moody, G.J., Thomas, J.D.R., Mulligan, D.C., Stoddart, J.F., and Zarzycki, R.J. (1988) *J. Chem. Soc., Faraday Trans. 2*, 319–324. (b) Coleman, A.W., Zhang, P., Ling, C.C., Parrot-Lopez, H., and Galon, H. (1992) *Carbohydr. Res.*, **224**, 307–309.
17. (a) Casu, B., Reggiani, M., Gallo, G.G., and Vigevani, A. (1968) *Tetrahedron*, **24**, 803–821. (b) Boger, J., Corcoran, R.J., and Lehn, J.-M. (1978) *Helv. Chim. Acta*, **61**, 2190–2218.
18. Sato, T., Nakamura, H., Ohno, Y., and Endo, T. (1990) *Carbohydr. Res.*, **199**, 31–35.
19. Angibeaud, P. and Utille, J.-P. (1991) *Synthesis*, **9**, 737–738.
20. Guitet, M., Adam de Beaumais, S., Vauzeilles, B., Blériot, Y., Zhang, Y., Ménand, M., and Sollogoub, M. (2012) *Carbohydr. Res.*, **356**, 278–281.
21. Hapiot, F., Tilloy, S., and Monflier, E. (2006) *Chem. Rev.*, **106**, 767–781.
22. (a) Melton, L.D. and Slessor, K.N. (1971) *Carbohydr. Res.*, **18**, 29–37. (b) Tabushi, I., Shimizu, N., Sugimoto, T., Shiozuka, M., and Yamamura, K. (1977) *J. Am. Chem. Soc.*, **99**, 7100–7102. (c) Takahashi, K., Hattori, K., and Toda, F. (1984) *Tetrahedron Lett.*, **25**, 3331–3334.
23. (a) Martin, K.A. and Czarnik, A.W. (1994) *Tetrahedron Lett.*, **35**, 6781–6782. (b) Huff, J.B. and Bieniarz, C. (1994) *J. Org. Chem.*, **59**, 7511–7516. (c) Yoon, J., Hong, S., Martin, K.A., and Czarnik, A.W. (1996) *J. Org. Chem.*, **60**, 2792–2795.
24. Hanessian, S., Benalil, A., and Laferriere, C. (1995) *J. Org. Chem.*, **60**, 4786–4797.
25. (a) Fügedi, P. and Nánási, P. (1988) *Carbohydr. Res.*, **175**, 173–181. (b) Chen, Z., Bradshaw, J.S., and Lee, M.L. (1996) *Tetrahedron Lett.*, **37**, 6831–6834.
26. Cornwell, M.J., Huff, J.B., and Bieniarz, C. (1995) *Tetrahedron Lett.*, **36**, 8371–8374.
27. Ueno, A. and Breslow, R. (1982) *Tetrahedron Lett.*, **23**, 3451–3454.
28. (a) Fujita, K., Yamamura, H., Matsunaga, A., Imoto, T., Mihashi, K., and Fujioka, T. (1986) *J. Am. Chem. Soc.*, **108**, 4509–4513. (b) Yamamura, H., Nagaoka, H., Saito, K., Kawai, M., Butsugan, Y., Nakajima, T., and Fujita, K. (1993) *J. Org. Chem.*, **58**, 2936–2937.
29. (a) Ueno, A., Moriwaki, F., Azuma, A., and Osa, T. (1989) *J. Org. Chem.*, **54**, 295–299. (b) Fujita, K., Yamamura, H., and Imoto, T. (1991) *Tetrahedron Lett.*, **32**, 6737–6740. (c) Fujita, K., Yamamura, H., Tah, Y., Imoto, T., Koga, T., Fujioka, T., and Mihashi, K. (1990) *J. Org. Chem.*, **55**, 877–880.
30. (a) Fujita, K., Matsunaga, A., Yamamura, H., and Imoto, T. (1988) *Chem. Lett.*, 1947–1950. (b) Fujita, K., Ishizu, T., Oshiro, K., and Obe, K. (1989) *Bull. Chem. Soc. Jpn.*, **62**, 2960–2962.
31. Fujita, K., Matsunaga, A., and Imoto, T. (1984) *J. Am. Chem. Soc.*, **106**, 5740–5741.
32. Tanimoto, T., Sakaki, T., and Koizumi, K. (1993) *Chem. Pharm. Bull.*, **41**, 866–869.
33. Tanimoto, T., Sakaki, T., Iwanaga, T., and Koizumi, K. (1994) *Chem. Pharm. Bull.*, **42**, 385–387.
34. (a) Tabushi, I., Nabeshima, T., Fujita, K., Matsunaga, A., and Imoto, T. (1985) *J. Org. Chem.*, **50**, 2638–2643. (b) Breslow, R., Canary, J.W., Varney, M., Waddell, S.T., and Yang, D. (1990) *J. Am. Chem. Soc.*, **112**, 5212–5219.
35. Yuan, D.Q., Immel, S., Koga, K., Yamaguchi, M., and Fujita, K. (2003) *Chem. Eur. J.*, **9**, 3501–3506.
36. Armspach, D., Poorters, L., Matt, D., Benmerad, B., Balegroune, F., and

Toupet, L. (2005) *Org. Biomol. Chem.*, **3**, 2588–2592.
37. Tabushi, I., Kuroda, Y., Yokota, K., and Yuan, L.C. (1981) *J. Am. Chem. Soc.*, **103**, 711–712.
38. Tabushi, I., Yamamura, K., and Nabeshima, T. (1984) *J. Am. Chem. Soc.*, **106**, 5267–5270.
39. Koga, K., Yuan, D.Q., and Fujita, K. (2000) *Tetrahedron Lett.*, **41**, 6855–6857.
40. Engeldinger, E., Armspach, D., and Matt, D. (2003) *Chem. Rev.*, **103**, 4147–4173.
41. Teranishi, K. (2000) *Chem. Commun.*, 1255–1256.
42. Teranishi, K. (2000) *Tetrahedron Lett.*, **41**, 7085–7088.
43. Teranishi, K. (2001) *Tetrahedron Lett.*, **42**, 5477–5480.
44. (a) Sakairi, N. and Kuzuhara, H. (1993) *Chem. Lett.*, 2077–2080. (b) Sakairi, N., Nishi, N., Tokura, S., and Kuzuhara, H. (1996) *Carbohydr. Res.*, **291**, 53–62.
45. Matsuoka, K., Shiraishi, Y., Terunuma, D., and Kuzuhara, H. (2001) *Tetrahedron Lett.*, **42**, 1531–1533.
46. Gramage-Doria, R., Rodriguez-Lucena, D., Armspach, D., Egloff, C., Jouffroy, M., Matt, D., and Toupet, L. (2011) *Chem. Eur. J.*, **17**, 3911–3921.
47. Tabushi, I., Yuan, L.C., Shimokawa, K., Yokota, K., Mizutani, T., and Kuroda, Y. (1981) *Tetrahedron Lett.*, **22**, 2273–2276.
48. Tabushi, I., Nabeshima, T., Kitaguchi, H., and Yamamura, K. (1982) *J. Am. Chem. Soc.*, **104**, 2017–2019.
49. Yuan, D.-Q., Kitagawa, Y., Fukudome, M., and Fujita, K. (2007) *Org. Lett.*, **9**, 4591–4594.
50. Yu, H., Yuan, D.-Q., Makino, Y., Fukudome, M., Xie, R.-G., and Fujita, K. (2006) *Chem. Commun.*, 5057–5059.
51. Atsumi, M., Izumida, M., Yuan, D.Q., and Fujita, K. (2000) *Tetrahedron Lett.*, **41**, 8117–8120.
52. Yuan, D.Q., Yamada, T., and Fujita, K. (2001) *Chem. Commun.*, 2706–2707.
53. Petrillo, M., Marinescu, L., Rousseau, C., and Bols, M. (2009) *Org. Lett.*, **11**, 1983–1985.
54. Jouffroy, M., Gramage-Doria, R., Armspach, D., Matt, D., and Toupet, L. (2012) *Chem. Commun.*, 6028–6030.
55. Boger, J., Brenner, D.G., and Knowles, J.R. (1979) *J. Am. Chem. Soc.*, **101**, 7630–7631.
56. Ling, C.C., Coleman, A.W., and Miocque, M. (1992) *Carbohydr. Res.*, **223**, 287–291.
57. Armspach, D. and Matt, D. (1998) *Carbohydr. Res.*, **310**, 129–133.
58. Poorters, L., Armspach, D., and Matt, D. (2003) *Eur. J. Org. Chem.*, 1377–1381.
59. Heck, R., Jicsinsky, L., and Marsura, A. (2003) *Tetrahedron Lett.*, **44**, 5411–5413.
60. (a) Hengge, A.C., Tobin, A.E., and Cheland, W.W. (1995) *J. Am. Chem. Soc.*, **117**, 5919–5926. (b) Hamasaki, K. and Ueno, A. (1995) *Chem. Lett.*, 859–860.
61. Yuan, D.Q., Kitagawa, Y., Aoyama, K., Douke, T., Fukudome, M., and Fujita, K. (2007) *Angew. Chem. Int. Ed.*, **46**, 5024–5027.
62. Sollogoub, M., Das, S.K., Mallet, J.-M., and Sinaÿ, P. (1999) *C.R. Acad. Sci., Sér. IIc*, **2**, 441–448.
63. Lecourt, T., Herault, A., Pearce, A.J., Sollogoub, M., and Sinaÿ, P. (2004) *Chem. Eur. J.*, **10**, 2960–2971.
64. Kuranaga, K., Ishihara, S., Ohtani, N., Satake, M., and Tachibana, K. (2010) *Tetrahedron Lett.*, **51**, 6345–6348.
65. Pearce, A.J. and Sinaÿ, P. (2000) *Angew. Chem. Int. Ed.*, **39**, 3610–3612.
66. Rawal, G.K., Rani, S., and Ling, C.-C. (2009) *Tetrahedron Lett.*, **50**, 4633–4636.
67. Rawal, G.K., Rani, S., Ward, S., and Ling, C.-C. (2010) *Org. Biomol. Chem.*, **8**, 171–180.
68. Ghosh, R., Zhang, P., Wang, A., and Ling, C.-C. (2012) *Angew. Chem. Int. Ed.*, **51**, 1548–1552.
69. Xiao, S., Yang, M., Sinaÿ, P., Blériot, Y., Sollogoub, M., and Zhang, Y. (2010) *Eur. J. Org. Chem.*, 1510–1516.
70. Roizel, B.D., Baltaze, J.P., and Sinaÿ, P. (2002) *Tetrahedron Lett.*, **43**, 2371–2373.
71. Xiao, S., Zhou, D., Yang, M., Sinaÿ, P., Sollogoub, M., and Zhang, Y. (2011) *Tetrahedron Lett.*, **52**, 5273–5276.
72. Luo, X., Chen, Y., Huber, J.G., Zhang, Y., and Sinaÿ, P. (2004) *C.R. Chim.*, **7**, 25–28.
73. Chen, Y., Huber, J.G., Zhang, Y., and Sinaÿ, P. (2005) *C.R. Chim.*, **8**, 27–30.

74. Guieu, S. and Sollogoub, M. (2008) *Angew. Chem. Int. Ed.*, **47**, 7060–7063.

75. Bistri, O., Sinaÿ, P., Jiménez Barbero, J., and Sollogoub, M. (2007) *Chem. Eur. J.*, **13**, 9757–9774.

76. Bistri, O., Sinaÿ, P., and Sollogoub, M. (2005) *Tetrahedron Lett.*, **46**, 7757–7760.

77. (a) Bistri, O., Sinaÿ, P., and Sollogoub, M. (2006) *Chem. Commun.*, 1112–1114. (b) Bistri, O., Sinaÿ, P., and Sollogoub, M. (2006) *Chem. Lett.*, 534–535. (c) Bistri, O., Sinaÿ, P., and Sollogoub, M. (2006) *Tetrahedron Lett.*, **47**, 4137–4139.

78. Guieu, S. and Sollogoub, M. (2008) *J. Org. Chem.*, **73**, 2819–2828.

10
Design and Synthesis of GM1 Glycomimetics as Cholera Toxin Ligands

José J. Reina and Anna Bernardi

10.1
Introduction

GM1 belongs to the family of gangliosides, glycosphingolipids that contain sialic acid (NeuAc, neuraminic acid) residues in their oligosaccharidic head-group. They are amphiphatic molecules constituted by an oligosaccharide head of variable length and complexity, connected to a ceramide lipid anchor [1, 2] (Figure 10.1). Gangliosides are ubiquitous components of mammalian cell membranes, but they are particularly abundant in the nervous system and, within the nervous system, they are present at high levels in neurons [3]. The main location of gangliosides is the outer layer of the plasma membrane, where their head-groups are exposed at the cell surface, while the ceramide hydrophobic moieties are inserted into the membrane's external layer [4]. Over the years, gangliosides have been implicated in fundamental cellular processes such as growth, differentiation, and adhesion [5]. Recently, progresses have been made in understanding how they exert their effects on cell behavior through participation in cell signaling pathways [6].

The GM1 ganglioside has important physiological properties and impacts neuronal plasticity and repair mechanisms as well as the release of neurotrophins in the brain [7]. Besides its function in the physiology of the brain, GM1 acts as the specific cell membrane receptor for both cholera toxin (CT) and the *Escherichia coli* heat-labile enterotoxins (LTs) in the small intestine of humans. It is well established that CT is responsible for the symptoms presented by patients infected with *Vibrio cholerae* and exerts its action after adhesion to intestinal epithelial cells via the interaction with the oligosaccharidic head-group of GM1 (GM1-os, GM1 oligosaccharide) [8, 9]. Thus, ligands antagonizing this interaction may have therapeutic value. In addition, the CT/GM1-os interaction is known in atomic detail [10] and high quality thermodynamic data have been obtained [11]. Thus the CT/GM1-os pair has been used by us [12] and others [10c, 13] as an ideal candidate for research projects directed toward the design and synthesis of mimics of oligosaccharides. Here, while reviewing the rationale behind the design, we will mostly detail our synthetic approaches toward the synthesis of GM1 glycomimetics.

Modern Synthetic Methods in Carbohydrate Chemistry: From Monosaccharides to Complex Glycoconjugates, First Edition. Edited by Daniel B. Werz and Sébastien Vidal.

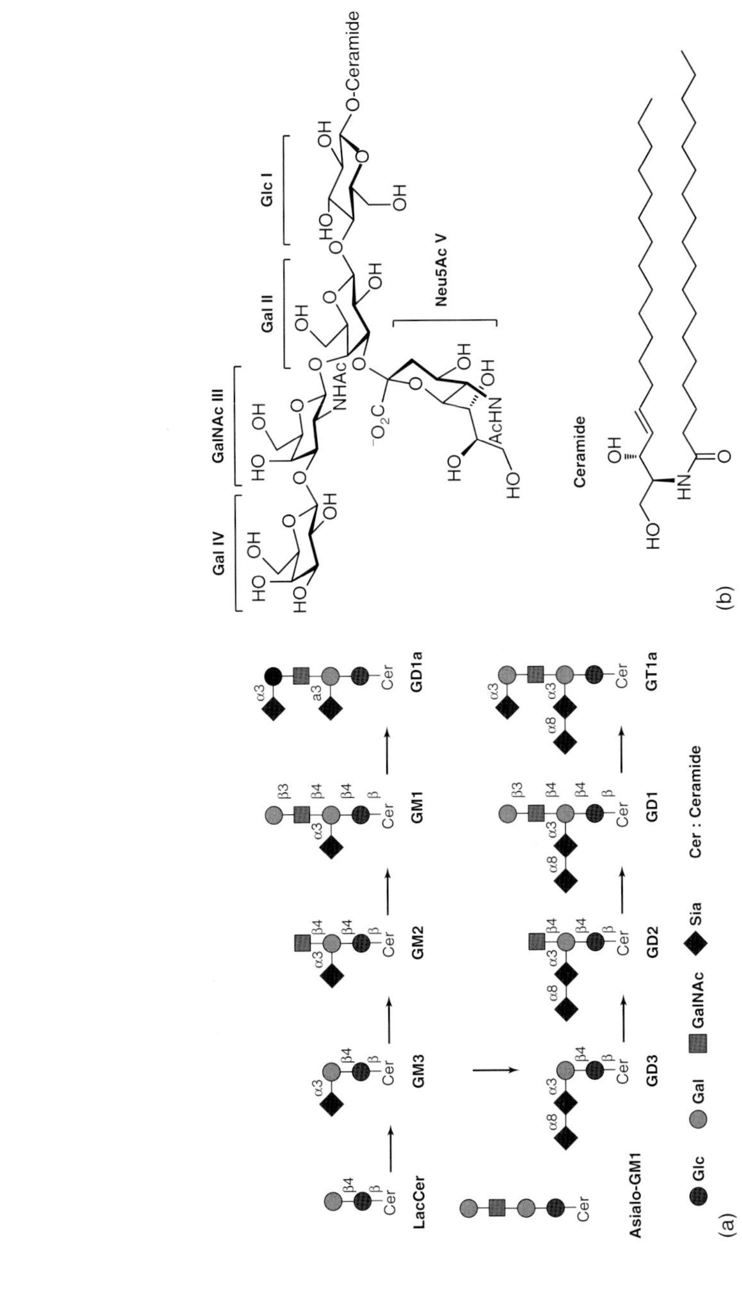

Figure 10.1 (a) Schematic representation of the most common gangliosides. (b) Structure of GM1 ganglioside head-group.

10.2
Cholera Toxin and Its Specific Membrane Receptor, the GM1 Ganglioside

CT belongs to the AB_5 bacterial toxins family, which includes CT itself and the *E. coli* LTs, LT-I and LT-II, among others. The structure and functions of the AB_5 toxins have been reviewed in detail on several occasions [10]. They are named after their particular architecture that consists of a single catalytically active component, A (~27.4 kDa; 240 amino acids) and a nontoxic pentamer of identical B subunits (B_5) (~11.6 kDa each B subunit; 103 amino acids) [10a, 14, 15]. The A subunit and the B pentamer are not cytotoxic individually, but the complete AB_5 holotoxin is required for intoxication. In the case of CT and of the highly homologous LTs, the B pentamer is responsible for binding to GM1 on the external membrane of intestinal epithelial cells. This event initiates the threatening action of CT. The interaction of the oligosaccharidic head-group of ganglioside GM1 (Galβ1-3GalNAcβ1-4(NeuAcα2–3)Galβ1-4Glcβ1-OH, GM1-os) with the cholera toxin B pentamer (CTB) is depicted in Figure 10.2. It is interesting that the binding capability of the B pentamer to cell surface receptors is retained even in the absence of the A subunit.

Several high-resolution X-ray structures of AB_5 toxins with or without bound ligands are available [16–22]. Binding data have been obtained through a variety of biophysical techniques, including nuclear magnetic resonance (NMR) [23], solid-phase and thin layer chromatography (TLC) overlay assay [24, 25], surface plasmon resonance (SPR) [26], fluorescence spectroscopy [27, 28], flow cytometry [29], atomic force microscopy [30], and isothermal titration calorimetry (ITC) [27, 31]. Given the importance of the processes that they promote, the complexes formed between gangliosides and AB_5 toxins have been studied extensively at different levels. For basic research, they have offered a paradigmatic model for studying the structural and thermodynamic basis of protein–carbohydrate interactions; and, for medicinal

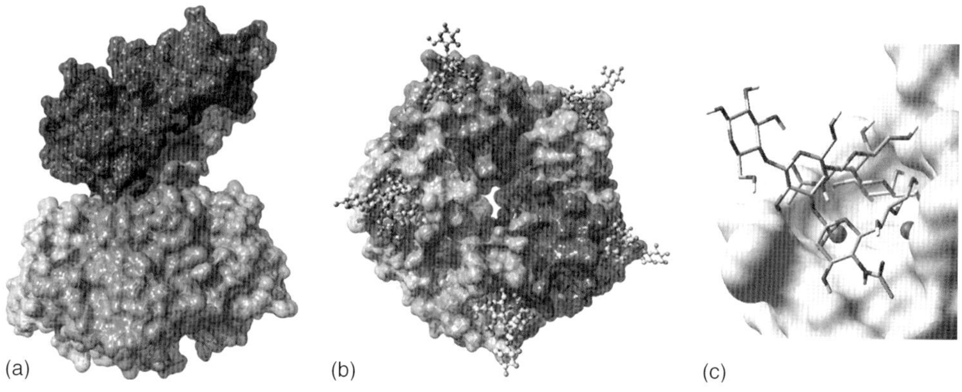

Figure 10.2 (a) Holotoxin CT as the AB_5 assembly, (b) pentamer B complexed with five copies of GM1-os, and (c) X-ray structure of the CTB:GM1-os complex: detail of the binding site.

chemistry, they provide key insights for the structure-based design of ligands that may be used to treat the diseases caused by AB_5 bacterial toxins, such as traveler's diarrhea and cholera.

10.2.1
Interaction of Cholera Toxin and GM1-os

The B pentamer of CT (CTB) interacts with the soluble, monovalent oligosaccharide portion of GM1 (GM1-os) with strong affinity, the binding process is weakly cooperative. The dissociation constant for the monovalent interaction of one GM1-os with one B pentamer binding site has been evaluated by ITC and found to be 43 nM [11], the highest affinity for a carbohydrate–protein interaction described to date. On cell membranes, this initial binding event is further amplified by the multivalent interaction of the B_5 pentamer with multiple copies of GM1-os presented at high concentration in lipids rafts at the cell surface [29].

A high-resolution (1.25 Å) X-ray structure of the CTB/GM1-os complex was reported [18]. A view of the binding site with bound GM1-os is shown in Figure 10.2c. The crystal structure shows a bidentate interaction of the branched GM1-os pentasaccharide, which has been described as a "two fingers grip": the first finger is a sialic acid "thumb," which grazes the surface of the protein and the second one is a Galβ(1–3)GalNAc "forefinger," which penetrates more deeply into the binding site. Most of the contacts are given by the "finger" tips: in terms of buried protein surface, the terminal Gal and NeuAc residues contribute 39 and 43% of the intermolecular contacts, the rest and minor part of protein surface is buried by GalNAc. A comparison between the bound GM1-os and previously reported NMR-based solution structure of GM1-os [32] shows that this oligosaccharide is highly preorganized for interaction with CTB. Such preorganization appears to be the source of the unusually high binding affinity of the CTB/GM1-os pair which we mentioned earlier. Indeed, it has been also observed that all of the monosaccharide or disaccharide fragments of GM1-os bind to CTB much more weakly than the whole molecule. For example, the K_d for galactose binding is only 15 mM, which is improved by only a factor of 2 in the case of the full Gal–GalNAc forefinger. The other important binding determinant, NeuAc, binds even more weakly to the protein ($K_d \approx 200$ mM) [11]. A thermodynamic analysis of these ITC binding data has shown that high affinity and selectivity of CTB/GM1-os interaction originates mainly from the conformational preorganization of the branched GM1 pentasaccharide, rather than through the cooperativity of the terminal residues (galactose and sialic acid) in the oligosaccharide [11]. The terminal galactose residue in the "forefinger" binds to the CTB binding site very specifically. The pyranose ring of this galactose is stacked on top of Trp88 (CH/π interaction) and forms an extensive hydrogen bond network with Asn90, Lys91, Glu51, and Gln61 residues from CTB. The galactose-binding cavity is deep and shielded from the solvent. On the contrary, the rest of the toxin's binding site is shallow and solvent exposed. The sialic acid residue represents the second important moiety of GM1 required for recognition of CT. In fact, asialo-GM1 (Figure 10.1) binds to CT with much lower affinity.

The sugar ring of sialic acid makes hydrophobic interaction with CT Tyr12, and its carboxyl group forms a water-mediated hydrogen bond with Trp88.

10.3 Rational Design of GM1-os Mimics as Cholera Toxin Inhibitors and Synthesis of First-Generation Ligands

Preventing the adhesion of CT to cell surfaces is one of the therapeutic strategies that can be adopted to reduce the severity of the symptoms after *V. cholerae* infection. This has been attempted, initially, using GM1 or GM1-os themselves. The synthesis of the full ganglioside is very complex and labor-intensive [33, 34]. The ganglioside can be extracted from the brain of a cow or a pig and the pentasaccharide is obtained by controlled hydrolysis. More recently, large-scale production in *E. coli* has been reported [35]. The main drawback of using GM1 therapeutically is that it can induce cholera sensitivity in cells normally lacking the receptor.

Some years ago, we became interested in the possibility of designing functional and structural mimics of GM1-os, which can function as soluble ligands of CT and prevent its adhesion to the intestine. At the time, very few examples were available in the literature for the rational design of oligosaccharide mimetics. The most notable ones came from the Ernst's group (then at Ciba-Geigy Ltd., Basel (CIBA)), who had begun to show the importance of conformational preorganization and of lipophilic interactions in the design of E-selectin ligands [36]. Our attention was attracted by NMR and theoretical studies of the conformational behavior of GM1-os and other ganglioside head-groups (e.g., GM2, GM3, and asialo-GM1). These studies were suggesting that 3,4-branching at Gal-II residue (Figure 10.1) acts as a conformational lock and blocks the two GM1-os "fingers" in the proper orientation for optimal interaction with CT [37, 38]. From the above structure-based hypothesis, we designed our first mimic **1** (psGM1, pseudo-GM1; Figure 10.3) by replacing this residue with the enantiomerically pure *cis*-cyclohexane diol DCCHD

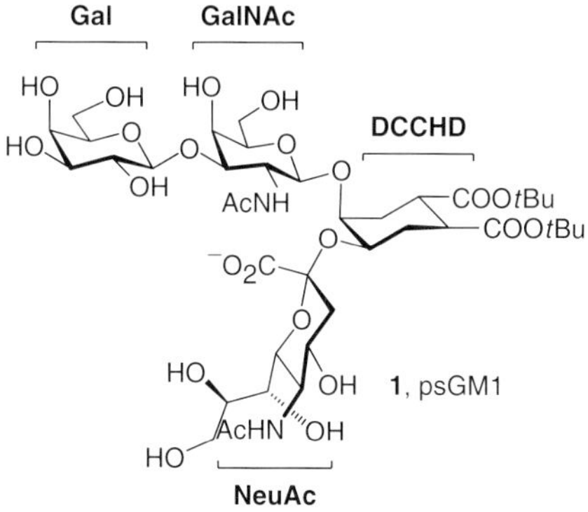

1, psGM1

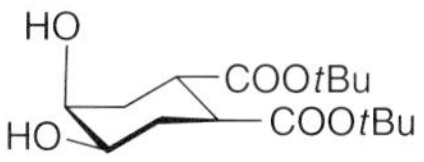

4, DCCHD

Figure 10.3 Structures of (a) GM1 mimic psGM1 (**1**) and (b) DCCHD **4**.

(dicarboxy-cyclohexane-diols) **4**, which possesses the same absolute and relative configuration of natural galactose and is locked in a single-chair conformation by the two carboxy groups [39]. The use of cyclohexanediols as core sugar mimetics is an attractive strategy that can simplify the molecular structure of bioactive oligosaccharides and produce functional analogs that may be easier to synthesize and of increased metabolic stability [12]. So, apart from the therapeutic potential of the work in the fight against cholera, a main driving force of this research was an interest in developing a general approach to the design of glycomimetic molecules.

A computational protocol initially allowed validation of the working hypothesis [40]. The conformation of **1** was studied both for the isolated molecule and in the binding pocket of LT, used as a model for CT, and the predicted three-dimensional (3D) structures were compared with those described for GM1. The calculation predicted that **1** and GM1-os would share the same conformation and that the LT:mimic complex would feature all the expected intermolecular interactions. The superimposition of the computer model of the LT:mimic complex with the X-ray structure of CT-GM1 showed that the mimic and GM1 adopt a common disposition in the toxin-binding pocket.

The diol **4** had never been described in the literature before we used it in the synthesis of **1**. Initially, we tried to synthesize it using an enantioselective Diels–Alder reaction between a chiral fumarate equivalent **5** (Scheme 10.1) and butadiene **6**, followed by double-bond dihydroxylation [41]. However, a more effective protocol for large-scale synthesis could finally be based on the procedure shown in Scheme 10.2, leading to enantiomerically pure (1*S*,2*S*)-cyclohex-4-ene-1,2-dicarboxylic acid **7** [42]. Here, a Bölm desymmetrization of tetrahydrophthalic anhydride **8** using quinine leads to monoester **9** in 89% ee. The monoester **9**, in turn, can be epimerized to the trans isomer **10** which is hydrolyzed to obtain **7**, after crystallization. This intermediate is the starting material for the synthesis of **4** and of other conformationally constrained DCCHD to be used as monosaccharide mimics [43, 44]. Our current best protocol for the large-scale synthesis of **7** is reported at the end of this chapter.

With this diacid in hand, the synthesis of **4** can be completed by reaction with dimethylformamide di-*tert*-butyl acetal to afford the bis *tert*-butyl ester, followed by dihydroxylation (cat. $OsCl_3$ and Me_3NO) [41].

The synthesis of **1** from **4** was adapted from previous synthesis of GM1-os and performed as shown in Scheme 10.3. The α-sialylation of **4** took advantage of the markedly different reactivity of equatorial and axial hydroxyl groups typically observed in the sialylation of Gal [45] and could be performed regioselectively at the equatorial position using phosphite **11** (Scheme 10.3). The desired α-sialoside **12** could be purified by flash chromatography in acceptable yields and was submitted to glycosylation by the Galβ(1–3)GalNAc donor **13** (Scheme 10.3). It is well known that 2-acetamido sugars are poor glycosyl donors [46]; however, **13** is readily accessible [47] and allows to introduce the desired 2-acetamido functionality directly in the reaction product, thus avoiding further manipulation of the oligosaccharide and possibly compensating for low yields in the glycosylation reaction. Indeed,

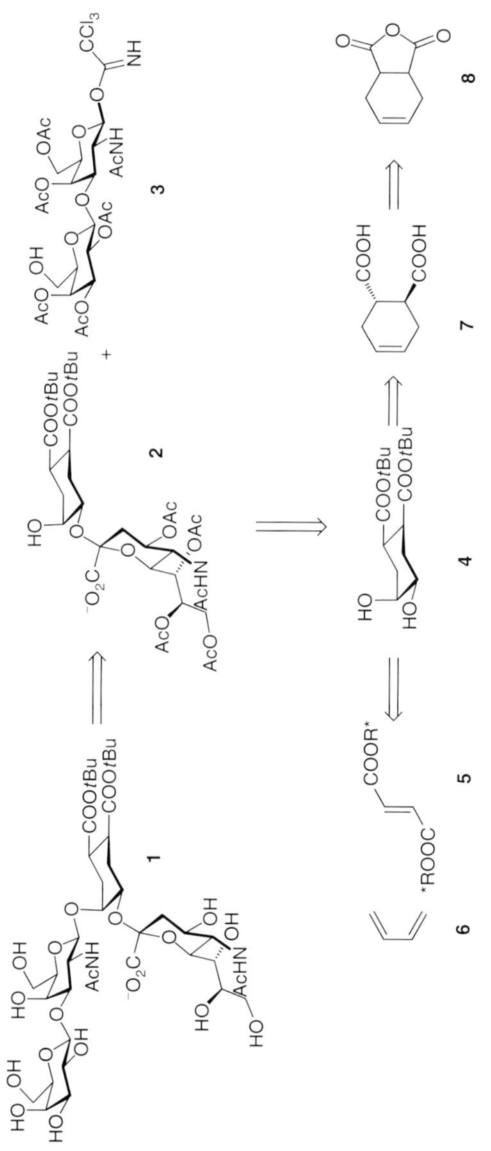

Scheme 10.1 Retrosynthetic analysis of psGM1 **1**.

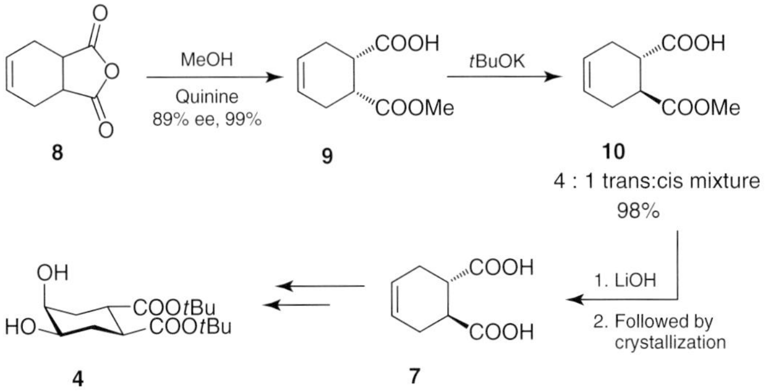

Scheme 10.2 Enantioselective synthesis of the DCCHD derivative **4**.

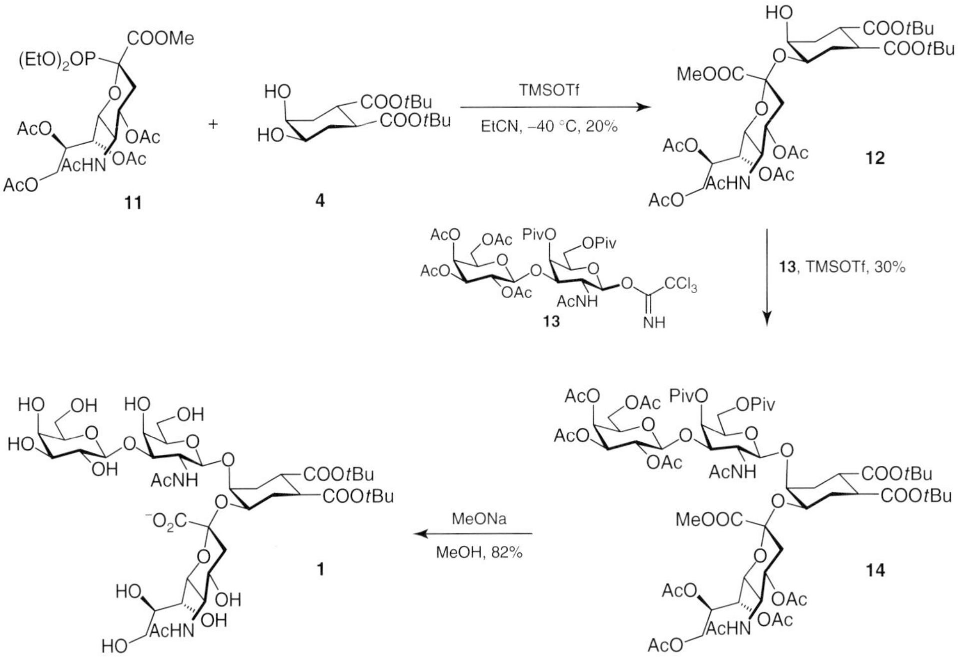

Scheme 10.3 Synthesis of pseudo-GM1 **1**.

using an excess of **13** (2.3 equiv) and a catalytic amount of trimethylsilyl trifluoromethanesulfonate (TMSOTf) in refluxing CH_2Cl_2, the pseudotetrasaccharide **14** was isolated in 30% yield. Finally, MeONa/MeOH trans-esterification gave the target pseudo-GM1 **1** [41]. This compound, as predicted, turned out to share the 3D structure and CT-binding ability of GM1-os, thus validating the design principle and the use of DCCHD **4** as a mimic of a 3,4-disubstituted Gal residue [39].

With a very similar approach, but using the Galβ(1–3)GlcNAc donor **15** (Scheme 10.4) we also synthesized the pseudo-GM1 analog **16**, where a GlcNAc residue replaces the native GalNAc [48]. Also in this case, the design process involved preliminary simulation of the ligand conformation and of its interaction with the protein [49] and the computational predictions were borne out by the experimental results. It was shown that, as expected, neither the 3D shape of the ligand nor its binding mode and affinity for CT are significantly modified passing from **1** to **16**.

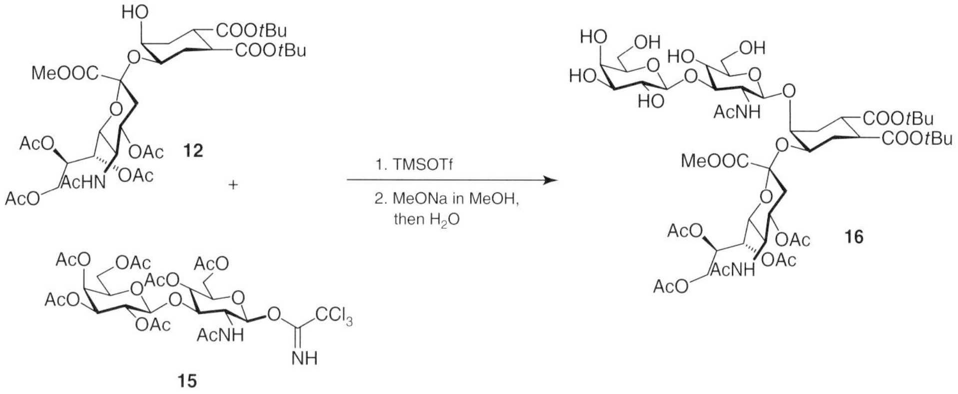

Scheme 10.4 Synthesis of the GlcNAc analog **16** of pseudo-GM1.

Thus, the scaffold-replacement approach afforded two GM1 mimics that share the exceptional CT-binding properties of the natural template, while reducing the structural complexity and the sugar-like character of GM1-os.

Further simplification of the pseudo-GM1 structure of **1** and **16** addressed synthetic drawbacks and sought to reduce the structural complexity of the ligands. The main bottleneck in the synthesis of the ganglioside mimics described above is the α-sialylation of diol **4** to afford **12**. Therefore, a further simplification of the pseudo-GM1 structure was envisaged based on replacement of the sialic acid residue with more treatable chemical entities.

10.3.1 Second-Generation Mimics of GM1 Ganglioside: Replacement of the Sialic Acid Moiety

A second generation of GM1 ganglioside mimics were prepared by substitution of the synthetically demanding α-NeuAc moiety with simple α-hydroxy acids [50, 51]. The series of compounds **17–21** (Scheme 10.5) were synthesized from **4** as shown in Scheme 10.5. Bu_2SnO-mediated monoalkylation of **4** with the appropriate nucleophiles **22–26** afforded the monoethers **27–31**. Glycosylation of the axial hydroxy group with the Galβ(βl → 3)GalNAc donor **13** was promoted with TMSOTf or triflic acid (TfOH), to give the pseudotrisaccharides **32–36**, from which standard removal of the protecting groups yielded **17–21**.

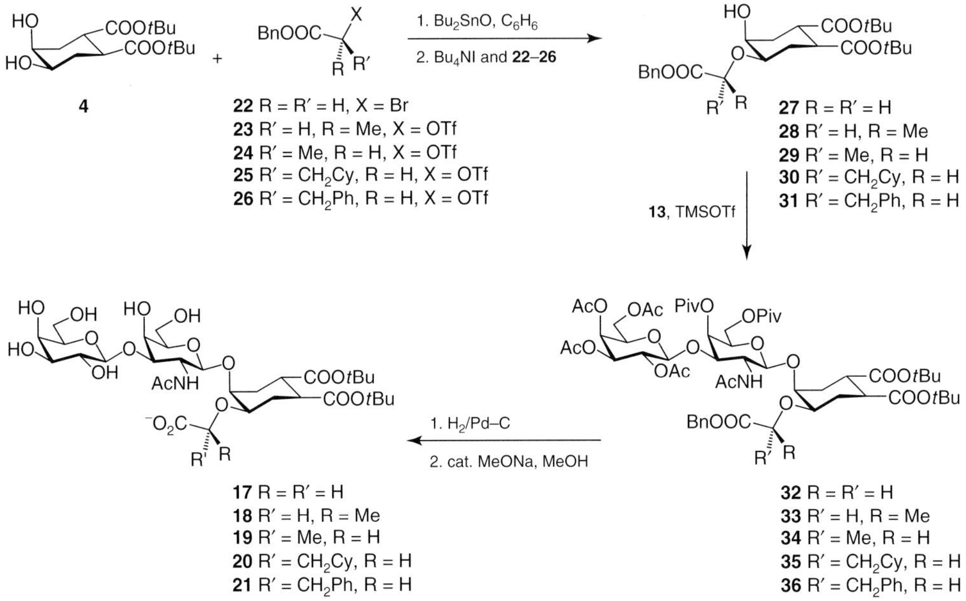

Scheme 10.5 Synthesis of the second generation of GM1 ganglioside mimics.

A similar approach had been used by Ernst in his design of sialyl Lewis-X (sLe^X) mimics [52]. Similar to GM1-os, the sLe^X tetrasaccharide contains a NeuAc(α2–3)Gal moiety at the nonreducing end. However, in the bioactive conformation of sLe^X, the NeuAc residue adopts a *gauche* orientation of the carboxy group relative to the galactose ring, whereas in the bioactive conformation of GM1-os, this orientation is *anti*. Accordingly, it was found that the NeuAc residue was best replaced by α-hydroxy acids of opposite configuration in the two situations, and particularly the (*R*)-configuration at the acid stereocenter was required for GM1-os mimics. For instance, within the series **17–19**, the (*R*)-lactic acid derivative **19** displays the strongest affinity for CTB. The dissociation constants determined by nonlinear regression analysis of fluorescence titration data are 667 μM for the glycolic acid derivative **17**, 1.1 mM for the (*S*)-lactic acid derivative **18**, and 190 μM for the (*R*)-epimer **19** [50]. CTB has a dissociation constant of about 40 and 81 mM for galactose and lactose, respectively, whereas on comparison, asialo-GM1 (Figure 10.1) showed no detectable binding to CT in TLC overlays. Thus, the carboxy group of **17–19** appears to have a measurable effect on the affinity of the artificial receptors for the toxin, and the (*R*)-configuration of the acid side-chain gives the best results in the series. Additional lipophilic groups in the side-chain, as in compounds **20** and **21**, were found to increase affinity for CT. The dissociation constant of 10 μM measured for **21** represents a one order of magnitude improvement over the parent ligand **19**.

These results were interpreted using modeling and NMR data generated and analyzed using a combination of different approaches [53, 54]. Initial modeling studies performed for compounds **17–19** suggested that, although more flexible than pseudo-GM1 **1**, these compounds should be able to simultaneously fit the galactose and the carboxy-binding sites of CT using low-energy conformations. Indeed, saturation transfer difference (STD) studies showed that all ligands interact with CT using the Gal residue and the acid group, while the DCCHD moiety is not involved in the process to a significant extent. In the free state, the three molecules were found to be rather flexible, especially in the hydroxy acid region. However, there is a process of conformational selection upon binding to CTB, and the selected conformation fits the galactose-binding pocket while placing the carboxy group in the carboxylate-binding region. Together with the stereocenter configuration, this, in turn, defines the position of the alkyl substituents in the complexes. For the (*R*)-lactic acid derivative **19** (Figure 10.4a), the bound conformation orients the methyl group toward a hydrophobic area located in the vicinity of the sialic acid side-chain binding region of the CT:GM1 complex.

Ligand **20** (Figure 10.4b), with the same side-chain configuration as **19**, produces a limited increase of van der Waals contacts in this region, which explains the modest affinity improvement over **19** (K_d 45 vs 190 μM). In fact, the STD spectrum of the CTB:**20** complex indicated marginal interactions of the cyclohexyl moiety with the protein and dynamic simulations also showed a high mobility of the cyclohexyl ring, which appears to flip in and out of the binding site. In contrast, the phenyl derivative **21** (Figure 10.4c) was found to be conformationally constrained by stacking interactions between the phenyl ring and the GalNAc residue. A single conformation was identified in the free state of the ligand, and was retained in the bound state. The aromatic protons of **21** gave STD signals corresponding

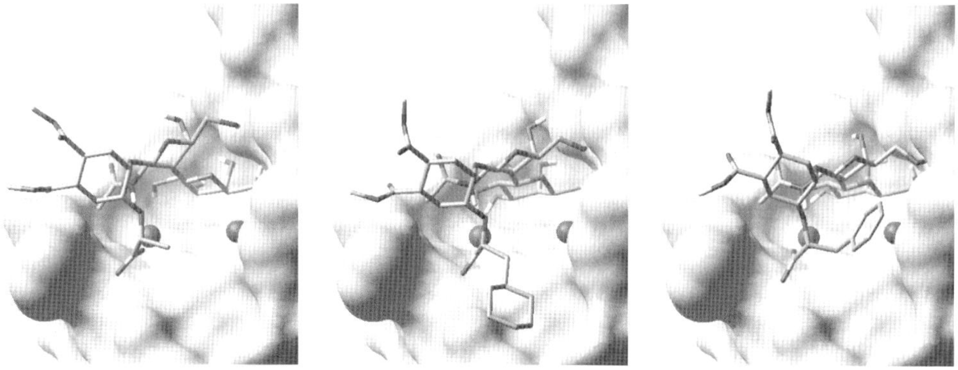
(a) (b) (c)

Figure 10.4 Suggested binding modes of ligands (a) (*R*)-epimer **19**, (b) ligand **20**, and (c) ligand **21** in CT-binding site.

to an intimate contact with the protein. The 3D model of the CT:**21** complex (Figure 10.4c) obtained as the best fit with the NMR data shows that the phenyl ring establishes extensive contacts with the hydrophobic patch in the protein, which are maintained throughout extensive molecular dynamics simulations. Although the importance of carbohydrate–aromatic interactions for the molecular recognitions of oligosaccharides by proteins has been well documented [55–58], this case represented the first clear-cut evidence of the relevance of this type of interactions in determining a ligand conformation. The aromatic-sugar interaction in **21** preorganizes the molecule in a conformation that allows optimal interaction of the carboxy group in the carboxylate-binding region of CT. A similar conformation appears to be attained by **20** in the bound state, but it has to be selected from a pool of different rotamers that are simultaneously present in solution. Thus, both the preorganization effect and a more efficient van der Waals interaction between the side-chain substituent and the protein appear to concur in determining the higher activity of **21** (K_d 10 μM).

Despite the somewhat limited affinity values obtained with these simplified ligands, their synthetic accessibility and the versatility of the DCCHD scaffold allowed us (in collaboration with other groups) to develop polyvalent versions, which finally afforded impressive affinity enhancements. A modular design, based on four different "blocks" (the ligand, the scaffold, the linker, and the spacers, Figure 10.5) in which all the different components of the polyvalent ligand can be easily modified, was adopted. The polyvalent scaffolds used were either dendrons [47] or calixarenes [59]. The length of the linkers was adjusted to allow simultaneous binding to two adjacent binding sites of the toxin.

The synthesis was made possible by the discovery that the two carboxy groups of DCCHD **4** can be differentiated, following an appropriate strategy, which is summarized in Scheme 10.6. Starting from diacid **7**, the bis-methyl ester diol **37** was synthesized as previously discussed. Monohydrolysis of the diester was obtained in good yield (80%) and with complete selectivity using 0.07 M NaOH in H_2O. The free acid was transformed into the *tert*-butyl ester using *N*,*N*-diisopropyl-*O*-*tert*-butyl isourea **38**, and the resulting diol **39** was subjected to the usual alkylation/glycosylation sequence under optimized conditions, yielding the pseudo-trisaccharide **41**, which was finally transformed with trifluoroacetic acid (TFA) into the required acid **42** [47]. Thus, the stage was set to achieve selective condensation with polyamine scaffolds.

Dendrimers up to a G3 generation (octavalent) were synthesized based on the 3,5-di-(2-aminoethoxy)-benzoic acid branching unit and tested in an SPR competition assay with immobilized asialofetuin. Strong multivalency effects were observed, up to a factor of 400, which reached the limit of the assay [47]. Even more spectacular enhancements were obtained with a divalent ligand, prepared by tethering two units of **42** onto a functionalized calix[4]arene. A 3800-fold (1900-fold per sugar mimic) affinity enhancement was measured by fluorescence spectroscopy, thus reaching the same potency of GM1-os [59].

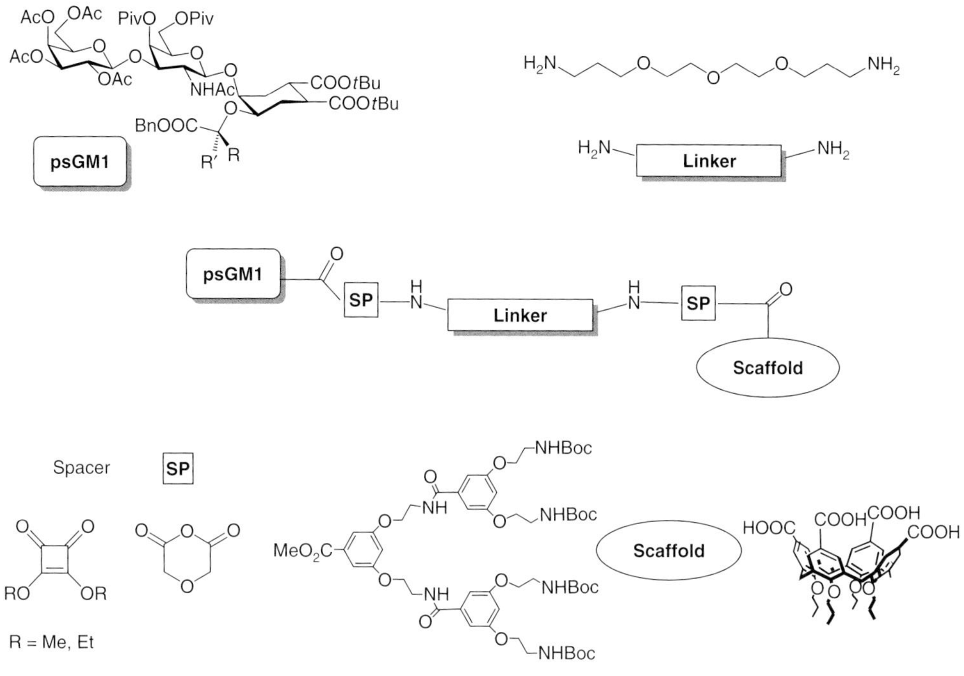

Figure 10.5 Modular design of polyvalent pseudo-GM1 ligands.

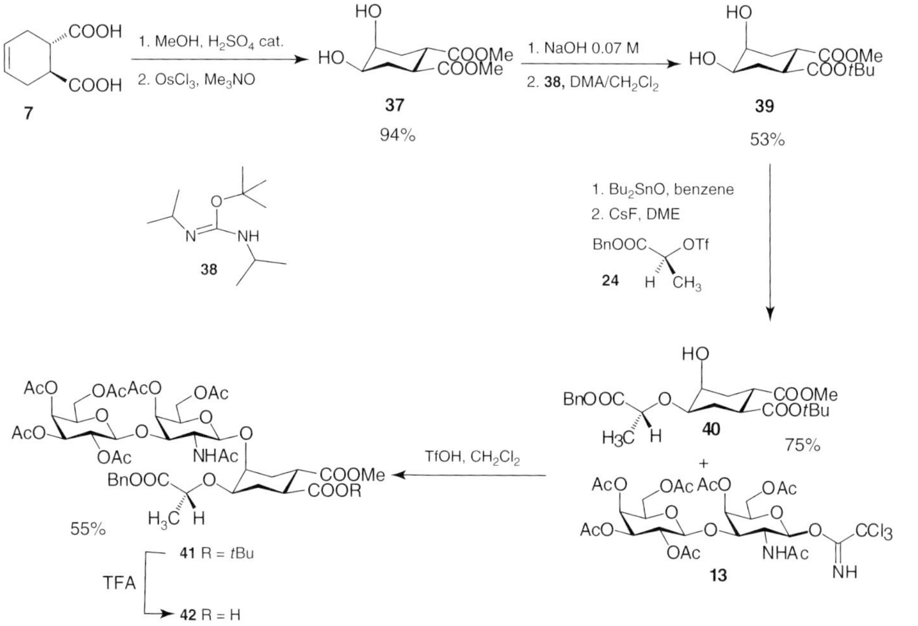

Scheme 10.6 Synthesis of the monovalent ligand **42**.

10.4 Third Generation of GM1 Ganglioside Mimics: Toward Nonhydrolyzable Cholera Toxin Antagonists

The GM1 ganglioside mimics described above, as well as other known antagonists of CT binding such as *meta*-nitrophenylgalactoside (MNPG) and its derivatives [13a, 21], still contain enzymatically labile *O*-glycosidic linkages that limit their potential applications *in vivo*. Furthermore, the synthetic methods used to connect the pharmacophoric sugar moieties in psGM1 are those of traditional carbohydrate chemistry, which are often laborious and low-yielding procedures. For these reasons, we sought to develop a third generation of metabolically stable and synthetically more accessible mimics of GM1-os. We set our efforts toward the synthesis of a library of bifunctional compounds of the general formula shown in Scheme 10.7. The target molecules contain a galactose and a sialic acid moiety connected through a linker. By taking advantage of the ready availability of sialic acid azide **43** [60], a terminal alkyne could be installed on the linker and a triazole used to connect the two sugar residues (Scheme 10.7). The library design relied on combinatorial selection of linear alkyne linkers and of galactose-replacing fragments, with the following two constraints: (i) *O*-glycosides were excluded to stabilize the constructs against the activity of hydrolytic enzymes and (ii) the functionalization of the galactose ring (X group in Scheme 10.7) was chosen to allow a facile conjugation to the putative linkers. Selection of the appropriate linker and identification of the best bidentate ligand was achieved by synthesizing

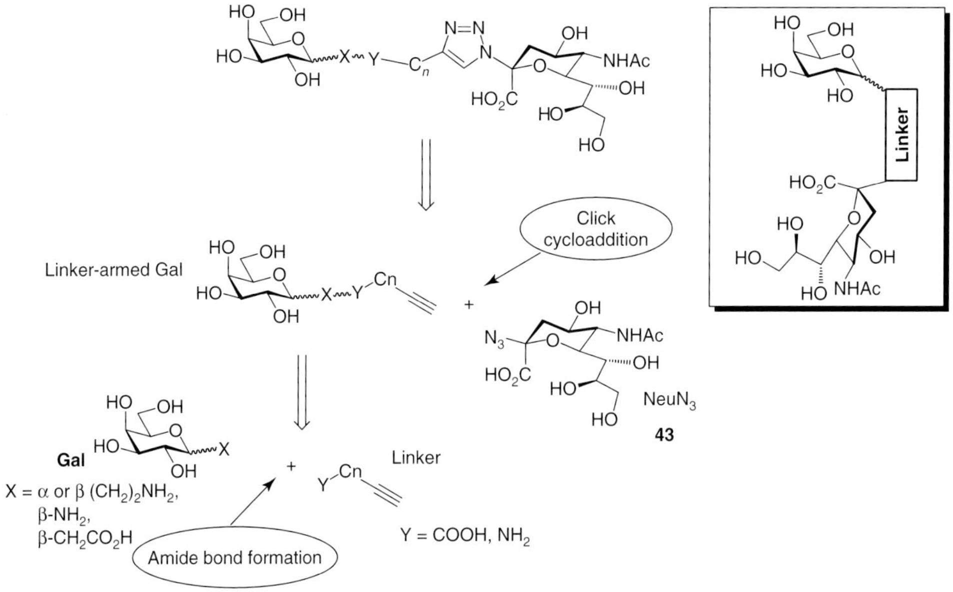

Scheme 10.7 General structure of nonhydrolyzable GM1-os mimics and general synthetic strategy for the library.

a small library of compounds and testing their CTB affinity by weak affinity chromatography (WAC) [61].

The four functionalized Gal fragments used were (Scheme 10.8): β-galactosyl amine **44**, α- and β-*C*-aminoethyl galactosides **45** and **46**, respectively, and β-carboxylic acid **47**. All of these fragments can be connected by amide bonds to the alkyne linkers to set the stage for the final "click" sialylation step. β-Galactosyl amine **44** is a well-known and readily available compound [62]. The remaining three *C*-galactose derivatives were obtained from easily accessible α-*C*-allyl galactoside **48** [63] via α-*C*-galactosyl aldehyde **49** (Scheme 10.8).

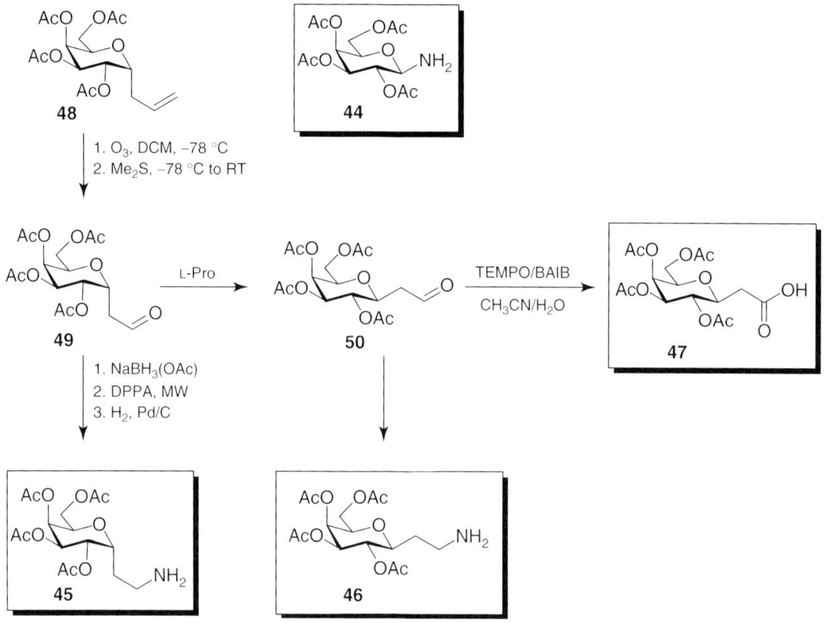

Scheme 10.8 The functionalized Gal fragments **44** and **45–47** used in the synthesis of third-generation GM1 mimics.

The crucial step of the synthesis of β-*C*-galactosides **46** and **47** (Scheme 10.8) was the inversion of configuration of **49** to the β-*C*-galactosyl aldehyde **50**, which was achieved by using L-proline catalysis, as recently described by Massi and Dondoni [64]. This reaction is believed to occur through an intermediate enamine **I** (Scheme 10.9), which promotes ring opening of the sugar via β-elimination to afford an acyclic α,β-unsaturated iminium ion **II**. This, in turn, undergoes intramolecular conjugate addition (hetero-Michael) through intermediate **III**, and the proline catalyst is released via hydrolysis. The less hindered and more stable β-*C*-glycosylmethyl carbonyl derivative **50** would then result as the thermodynamic product.

This process results in good yields under microwave (MW)-assisted conditions described in detail in Section 10.6. However, scale-up of this step was complicated by

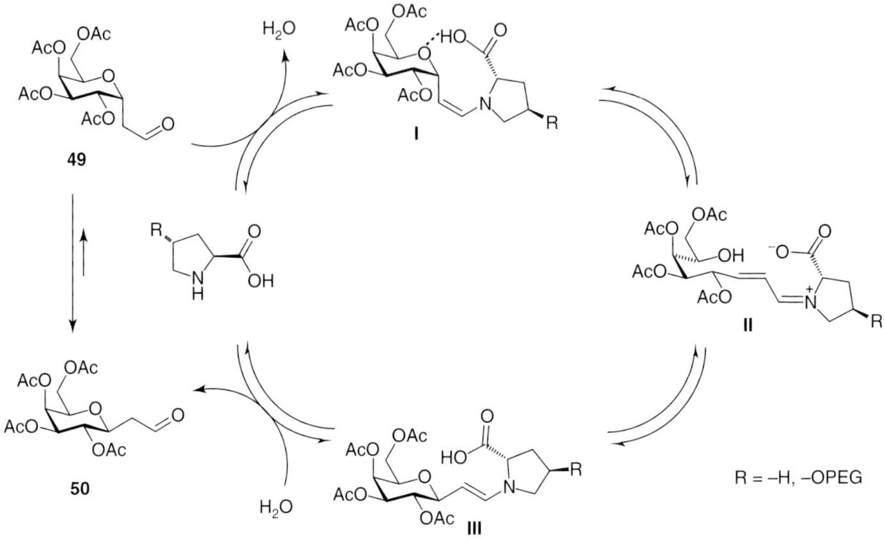

Scheme 10.9 Anomerization of *C*-glycosides with the Massi–Dondoni protocol.

proline-catalyzed auto-condensation of **50** during workup. The problem was solved using poly(ethylene glycol) (PEG)-supported proline [65] as the catalyst, which can be precipitated from the reaction mixture before workup. With this modification of the reported procedure, yields improved from moderate to excellent and purification issues reported in the original paper were resolved (the reaction was performed on up to 1.5 g of aldehyde **49** with 99% yield of **50**). The optimized gram-scale sequence leading from α-methyl galactoside to aldehydes **49** and **50** is described in Section 10.6.

Oxidation of **50** (Scheme 10.8) with 2,2,6,6-tetramethylpiperidine-*N*-oxyl (TEMPO) and bis[acetoxy(iodo)]benzene (BAIB) in a 1 : 1 mixture of acetonitrile and water [66], afforded acid **47** in good yields. The two amines **45** and **46** were synthesized from **49** and **50**, respectively, via the corresponding azides (Scheme 10.8). Functionalization of **44–47** with the ω-alkyne linkers (Scheme 10.10) yielded the 11 linker-armed Gal fragments **54–64** used in the library.

Finally, Cu-catalyzed (click) cycloaddition of alkynes **54–64** with fully acetylated sialyl azide **43** [60] under Sharpless reaction conditions ($CuSO_4$, sodium ascorbate) [67] was performed (Scheme 10.11), and complete deprotection of the coupling products gave the target bidentate adducts **65–75**. These compounds can be grouped into three general families of ligands according to the anomeric composition and configuration of the galactose moiety. For clarity, in the rest of the text, they will be identified as β-Gal-N (compounds **65–67**), α-Gal-C (**68–70**), and β-Gal-C (**71–75**).

The ligands were ranked by WAC [68], using a high-performance liquid chromatography (HPLC) column with immobilized CTB and MNPG as a reference. Under the conditions of the assay, the retardation of any given ligand on the

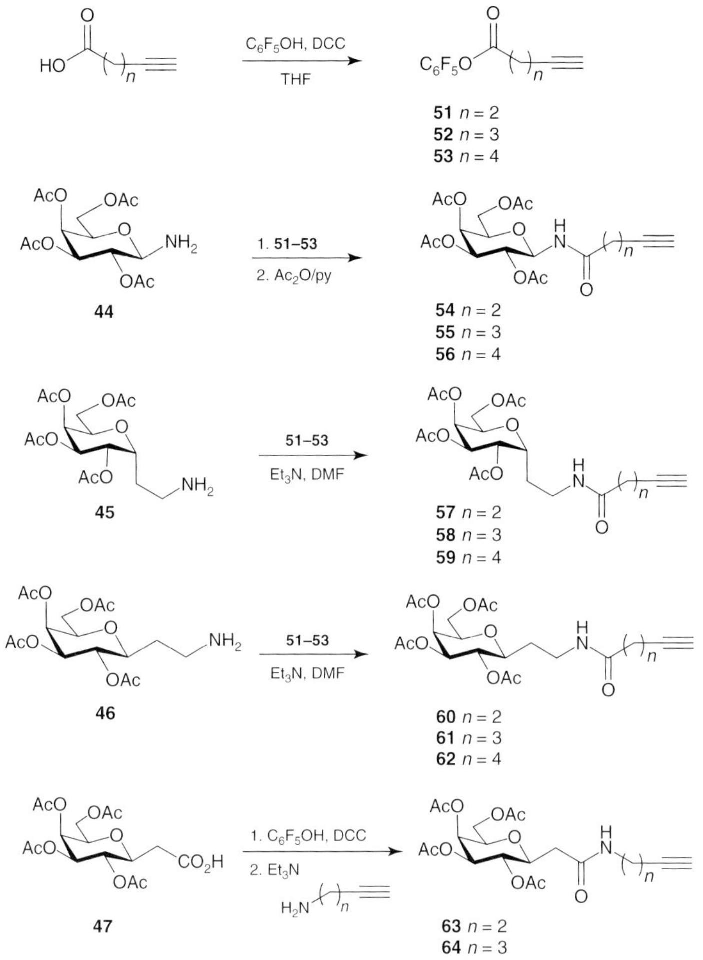

Scheme 10.10 Synthesis of linker-armed galactose fragments.

column is directly related to the affinity of the interaction and allows evaluating the dissociation constant of the ligand (K_d) by calibrating the retardation against a compound of known affinity. By varying the mobile phase in the WAC analysis, the pH dependence of the interaction could also be studied. Interestingly, the interaction with ligands **68–73** was found to be rather pH-dependent, with a maximum retardation at pH 6. MNPG binding was less pH-dependent and displayed maximum affinity at pH 7. From a structural point of view, WAC analysis of **65–75** showed that all β-Gal-N ligands **65–67** as well as **74** and **75** in the β-Gal-C-linked group exhibited negligible affinity. On the contrary, three compounds in the β-Gal-C family (**71–73**) and compound **69** in the α-Gal-C family displayed affinities one order of magnitude higher than the revelation threshold. These values may be

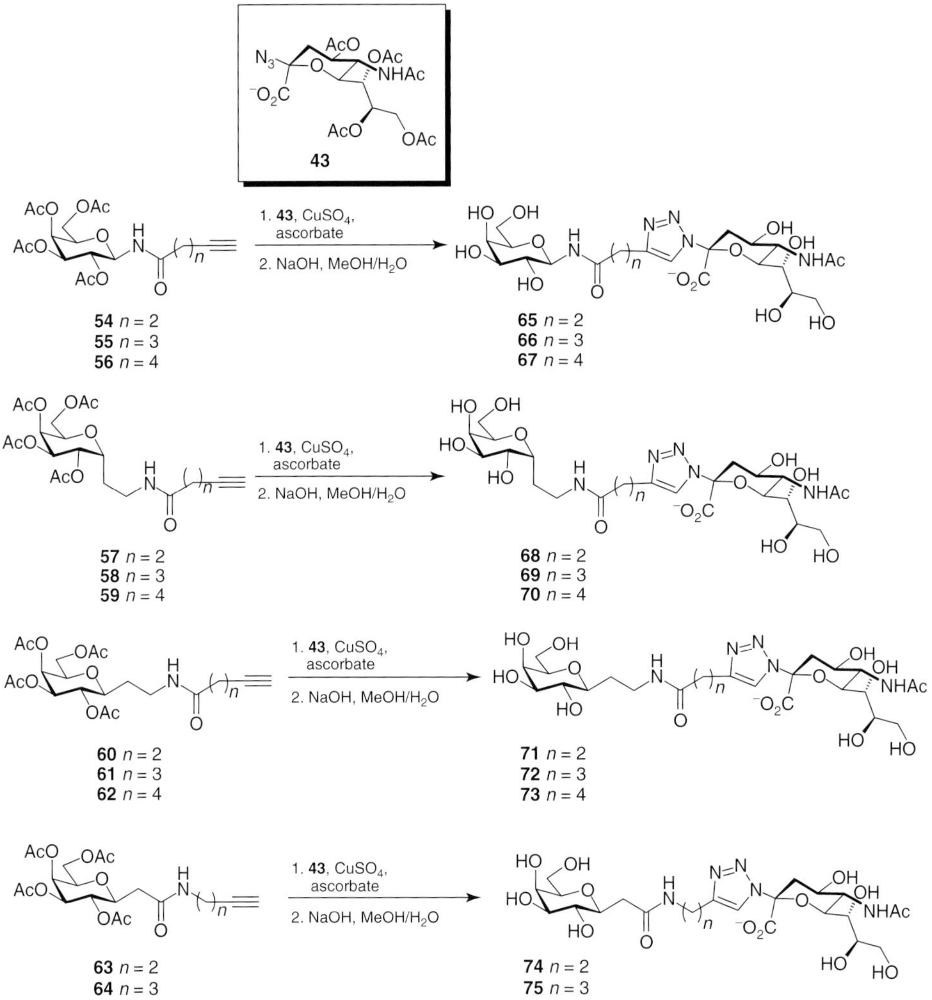

Scheme 10.11 Synthesis of 11 ligands of the primary library.

assumed to result from simultaneous interaction of the two sugar fragments with the toxin, which apparently is optimally allowed in this series by the framework of ligand **71** (C5 linker chain from β-Gal-C framework). Indeed, STD NMR spectroscopy experiments allowed to observe binding events between CT and ligand **71** and showed clear signals corresponding to the galactose (Gal-H_2 at 3.33 ppm) and sialic acid (NeuAc-H_5 at 3.89 ppm) fragments, thus confirming that **71** operates as a bidentate ligand.

In order to obtain compounds with the same framework as **71**, but be capable of further derivatization and conjugation to multivalent aglycons, the library was expanded to include molecules featuring an additional branching point on

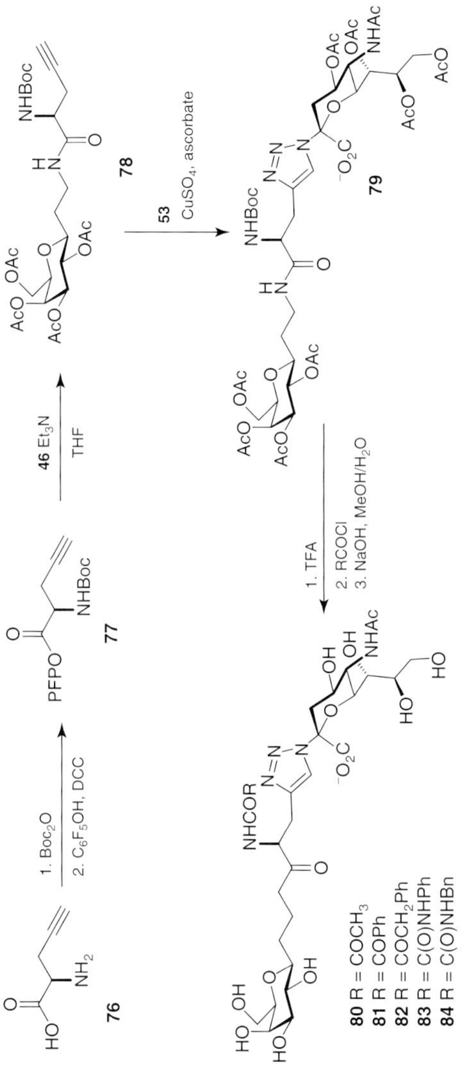

Scheme 10.12 Synthesis of the functionalized divalent ligands **80–84**.

the pentynoic acid backbone. This was achieved using commercially available propargyl glycine **76** (Scheme 10.12) as the linker precursor. After Boc protection, the pentafluorophenyl ester **77** was prepared and allowed to react with β-aminoethyl galactoside **46** to give alkyne **73**, which underwent click cycloaddition with sialyl azide **53** to afford the functionalized divalent ligand **79**.

With this approach, a group of 10 compounds **80–84** (Scheme 10.12, four pairs of epimers at the linker's stereocenter) were prepared, starting from either (*S*)- or (*R*)-propargylglycine and derivatizing **79** with different groups. Lipophilic substituents were mostly examined, in an effort to exploit a known lipophilic patch near the NeuAc side-chain binding region of CT [50, 54, 69]. Although a loss of affinity was observed upon branching of the linker, WAC analysis showed that the (*R*)-phenylurea (*R*)-**83** and the (*R*)-phenylacetamide (*R*)-**82** were interesting ligands, with K_d values of 1.2 and 0.8 mM, respectively.

The cornerstone of the approach described above for the synthesis of nonhydrolyzable CT ligands consists of tethering carbohydrate epitopes that are known to interact with the CTB binding site, galactose and sialic acid, to a properly designed linker through nonhydrolyzable, non-*O*-glycosidic bonds. The Gal epitope was introduced as one of the four simple *C*- or *N*-galactosides. The NeuAc residue was connected through a triazole spacer, starting from the known sialyl azide. Simple linear linkers were used to connect the sugar fragments. The results showed that only some appropriate combinations of fragments led to a measurable improvement of affinity over the individual epitopes. In particular, a group of molecules was identified that displayed affinities with at least one order of magnitude higher than galactose. Such values may be assumed to result from simultaneous interaction of the two sugar fragments with the toxin. The linear linker developed for the third generation of GM1 ganglioside mimetics can be used as a platform for rational ligand design directed toward the stabilization of the bound conformation of the ligand. Some of the most active molecules identified in this third series of GM1 mimetics also feature a point of further derivatization that can be used for conjugation with polyvalent aglycons. Even though there was either no, or negligible, affinity gained by linker derivatization, this approach opens the way to the creation of polyvalent constructs that may be used to block the toxin in a therapeutically relevant context. Interesting findings, such as pH dependence of the binding affinity of the ligands and the structural preference of the CTB binding site toward one of the two diastereomeric forms of the ligand were assessed with WAC, which will be useful for further development of inhibitors of CT binding.

10.5 Conclusions

The work summarized in this chapter elaborated on different strategies to design and synthesize mimics of oligosaccharides. In the first approach, the three-dimensional structure of GM1-os was analyzed and reproduced using an appropriately designed DCCHD as the key, rigidifying element. This strategy afforded an artificial ligand that optimally reproduced the biological activity of the natural one, while allowing ample room for further simplification and optimization. Of the various design lessons learned along this project, two stand out for their generality, that is, the potential of DCCHDs as lipophilic replacements of monosaccharides and the ability of aromatic-sugar interactions to control the conformational behavior of small molecules, as seen in ligand **21**.

In the second approach, the epitopic fragments of GM1-os were connected avoiding glycosidic bonds and with minimal design of the linker. This work relied on the optimization of large-scale synthesis of *C*- and *N*-glycosides and on the use of high yielding "click" chemistry reactions. Despite their lower affinity relative to GM1-os, the resulting molecules represent highly accessible and metabolically stable targeting devices for the preparation of multivalent constructs.

10.6 Experimental Section

10.6.1 Multigram-Scale Synthesis of (1*S*, 2*S*)-Cyclohex-4-ene-1,2-dicarboxylic acid 7

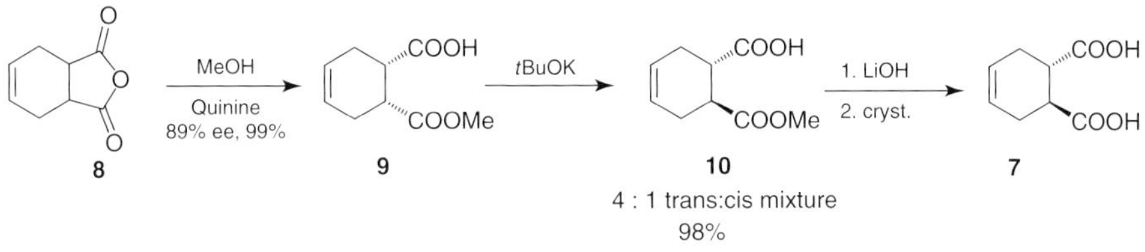

10.6.1.1 Synthesis of (1*S*, 2*R*)-Cyclohex-4-ene-1,2-carboxylic acid monomethylester 9

Using Bölm's procedure [70], tetrahydrophthalic anhydride **8** (5 g, 32.9 mmol, 1 equiv) and quinine (11.8 g, 36.4 mmol, 1.1 equiv) were dissolved in a 1 : 1 mixture of dry toluene (75 ml) and dry CCl_4 (75 ml) and cooled to −50 °C, under N_2. MeOH (4 ml, 98.7 mmol, 3 equiv) was added dropwise, under vigorous stirring and the suspension was stirred at −50 °C for 24 h, during which it turned into a glassy solid. The solvent was evaporated, the residue dissolved in AcOEt, and the organic phase extracted three times with 6N HCl. The organic solvent was dried with Na_2SO_4 and evaporated to yield 5.9 g of the *cis*-monomethylester **9** (98%), which was used without further purification. To recover quinine: NaOH is added to the water phase (pH 12). The resulting solid is filtered, then redissolved in CH_2Cl_2, the solution is washed with water, and the organic phase evaporated to dryness.

10.6.1.2 *cis–trans* Equilibration of the Monomethylester: Synthesis of 10

To a stirred solution of *t*BuOK (5.24 g, 46.7 mmol, 1.5 equiv) in dry tetrahydrofuran (THF) (27 ml), a solution of **9** (5.73 g, 31.1 mmol, 1 equiv) in THF (67 ml) was added under N_2 at 0 °C. The solution was stirred at room temperature for 1 h (the extent of equilibration can be checked by ^{1}H NMR spectroscopy, $CDCl_3$ by integration of the multiplets at 2.9 ppm (2H, H1, and H2 of the *trans*-isomer **10**) and 3.1 ppm (2H, H1, and H2 of the *cis*-isomer **9**)) then concentrated under reduced pressure to about one-third of the original volume and 6N HCl was added to pH 1. The aqueous phase was extracted with Et_2O, the organic phases dried with Na_2SO_4 and the solvent evaporated, to yield 5.67 g of crude 4 : 1 **10**:**9** mixture, as evaluated by ^{1}H NMR spectroscopy.

If the reaction conditions are forced (longer time, higher concentration, or temperature), some trans-esterification also takes place, yielding variable amounts of a mono *tert*-butylester, which can be hydrolyzed to the diacid in quantitative yield by treatment of the crude with TFA.

10.6.1.3 **Synthesis of (1*S*,2*S*)-Cyclohex-4-ene-1,2-dicarboxylic acid 7**

To a solution of a 4 : 1 mixture of the *trans*- and *cis*-monomethylesters **10** and **9** (1 g, 5.4 mmol, 1 equiv) in 3 : 1 MeOH:H_2O (8 ml) LiOH·H_2O (863 mg, 20.6 mmol, 3.8 equiv) was added. The reaction can be monitored by TLC (silica gel, eluant: hexane/AcOEt/AcOH 50/50/3, $KMnO_4$ detection. **10**: R_f 0.45; **7**: R_f 0.2). The solution was stirred at room temperature for 2 h, and when concentrated under reduced pressure before adding 6N HCl, pH 1 was reached. The aqueous phase was extracted with AcOEt, the organic solvent dried with Na_2SO_4 and evaporated to yield 870 mg of a 4 : 1-mixture of *trans*- and *cis*-diacids (95%). This mixture (870 mg total, containing 174 mg of *cis*-diacid, 1.02 mmol, 1 equiv) was dissolved in dry toluene (9 ml) and Ac_2O (112 μl, 1.02 mmol, 1 equiv) was added under N_2. The solution was stirred at 80 °C for about 2 h (formation of the *cis*-anhydride **8** is monitored by ^{1}H NMR spectroscopy: 3.40 (m, 1H), 6.00 (m, 2H)), then the solvent was evaporated, and the residue crystallized from benzene (2 ml) to yield 555 mg of pure **7** (80%).

The enantiomeric excess of **7** can be improved by recrystallization of its quinine salt, using the following procedure: to a solution of **7** (74% ee, 5.12 g, 30.1 mmol, 1 equiv) in MeOH (36 ml) was added a solution of quinine (12.4 g in 12.8 ml of MeOH, 38.2 mmol, 1.3 equiv) and the solution was stirred at 40 °C for 10 min. Then, the solvent was evaporated, the residue redissolved in AcOEt (25 ml) and heated under reflux for 10 min. After cooling to room temperature, the mixture was filtered to eliminate the excess of quinine. The filtrate was evaporated and the solid (15.65 g) crystallized from MeOH (about 20 ml) to obtain the quininium salt as a white solid.

To recover the acid, the quininium salt was dissolved in a 10% HCl solution in water (500 ml) and extracted with AcOEt (3× 240 ml). The combined organic phases were washed with 10% HCl (2× 50 ml), brine (2× 40 ml), and H_2O (40 ml). The organic phase was dried with Na_2SO_4 and the solvent was evaporated to obtain the enantiomerically pure (1*S*,2*S*)-diacid **7** (3.7 g, 82%). Quinine was recovered by treatment of the acidic water solution with NaOH. The solid was filtered and washed with H_2O.

10.6.2
Synthesis of α- and β-2,3,4,6-tetra-*O*-Acetyl-1-*C*-(2-oxo-ethyl)-D-galactopyranose 49 and 50

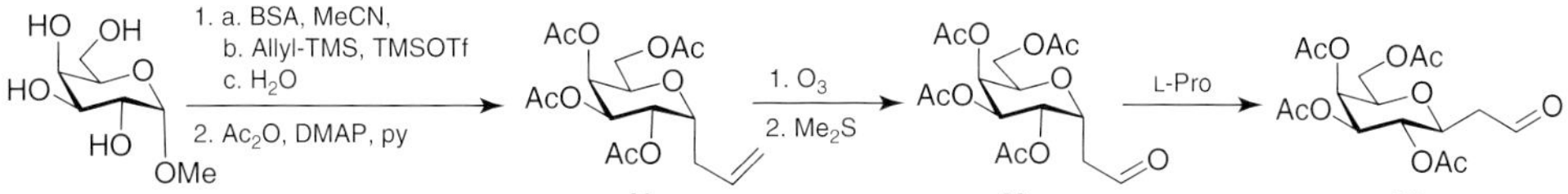

48 **49** **50**

10.6.2.1 Synthesis of 2,3,4,6-tetra-*O*-Acetyl-1-*C*-allyl-α-D-galactopyranose 48

To a solution of methyl-α-D-galactopyranoside (2.02 g, 10.4 mmol) in dry CH_3CN (20 ml) under N_2 atmosphere, bis-trimethylsilylacetamide (BSA) (7.6 ml, 31.2 mmol) was added and the solution was stirred at 80 °C until the mixture was clear and uniform [64]. Then, the solution was cooled to room temperature and allyl-TMS (trimethylsilyl 8.3 ml, 52 mmol) was added, followed by TMSOTf (9.6 ml, 52 mmol). The reaction was stirred for 18 h and was worked up by adding ice-cold deionized H_2O (12 ml) and stirring for 10 min. Then, the solution was neutralized with Et_3N and the solvent was evaporated under reduced pressure.

The crude was stirred in 2 : 1 (v/v) pyridine:acetic anhydride mixture (15 ml) for 18 h. The reaction was quenched with ice-cold water (200 ml) and, after 10 min stirring, it was diluted with CH_2Cl_2 (200 ml). The phases were separated and the organic phase was washed with 1N HCl (200 ml) and brine (200 ml), dried over Na_2SO_4 and the solvent was evaporated. The reaction crude was purified by automatic flash chromatography (silica gel, eluant: hexane/AcOEt, R_f 0.5) to obtain **48** (3.43 g, 94%) as a colorless oil.

10.6.2.2 Synthesis of 2,3,4,6-tetra-*O*-Acetyl-1-*C*-(2-oxo-ethyl)-α-D-galactopyranose 49

Ozone was bubbled into a solution of *C*-allylated α-D-galactose penta-*O*-acetate **48** (1.98 g, 5.3 mmol) in dry dichloromethane (40 ml) under nitrogen at −78 °C until the solution turned blue (30 min). Nitrogen was then bubbled through the solution until it turned colorless, Me_2S was added (8 ml) and the solution was warmed to room temperature. Then, the solution was stirred overnight at 40 °C, the solvent was evaporated and the reaction crude was purified by automatic flash chromatography to obtain **49** (1.91 g, 96%) as a colorless oil.

10.6.2.3 Synthesis of 2,3,4,6-tetra-*O*-Acetyl-1-*C*-(2-oxo-ethyl)-β-D-galactopyranose 50

L-Proline PEG conjugate [61, 65] (2.27 g, 0.1 equiv) was added to a stirred solution of **48** (1.58 g, 4.22 mmol) in MeOH (11 ml) at 0 °C. The mixture was sonicated and then subjected to MW irradiation for 4.5 h at a controlled temperature of 55 °C (constant power 13 W, cooling by compressed air). The reaction was monitored by ^{1}H NMR spectroscopy (H_1-Galα: m, $\delta = 4.86$ ppm; H_1-Galβ: m, $\delta = 4.00$ ppm). When the reaction is complete, the reaction mixture was allowed to cool to room temperature, Et_2O was added to precipitate the proline–PEG conjugate, which was filtered and recovered. The filtrate was evaporated and the resulting solid was purified by automated flash chromatography (hexane/AcOEt, 1 : 1) to afford pure β-anomer **50** (1.57 g, 99%).

Acknowledgments

J. J. R. was supported by a Marie Curie Intra-European Fellowship within Seventh EU Framework program (PIEF-2009-GA-251763).

List of Abbreviations

Ac	Acetyl
Asn	Asparagine
BAIB	Bis[acetoxy(iodo)]benzene
BSA	Bis-trimethylsilylacetamide
CT	Cholera toxin
CTB	Cholera toxin B pentamer
Cy	Cyclohexyl
DCCHD	Dicarboxy-cyclohexane-diols
DMA	Dimethylacetamide
DME	Dimethoxyethane
DPPA	Diphenylphosphoryl azide
Et	Ethyl
Gal	Galactose
GalNAc	*N*-Acetyl-galactosamine
Glc	Glucose
GlcNAc	*N*-Acetyl-glucosamine
Gln	Glutamine
Glu	Glutamic acid
GM1	Galβ1-3GalNAcβ1-4(NeuAcα2–3)Galβ1-4Glcβ1-*O*-Ceramide
GM1-os	GM1 oligosaccharide
HPLC	High-performance liquid chromatography
ITC	Isothermal titration calorimetry
LT	Heat-labile toxin of *E. coli*
Lys	Lysine
Me	Methyl
MNPG	*meta*-Nitrophenylgalactoside
MW	Microwaves
NeuAc	Neuraminic acid (Sialic acid)
NMR	Nuclear magnetic resonance
NOESY	Nuclear overhauser effect spectroscopy
PEG	Poly(ethylene glycol)
PFP	Pentafluorophenyl
Piv	Pivaloyl
psGM1	Pseudo-GM1

py	Pyridine
sLe^x	Sialyl Lewis-X
SPR	Surface plasmon resonance
STD	Saturation transfer difference
*t*Bu	*tert*-Butyl
TCA	Trichloroacetimidate
TEMPO	2,2,6,6-Tetramethylpiperidine-*N*-oxyl
Tf	Trifluoromethanesulfonate
TFA	Trifluoroacetic acid
THF	Tetrahydrofuran
TLC	Thin layer chromatography
TMS	Trimethylsilyl
TMSOTf	Trimethylsilyl trifluoromethanesulfonate
Tr-NOESY	Transferred nuclear Overhauser effect spectroscopy
Trp	Tryptophan
Tyr	Tyrosine
WAC	Weak affinity chromatography

References

1. Hakomori, S.I. (1983) in *Sphingolipid Biochemistry* (eds J.N. Kanfer and S.I. Hakomori), Plenum Press, New York, pp. 1–165.
2. Yu, R.K., Tsai, Y.-T., Ariga, T., and Yanagisawa, M. (2011) *J. Oleo Sci.*, **60**, 537–544.
3. Posse de Chaves, E. and Sipione, S. (2010) *FEBS Lett.*, **584**, 1748–1759.
4. Tettamanti, G. (2004) *Glycoconjugate J.*, **20**, 301–317.
5. Hakomori, S. (2000) *Glycoconjugate J.*, **17**, 627–647.
6. (a) Allende, M.L. and Proia, R.L. (2002) *Curr. Opin. Struct. Biol.*, **12**, 587–592. (b) Lopez, P.H. and Schnaar, R.L. (2009) *Curr. Opin. Struct. Biol.*, **19**, 549–557.
7. (a) Mocchetti, I. (2005) *Cell. Mol. Life Sci.*, **62**, 2283–2294. (b) Duchemin, A.M., Ren, Q., Mo, L.L., Neff, N.H., and Hadjiconstantinou, M. (2002) *J. Biol. Chem.*, **81**, 686–707.
8. Pieters, R.J. and Liskamp, R.M.J. (2008) *Anti-Infect. Agents Med. Chem.*, 7, 193–200.
9. Ivarsson, M.E., Leroux, J.-C., and Castagner, B. (2012) *Angew. Chem. Int. Ed.*, **51**, 4024–4045.
10. (a) Merrit, E.A. and Hol, W.G.J. (1995) *Curr. Opin. Struct. Biol.*, **5**, 165–171. (b) Fan, E.K., Merritt, E.A., Verlinde, C.L.M.J., and Hol, W.G.J. (2000) *Curr. Opin. Struct. Biol.*, **10**, 680–686. (c) Fan, E.K., O'Neal, C.J., Mitchell, D.D., Robien, M.A., Zhang, Z.S., Pickens, J.C., Tan, X.J., Korotkov, K., Roach, C., Krumm, B., Verlinde, C.L.M.J., Merritt, E.A., and Hol, W.G.J. (2004) *Int. J. Med. Microbiol.*, **294**, 217–223. (d) Beddoe, T., Paton, A.W., Le Nours, J., Rossjohn, J., and Paton, J.C. (2010) *Trends Biol. Sci.*, **35**, 411–418.
11. Turnbull, W.B., Precious, B.L., and Homans, S.W. (2004) *J. Am. Chem. Soc.*, **126**, 1047–1054.
12. Cheshev, P. and Bernardi, A. (2008) *Chem. Eur. J.*, **14**, 7434–7441.
13. (a) Liu, J., Begley, D., Mitchell, D.D., Verlinde, C.L.M.J., Varani, G., and Fan, E. (2008) *Chem. Biol. Drug Des.*, **71**, 408–419. (b) Tran, H.A., Kitov, P.I., Paszkiewicz, E., Sadowska, J.M., and Bundle, D.R. (2011) *Org. Biomol. Chem.*, **9**, 3658–3671.
14. Lai, C.-Y. (1977) *J. Biol. Chem.*, **252**, 7249–7256.

15. Spangler, B.D. (1992) *Microbiol. Rev.*, **56**, 622–647.
16. Merritt, E.A., Sarfaty, S., van den Akker, F., L'Hoir, C.L., Martial, J.A., and Hol, W.G.J. (1994) *Protein Sci.*, **3**, 166–175.
17. Kuhn, P., Sarfaty, S., Erbe, J.L., Holmes, R.K., and Hol, W.G.J. (1998) *J. Mol. Biol.*, **282**, 1043–1059.
18. Merritt, E.A., Sarfaty, S., Feil, I.K., and Hol, W.G.J. (1997) *Structure*, **5**, 1485–1499.
19. Fan, E., Merritt, E.A., Zhang, Z., Pickens, J.C., Roach, C., Ahn, M., and Hol, W.G.J. (2001) *Acta Crystallogr.*, **D57**, 201–212.
20. van den Akker, F., Steensma, E., and Hol, W.G.J. (1996) *Protein Sci.*, **5**, 1184–1188.
21. Pickens, J.C., Merritt, E.A., Ahn, M., Verlinde, C.L.M.J., Hol, W.G.J., and Fan, E.K. (2002) *Chem. Biol.*, **9**, 215–224.
22. Merritt, E.A., Sixma, T.K., Kalh, K.H., van Zanten, B.A.M., and Hol, W.G.J. (1994) *Mol. Microbiol.*, **13**, 745–753.
23. Yung, A., Turnbull, W.B., Kalverda, A.P., Thompson, G.S., Homans, S.W., Kitov, P., and Bundle, D.R. (2003) *J. Am. Chem. Soc.*, **125**, 13058–13062.
24. Fukuta, S., Magnani, J.L., Twiddy, E.M., Holmes, R.K., and Ginsburg, V. (1988) *Infect. Immun.*, **56**, 1748–1753.
25. Angströn, J., Teneberg, S., and Karlsoon, K.-A. (1994) *Proc. Natl. Acad. Sci. U.S.A.*, **91**, 11859–11863.
26. Kuziemko, G.M., Stroh, M., and Stevens, R.C. (1996) *Biochemistry*, **35**, 6375–6384.
27. Schön, A. and Freire, E. (1989) *Biochemistry*, **28**, 5019–5024.
28. Mertz, J.A., McCann, J.A., and Picking, W.D. (1996) *Biochim. Biophys. Res. Commun.*, **266**, 140–144.
29. Lauer, S., Goldstein, B., Nolan, R.L., and Nolan, J.P. (2002) *Biochemistry*, **41**, 1742–1751.
30. Cai, X.-E. and Yang, J. (2003) *Biochemistry*, **42**, 4028–4034.
31. Masserini, M., Freire, E., Palestini, P., Calappi, E., and Tettamanti, G. (1992) *Biochemistry*, **31**, 2422–2426.
32. Acquotti, D., Poppe, L., Dabrowski, J., von der Lieth, C.W., Sonnino, S., and Tettamanti, G. (1990) *J. Am. Chem. Soc.*, **112**, 7772–7778.
33. Velter, I.A., Politi, M., Podlipnik, C., and Nicotra, F. (2007) *Mini-Rev. Med. Chem.*, **7**, 159–170.
34. Sigumoto, M., Numata, M., Koike, K., Nakahara, Y., and Ogawa, T. (1986) *Carbohydr. Res.*, **156**, c1–c5.
35. (a) Antoine, T., Priem, B., Heyraud, A., Greffe, L., Gilbert, M., Wakarchuk, W.W., Lam, J.S., and Samain, E. (2003) *ChemBioChem*, **4**, 406–412. (b) Pukin, A.V., Weijers, C.A.G.M., van Lagen, B., Wechselberger, R., Sun, B., Gilbert, M., Karwaski, M.-F., Florack, D.E.A., Jacobs, B.C., Tio-Gillen, A.P., van Belkum, A., Endtz, H.P., Visser, G.M., and Zuilhof, H. (2008) *Carbohydr. Res.*, **343**, 636–650.
36. Binder, F.P.C., Lemme, K., Preston, R.C., and Ernst, B. (2012) *Angew. Chem. Int. Ed.*, **51**, 7327–7331.
37. Brocca, P., Bernardi, A., Raimondi, L., and Sonnino, S. (2000) *Glycoconjugate J.*, **17**, 283–299 and references therein.
38. Bernardi, A., Arosio, D., and Sonnino, S. (2002) *Neurochem. Res.*, **27**, 539–545.
39. Bernardi, A., Checchia, A., Brocca, P., Sonnino, S., and Zuccotto, F. (1999) *J. Am. Chem. Soc.*, **121**, 2032–2036.
40. Bernardi, A., Raimondi, L., and Zuccotto, F. (1997) *J. Med. Chem.*, **40**, 1855–1865.
41. Bernardi, A., Boschin, G., Checchia, A., Lattanzio, M., Manzoni, L., Potenza, D., and Scolastico, C. (1999) *Eur. J. Org. Chem.*, 1311–1317.
42. Bernardi, A., Arosio, D., Dellavecchia, D., and Micheli, F. (1999) *Tetrahedron-Asymmetry*, **40**, 3403–3407.
43. Bernardi, A., Arosio, D., Manzoni, L., Micheli, F., Pasquarello, S., and Seneci, P. (2001) *J. Org. Chem.*, **66**, 6209–6216.
44. Mari, S., Posteri, H., Marcou, G., Potenza, D., Micheli, F., Cañada, F.J., Jiménez-Barbero, J., and Bernardi, A. (2004) *Eur. J. Org. Chem.*, 5119–5125.
45. Okamoto, K. and Goto, T. (1990) *Tetrahedron*, **46**, 5835–5857.
46. Banoub, J., Boullanger, P., and Lafont, D. (1992) *Chem. Rev.*, **92**, 1167–1195.
47. Arosio, D., Vrasidas, I., Valentini, P., Liskamp, R.M.J., Pieters, R.J., and Bernardi, A. (2004) *Org. Biomol. Chem.*, **2**, 2113–2124.
48. Bernardi, A., Arosio, D., Manzoni, L., Monti, D., Posteri, H., Potenza, D.,

Mari, S., and Jiménez-Barbero, J. (2003) *Org. Biomol. Chem.*, **1**, 785–792.
49. Bernardi, A., Galgano, M., Belvisi, L., and Colombo, G. (2001) *J. Comp-Aided Mol. Des.*, **15**, 117–128.
50. Bernardi, A., Carrettoni, L., Grosso Ciponte, A., Monti, D., and Sonnino, S. (2000) *Bioorg. Med. Chem. Lett.*, **10**, 2197–2200.
51. Arosio, D., Baretti, S., Cattaldo, S., Potenza, D., and Bernardi, A. (2003) *Bioorg. Med. Chem. Lett.*, **13**, 3831–3834.
52. Kolb, H.C. and Ernst, B. (1997) *Chem. Eur. J.*, **3**, 1571–1578.
53. Bernardi, A., Potenza, D., Capelli, A.M., García-Herrero, A., Cañada, F.J., and Jiménez-Barbero, J. (2002) *Chem. Eur. J.*, **8**, 4597–4612.
54. Bernardi, A., Arosio, D., Potenza, D., Sanchez-Medina, I., Mari, S., Canada, F.J., and Jiménez-Barbero, J. (2004) *Chem. Eur. J.*, **10**, 4395–4406.
55. Vyas, N.K. (1991) *Curr. Opin. Struct. Biol.*, **1**, 732–740.
56. Quiocho, F.A. (1993) *Biochem. Soc. Trans.*, **21**, 442–448.
57. Vandenbussche, S., Díaz, D., Fernández-Alonso, M.C., Pan, W., Vincent, S.P., Cuevas, G., Cañada, F.J., Jiménez-Barbero, J., and Bartik, K. (2008) *Chem. Eur. J.*, **14**, 7570–7578and references therein.
58. Laughrey, Z.R., Kiehna, S.E., Riemen, A.J., and Waters, M.L. (2008) *J. Am. Chem. Soc.*, **130**, 14625–14633and references therein.
59. Arosio, D., Fontanella, M., Baldini, L., Mauri, L., Bernardi, A., Casnati, A., Sansone, F., and Ungaro, R. (2005) *J. Am. Chem. Soc.*, **127**, 3660–3661.
60. Tropper, F.D., Anderson, F.O., Braun, S., and Roy, R. (1992) *Synthesis*, **1992**, 618–620.
61. Cheshev, P., Morelli, L., Marchesi, L., Podlipnik, C., Bergström, M., and Bernardi, A. (2010) *Chem. Eur. J.*, **16**, 1951–1967.
62. Likhosherstov, L.M., Novikova, O.S., Derevitskaya, V.A., and Kochetkov, N.K. (1986) *Carbohydr. Res.*, **146**, C1–C5.
63. Bennek, J. and Gray, G. (1987) *J. Org. Chem.*, **52**, 892–896.
64. Massi, A., Nuzzi, A., and Dondoni, A. (2007) *J. Org. Chem.*, **72**, 10279–10282.
65. Benaglia, M., Celentano, G., and Cozzi, F. (2001) *Adv. Synth. Catal.*, **343**, 171–173.
66. Epp, J.B. and Widlanski, T.S. (1999) *J. Org. Chem.*, **64**, 293–295.
67. Rostovtsev, V.V., Green, L.G., Fokin, V.V., and Sharpless, K.B. (2002) *Angew. Chem. Int. Ed.*, **41**, 2596–2599.
68. Bergström, M., Liu, S., Kiick, K., and Ohlson, S. (2009) *Chem. Biol. Drug Des.*, **73**, 132–141.
69. Minke, W.E., Hong, F., Verlinde, C.L.M.J., Hol, W.G.J., and Fan, E. (1999) *J. Biol. Chem.*, **274**, 33469–33473.
70. Bölm, C., Gerlach, A., and Dinter, C.L. (1999) *Synlett*, 195–196.

11
Novel Approaches to Complex Glycosphingolipids

Hiromune Ando, Rita Pal, Hideharu Ishida, and Makoto Kiso

11.1
Introduction

The surfaces of eukaryotic cells are covered by a carbohydrate layer, which is called *glycocalyx*. The glycocalyx is composed of glycoproteins, glycolipids, and glycosaminoglycans. Glycosphingolipids (GSLs), which are focused on in this chapter, are a major family of glycolipids as well as glycoglycerolipids. Commonly, GSLs are composed of a lipid part, which is called *ceramide*, and a glycan part (monosaccharide or oligosaccharide). In the outer leaflet of the plasma membrane, GSLs are anchored by their lipid moieties through hydrophobic interactions with other membrane molecules, presenting their glycan moieties to outer milieu. Owing to this positioning of glycan moieties, GSLs are involved in various biological processes through glycan–protein or glycan–glycan interactions, so-called trans interaction, such as cell-type specific adhesion, bindings of toxins, viruses, and bacteria to host cells. In addition, GSLs can modulate the functions of protein receptors in the same plasma membrane via glycan–protein interactions (cis interaction), thereby participating in the regulation of embryogenesis and neuronal cell and leukocyte differentiation [1–3].

GSLs in vertebrates are classified into several families based on the structures of glycan moieties such as cerebrosides, sulfatides, ganglio-series, lacto-series, neolacto-series, globo-series, and isoglobo-series. In the biosynthetic pathway, except cerebrosides and sulfatides, the glycan sequences of all families can be extended by sialic acid residue(s) to be transformed into acidic GSLs, which were called *gangliosides*. The glycan sequences of the ganglio-series only have several sites to be elongated with a sialic acid, disialic acid, or trisialic acid, thereby displaying most diverse and unique glycan structures among vertebrate gangliosides. Sialic acids are structurally distinguished from other monosaccharides by the C1 carboxy group and the glycerol chain extending from C6, being defined as a diverse family of monosaccharides derived from 3-deoxy-non-2-ulosonic acid. The three major sialic acids found in nature are *N*-acetyl (Neu5Ac) and *N*-glycolyl (Neu5Gc) derivatives of neuraminic acid (5-amino-3,5-dideoxy-D-glycero-D-galacto-non-2-ulosonic acid) and KDN (3-deoxy-D-glycero-D-galacto-non-2-ulosonic acid). They are sometimes

Modern Synthetic Methods in Carbohydrate Chemistry: From Monosaccharides to Complex Glycoconjugates,
First Edition. Edited by Daniel B. Werz and Sébastien Vidal.

further modified at their hydroxyl groups by acetylation, sulfonylation, methylation, and so on, being converted into over 50 congeners. Within the glycan chains of gangliosides present in the outer leaflet of the cell membrane, sialic acids typically occupy the distal end of glycan chains through α(2,3)- and/or α(2,6)-linkages with galactose, *N*-acetyl galactosamine, glucose, or *N*-acetyl glucosamine, and through α(2,8)- or α(2,4)-linkages with another neuraminic acid residue, facing extracellular milieu. The outermost positioning of sialic acids makes gangliosides also biologically distinguishable from other GSLs. Owing to their special biological aspect and structural features, gangliosides have been centered in the synthetic and biological study of GSLs [4].

On the other hand, GSLs in invertebrates have quite different glycan moieties from those of vertebrates, which are classified as arthro-, mollu-, gala-, neogala-, spirometo-, and schisto-series. However, a variety of unique GSLs, which does not match with the classification, have also been identified from invertebrates. Furthermore, exceptionally from echinoderms, many kinds of very unique gangliosides have been identified.

For detailing the known functions of GSLs in the cell membrane or uncovering new biological functions of GSLs, the supply of homogenous GSLs is an essential subject. This chapter focuses on recent advances in the synthesis of complex GSLs, especially complex gangliosides.

11.2
Syntheses of Complex Glycans of Gangliosides

Because the structural complexity of ganglioside relies mainly on the glycan moiety, considerable efforts and time of carbohydrate chemists have been spent to establish an efficient and powerful method for synthesizing intricate ganglioside glycans [5]. The major difficulty in the synthesis of ganglioside glycans will be the stereoselective formation of α-glycoside of sialic acid. α-Glycosides of sialic acid can be synthesized by the reaction of oxocarbenium ion generated from sialic acid donor with the hydroxyl of glycosyl partner. However, the issue of coupling yield and stereoselectivity is much more complicated by the special structural features of sialic acid. First, the electron-withdrawing carboxyl group on anomeric center makes the tertiary oxocarbenium ion intermediate unstable and susceptible to 2,3-elimination in collaboration with its 3-deoxy structure. Second, because of the deoxy structure, no neighboring functionality at C3 position is available, assisting the formation of the alternative thermodynamically more stable β-glycoside. Also the glycerol moiety branching from C6 position results in more steric hindrance to the anomeric carbon than in common hexopyranosides. If a glycosidation of a sialyl donor is not stereoselective and affords an anomeric mixture, the complete separation of the anomeric isomers will also be troublesome and time consuming. Furthermore, in the approach to highly sialylated glycans, such as those having an α(2,8)- or α(2,4)-linked disialic acid, the sialylation of the less-reactive C8 or C4 hydroxyl group of sialic acid will be more difficult. However, these difficulties have been surmounted with the advent of new chemistry for sialylations [6].

11.2.1
Glycan Moiety of Ganglioside Hp-s6 (Hp-s6 Glycan)

The glycan part of Hp-s6 contains a tandem of partially modified sialic acid linking through an α(2,8)-linkage, which is the most difficult glycosidic linkage to synthesize. To boost the reactivity of the C8 hydroxyl group of sialic acid, the formation of 1,5-lactam has proven to be effective by our research group, which is probably due to the absence of hydrogen bonding with an amide group [7]. In the synthesis of the Hp-s6 glycan **8** [7a], a set of phenylthioglycosides of sialic acid was utilized to construct the tandem of sialic acid; *N*-Troc-sialyl thioglycoside **2** and 1,5-lactamized sialyl thioglycoside **3** (Scheme 11.1). Because the phenylthio group at the bridgehead anomeric center of **3** was unreactive because of Bredt's rule, **3** could serve as glycosyl acceptor in the coupling with **2** to afford α(2,8)-linked disialic acid derivative **4** with complete stereoselectivity. Next, the locked-up phenylsulfenyl group was converted into an active state by opening the lactam ring. For the purpose of 8-*O*-sulfonation in the final stage, an 8-hydroxy derivative was produced through selective deprotection of the Troc group and subsequent 8O to 5N migration of the acetyl group, and the resulting hydroxyl group was protected with a levulinoyl (Lev) group, giving **5**. The glycosylation of glucosyl acceptor **6** with **5** afforded trisaccharide **7**, which was converted into the target molecule **8** via selective removal of the Lev group, sulfonation and global deprotection.

11.2.2
Glycan Moiety of Ganglioside HPG-7 (HPG-7 Glycan)

Ganglioside HPG-7 **15** was isolated from the sea cucumber, *Holothuria pervicax*, together with a number of other new ganglioside species [8]. Most characteristic structural feature of HPG-7 glycan **15** is an unusual trisialic acid residue embedded within the glycan moiety that exhibited different inter-residual linkages between the sialic acids. For assembling the glycan part, suitably differentiated sialyl units were utilized (Scheme 11.2) [9]. A 1,5-lactamized sialyl glucose derivative **9** was employed as a glycosyl acceptor for the construction of Neuα(2,4)Neu linkage, and as its glycosyl partner and the precursor of 5-amino-sialyl unit for amide formation, 5-azido-sialic acid derivative **10** [10] was chosen. For designing the Fucα(1,4)Neu substructure at the terminus, *N*-Troc-sialic acid derivative **12** was utilized in the glycosylation with Fuc donor **13** because a Troc group at the C5 position of Neu is known to enhance the reactivity of not only the corresponding glycosyl donor but also that of the adjacent C4 hydroxyl group [11]. Among three hydroxyl groups within the 1,5-lactam sialyl residue of **9**, the least hindered C4 hydroxyl group was predominantly glycosylated with the sialyl donor **10**, yielding trisaccharide in 50%. Then, the resulting trisaccharide was converted into a lightly protected derivative

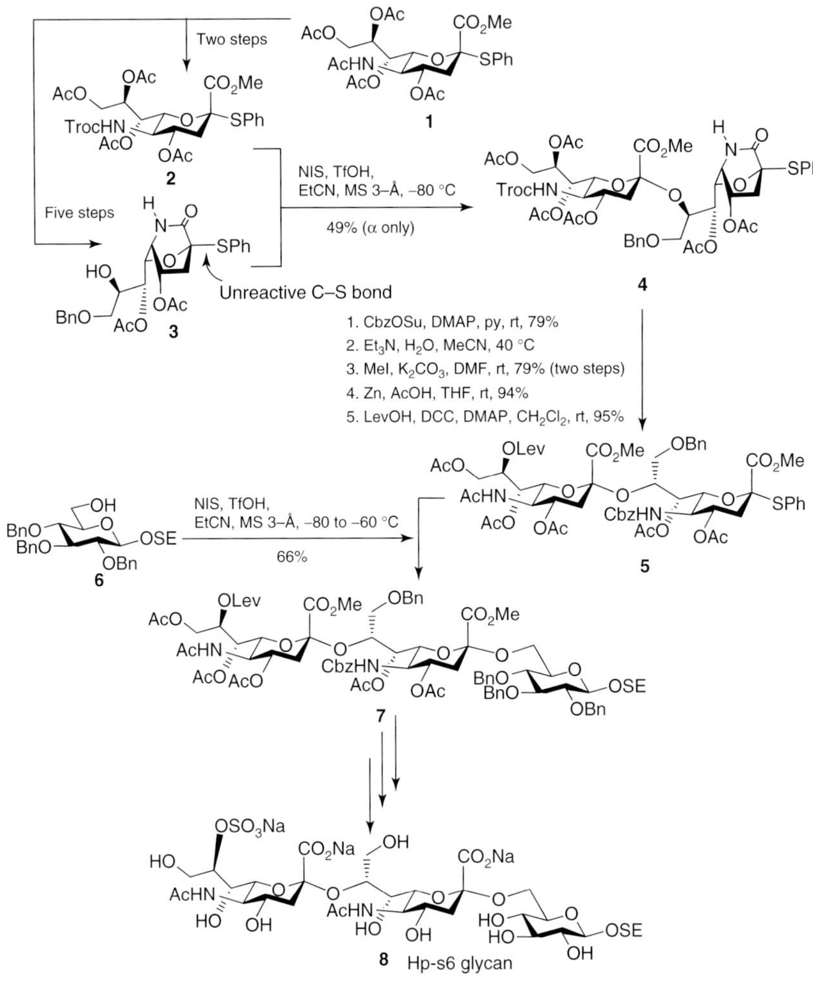

Scheme 11.1 Synthesis of glycan moiety of ganglioside Hp-s6.

11 in order to minimize steric hampering to C5 amino group during the coupling reaction with carboxylic acid unit **14**. The final coupling of **11** and **14** provided the framework of HPG-7 glycan in high yield (83%), which was successfully converted into the target molecule **15**. By applying a similar approach to that of the HPG-7 glycan synthesis, the glycan moiety of ganglioside HPG-1 was also synthesized [12].

11.2.3
Glycan Moiety of Ganglioside AG-2 (AG-2 Glycan)

Ganglioside AG-2, which was isolated from the starfish *Acanthaster planci*, has also an unusual glycan sequence (pentasaccharide) involving an inner sialic acid

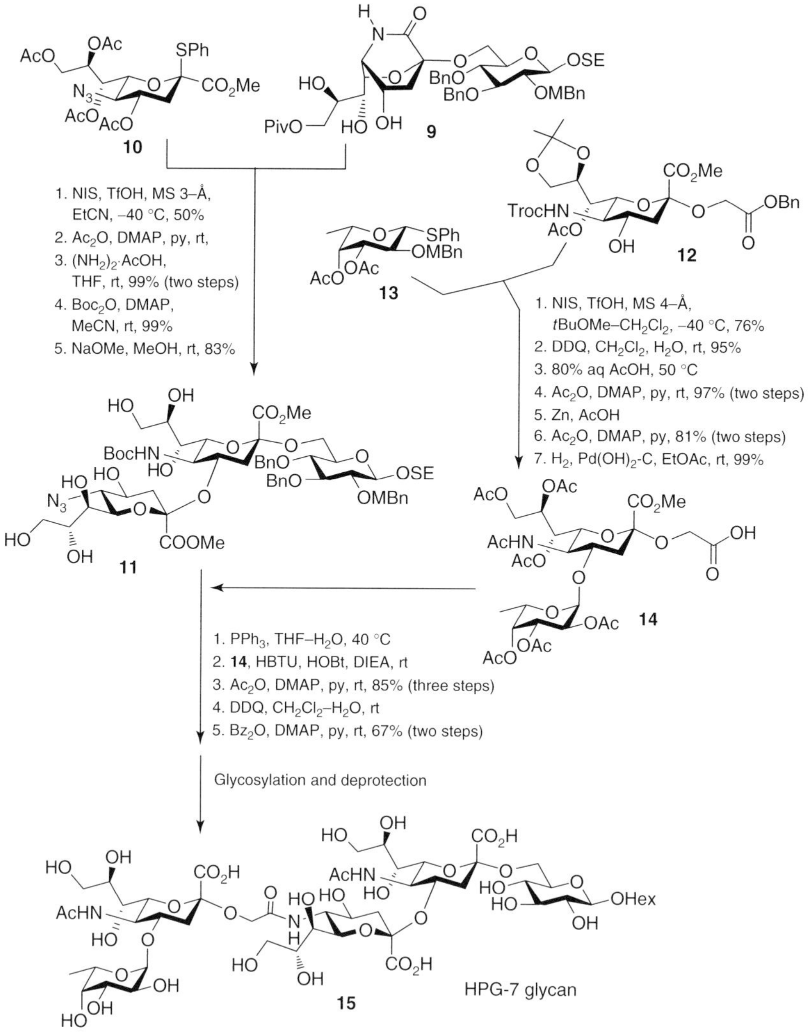

Scheme 11.2 Synthesis of glycan moiety of ganglioside HPG-7.

residue linked with a galactofuranose residue at the C4 position. A focal point in the synthesis of AG-2 glycan is to establish the linkage of Galα(1,4)-Neu. To achieve this glycoside linkage, Hanashima *et al.* [13] utilized an expedient sialic acid building block **16**, which carries an oxazolidinone and di-*tert*-butylsilylene double-lock on the central pyranose ring (Scheme 11.3). The sialic acid building block **16** could serve as a highly effective glycosyl donor largely because of the torsional effect by the two external rings. After the coupling of **16** with a lactose acceptor **17**, the oxazolidinone ring was selectively removed to retrieve a free hydroxyl group at the C4 position, giving trisaccharide acceptor **18**. Then, the C4 hydroxyl group was

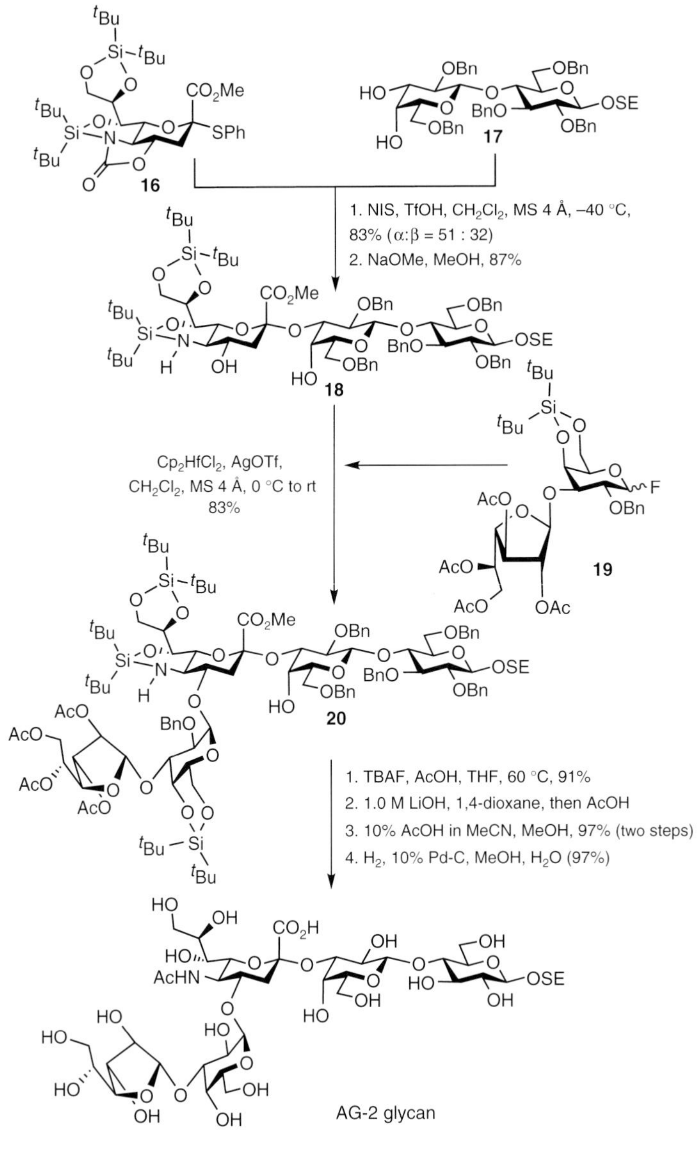

Scheme 11.3 Synthesis of glycan moiety of ganglioside AG-2.

glycosylated with Gal*f*α(1,3)Gal donor **19** to produce pentasaccharide **20** in 83% yield with predominant α-selectivity, which was brought about by the steric effect of the 4,6-DTBS moiety [14]. This result demonstrated that the C4 hydroxyl group exposed by the 5,7-*N*,*O*-DTBS (di-*tert*-butylsilylene) ring possessed high reactivity rather than that shielded by C5 acetamide group. Finally, the synthesis of AG-2 glycan was completed by the reaction sequence for global deprotection.

11.2.4
Glycan Moiety of Ganglioside GP1c (GP1c Glycan)

Ganglioside GP1c, which is a member of ganglio-series gangliosides, possesses a highly complicate glycan moiety composed of a tetrasaccharide (Galβ(1,3)GalNAcβ(1,4)Galβ(1,4)Glc) that carries α(2,8)-linked di- and trisialic acid units. The congestion of highly sialylated substructures at the middle galactose residue defied any challenge of its chemical synthesis. A breakthrough came with the development of a highly stereoselective and efficient method for synthesizing α(2,8)-linked oligosialic acid, which utilized 5*N*,4*O*-oxazolidinone sialic acid donor and acceptor [15]. It was revealed that 5*N*,4*O*-oxazolidinone sialic acid donor **21** could provide very high α-selectivity during coupling reaction, and the C8 hydroxyl group in the 5*N*,4*O*-oxazolidinone sialic acid acceptor was very reactive. Taking advantage of the prominent features of the oxazolidinone sialyl donor **21** and acceptor **22**, Tanaka *et al.* [16] were able to fabricate a trisialyl Gal–Glc acceptor **26** and a disialyl Gal–GalN donor **24** in good yields, respectively (Scheme 11.4). Then, the (4+5)-coupling was successfully carried out to deliver the framework of GP1c glycan **27**, which underwent the manipulations for deprotection, furnishing GP1c glycan.

11.3
Total Syntheses of Complex Gangliosides

11.3.1
Synthesis of Ceramide Moiety

Ceramide is composed of a sphingoid base (a long chain chiral amino alcohol) linked with a fatty acid through an amide linkage. Sphingoid bases are of three general chemical types: sphingosine, sphinganine, and phytosphingosine. The most common sphingoid base in ceramide of mammals contains a C4–C5 double bond in the *trans*-D-erythro configuration. Since Shapiro and Segal [17] first reported the synthesis of the racemic sphingosine in 1954, the syntheses of various types of mammalian sphingosine and ceramide have been reported by many research groups. The earlier successful syntheses of the most common *trans*-D-erythro-sphingosine mainly utilized a monosaccharide (D-Glc [18], D-Gal [19, 20], and D-Xyl [20]) as a chiral template of two stereocenters and constructed the *trans*-olefin by stereoselective Wittig reaction (Scheme 11.5). This strategy was also applied in the syntheses of saturated sphinganines and phytosphingosines. Recently, Murakami *et al.* [21] developed a more efficient synthetic method for the *trans*-D-erythro-sphingosine, which employed a highly diastereoselective carbon elongation of Garner's aldehyde **33** with 1-alkenyl-zirconocene chloride **34**.

11.3.2
Glucosyl Ceramide Cassette Approach

As we discussed earlier, a major difficulty in the total synthesis of complex ganglioside lies in the construction of its sialyl glycan moiety, so that considerable

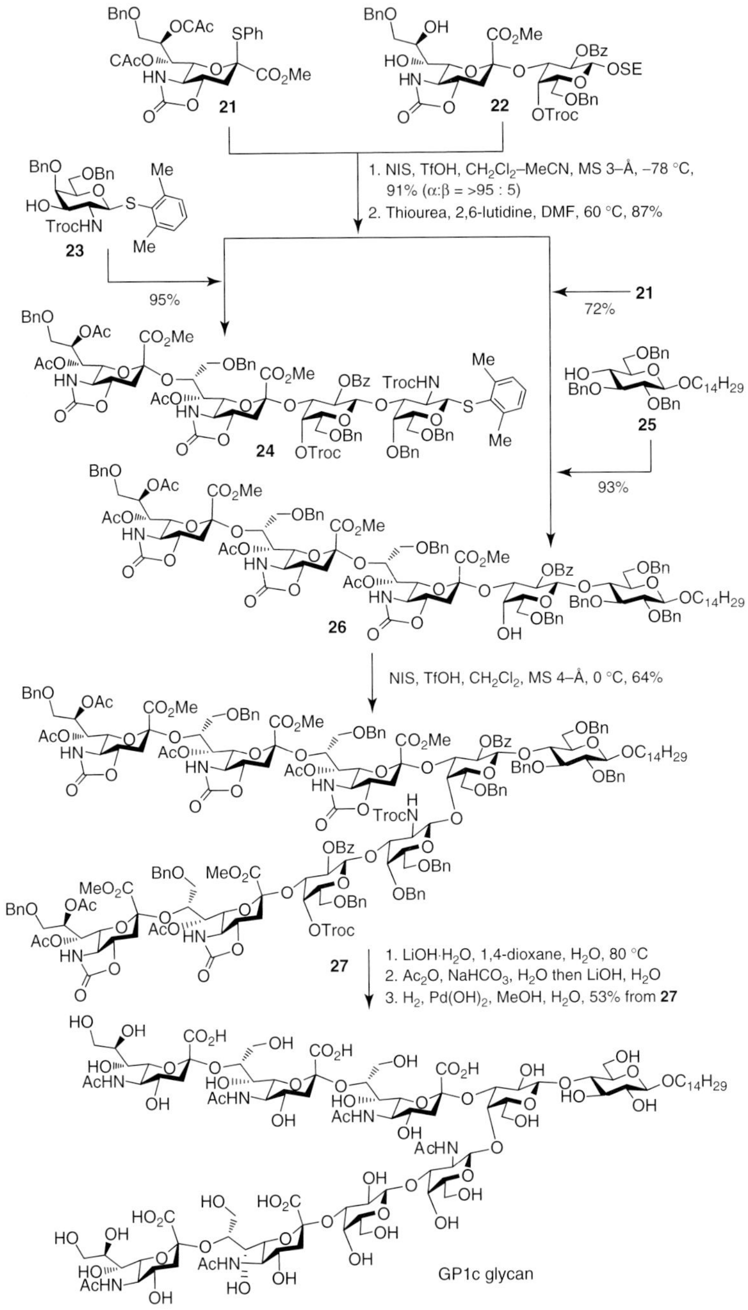

Scheme 11.4 Synthesis of glycan moiety of ganglioside GP1c.

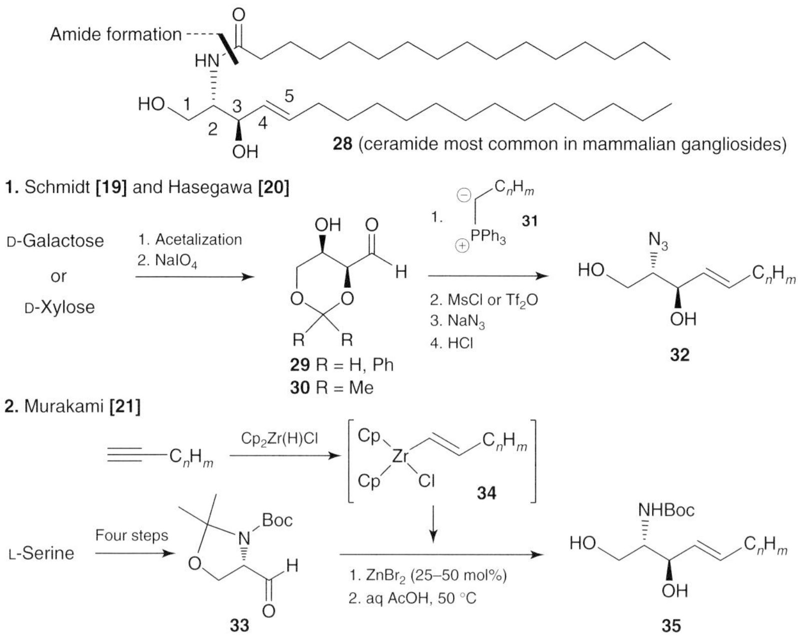

Scheme 11.5 Structure of the most common ceramide in mammalian gangliosides and synthetic methods to access *trans*-D-erythro-sphingosine derivatives.

efforts have been paid to establish an expedient method for the sialylation reaction. However, there is a parallel difficulty in combining the glycan moiety with the ceramide moiety. The first synthesis of GSLs dates back to the synthesis of dihydro-cerebrosides by Shapiro and Flowers [22], which was followed by the first synthesis of ganglioside GM3 reported by Ogawa *et al.* [23]. In both cases, the frameworks of glycosyl ceramides were assembled by the coupling of glycan donor and the C1 hydroxyl group of ceramide. Although this strategy was employed in several ganglioside syntheses, the coupling yields tend to diminish as the length of the glycosyl donor increases, which was mainly attributable to the attenuated likelihood of the attack of the C1 hydroxyl group within ceramide to an oxocarbenium ion generated from glycosyl donor owing to self-aggregation of ceramide. To circumvent the direct connection of the ceramide to the glycan building block, Schmidt [24] incorporated 2-azido-sphingosine into glycan, and constructed ceramide moiety via selective reduction of the azide group and subsequent coupling with fatty acid (Scheme 11.6). This stepwise approach was applied in a broad spectrum of ganglioside syntheses by many research groups. However, according to the reported results, the yields of the coupling reactions of glycans and 2-azido-sphingosine derivatives altered depending on the structures of glycan donors, ranging from 25 to 92%. Therefore, the conjugation of glycan moiety and ceramide moiety was a major difficulty in the synthesis of complex gangliosides.

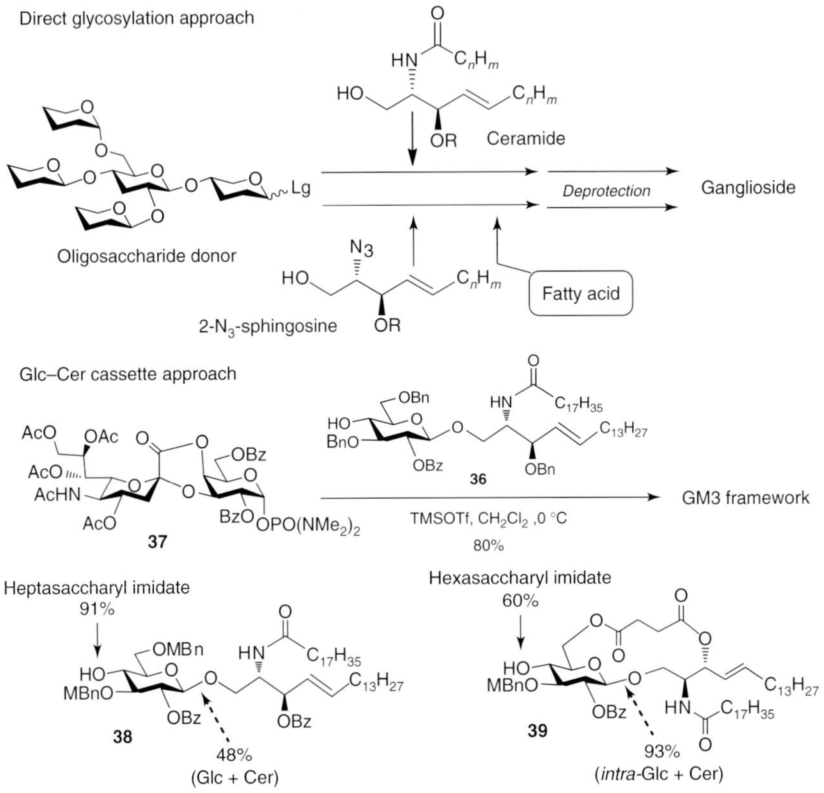

Scheme 11.6 Synthetic approaches to oligosaccharyl ceramide.

In 2000, Hashimoto *et al.* [25] reported the glycosylation of C4′ hydroxyl group of glucosyl ceramide (GC) with sialyl galactosyl donor in high yield in the synthesis of ganglioside GM3. Thus, they performed the glycosylation of GC acceptor **36**, where the C3 and C6 hydroxyl groups were protected with benzyl groups and sialyl galactosyl phosphate donor **37** at 0 °C in CH_2Cl_2 to give the framework of ganglioside GM3 in 80%. Afterward, the GC acceptor **36** was modified into a 3,6-methoxybenzylated derivative **38**, which was successfully applied in the synthesis of ganglioside GQ1b [26]. The use of MBn group benefitted GC acceptor by boosting the reactivity of the C4 hydroxyl group with its electron-donating nature and its selective removal under mild acidic conditions. In the synthesis of GQ1b, GC acceptor **38** was glycosylated with heptasaccharyl donor in 91% yield, while **36** was glycosylated in 73% yield. The MBn groups were then removed by action of trifluoroacetic acid (TFA) in CH_2Cl_2 at 0 °C without affecting the chemical structure of the intricate molecule. Furthermore, to improve the glycosylation of the C1 hydroxyl group of ceramide with a glucosyl donor, an intramolecular glycosylation using a Glc donor tethered with Cer was examined [27]. The best yield of the intramolecular glycosylation was obtained when promoted by dimethyl(methylthio)sulfonium triflate (DMTST), yielding a cyclized GC **39** in 93%. The cyclized GC acceptor **39** was proven to be as reactive as GC **38** by the results of the coupling reaction with oligosaccharide donors. The efficacy of **39** was exemplified in the syntheses of highly branched gangliosides. The above-mentioned approaches using GC acceptors are called *GC cassette approach*. Since mammalian gangliosides and other mammalian GSLs share a common substructure at their tails, Galβ(1,4)Glcβ(1,1)Cer, GC cassette approach would be a promising option for the syntheses of mammalian GSLs.

11.3.3
Total Synthesis of Ganglioside GQ1b

Ganglioside GQ1b is the most complex species of the b-series gangliosides. This molecule is present abundantly in the mammalian central nervous system and involved in diverse biological events such as neurite extension, toxin binding, cell adhesion, cell growth, and so on. The first total synthesis of GQ1b was achieved in 1994 by our research group [28], and in 2009, it was synthesized according to a renewed synthetic strategy designed for a large-scale preparation, where the above-mentioned GC cassette approach was first exploited [26] (Scheme 11.7). In the renewed synthesis, the trisaccharide building unit **40**, Neuα(2,8)Neuα(2,3)Gal, was synthesized by the glycosylation of galactosyl acceptor with disialyl donor, and **40** was also converted into the suitably designed glycosyl donor **41**. Next, **40** was elongated by stepwise glycosylation with GalN donor **42** and disialyl Gal donor **41** to afford a branched heptasaccharide, which was further transformed into an imidate donor. The final coupling with GC cassette **38** was quite efficient, providing a high yield of a protected GQ1b **44** (91%). It is noteworthy that the first attempt at this synthetic route produced 80 mg of homogenous GQ1b.

11.3.4
Total Synthesis of Ganglioside GalNAc-GD1a

In the human body, ganglioside GalNAc-GD1a is localized in the ventral spinal root and closely associated with the development of Guillain–Barré Syndrome, which is the most frequent cause of acute flaccid paralysis worldwide. The structure of ganglioside GalNAc-GD1a features a highly branched glycan moiety, which is composed of a tandem of trisaccharide called *GM2-core*, Neu5Acα(2,3){GalNAcβ(1,4)}Gal. To approach ganglioside GalNAc-GD1a in a convergent manner, the molecule was fragmented into a hexasaccharide donor and a GC cassette [29]. The hexasaccharide donor was further disconnected at the linkage between Gal and GalNAc residues, providing GM2-core donor **49** and acceptor **48** (Scheme 11.8). The synthesis of the GM2-core units profited from the use of a Neu5Troc-Gal unit **45**, which can be produced on a large scale because it can be purified by crystallization [11a, 30]. Thus, the synthesis of GM-core units started from the conversion of the Neu5Troc-Gal **45** into N-Ac derivative **46**, and the glycosylation of **46** with GalNTroc donor **47** produced a GM2-core intermediate.

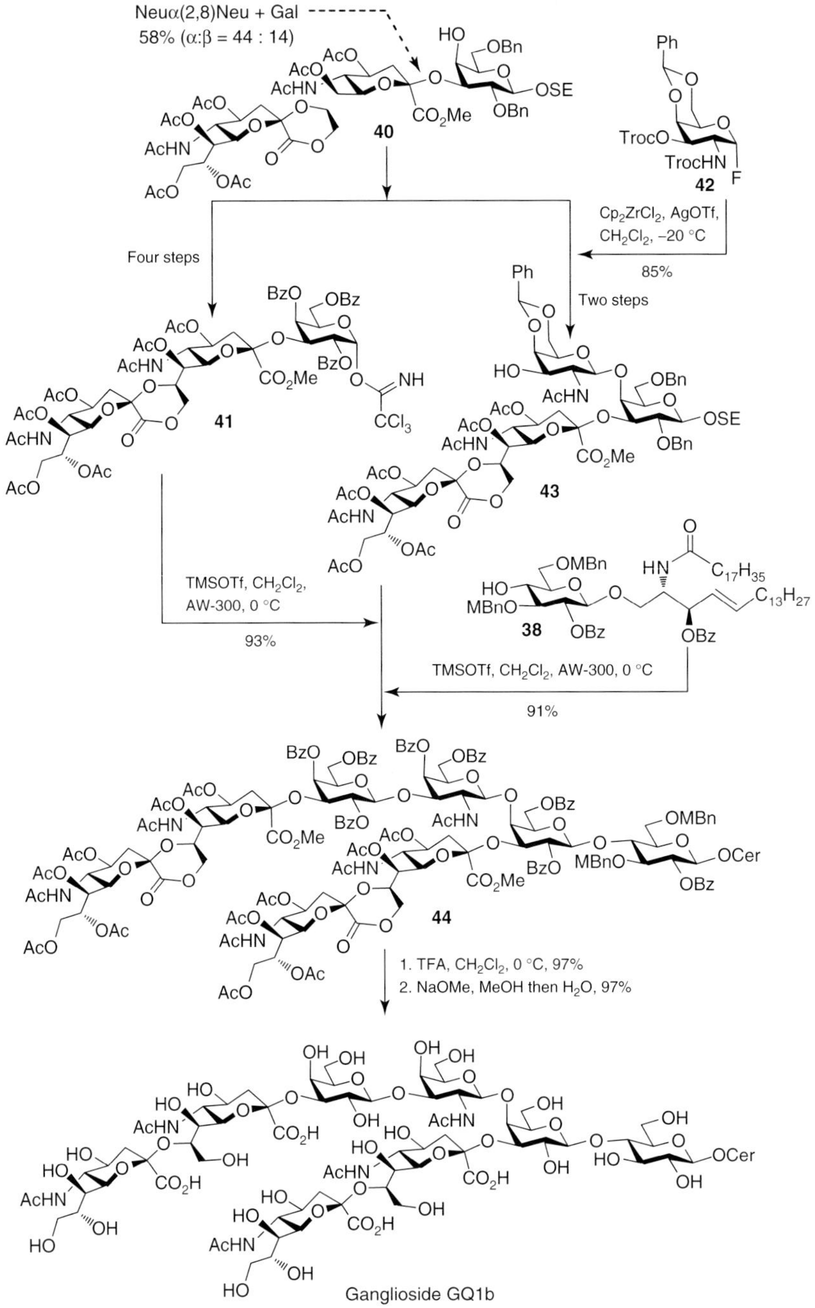

Scheme 11.7 Total synthesis of ganglioside GQ1b.

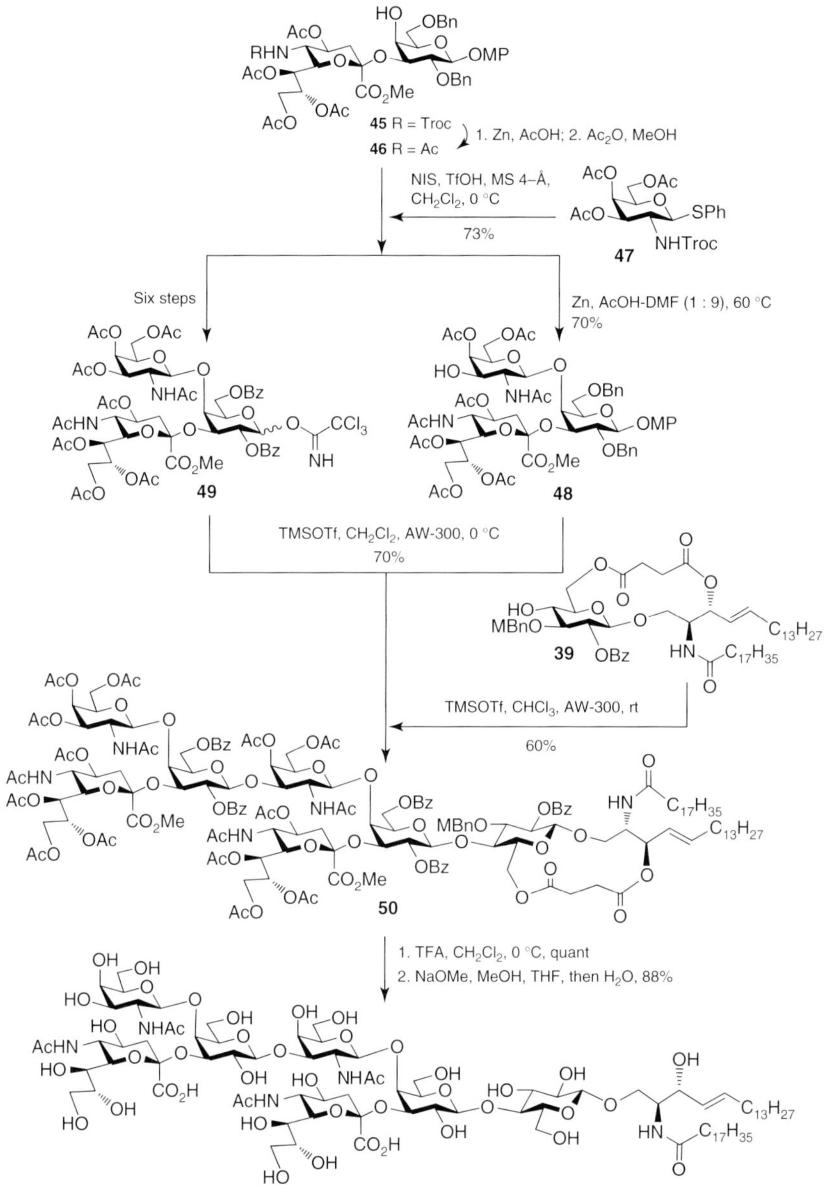

Scheme 11.8 Total synthesis of ganglioside GalNAc-GD1a.

To transform the intermediate into a glycosyl acceptor, the C3 hydroxyl group was liberated by migration of the acetyl group from O to N, which concomitantly occurred with deprotection of the Troc group with zinc in acetic acid, thereby yielding GM2-core acceptor **48** in 70% yield. On the other hand, the GM2-core intermediate was also converted into a glycosyl donor **49** through conventional manipulations of protecting groups. Next, the dimerization of GM2-core unit using **48** and **49** was successfully carried out to furnish a hexasaccharide in 70% yield, which was followed by conversion into the corresponding glycosyl donor. In this synthesis, the cyclized GC acceptor **39** was employed for the glycosylation with the hexasaccharyl donor. As a result, the framework of ganglioside GalNAc-GD1a **50** was produced in 60% yield. Finally, the global deprotection in a conventional manner delivered 20 mg of ganglioside GalNAc-GD1a.

The key unit **49** was also used in the syntheses of highly branched gangliosides that contain GM2-core sequence at the tail parts of their glycan moieties, gangliosides X1 [31] and X2 [32].

11.3.5 Total Synthesis of Ganglioside LLG-3

Ganglioside LLG-3 was identified in the starfish *Linckia laevigata*. Similar to other echinodermatous gangliosides, this molecule also caused neurogenesis on PC-12 cells in the presence of neuron growth factor (NGF). This activity was the second most potent among the 15 echinoderm gangliosides examined. Thus, LLG-3 is of considerable interest in view of medicinal chemistry. Ganglioside LLG-3 contains a modified disialic acid residue, 8-*O*-Me-Neu5Acα(2,11)Neu5Gc, which is presumably responsible for neurogenic activity, and a ceramide which is composed of 3,4-dihydroxysphingosine (phytosphingosine) and an α-hydroxy fatty acid. So far, the synthesis of ganglioside LLG-3 was succeeded by a chemical method and a chemo-enzymatic method.

11.3.5.1 Chemical Synthesis

In the chemical synthesis of ganglioside LLG-3 performed by our research group, the characteristic modified disialic acid was accessed with a high degree of efficiency from a common Neu5Troc donor **2** [33] (Scheme 11.9). As mentioned in the previous section, the Neu5Troc donor **2** was utilized in the production of a Neu5Troc-Gal **45**, which was converted into 5-NH_2-Neu **54** through selective removal of the Troc group. On the other hand, 8-*O*-methyl-Neu glycolate derivative **51** could be prepared from **2** in a relatively short sequence including the 8-O to 5-N migration of the acetyl group after deprotection of the Troc group as a key step. Next, the carboxylic acid unit **52** and the amine unit **54** underwent a coupling reaction mediated by EDC·HCl and HOBt (EDC, 1-ethyl-3-(3-dimethylaminopropyl)carbodiimide; HOBt, 1-hydroxybenzotriazole), providing a high yield of trisaccharide. For the synthesis of the ceramide moiety, the (*R*)-2-hydroxytricosanoic acid derivative was assembled from oxirane unit and icos-1-yne, and coupled with the phytosphingosine, which was synthesized from 1,2 : 4,6-diacetone-D-mannose, to yield a phytoceramide. To

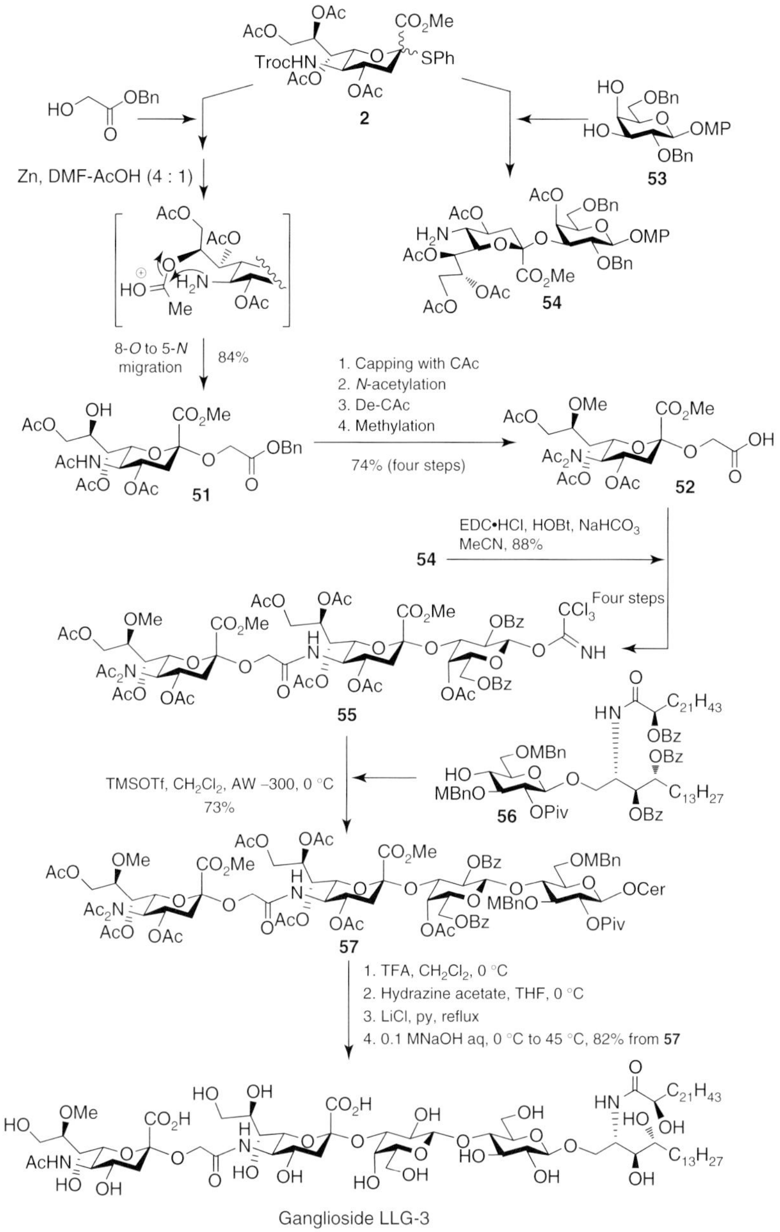

Scheme 11.9 Total synthesis of ganglioside LLG-3.

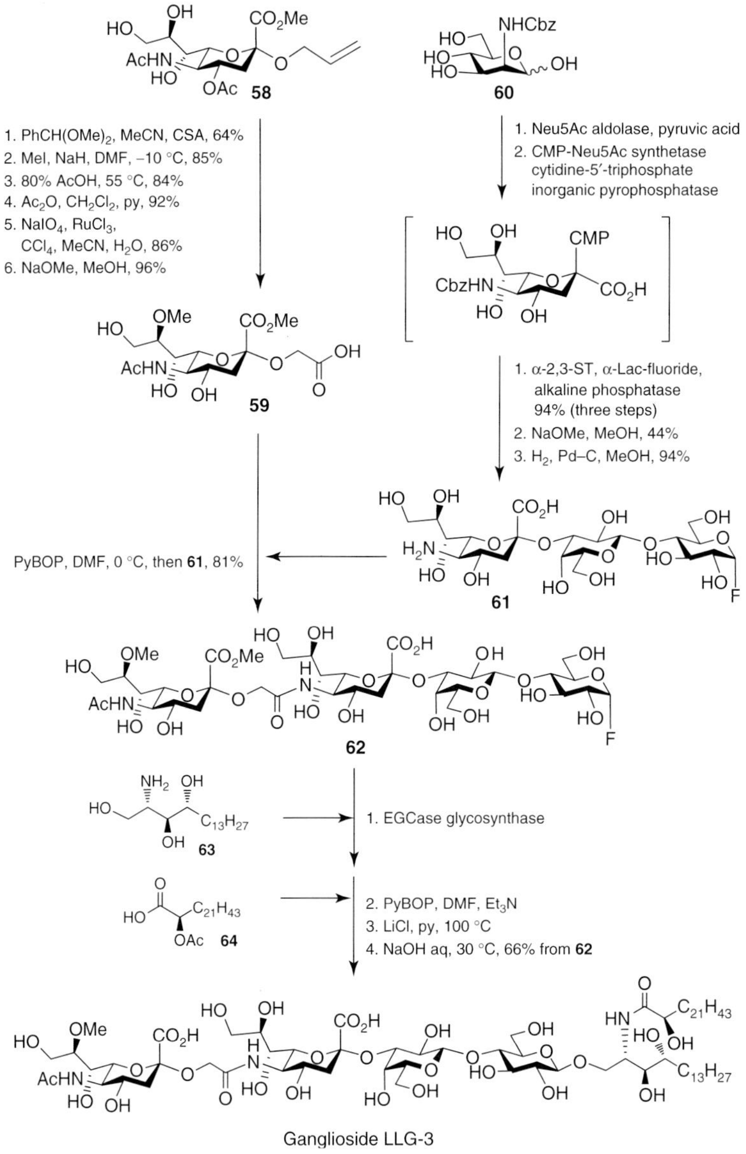

Scheme 11.10 Chemo-enzymatic synthesis of ganglioside LLG-3.

avoid great loss of the full-length glycan part because of low efficiency of coupling with the ceramide, the GC cassette approach was employed for the synthesis of LLG-3. Thus, the trisaccharide donor **55** was coupled with a GC cassette **56** to furnish the framework of ganglioside LLG-3 **57** in high yield. Finally, compound **57** was successfully deprotected, thereby delivering ganglioside LLG-3.

11.3.5.2 Chemo-Enzymatic Synthesis

In the chemo-enzymatic synthesis of LLG-3 performed by Withers *et al.* [34], the glycoside of 8-*O*-Me-Neu5Ac with glycolic acid **59** was synthesized from Neu5Ac derivative **58** by a chemical method because of the absence of a sialyltransferase capable of sialyltransfer to the hydroxyl group of the glycosyl amide in the penultimate sialic acid residue and an 8-*O*-methyltransferase for NeuAc (Scheme 11.10). The coupling partner of **59**, sialyllactosyl fluoride carrying an amine group at the C5 position of Neu residue (**61**) was produced via sequential enzymatic reactions (N-protected Neu formation, CMP-Neu production (CMP, cytidine monophosphate), and sialylation of lactosyl fluoride) using Neu5Ac aldolase, CMP-Neu5Ac synthetase, and α-2,3-sialyltrasferase. Then, the glycan moiety of LLG-3 **62** was synthesized by the dehydrative condensation of **59** and **61**. Finally, the glycan moiety **62** served as a glycosyl donor in the enzymatic glycosylation of sphingosine **63** catalyzed by EGCase glycosynthase to produce lyso-LLG-3, which was then coupled with a fatty acid **64** and fully deprotected, delivering LLG-3.

11.4 Conclusion and Outlook

The parallel problems, low efficiencies of sialylation and the introduction of ceramide into the glycan part, impeded the access to complex gangliosides. The low efficiency of sialylation has been greatly improved by developing the highly reactive C5-modified sialic acid donors, which were successfully applied for the synthesis of sialic acid-enriched glycans. The GC cassette has proven to be a reliable synthetic unit of choice for combining an intricate glycan and a lipid part. For the purpose of advancing glycobiology of GSLs with increasing speed, we must elaborate the synthetic methodology in view of rapid mass production of GSLs. In this context, the innovation of purification technique will also be necessitated for this purpose.

11.5 Experimental Section

11.5.1 Synthesis of *N*-Troc Sialyl Donor 2

Special precautions: Do not alkalize the reaction mixture of the deacetylation at the workup to prevent the formation of 4,5-oxazolidinone derivative. It is noteworthy that the resulting *N*-hydroxysuccinimide (SuOH) should be extracted with water from the reaction mixture because SuOH is hard to remove by chromatographic separation after the acetylation.

Procedure: Add dry methanol (343 ml) to a flask containing phenylthioglycoside of 4,7,8,9-tetraacetyl-Neu5Ac (20.0 g, 34.3 mmol), and then add methanesulfonic acid (6.7 ml, 102.9 mmol). Leave the solution refluxing for 21 h. The reaction progress can be monitored by thin layer chromatography (TLC) analysis ($CHCl_3$:MeOH = 3 : 1). Concentration of the reaction mixture gives the syrupy residue, and exposes it to vacuum for 12 h. After carefully adding 1 M $NaHCO_3$ (103 ml) to the crude residue in an ice water bath (*Caution*: exothermic because of neutralization), add a solution of TrocOSu (19.9 g, 68.6 mmol) in 1,4-dioxane (103 ml) and leave the mixture stirring for 1 h at ambient temperature. Then, remove the solvents by co-evaporation with ethanol, and expose the obtained crude residue to vacuum to dryness. Add dry pyridine (343 ml) to the dried residue and cool the mixture in an ice water bath. Drop acetic anhydride (26.0 ml, 274 mmol) to the mixture over 30 min, and add 4-dimethylaminopyridine (42 mg, 0.343 mmol). Leave the mixture stirring at ambient temperature for 12 h. The reaction can be monitored by TLC analysis ($CHCl_3$:MeOH = 30 : 1). Then to quench the reaction, add methanol to the reaction mixture in ice water bath. Co-evaporate with toluene, dilute the resulting residue with ethyl acetate, and wash with 2 M HCl. The aqueous layer should be washed at least twice with ethyl acetate. Then wash the combined organic layer with saturated $NaHCO_3$(aq) and brine, and dry over Na_2SO_4. Filtration and evaporation give the crude material. Chromatographic purification can be conducted on silica gel (AcOEt:Hex = 1 : 4) to afford the pure title compound **2**.

11.5.2
Synthesis of *N*-Troc Sialyl Galactoside 45

Note: This reaction produces *N*-Troc sialyl galactoside **45**, which can serve as a key common intermediate for the synthesis of gangliosides. Namely, the objective product in this reaction can function as a good coupling partner to galactosaminyl donor and galactosyl galactosaminyl donor, thus producing GM2 core and GM1 core sequences, respectively. By going through the manipulations of the protecting groups including debenzylation, benzoylation, and de-*O*-methoxyphenylation, the formation of trichloroacetimidate at the anomeric position, the product **45** can be converted into the corresponding donor, which has been widely utilized as a terminal sialyl galactose unit in the syntheses of gangliosides. The strong merit of this reaction is that the α-isomer can be crystallized from the anomeric mixture, thereby facilitating the large-scale production of this unit.

Procedure: Add dry EtCN (47 ml) to a flask containing Neu5Troc donor (**2**; 4.0 g, 5.58 mmol) and galactosyl acceptor (**53**; 1.74 g, 3.72 mmol) under Ar atmosphere, add 3 Å molecular sieves (1.8 g), and stir the mixture for 1 h at ambient temperature. Then, cool the mixture down to −50 °C. To the stirring mixture, add *N*-iodosuccinimide (NIS; 1.88 g, 8.37 mmol) and triflic acid (TfOH; 74 μl, 0.84 mmol) (*Caution*: Add the solution of TfOH in EtCN dropwise in case of large-scale reaction (>1.0 g) because it is exothermic.) at −50 °C, and leave for 30 min. The reaction progress can be monitored by TLC analysis (AcOEt:PhMe = 1 : 3). Then, add excess saturated $NaHCO_3$ (aq) to the reaction mixture at 0 °C and stir the suspension vigorously, filter through a pad of celite, washing with AcOEt thoroughly, and combine the filtrate and washings. Wash the combined solution with saturated Na_2SO_4 (*Caution*: Add until the yellow or wine red organic phase turns colorless.) and brine, and dry over Na_2SO_4. Filtration and evaporation give the crude material. The crude α,β-mixture of NeuTroc(2,3)Gal, which is prepurified by chromatography on short silica gel pad (AcOEt:PhMe = 1 : 4), can be crystallized from AcOEt–Hex to afford pure α-isomer **45** (2.0 g, 52%) as crystalline compound.

List of Abbreviations

Ac	Acetyl
Bn	Benzyl
Boc	*tert*-Butoxycarbonyl
Bz	Benzoyl
CAc	Chloroacetyl
Cbz	Benzyloxycarbonyl
Cer	Ceramide
CMP	Cytidine monophosphate
Cp	Cyclopentadinenyl
DCC	*N,N′*-Dicyclohexylcarbodiimide
DDQ	2,3-Dichloro-5,6-dicyano-*p*-benzoquinone
DIEA	*N,N*-Diisopropylethylamine
DMAP	4-(Dimethylamino)pyridine
DMF	*N,N*-Dimethylformamide
DMTST	Dimethyl(methylthio)sulfonium triflate
EDC	1-Ethyl-3-(3-dimethylaminopropyl)carbodiimide
Gal	Galactose
GalNAc	*N*-Acetylgalactosamine
Glc	Glucose
HBTU	2-(1*H*-Benzotriazole-1-yl)-1,1,3,3-tetramethyluronium hexafluorophosphate
HOBt	1-Hydroxybenzotriazole
Hex	*n*-Hexyl
Lac	Lactose
Lev	Levulinoyl (=4-oxopentanoyl)
MBn	*p*-Methoxybenzyl
Me	Methyl
MP	*p*-Methoxyphenyl
MS	Molecular sieves
Ms	Methanesulfonyl
Neu	Neuraminic acid (sialic acid)
NGF	Neuron growth factor
NIS	*N*-Iodosuccinimide
Piv	Pivaloyl (*tert*-butylcarbonyl)
py	Pyridine
PyBOP	1-Benzotriazolyloxy-tris(pyrollidino)phosphonium hexafluorophosphate

SE	2-(Trimethylsilyl)ethyl
ST	Sialyltransferase
Su	Succinimidyl
TBA	tetra-*n*-Butylammonium
TMS	Trimethylsilyl
Tf	Trifluoromethanesulfonyl
TFA	Trifluoroacetic acid
THF	Tetrahydrofuran
Troc	2,2,2-Trichloroethoxycarbonyl
Xyl	Xylose

References

1. Kolter, T. and Sandhoff, K. (1999) *Angew. Chem. Int. Ed.*, **38**, 1532–1568.
2. Hakomori, S.-I. and Handa, K. (2000) in *Glycosphingolipid Microdomains in Signal Transduction, Cancer, and Development in Carbohydrates in Chemistry and Biology*, Vol. 4 (eds B. Ernst, G.W. Hart, and P. Sinäy), Wiley-VCH Verlag GmbH, Weinheim, pp. 771–781.
3. Schnaar, R.L., Suzuki, A., and Stanley, P. (2009) in *Essentials of Glycobiology*, 2nd edn (eds A. Varki, R.D. Cummings, J.D. Esko, H.H. Freeze, P. Stanley, C.R. Bertozzi, G.W. Hart, and M.E. Etzler), Cold Spring Harbor Laboratory Press, Cold Spring Harbor, NY, pp. 129–141.
4. Angata, T. and Varki, A. (2002) *Chem. Rev.*, **102**, 439–469.
5. (a) Hasegawa, A. and Kiso, M. (1992) in *Carbohydrates Synthetic Methods and Application in Medicinal Chemistry* (eds H. Ogura, A. Hasegawa, and T. Suami), Kodansha/Wiley-VCH Verlag GmbH, Tokyo, Weinheim, pp. 243–266. (b) Hasegawa, A. and Kiso, M. (1997) in *Preparative Carbohydrate Chemistry* (ed. S. Hansessian), Marcel Dekker, New York, pp. 357–379. (c) Ishida, H. (2000) in *Glycolipid Synthesis in Carbohydrates in Chemistry and Biology*, Vol. 1 (eds B. Ernst, G.W. Hart, and P. Sinäy), Wiley-VCH Verlag GmbH, Weinheim, pp. 305–317.
6. (a) Boons, G.-J. and Demchenko, A.V. (2000) *Chem. Rev.*, **100**, 4539–4565. (b) Kiso, M., Ishida, H., and Ito, H. (2000) in *Special Problems in Glycosylation Reactions: Sialidations in Carbohydrates in Chemistry and Biology*, Vol. 1 (eds B. Ernst, G.W. Hart, and P. Sinäy), Wiley-VCH Verlag GmbH, Weinheim, pp. 345–365. (c) Ando, H. and Imanura, A. (2004) *Trends Glycosci. Glycotechnol.*, **16**, 293–303. (d) De Meo, C. and Priyadarshani, U. (2008) *Carbohydr. Res.*, **343**, 1540–1552. (e) Ando, H. and Kiso, M. (2008) in *Selective α-Sialylation* (eds B. Fraser-Reid, K. Tatsuta, and J. Thiem), Springer-Verlag, Berlin, Heidelberg, pp. 1313–1359. (f) Hanashima, S. (2011) *Trends Glycosci. Glycotechnol.*, **23**, 111–121.
7. (a) Ando, H., Koike, Y., Koizumi, S., Ishida, H., and Kiso, M. (2005) *Angew. Chem. Int. Ed.*, **44**, 6759–6763. (b) Ando, H., Shimizu, H., Katano, Y., Koizumi, S., Ishida, H., and Kiso, M. (2006) *Carbohydr. Res.*, **341**, 1522–1532.
8. Yamada, K., Harada, Y., Miyamoto, T., Isobe, R., and Higuchi, R. (2000) *Chem. Pharm. Bull.*, **48**, 157–159.
9. Iwayama, Y., Ando, H., Tanaka, H.-N., Ishida, H., and Kiso, M. (2011) *Chem. Commun.*, **47**, 9726–9728.
10. Yu, C.-S., Niikura, K., Lin, C.-C., and Wong, C.-H. (2001) *Angew. Chem. Int. Ed.*, **40**, 2900–2903.
11. (a) Ando, H., Koike, Y., Ishida, H., and Kiso, M. (2003) *Tetrahedron Lett.*, **44**, 6883–6886. (b) Tanaka, H., Adachi, M., and Takahashi, T. (2005) *Chem. Eur. J.*, **11**, 849–862. (c) Tanaka, H., Nishiura, Y., Adachi, M., and Takahashi, T. (2006) *Heterocycles*, **67**, 107–112.

12. Shimizu, H., Iwayama, Y., Imamura, A., Ando, H., Ishida, H., and Kiso, M. (2011) *Biosci. Biotechnol. Biochem.*, **75**, 2079–2082.
13. Hanashima, S., Yamaguchi, Y., Ito, Y., and Sato, K.-I. (2009) *Tetrahedron Lett.*, **50**, 6150–6153.
14. Imamura, A., Kimura, A., Ando, H., Ishida, H., and Kiso, M. (2006) *Chem. Eur. J.*, **12**, 8862–8870.
15. Tanaka, H., Nishiura, Y., and Takahashi, T. (2006) *J. Am. Chem. Soc.*, **128**, 7124–7125.
16. Tanaka, H., Nishiura, Y., and Takahashi, T. (2008) *J. Am. Chem. Soc.*, **130**, 17244–17245.
17. Shapiro, D. and Segal, K.H. (1954) *J. Am. Chem. Soc.*, **76**, 5894–5895.
18. Koike, K., Nakahara, Y., and Ogawa, T. (1984) *Glycoconj. J.*, **1**, 107–109.
19. Schmidt, R.R. and Zimmermann, P. (1986) *Tetrahedron Lett.*, **27**, 481–484.
20. Kiso, M., Nakamura, A., Tomita, Y., and Hasegawa, A. (1986) *Carbohydr. Res.*, **158**, 101–111.
21. Murakami, T. and Furusawa, K. (2002) *Tetrahedron*, **58**, 9257–9263.
22. Shapiro, D. and Flowers, H.M. (1959) *J. Am. Chem. Soc.*, **81**, 2023–20244.
23. Sugimoto, M. and Ogawa, T. (1985) *Glycoconj. J.*, **2**, 5–9.
24. Zimmermann, P., Bommer, R., Bare, T., and Schmidt, R.R. (1988) *J. Carbohydr. Chem.*, **7**, 435–452.
25. Sakamoto, H., Nakamura, S., Tsuda, T., and Hashimoto, S. (2000) *Tetrahedron Lett.*, **41**, 7691–7695.
26. Imamura, A., Ando, H., Ishida, H., and Kiso, M. (2009) *J. Org. Chem.*, **74**, 3009–3023.
27. (a) Fujikawa, K., Imamura, A., Ishida, H., and Kiso, M. (2008) *Carbohydr. Res.*, **343**, 2729–2734. (b) Fujikawa, K., Nohara, T., Imamura, A., Ando, H., Ishida, H., and Kiso, M. (2010) *Tetrahedron Lett.*, **51**, 1126–1130.
28. Ishida, H.-K., Ishida, H., Kiso, M., and Hasegawa, A. (1994) *Tetrahedron: Asymmetry*, **5**, 2493–2512.
29. Fujikawa, K., Nakashima, S., Konishi, M., Fuse, T., Komura, N., Ando, T., Ando, H., Yuki, N., Ishida, H., and Kiso, M. (2011) *Chem. Eur. J.*, **17**, 5641–5651.
30. Fuse, T., Ando, H., Imamura, A., Sawada, N., Ishida, H., Kiso, M., Ando, T., Li, S.-C., and Li, Y.-T. (2006) *Glycoconj. J.*, **23**, 329–343.
31. Nakashima, S., Ando, H., Imamura, A., Yuki, N., Ishida, H., and Kiso, M. (2011) *Chem. Eur. J.*, **17**, 588–597.
32. Nakashima, S., Ando, H., Saito, R., Tamai, H., Ishida, H., and Kiso, M. (2012) *Chem. Asian J.*, **7**, 1041–1051.
33. Tamai, H., Ando, H., Tanaka, H.-N., Hosoda-Yabe, T., Yabe, T., Ishida, H., and Kiso, M. (2011) *Angew. Chem. Int. Ed.*, **50**, 2330–2333.
34. Rich, J.R. and Withers, S.G. (2012) *Angew. Chem. Int. Ed.*, **51**, 8640–8643.

12
Chemical Synthesis of GPI Anchors and GPI-Anchored Molecules

Ivan Vilotijevic, Sebastian Götze, Peter H. Seeberger, and Daniel Varón Silva

12.1
Introduction

Over a quarter of all known proteins are associated with cellular membranes. While many of these proteins establish contact with the lipid bilayer via hydrophobic peptide fragments, others associate with the membranes only transiently, after appropriate posttranslational modification. These posttranslational modifications can be structurally simple like palmitoylation, myristoylation, or prenylation of cysteine and serine residues, or complex like the covalent attachment of a glycosylphosphatidylinositol (GPI) glycolipid to the C-terminus of a protein [1, 2].

The GPIs started being recognized a unique tool for membrane anchoring after first observations that bacterial phosphatidylinositol phospholipase C releases alkaline phosphatase, 5′-nucleotidase, and acetylcholinesterase from various tissues in soluble form suggesting that these proteins are localized at the periphery of the cell membrane by a phosphatidylinositol-containing lipid [3, 4]. Further studies on these and other eukaryotic proteins led to the first full structural elucidation of a GPI molecule, the *Trypanosoma brucei* variant surface glycoprotein (VSG) GPI anchor [5, 6].

Analysis of the GPI structures characterized to date, quickly reveals the conserved α-Man-(1 → 2)-α-Man-(1 → 6)-α-Man-(1 → 4)-α-GlcN-(1 → 6)-myo-inositol core glycan structure (Figure 12.1) [7]. The conserved GPI glycan can be further decorated, in a species or tissue characteristic manner, with phosphodiesters and additional saccharide units. The most common saccharide modification is α-mannosylation at the C2 position of ManIII (ManIII refers to the third mannose residue of the GPI glycan as depicted in Figure 12.1, this notation is used throughout the chapter). A β-galactose residue can be present at the C3 position of ManII. The oligosaccharide branches at the C4 and occasionally at the C3 position of ManI introduce significant structural diversity to this family of glycolipids. The presence of a phospholipid at the C1 position of inositol is common for all GPIs. Owing to the late stage lipid remodeling in the biosynthesis of GPIs, the structures of the GPI lipids are diverse and may include diacylglycerols, alkylacylglycerols, lyso-alkylglycerols, or ceramides with chains

Modern Synthetic Methods in Carbohydrate Chemistry: From Monosaccharides to Complex Glycoconjugates, First Edition. Edited by Daniel B. Werz and Sébastien Vidal.

Origin	R1	R2	R3	R4	R5	R6	R7	R8	Lipid
L. major PSP	H	H	H	H	H	H	H	H	AAG
P. falciparum	±Man α	H	H	H	H	H	H	Acyl	DAG
T. cruzi NETNES	Man α	H	H	H	H	H	AEP	H	AAG
P. communis AGP	H	H	±Galβ	H	H	H	H	H	Ceramide
T. brucei VSG 117	H	H	H	$Gal_{2-4}\alpha$	H	H	H	H	DAG
T. brucei VSG 121	H	Galβ	H	$Gal_{2-4}\alpha$	H	H	H	H	DAG
T. gondii	H	H	±Glc-GalNAcβ	H	H	H	H	H	DAG
T. congolense VSG	H	H	Gal-GlcNAcβ	H	H	H	H	H	DAG
Rat brain Thy-1	±Manα	H	GalNAcβ	H	H	PEtN	H	H	n d
Hamster brain PrP^{Sc}	±Manα	H	±Sia-±Gal-GalNAcβ	H	H	PEtN	H	H	n d
Human CD5 2	±Manα	H	H	H	H	PEtN	H	±pal	DAG
Human sperm CD5 2	H	H	H	H	H	PEtN	H	pal.	AG
Human erythrocyte CD59	H	H	±GalNAcβ	H	PEtN	PEtN	H	pal.	AAG

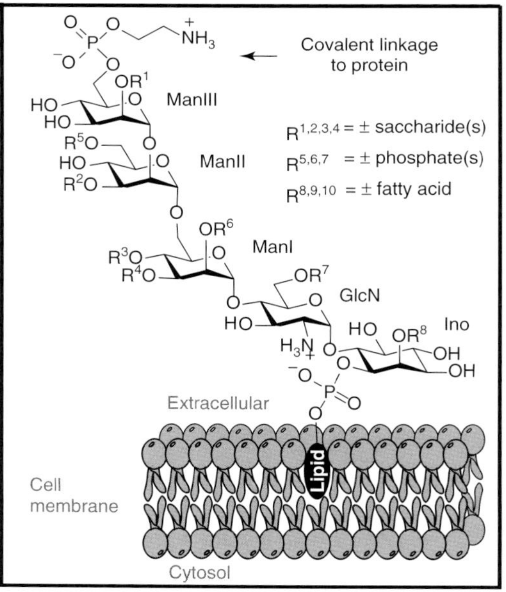

Figure 12.1 Conserved core structure of GPIs with possible substituents [10, 11].

of different length and varying degrees of unsaturation [8]. In addition to the phospholipid, a fatty acid ester is occasionally present at the C2 position of inositol. The phosphoethanolamine in C6 position of ManIII is present in all GPIs and it serves as the linker to the protein. Additional phosphoethanolamines can be found at the C2 position of ManI (characteristic for mammalian GPIs) and C6 position of ManII. The presence of an aminoethylphosphonate at C6 position of GlcN is a unique feature of the GPIs from *Trypanosoma cruzi* [9].

Although some parasitic protozoa express free GPIs on their surface [12], GPIs are normally linked to the C-terminus of a protein via the phosphoethanolamine unit of the ManIII and their primary biological role is to localize the attached protein to the outer surface of the plasma membrane bilayer [4]. The ability of GPIs to associate the anchored proteins with lipid rafts [13, 14] implicates them in diverse biological processes such as regulation of innate immunity [15, 16], protein trafficking [17, 18], and antigen presentation [5]. The complexity and diversity of GPI structures together with the conserved nature of the GPI core glycan and their ubiquity among eukaryotes are intriguing indications of additional functions that may be attributed to these glycolipids. Deciphering the biological functions of different GPIs hinges on the access to homogeneous GPI samples that are difficult to isolate from natural sources. Inspired by their elaborate structures and the specific synthetic challenges presented by GPIs, a number of research groups have developed synthetic routes and prepared various GPIs that have addressed this need and served as powerful tools for biological research over the past two decades [11, 19, 20].

Rather than presenting a comprehensive review of these routes, this chapter summarizes the achievements in the synthesis of GPIs from the pioneering efforts in the early 1990s to the recent development of the general strategies for synthesis of GPIs and glycosylphosphatidylinositol-anchored proteins (GPI-APs). We first discuss the challenges in the synthesis of GPIs and analyze the tools and strategies used to solve the specific problems. This is followed by a summary of the representative syntheses of structurally diverse GPI targets. A short analysis of the synthetic GPI-based tools for biological research serves as the introduction to the discussion on the current state of the art in the synthesis of GPI-anchored peptides and proteins.

12.2
Challenges in the Synthesis of GPIs

The work on the synthesis of GPIs started soon after the first complete structures of GPI glycolipids were disclosed in 1988 [6, 21]. A number of elegant synthetic approaches to GPI fragments and several total syntheses of structurally diverse GPIs [22, 23] have been disclosed over the last two decades (Figure 12.2). These syntheses have served as platforms for the development of new synthetic methodologies [24–26] and technologies [27–29]. The body of reported work on GPI synthesis

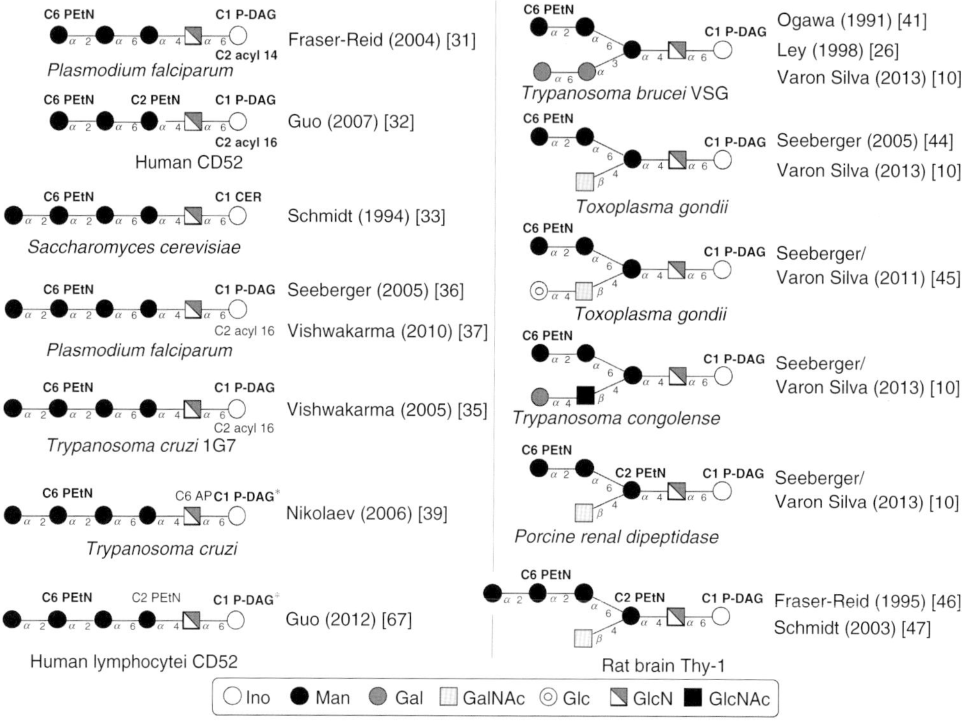

Figure 12.2 The list of completed GPI targets [10, 26, 30–47].

constitutes an excellent reference material that documents the challenges in the synthesis of these complex glycolipids and the synthetic tools that have been successfully used to overcome these challenges.

There are several strategic considerations and specific problems that a successful route to a GPI has to address [22, 23, 48]. A typical approach requires an assembly of the GPI molecule starting from the building blocks defined by their structural identity: monosaccharide, inositol, phospholipid, and phosphoethanolamine building blocks. The methods to construct the glycosidic bonds and to phosphorylate the GPI glycan are, therefore, the foundation for the design of any synthetic route to a GPI. The glycosylation and phosphorylation strategies in the syntheses of GPIs can be analyzed and designed independent of each other because they are executed at different stages of the synthesis with the glycan assembly occurring first in most of the reported routes. These two strategies are, however, intertwined by the use of orthogonal protecting groups necessary for both the glycan assembly and the regioselective phosphorylation of the GPI glycan.

Considering the number and arrangement of saccharide units (linear and branched) together with the number and possible positions of the phosphodiesters, the number of orthogonal protecting groups necessary to assemble the entire GPI molecule quickly becomes impractical. In order to reduce the required levels of orthogonal protection, an efficient protecting group strategy has to be general and common for both the glycosylations and the phosphorylations. The choice of permanent protecting group is also of great importance in the GPI synthesis as it restricts the type of reaction conditions that can be used and defines the requirements for the global deprotection that is not trivial because of the poor solubility profiles and amphiphilic character of the native GPIs.

As in any multistep synthesis, a convergent approach is preferred over the linear assembly of the building blocks. To reduce the number of operations on advanced intermediates, the precursors of phosphate groups are routinely introduced at a late stage of the synthesis as part of the simpler building blocks, for example, the (phospho)ethanolamine or the (phospho)lipid. The late stage formation of phosphodiesters also allows for more flexibility in the assembly of the GPI glycan, especially with respect to the use of acidic conditions often required for glycosylation reactions.

The most challenging task in the synthesis of GPIs is the construction of the GPI glycan. The GPI pseudooligosaccharides feature a linear backbone with three to four α-mannoses extending from the α-GlcN-(1 → 6)-myo-inositol pseudodisaccharide. It is, therefore, imperative that the glycosylation method used in the synthesis of GPI glycans produces α-configured mannosides in an efficient manner. The introduction of an α-configured glucosamine often requires an elaborate solution. Owing to the structural diversity of the saccharides outside of the core glycan, a number of synthetically challenging glycosidic linkages, α-galactose, α-glucose, or a sialic acid have to be constructed during the synthesis of various GPIs.

A typical synthetic route to a GPI target is modular and, therefore, formally convergent. It is important to emphasize the need for convergent approach in the synthesis of more complex, branched GPI glycans. Convergent routes offer more flexibility with respect to chemistries used to construct the subunits and allow for different glycosylation strategies to be employed for construction of challenging glycosidic linkages positioned in different subunits of the GPI molecule [10].

12.3 Tools for Synthesis of GPIs

Aside from the common problems in carbohydrate chemistry, synthesis of GPIs often requires creative solutions and special tools to address the specific challenges of the complex glycolipid synthesis. The problems in the GPI synthesis are intertwined but they can be classified in several categories according to the tools that have been used to address them: (i) the synthesis of building blocks, (ii) the assembly of GPI glycan by stereoselective glycosylations, (iii) the installation of phosphodiesters via phosphorylation reactions, and (iv) the strategic synthesis planning including the protecting group strategy and synthetic convergence.

12.3.1 Synthesis of Building Blocks

The building blocks required for the synthesis of GPI anchors are defined by the choice of glycosylation method and the choice of permanent and orthogonal protecting groups in the synthetic route. While the GPI monosaccharide building blocks occasionally call for novel synthetic approaches, they do not differ significantly from the typical building blocks used in the synthesis of other oligosaccharides. Their syntheses focus mostly on the selective protection of hydroxyls within the monosaccharide starting material and creation of the desired glycoside. A particularly rewarding strategy to prepare the GPI monosaccharide building blocks includes the synthesis of multiple building blocks starting from a common late stage intermediate. This strategy is commonly utilized for the synthesis of different mannose building blocks [30, 49].

In contrast to the common monosaccharide building blocks, synthesis of the myo-inositol building blocks features distinct challenges. Positions C3, C4, and C5 of the myo-inositol in natural GPIs are not substituted and position C2 only infrequently carries a fatty acid ester. A typical building block target is, therefore, an enantiopure, differentially protected myo-inositol with orthogonal protection in positions C1 and C6 (and sometimes C2). Such building blocks have been prepared either through desymmetrization and resolution of enantiomers of the appropriate myo-inositol derivatives or via *de novo* synthesis starting from common chiral pool materials.

myo-Inositol is a meso compound and, therefore, optically inactive. Upon introduction of the appropriate protecting groups, the molecule is desymmetrized and its enantiomers can be purified either by enzymatic resolution or resolution of diastereomers formed by coupling of the racemic mixture with optically pure partners, such as enantiopure carboxylic acids. These routes usually sacrifice half of the material containing the undesired enantiomer. Gou and coworkers have, for example, established an elegant enzymatic resolution strategy based on the ability of porcine pancreatic lipase to resolve enantiomers of protected myo-inositol propionic ester **1** (Scheme 12.1a). To avoid the loss of material, both the hydrolyzed product **2** and the remaining ester enantiomer (−)-**1** were transformed to the targeted building block **5** used in the synthesis of the sperm cluster of differentiation 52 (CD52) glycopeptide GPI anchor [50]. As in many similar routes, transformations of (−)-**1** and **2** to the appropriately protected building block make frequent use of the selective *O*-alkylation of the bis(dibutylstannylene) ketals [51] and in this specific case afford the building block **5** with chemically differentiated positions C1, C2, and C6.

An alternative approach to the synthesis of enantiopure myo-inositol building blocks is the *de novo* synthesis that usually starts from chiral pool materials [25, 54, 55]. Although there are no rules regarding the starting material [54], most of the approaches used in the synthesis of GPIs have relied on monosaccharides. A popular approach to myo-inositol building blocks starting from α-D-methylglucoside has been first developed by Fraser-Reid and later modified by the Seeberger

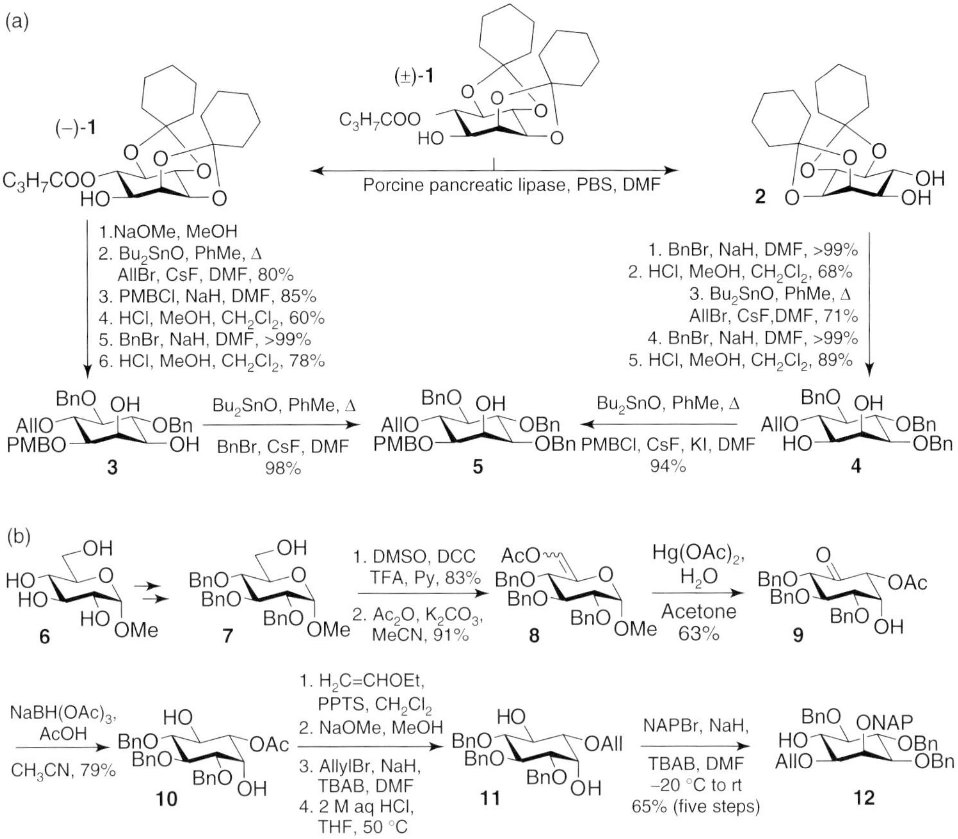

Scheme 12.1 (a) Enzymatic resolution [52, 50] and (b) *de novo* synthesis in the synthesis of myo-inositol building blocks [25, 53].

group (Scheme 12.1b). Installation of the permanent benzyl protecting groups is achieved in three simple steps early in the synthesis to produce **7**. Ferrier type-2 rearrangement and stereoselective reduction of the resulting ketone produces the inositol **10**. Protecting group exchange at the C1 and regioselective *O*-alkylation at the C2 position installs the required orthogonal protecting groups and completes the differentially protected myo-inositol building block **12**.

12.3.2 Glycosylation Strategy

Although there are no formal restrictions for the use of multiple glycosylation methods within one synthetic route to a GPI target, most of the reported syntheses rely on only one type of glycosylating agents and consequently use only one glycosylation method. This simplifies strategic synthesis planning and design of the protecting group strategy but also expands the list of requirements that the glycosylation method has to accommodate (synthesis of α-mannosides, construction of α-glucosamine linked to myo-inositol and any other challenges presented by the side branch oligosaccharide).

The glycosylation methods used to prepare the GPI glycan reflect, in many cases, the state of the art of the carbohydrates chemistry at the time of the report. Type of glycosylating agents that were successfully utilized in various GPI syntheses is indicative of the glycosylation methods and they include glycosyl halides, glycosyl *n*-pentenyl orthoesters, glycosyl chalcogenides, and glycosyl trichloroacetimidates. The general rules and restrictions for various activation methods [56] (for example, *N*-iodosuccinimide (NIS) activation of thioglycoside is not recommended for substrates containing alkenes or alkynes) generally hold up in the synthesis of GPI anchors and they will not be discussed in detail.

The common element of stereocontrol in α-mannosylation, regardless of the choice of the glycosylation method, is the presence of a participating group in C2 position of the glycosylation agent (although this is not always necessary). Construction of the notorious α-glycosidic linkage between the glucosamine and inositol, on the other hand, typically requires an elaborate solution and is consequently always created between the monosaccharide building blocks early in the synthesis. In fact, the α-GlcN-(1 → 6)-myo-inositol subunit is often treated as a single building block with respect to the assembly of the GPI glycan. The methods that have been used to construct the pseudodisaccharide vary from the direct glycosylation of glycosyl halides, which under appropriate conditions proceeds with good α-selectivity [57], to the installation of a mannose unit instead of a glucosamine followed by the S_N2 inversion of the C2 hydroxyl with a latent amine nucleophile such as an azide [30] (Scheme 12.2a,b). Well-documented influences of the protecting groups at C3, C4, and C6 positions on the stereoselectivity in glycosylations with 2-azido-2-deoxy-glucose building blocks [58] have also been exploited for the construction of the GPI pseudodisaccharide, although elaboration of the produced triacetate into a building block with the required arrangement of protecting groups may be lengthy (Scheme 12.2c).

Several syntheses of GPI glycans have profited from the methods for rapid construction of oligosaccharides such as the orthogonal glycosylation strategies [57, 61, 62] or the armed/disarmed glycosylating agent approach [26, 43]. These strategies are discussed with the specific synthetic routes in the following section.

12.3.3
Phosphorylation Strategies

Phosphorylation methods that are used in the synthesis of GPIs are adopted from the well-established protocols for the synthesis of oligonucleotides. The two commonly used methods are both based on phosphitylation that is followed by oxidation. Phosphitylation of an alcohol with H-phosphonates activated via transient formation of mixed anhydride with pivaloyl chloride followed by oxidation with iodine in wet pyridine is commonly used to prepare the phosphate salts (Scheme 12.3, introduction of the phospholipid phosphodiester). The reagents

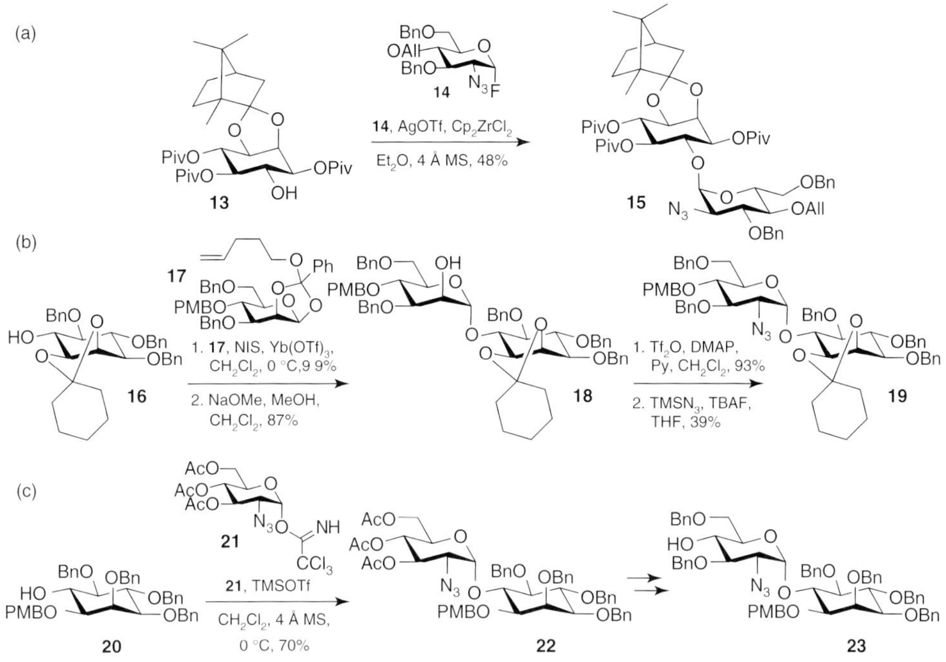

Scheme 12.2 (a–c) Synthetic approaches to the α-GlcN-(1 → 6)-myo-inositol GPI pseudodisaccharide [30, 57, 59, 60].

used in these reactions are cheap and easy to prepare, which makes this methodology versatile with respect to the synthesis of analogs. Another commonly used method to introduce the phosphodiesters is phosphitylation of a hydroxyl group in the GPI glycan with phosphoramidites activated with 1*H*-tetrazole followed by the oxidation with peroxides or peroxy acids (Scheme 12.3, introduction of the phosphoethanolamine). While these reactions proceed under generally milder conditions, the required reagents are expensive and difficult to prepare and/or to store. Methods based on phosphoramidites produce the protected phosphates that are generally easier to handle but more challenging to purify as they are normally produced as a mixture of phosphotriester diastereomers.

12.3.4
Strategic Synthesis Planning

In most of the reported synthetic routes to both linear and branched GPI structures, the target molecules are built from individual monosaccharides in a linear, stepwise manner (Figure 12.3). The oligomannoside segment of the molecule, in particular, has often been constructed in a linear manner using mannose building blocks with a participating group at C2 position to facilitate stereoselective α-glycosylation. Several convergent routes have also been developed. These routes are especially

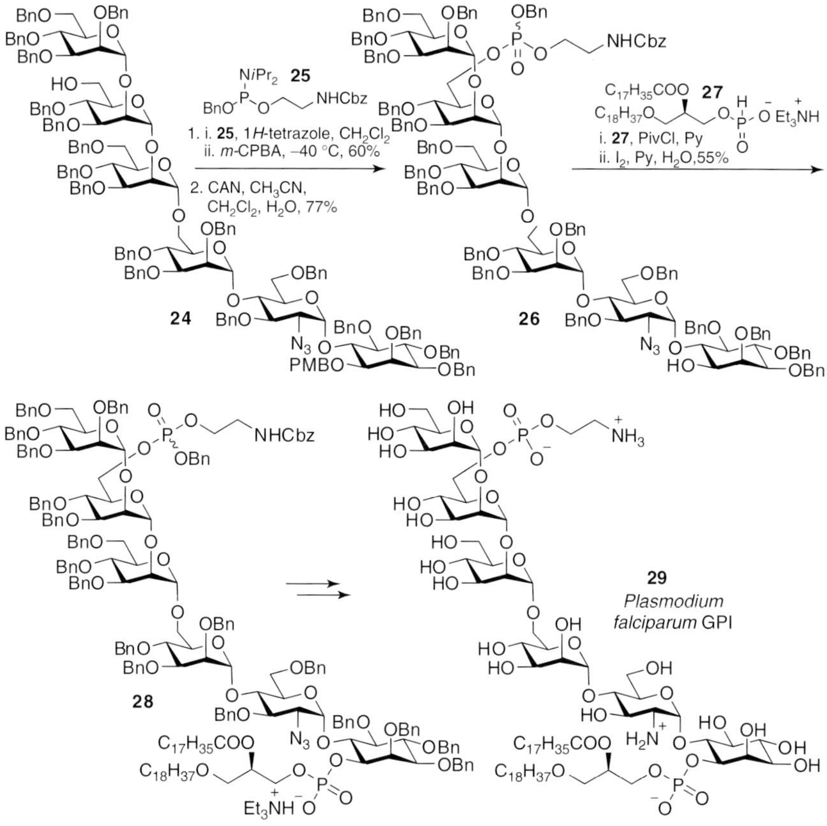

Scheme 12.3 Phosphorylation of the GPI glycan **24** using H-phosphonates and phosphoramidites [38].

suitable for the synthesis of branched GPIs (Figure 12.3). At the moment, however, synthesis of multiple branched GPI targets via glycosylation of the appropriately protected core glycan acceptor with various side branch oligosaccharide donors is not possible [10].

Convergent synthetic routes typically require more elaborate orthogonal protection patterns than linear syntheses. The well-designed protecting group strategy with the special emphasis on orthogonal protection is, thus, an important part in the design of a concise GPI synthesis. Considering the number of possible substituents on the GPI core glycan, one could expect that a convergent GPI synthesis requires prohibitively large number of orthogonal protecting groups to ensure both the convergent assembly of the branched glycan and the regioselective installation of the phosphodiesters. It was recently demonstrated that, when all known GPI structures are taken into account together with the identity and the arrangement of additional substituents of the core glycan, a single group for permanent

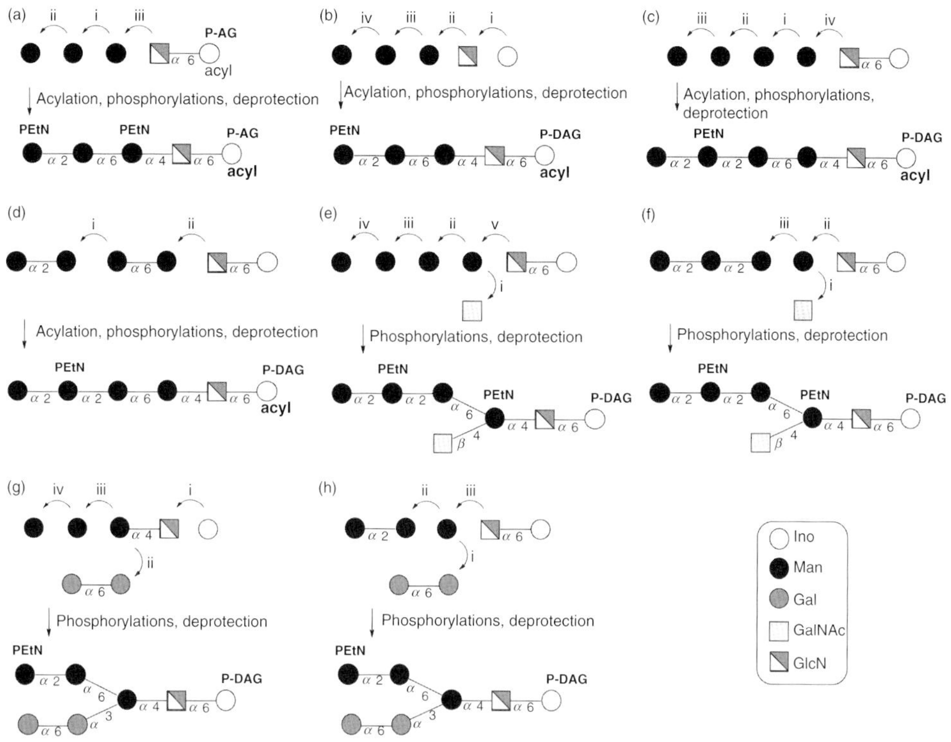

Figure 12.3 Summary of the linear (a,b,c,e) and convergent (d,f,g,h) strategies used to prepare various GPIs.

protection and four or less fully orthogonal protecting groups are sufficient to enable convergent synthesis of any natural GPI structure [10].

The choice of orthogonal protecting groups depends in the first line on the group chosen for permanent protection of the glycan. Although multiple permanent protecting groups can and have been used in the same route, such syntheses require multistep global deprotection protocols that increase the number of operations on advanced intermediates. Owing to their reliability and stability under a range of acidic and basic conditions, simple benzyl ethers are commonly used for permanent protection of the glycan. Conditions required for hydrogenolytic removal of benzyl ethers from advanced intermediates are simple and normally require minimal effort in work up and purification of the products. This is particularly suitable for the synthesis of GPIs that are notoriously difficult to handle due to their amphiphilic character and poor solubility. The requirement for hydrogenolytic conditions in the global deprotection, on the other hand, prevents the incorporation of unsaturated lipids into the synthetic GPIs prepared via routes that use benzyl ethers for permanent protection of the glycan. Common orthogonal protecting groups used in combination with benzyl ethers include silyl, allyl, 2-naphthylmethyl, and *p*-methoxybenzyl (PMB) ethers as well as benzoyl, acetyl, and levulinoyl esters.

In addition to the benzyl ethers, PMB ethers and benzoyl esters have been used as permanent protecting groups in the synthesis of GPIs. These protecting groups are characterized by the lower stability under strongly acidic or basic conditions and they put significant restrictions on the choice of conditions used in the synthesis of the GPI target. Global deprotection reactions to remove PMB ethers or benzoyl esters can be operationally challenging and require demanding purification protocols for the resulting materials due to the poor solubility profiles of the native GPIs and intermediates in these reactions. These routes, however, enable the synthesis of unsaturated GPI glycolipids that are otherwise not accessible [39, 63].

With the tools that are used in the synthesis of GPI glycolipids explained, the representative routes to the GPIs with linear and branched glycan cores can now be showcased with focus on the strategic synthesis planning and the assembly of building blocks. The synthetic routes are organized into two groups based on the complexity of the GPI glycan in the target molecule and each route demonstrates the use of a different type of building blocks and glycosylation method for the glycan assembly.

12.4 Synthesis of GPIs with Linear Glycan Core

12.4.1 Synthesis of the GPI from *Plasmodium falciparum* Using *n*-Pentenyl Orthoesters

The Fraser-Reid group has disclosed the synthesis of the *Plasmodium falciparum* GPI (**30**, Scheme 12.4) based on the use of *n*-pentenyl orthoester building blocks [30, 31, 64]. In addition to the conserved pseudopentasaccharide, the target GPI features a myristoyl ester at the C2 position of the inositol, diacylglycerol phosphate at the C1 of inositol, and the mandatory phosphoethanolamine at ManIII.

Fraser-Reid envisioned **30** via linear assembly of the glycan by way of α-selective mannosylations using *n*-pentenyl orthoester glycosylating agents **17**, **31–33** [49] and subsequent phosphorylation using phosphoramidites **34** and **25** (Scheme 12.4). Trityl group and cyclohexanone ketal were chosen for protection of the phosphorylation sites, while the temporary *tert*-butyldimethylsilyl (TBS), acyl and PMB groups enabled extension of the glycan. Building blocks **17**, **31–33** were prepared from a common precursor in just a few steps and the inositol building block **16** was prepared via a *de novo* synthesis [25]. The direct glycosylation to install an α-configured glucosamine was circumvented by the formation of the corresponding mannoside and inversion of the C2 alcohol with an azide as described in the previous section (Scheme 12.2b).

The synthesis commenced with the assembly of the glycan. Glycosylation of myo-inositol **16** with *n*-pentenyl orthoester **17** using ytterbium(III) triflate and

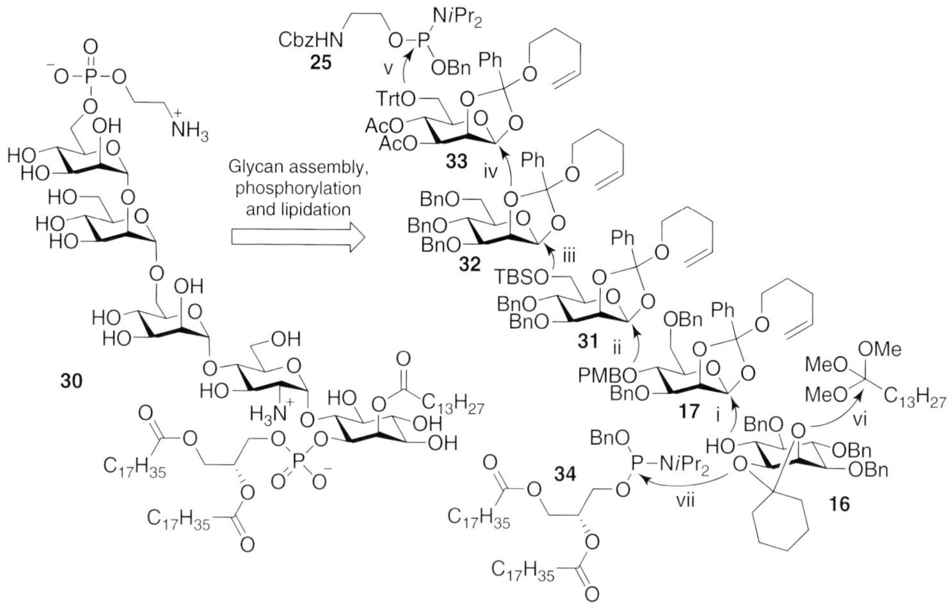

Scheme 12.4 Retrosynthetic analysis for the GPI **30** from *P. falciparum*, by the Fraser-Reid group. Small Roman numerals indicate the order of assembly.

NIS as promoters yielded the corresponding pseudodisaccharide in 99% yield as the pure α-isomer (Scheme 12.2b). Upon removal of the C2 benzoate and conversion of the alcohol to the corresponding triflate, the stereocenter at position C2 of the mannose in **18** was inverted using trimethylsilylazide in the presence of tetra-*n*-butyl-ammonium fluoride (TBAF) to afford the pseudodisaccharide **19**. Removal of the PMB ether from **19** under acidic conditions ($BF_3{\cdot}Et_2O$) unveiled the pseudodisaccharide alcohol **35**. Glycosylation of **35** with *n*-pentenyl orthoester **31** in the presence of NIS and $BF_3{\cdot}Et_2O$ produced the corresponding pseudotrisaccharide in 79% yield (Scheme 12.5). The exchange of benzoyl ester for benzyl ether in C2 position and the removal of the C6 silyl ether of ManI set the stage for glycosylation with *n*-pentenyl orthoester **32** in the presence of NIS and $BF_3{\cdot}Et_2O$ to form the desired tetrasaccharide in an excellent yield. Removing the benzoate to generate **37** followed by glycosylation using mannoside **33** produced the corresponding pseudopentasaccharide. Exchange of the acyl groups to benzyl ethers furnished the fully protected GPI glycan **38** in 55% yield over three steps starting from **37**.

Cleavage of the trityl ether from the ManIII with formic acid followed by the phosphitylation with phosphoramidite **25** and *meta*-chloroperoxybenzoic acid (*m*-CPBA) oxidation at −40 °C gave phosphotriester **39** as a 1 : 1 mixture of diastereomers. Upon removing the cyclohexanone ketal from the inositol, the corresponding diol was converted into an orthoester using the trimethyl orthoester of myristic acid. Hydrolysis of the orthoester in the presence of $Yb(OTf)_3$ installed the myristate ester

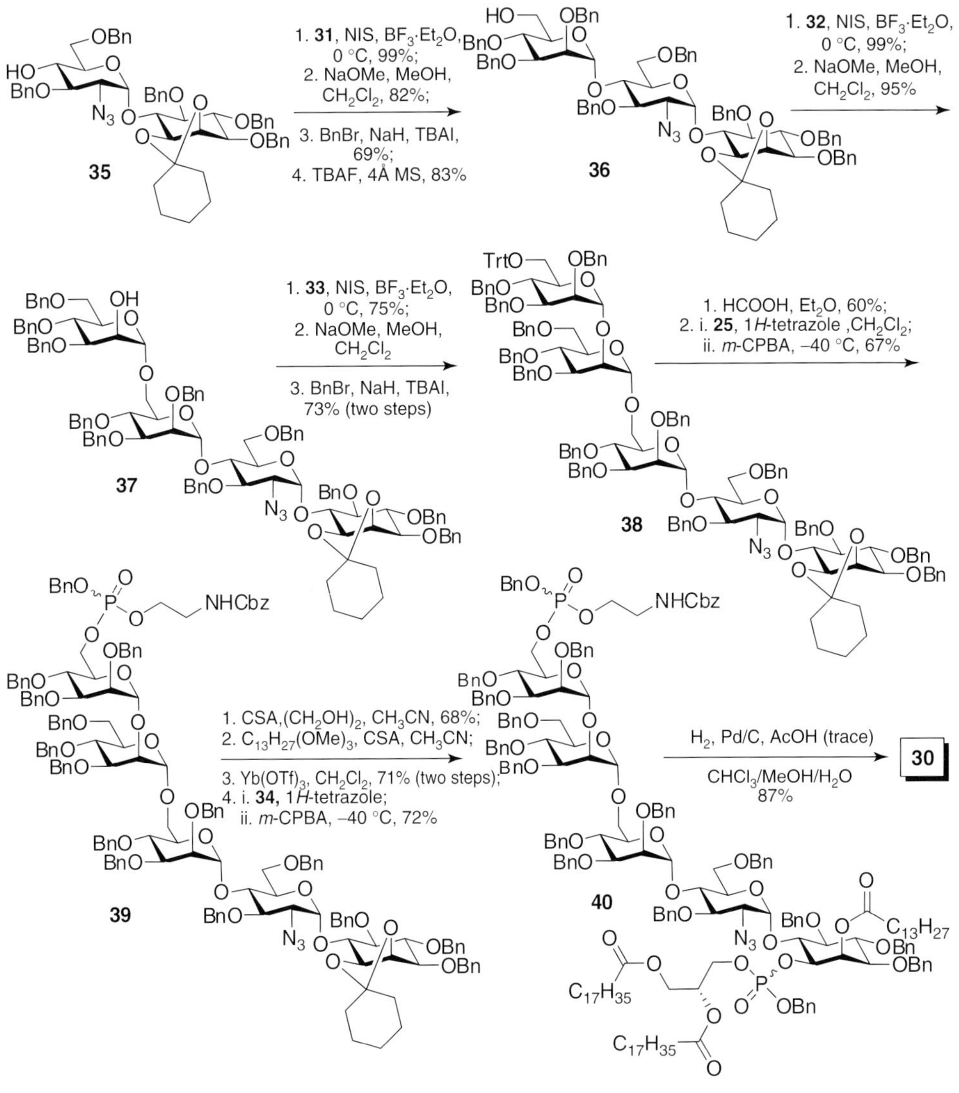

Scheme 12.5 Synthesis of the *P. falciparum* GPI **30** by the Fraser-Reid group [30].

at C2 position of myo-inositol in 71% yield and good selectivity with only 15% of the regioisomer formed. The C1 position of inositol was phosphitylated using phosphoamidite **34** in the presence of 1*H*-tetrazole and the corresponding phosphotriester was oxidized to the phosphate with *m*-CPBA to yield the fully protected GPI **40**. Hydrogenolysis over palladium on charcoal in chloroform–methanol–water mixture, with a single change of solvent to maintain the solubility of the intermediates delivered GPI **30** in 87% yield.

12.4.2
Synthesis of the GPI from *Saccharomyces cerevisiae* Using Trichloroacetimidates

Schmidt and colleagues [34] have reported the first synthesis of a GPI isolated from yeast (*Saccharomyces cerevisiae*). Similar to the previous synthesis, the target molecule (**41**, Scheme 12.6) features a linear core glycan extended with one α-mannose attached to the C2 position of ManIII. A distinguishing feature of this GPI is the presence of a ceramide instead of the usual glycerolipid.

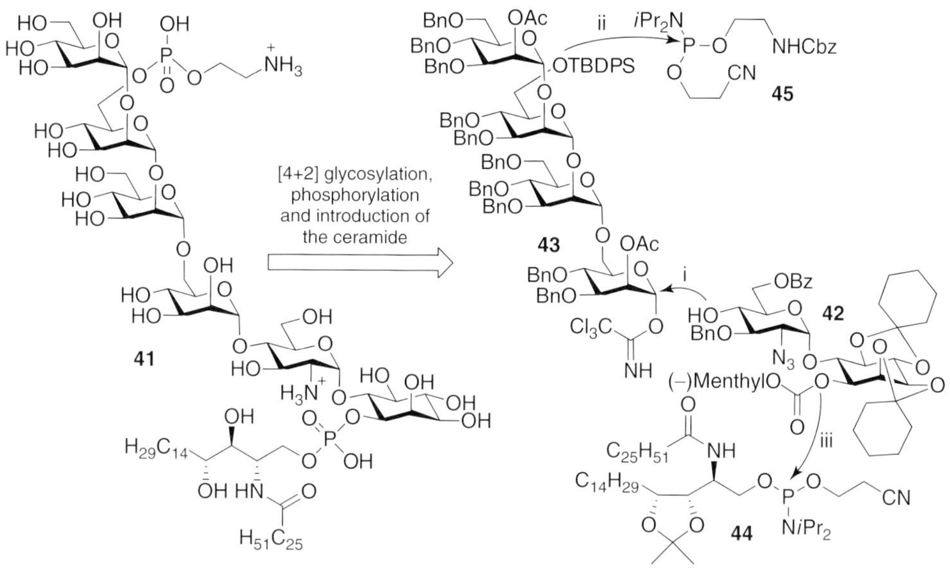

Scheme 12.6 Retrosynthetic analysis of the GPI anchor from *S. cerevisiae*, by the Schmidt group. Small Roman numerals indicate the order of assembly.

The key disconnections in the retrosynthetic analysis, the phosphodiester bonds, and the glycosidic bond between GlcN and ManI, define a convergent [4+2] approach to the GPI glycan [33, 34]. Phosphodiesters were envisioned from the protected GPI glycan and phosphoramidites **45** and **44** (prepared by a *de novo* approach). A sequence of highly stereoselective glycosylations with benzyl protected mannose trichloroacetimidate building blocks featuring a C2-participating group [56, 65] was planned for the synthesis of **43** [66]. The pseudodisaccharide **42** was envisioned to arise via glycosylation of the protected myo-inositol (−)-menthyl formate with peracetylated 2-azido-2-deoxy-glucose trichloroacetimidate **21** following the protecting group exchange as described in a similar synthesis in the previous section (Scheme 12.2c).

En route the tetramannoside **43**, building block **46** was glycosylated with trichloroacetimidate **47** in the presence of catalytic amounts of TMSOTf and subsequently deacetylated to produce dimannoside **48** in 81% yield over two steps (Scheme 12.7). Dimannoside **48** was then subjected to glycosylation with

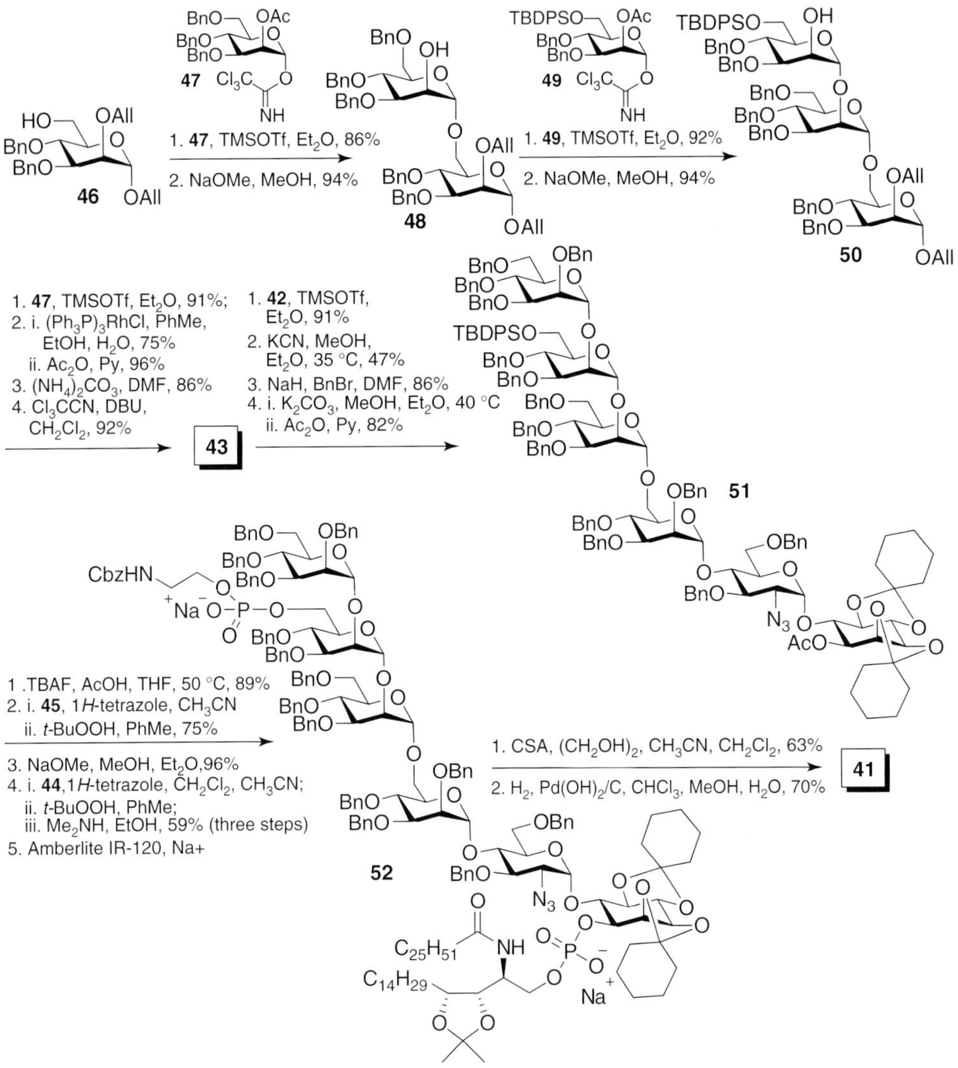

Scheme 12.7 Synthesis of the *S. cerevisiae* GPI **41** by the Schmidt group [33].

trichloroacetimidate **49** under the standard conditions followed by removal of the acetate to produce trimannoside **50** in 86% yield over two steps. Another TMSOTf promoted glycosylation utilizing building block **47** converted the trimannoside **50** to the corresponding tetramannoside. Exchange of allyl ethers for acetate esters was accomplished by removal of the allyl ethers via rhodium-catalyzed isomerization followed by hydrolysis of the resulting enol ethers and subsequent acetylation. Removal of the anomeric acetate and formation of trichloroacetimidate afforded the desired tetramannose building block **43** in 52% yield over five steps starting from **50**. The [4+2] glycosylation to combine **42** and **43** proceeds in an excellent yield and affords only the α-configured product owing to the neighboring group participation. Replacing the acetates with the benzyl ethers and the carbonate at the C1 position of inositol with an acetyl ester provided the appropriately protected glycan **51**.

In order to install the phosphodiester, the *tert*-butyldiphenylsilyl (TBDPS) group at ManIII was removed using TBAF and the corresponding alcohol was treated with phosphoamidite **45** and 1*H*-tetrazole to form the phosphite triester that was further oxidized with *tert*-butylhydroperoxide. Removal of acetate from inositol followed by phosphitylation with phytosphingosine building block **44** followed by oxidation of the resulting phosphite afforded the protected glycolipid **52** in 39% yield over four steps starting from **51**. Global deprotection of the acid-labile acetals with camphorsulfonic acid followed by hydrogenolysis over palladium gave GPI anchor **41** in 44% yield over two steps.

12.4.3
Synthesis of Unsaturated GPIs from *Trypanosoma cruzi*

GPI fraction purified from *T. cruzi* trypomastigote mucins has a level of proinflammatory activity that is comparable with bacterial lipopolysaccharide [9]. Given that neither the glycan structure alone nor the lyso-alkyl- or acyl-alkyl-phosphatidylinositol moieties are able to induce the production of NO, interleukin 12 (IL-12), and tumor necrosis factor-α (TNF-α) in macrophages, it is hypothesized that the full GPI structure, containing the unsaturated fatty acids, is necessary for the observed proinflammatory activity.

The synthesis of GPIs containing unsaturated lipids has been a significant challenge because of the common use of benzyl ethers for permanent protection of the glycan. When benzyl ethers are used, global deprotection requires hydrogenolytic conditions that also reduce the alkenes present in fatty acids. The first synthesis of a GPI containing an unsaturated lipid based on the use of base-labile [39] or acid-labile [22, 62] permanent protecting groups was reported by Nikolaev and coworkers. The target *T. cruzi* GPIs **53** and **54** (Scheme 12.8) feature the extended core pseudohexasaccharide, the 2-aminoethylphosphonic diester at the C6 position of GlcN, unique to *T. cruzi*, and either an oleyl or linoleyl ester.

The central retrosynthetic disconnections are the phosphate and phosphonate esters (Scheme 12.8). The GPI glycan was envisioned via a convergent [4+2] glycosylation strategy utilizing trichloroacetimidate building blocks. In order to address the problems associated with hydrogenolytic removal of benzyl ethers in the presence of unsaturated lipids, Nikolaev uses benzoyl esters for the permanent protection of tetrasaccharide **56** and an orthogonal combination of silyl ethers and cyclohexanone acetals for the pseudodisaccharide **55**.

The pseudohexasaccharide **61** was constructed via glycosylation of **55** with tetramannoside **56** in the presence of TMSOTf in 71% yield (Scheme 12.9). Selective removal of the triethylsilyl ether with TBAF uncovered the C6 position of GlcN. 1*H*-tetrazole-catalyzed esterification with phosphonic dichloride **59**, followed by methanolysis gave the corresponding phosphonic diester. Staudinger reduction

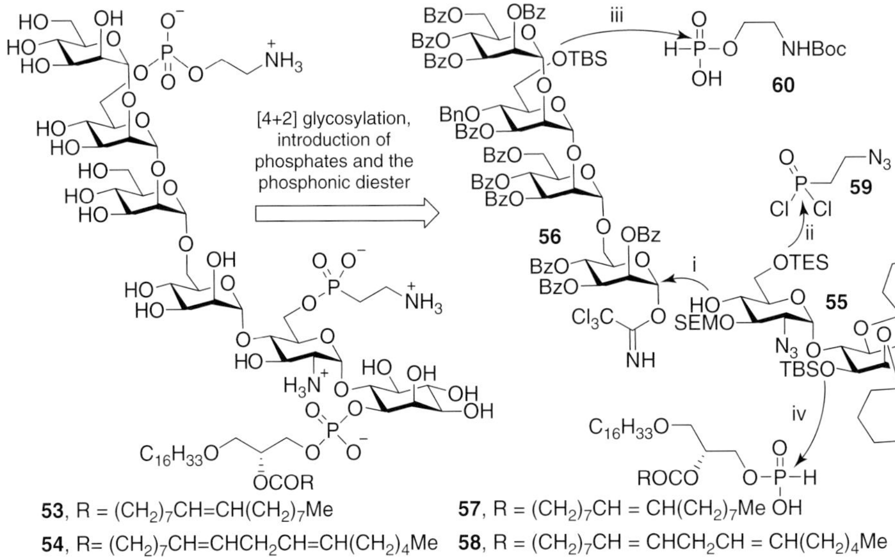

Scheme 12.8 Retrosynthetic analysis of GPI anchors **53** and **54** of *T. cruzi* by the Nikolaev group [39]. Small Roman numerals indicate the order of assembly.

of the azide and Boc-protection of the resulting amine followed by the removal of the primary TBS ether from C6 position of ManIII with 3HF·Et_3N produced **62** in 80% over three steps. The protected phosphoethanolamine was introduced via phosphitylation of **62** with H-phosphonate **60** followed by oxidation with iodine in wet pyridine. Removal of the TBS ether from the C1 position of inositol followed by phosphitylation with H-phosphonates **57** or **58** and oxidation with iodine introduced the desired lipid. Selective removal of methyl ester from the phosphonate effected by thiophenol, followed by the selective methanolysis of benzoyl esters in the presence of fatty acid esters and removal of the acid-labile 2-trimethylsilylethoxymethoxy (SEM) and acetal groups in aqueous trifluoroacetic acid (TFA) produced the desired unsaturated GPIs **53** and **54**. Electronic differences between the benzoyl and the fatty acid esters together with micelle formation that makes lipid esters less accessible are quoted as origins of the good selectivity observed in the global deprotection step.

A similar strategy toward unsaturated GPIs has been reported by the Guo group [63] in their syntheses of a simple GPI anchor containing dioleoylglycerol phosphate and the GPI of the human lymphocyte CD52 antigen [67], where PMB ethers are used for permanent protection of the glycan. Unlike benzyl ethers that are stable under a broad range of acidic and basic conditions, PMB ethers can be cleaved under acidic conditions. While this restricts the choice of glycosylation conditions and causes problems in the glycan assembly with respect to the purification of intermediates, it also enables the global deprotection under acidic conditions tolerated by the fatty acid esters and allows for the synthesis of unsaturated GPIs.

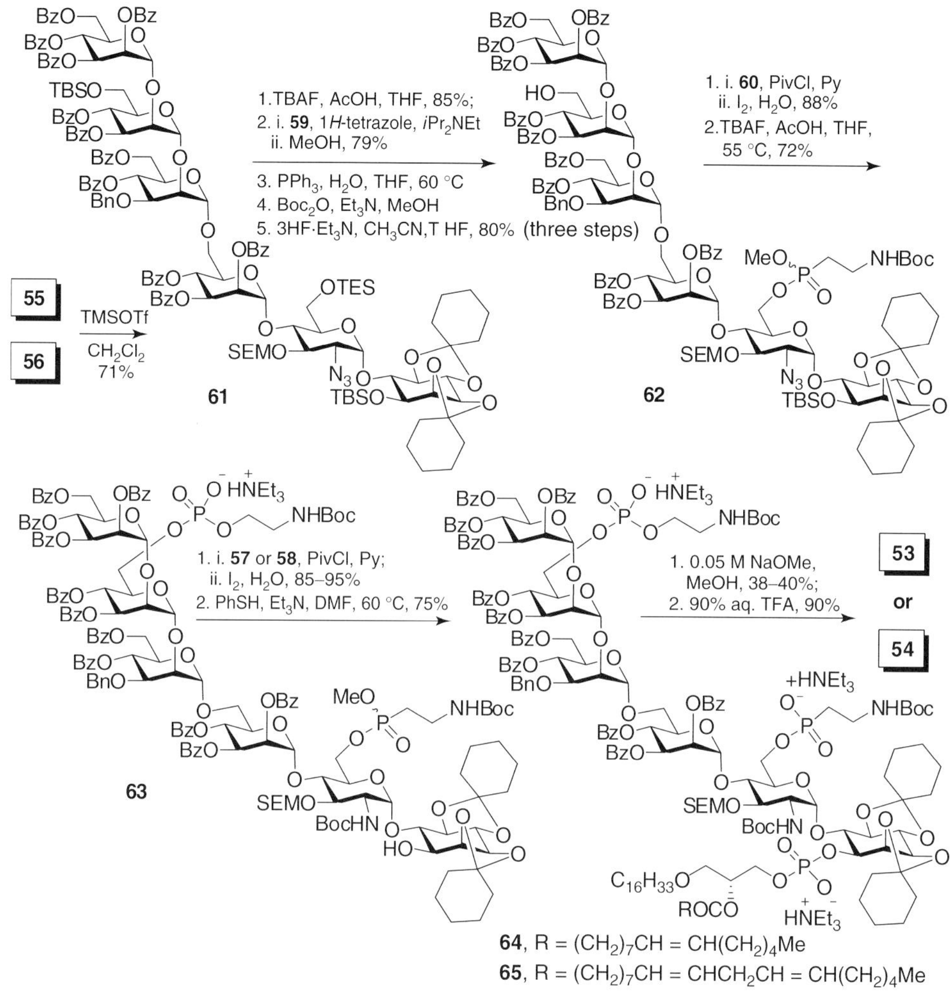

Scheme 12.9 Synthesis of GPIs **53** and **54** bearing unsaturated lipids, by Nikolaev and coworkers [39].

12.5 Synthesis of GPIs with Branched Glycan Core

12.5.1 Synthesis of the *Trypanosoma brucei* VSG GPI Using Glycosyl Halides

Coming only a few years after the first reports on the structure of GPIs, Ogawa's [40–42] work toward the GPI anchor from *T. brucei* was a grand endeavor in the synthesis of glycolipids and the first report of synthetic work toward a complete GPI structure. Besides the core glycan and the mandatory phosphodiesters, the *T. brucei*

GPI features a synthetically challenging side branch with up to four α-configured galactose units that extend from the C3 position of ManI (Scheme 12.10).

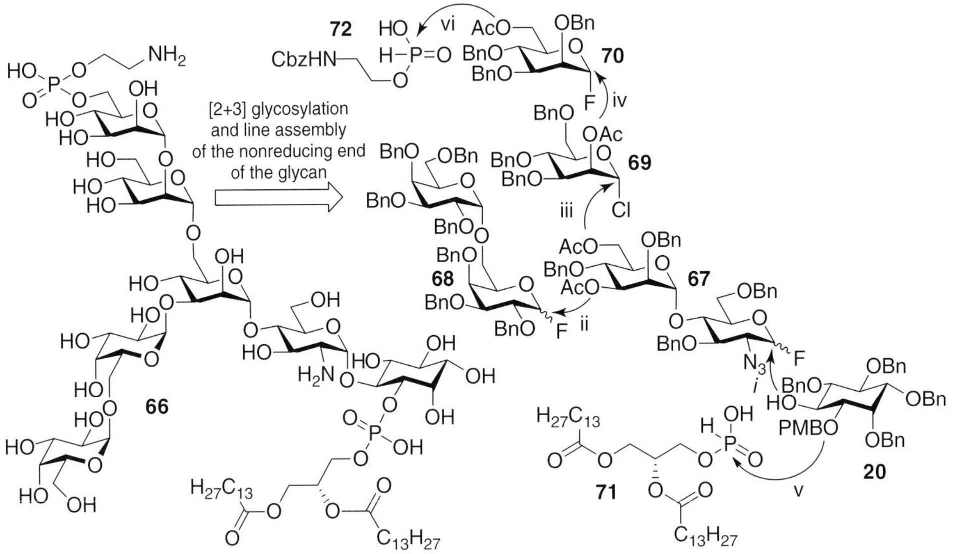

Scheme 12.10 Retrosynthetic analysis for the unnatural diastereomer of *T. brucei* GPI **66**. Small Roman numerals indicate the order of assembly.

In order to prepare the appropriate inositol building block, Ogawa and coworkers relied on a previously described protocol for resolution of myo-inositol enantiomers based on the formation of (*R*)-camphanic acid esters. Subsequent studies revealed that the diastereomers of the L- and D-myo-inositol esters of (*R*)-camphanic acid were wrongly assigned in the original paper [68, 69] which led to a revision of the structures prepared by Ogawa [70] and coworkers. Although the prepared material is a diastereomer of the GPI from *T. brucei* (differing in the absolute configuration of myo-inositol), this pioneering study deserves further discussion as it accurately reflects both the challenges associated with the synthesis of complex, branched glycolipids, and the state of the art of synthetic carbohydrate chemistry from over two decades ago [70, 71].

The first retrosynthetic disconnections were made at the phosphodiesters (Scheme 12.10). The core glycan structure was envisioned via a linear assembly of building blocks, adding one monosaccharide unit at a time via glycosylation using glycosyl halides. Because of the difficulties in the synthesis of α-galactosides, side branch is attached to the core glycan as a disaccharide at an opportune stage of the synthesis.

Glycosylation of glucosamine building block **73** with the *in situ* generated glycosylbromide derived from **74** in the presence of $CuBr_2$, Bu_4NBr, and AgOTf produced the desired α-linked disaccharide in 90% yield (Scheme 12.11) [72]. Removal of the TBS group from the mixed silyl acetal followed by (diethylamino)sulfur trifluoride

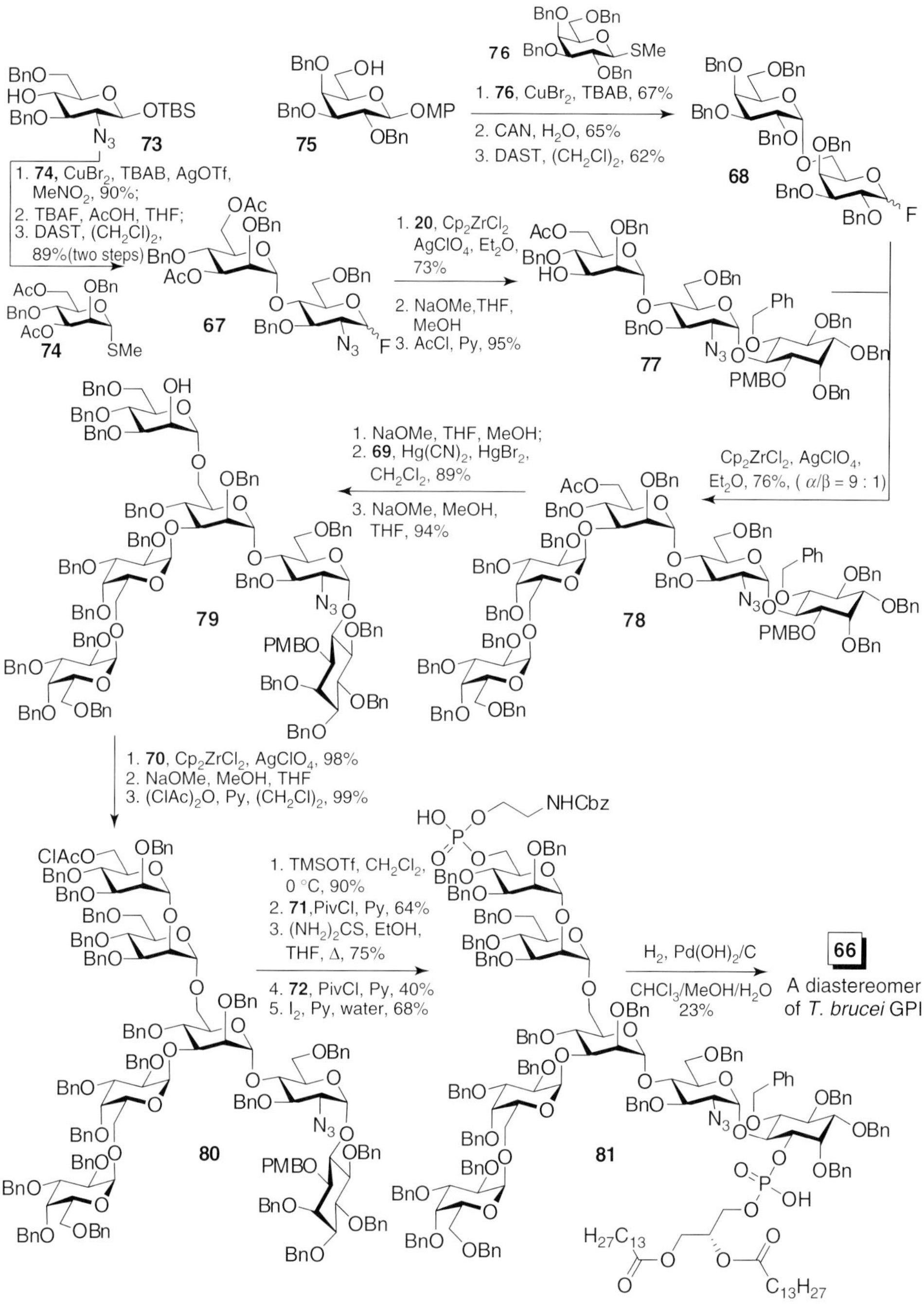

Scheme 12.11 Synthesis of the GPI **66** related to the *T. brucei* VSG GPI by the Ogawa group [40–42].

(DAST) promoted fluorination afforded the glycosyl fluoride **67** as a mixture of anomers. Cp_2ZrCl_2- and $AgClO_4$-promoted glycosylation of inositol (+)-**20** with the fluoride **67** afforded the corresponding pseudotrisaccharide in 93% yield but with only moderate selectivity for the desired anomer (α:β = 3.7 : 1). Cleavage of the two acetate esters followed by the selective acetylation of the primary alcohol provided the alcohol **77**.

For the synthesis of digalactoside **68**, building block **75** was glycosylated with thioglycoside **76** in the presence of $CuBr_2$ and tetra-*n*-butylammonium bromide (TBAB) in 77% yield with 6.7 : 1 selectivity favoring the α-anomer. Oxidative removal of the *p*-methoxyphenyl ether followed by fluorination of the resulting lactol with DAST produced **68** as a mixture of anomers. Glycosylation of trisaccharide **77** with fluoride **68** was effected by Cp_2ZrCl_2 and $AgClO_4$ to give pseudopentasaccharide **78** in 76% yield and with good selectivity favoring the desired α-isomer (α:β = 9 : 1).

Pseudopentasaccharide obtained by acetate removal from **78** was glycosylated with chloride **69** using mercury(II)cyanide as a promoter to yield the α-linked product in 89% yield (Scheme 12.11). Removal of the acetate unveiled the C2 hydroxyl of ManII (**79**, Scheme 12.11) for the subsequent glycosylation with fluoride **70** in the presence of Cp_2ZrCl_2 and $AgClO_4$ in diethylether. The fully protected glycan was obtained in 99% yield with the α-anomer as the major product (α:β > 15 : 1).

Owing to the problems associated with the formation of phosphodiester in the presence of acetate and problems with removal of the PMB ether in the presence of a phosphinic diester, an exchange of protecting groups at the C6 position of ManIII from acetate to chloroacetate was required. With chloroacetate in place, the PMB ether in **80** was removed under acidic conditions (TMSOTf) to produce the corresponding alcohol. Phosphitylation with the *in situ* prepared mixed anhydride of H-phosphonate **71** and pivaloyl chloride yielded the lipidated glycan. Chloroacetate was cleaved under mild conditions using thiourea in ethanol and the second phosphitylation/oxidation sequence was carried out with Cbz-protected phosphoethanolamine **72** to furnish bisphosphate **81**. Global deprotection of benzyl ethers under hydrogenolytic conditions produced **66**, an unnatural diastereomer of *T. brucei* VSG GPI.

12.5.2
Synthesis of *T. brucei* VSG GPI from Chalcogenide Glycosides of Finely Tuned Reactivity

The first synthesis of the *T. brucei* VSG GPI was accomplished by the Ley group [26, 43] in 1998. Similarly to Ogawa's retrosynthesis, Ley makes initial disconnection at the phosphodiesters. For the synthesis of the GPI glycan, a highly convergent approach with disconnections around ManI was outlined to come at the central mannose building block and three disaccharide fragments of similar complexity. The glycosylation plan was based on the orthogonal glycosylation strategies [73] and the concept of armed and disarmed glycosylating agents [74] that includes thio- and selenoglycosides. An elaborate protecting group pattern was designed and optimized to serve both the orthogonal protection of the glycan and the reactivity tuning of glycosyl donors predominantly through electronic effects of butane diacetal, chloroacetates, TBS ethers, allyl ethers, and benzyl ethers (Scheme 12.12a) [75].

Two building blocks that carry anomeric leaving groups of the same type can take part in a productive glycosylation reaction if their reactivities are sufficiently different. For example, selenoglycoside **84** (Scheme 12.12a) is deactivated by disarming protecting groups and it can be glycosylated with selenoglycoside **82** which is activated by arming protecting groups. Homodimers or polymers of **84** are not formed in such reaction as long as sufficient amounts of the more reactive glycoside **82** are present in the reaction mixture [76]. This concept eliminates functional group interconversions from the glycan assembly phase, for example, conversion of an allyl glycoside to a trichloroacetimidate, and reduces the number of synthetic operations on advanced intermediates and thus increases the efficiency of the route.

Construction of the protected GPI glycan **94** commenced with the glycosylation of galactosyl acceptor **84** with selenoglycoside **82** (Scheme 12.12b), which was selectively activated in the presence of NIS and catalytic amounts of triflic acid (TfOH), to afford disaccharide **87** in 71% and the corresponding β-isomer in 14% yield. Digalactoside **87** was directly used for methyl trifluoromethanesulfonate (MeOTf) promoted glycosylation of mannoside **88**. The cleavage of TBS ether from the resulting alcohol afforded alcohol **91** in 67% yield over two steps.

Dimannoside **86** was prepared via glycosylation of mannoside **85** with selenoglycoside **83** in the presence of NIS and triflic acid in 87% yield, as a single anomer. The MeOTf-promoted [2+3] glycosylation of **91** with **86** in excess quantities furnished pentasaccharide **92** in 75% yield with good α-selectivity. Pseudodisaccharide **93**, prepared by glycosylation of **90** with bromide **89** in the presence of TBAB followed by removal of the TBS ether, was glycosylated with pentasaccharide thioglycoside **92** in the presence of NIS and catalytic amounts of TfOH to afford, in 51% yield, the protected GPI glycan **94**, assembled in only eight synthetic operations starting from the monosaccharide building blocks.

The final steps of the synthesis included removal of the remaining orthogonal protecting groups and regioselective introduction of phosphates by the well-established phosphitylation/oxidation protocols using phosphoramidites **25** and **95** (Scheme 12.12b). As a trade-off for the use of multiple protecting groups, the global deprotection required optimization of a three step sequence. The fully protected GPI **96** was first submitted to palladium-catalyzed hydrogenolysis followed by the selective deacylation of chloroacetates in the presence of fatty acid esters with hydrazinedithiocarbonate and hydrolysis of the remaining acetal groups with TFA in water to provide the *T. brucei* VSG GPI **97**.

12.5.3
A General Synthetic Strategy for the Synthesis of Branched GPIs

The reported GPI syntheses are typically target-oriented and not readily amenable to modifications that would produce other GPIs or unnatural GPI analogs. Systematic biochemical and biophysical studies to decipher the functions of GPIs require

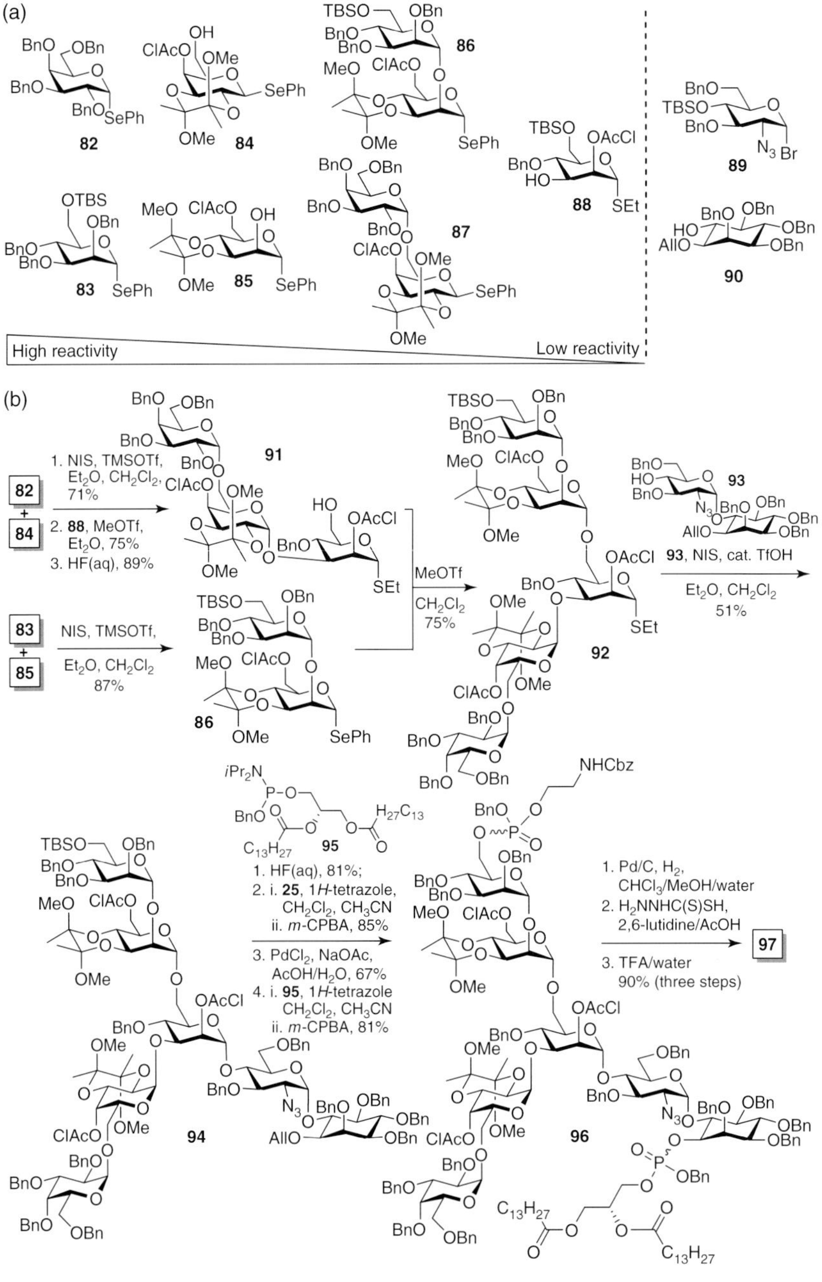

Scheme 12.12 (a) Structures and reactivity of the building blocks for the synthesis of *T. brucei* GPI. (b) Total synthesis of the GPI **97** from *T. brucei* by the Ley group [26, 43].

access to a structurally diverse set of these glycolipids that would be difficult to obtain using the traditional target-oriented routes. In order to address this problem, the Varon Silva group has developed a general convergent synthetic route to GPI glycolipids based on interchangeable common building blocks [10, 45].

Targeting all known GPIs, retrosynthesis is based on the analysis of the substitution patterns around the conserved pseudopentasaccharide in the characterized GPI structures and the frequency with which individual modifications occur in this set. The initial disconnections at the phosphodiesters emulate other reported GPI syntheses and define the requirement for orthogonal protection of the glycan (Scheme 12.13). Further disconnections of the GPI glycan around the ManI, similar to those used by Ley [26] and Schmidt [60], produce four sets of building blocks each of which can be used for synthesis of multiple GPI structures that are of interest for biological studies. To allow for more flexibility in the choice of reaction conditions in this general route, benzyl ethers were selected for permanent protection of the glycans and trichloroacetimidate glycosylating agents were selected as the basic building blocks for construction of the GPI glycan. Four levels of orthogonal protection (allyl ether, levulinoyl ester, silyl ether, and a 2-naphthylmethyl ether) are

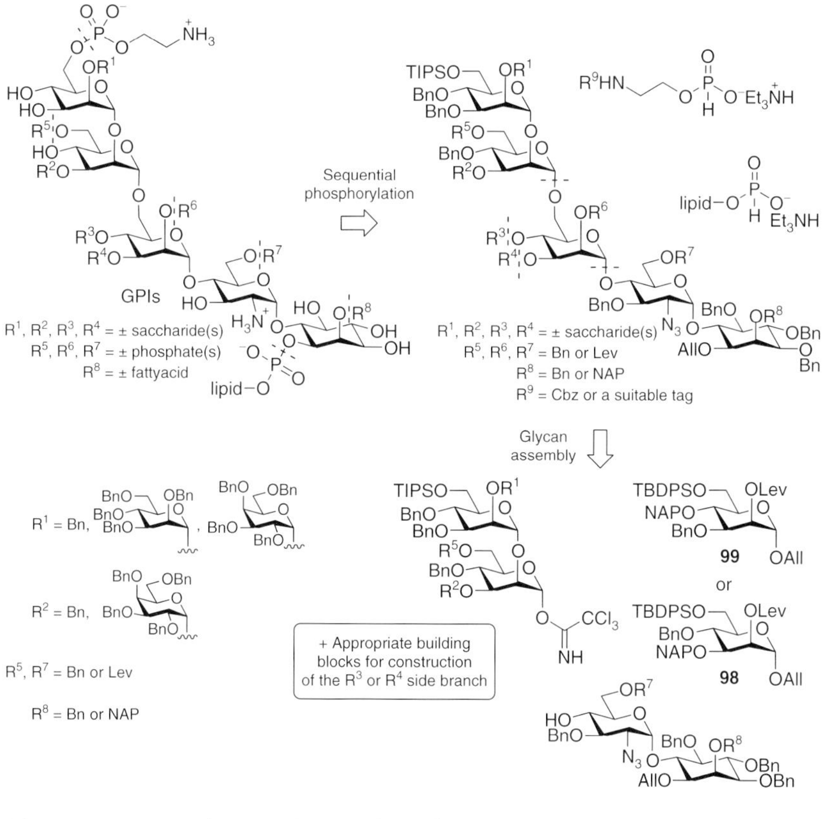

Scheme 12.13 General retrosynthetic analysis of the characterized GPIs [10].

sufficient to enable the synthesis of any phosphorylation pattern and the assembly of any glycan found in the natural GPI anchors characterized to date.

Initial optimization of the general route was performed during the synthesis of porcine renal dipeptidase GPI anchor. The target molecule features the pseudopentasaccharide core, two mandatory phosphodiesters, the common β-GalNAc side branch at C4 position of ManI, and an additional phosphoethanolamine unit at C2 position of ManI.

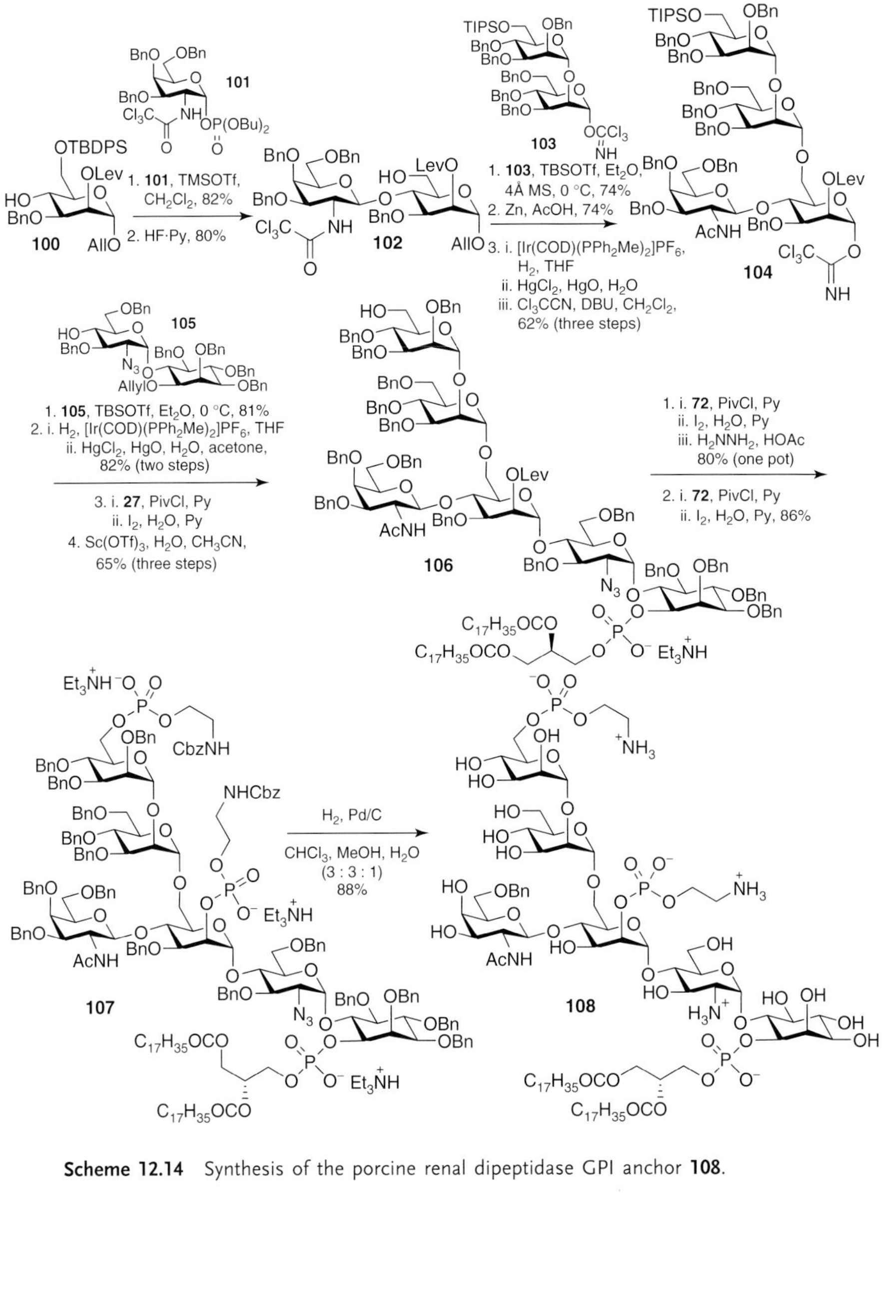

Scheme 12.14 Synthesis of the porcine renal dipeptidase GPI anchor **108**.

The assembly of the requisite glycan commenced with glycosylation of alcohol **100** with phosphate **101** to produce the disaccharide **102** (Scheme 12.14). Removal of TBDPS ether from C6 position of ManI with HF·Py followed by [2+2] glycosylation of the corresponding alcohol with dimannoside **103** provided the desired tetrasaccharide. Reduction of the trichloroacetamide with zinc in acetic acid at elevated temperatures and exchange of allyl glycoside for the corresponding trichloroacetimidate via isomerization with *in situ* prepared iridium hydride catalyst, mild acid hydrolysis in the presence of mercury salts and treatment with trichloroacetonitrile in the presence of 1,8-diazabicyclo[5.4.0]undec-7-ene (DBU) yielded imidate **104**. [4+2] glycosylation was accomplished by activation of **104** with catalytic amounts of *tert*-butyldimethylsilyl trifluoromethanesulfonate (TBSOTf) in the presence of pseudodisaccharide **105** at 0 °C in 81% yield (Scheme 12.14). Removal of the allyl ether, phosphitylation/oxidation using H-phosphonate **27** and acidic hydrolysis of the silyl group delivered monophosphate **106** in 65% over three steps. A one pot protocol consisting of the phosphitylation using H-phosphonate **72**, the subsequent oxidation with iodine in wet pyridine and the hydrazinolysis of the levulinoyl ester gave access to the bisphosphate. Another phosphitylation/oxidation using H-phosphonate **72** installed the phosphoethanolamine at C2 position of ManI and afforded the fully protected GPI **107** which underwent clean hydrogenolysis to furnish **108**, the triphosphorylated GPI anchor of the porcine renal dipeptidase in 88% yield.

This synthetic route has been further optimized to improve the yields in the glycan assembly process and has been successfully utilized to prepare the *Toxoplasma gondii* GPI anchor, the low-molecular-weight antigen of *T. gondii*, *Trypanosoma congolense* VSG GPI anchor, and *T. brucei* VSG 117 GPI anchor [10, 45]. This method facilitates high synthetic throughput, which can provide pure GPIs and derivatives for biological assays.

12.6 GPI Derivatives for Biological Research

One goal of the synthetic efforts directed toward GPIs is to enable investigation of their functional roles and biophysical properties [14, 77]. In addition to native GPIs, such studies often require GPI analogs that are traceable or tractable and modified GPIs that can be immobilized on a surface or conjugated to other molecules. To prepare such molecules, it is often sufficient to introduce a chemical tag for further functionalization or conjugation.

The Seeberger laboratory was the first to use GPI structures immobilized on various surfaces with the goal of identifying their binding partners and characterizing these interactions. When amines of a GPI molecule are used as linkers for immobilization on surfaces, orientation in which the molecules are presented is significantly different from that on the cell surface. To create a system in which the presentation of GPI molecules mimics the presentation of these molecules on the surface of a lipid bilayer, the Seeberger group linked a 6-mercaptohexan-1-ol to the C1 position of inositol as a phosphodiester instead of a lipid and used the free thiol to immobilize these structures on chips and conjugate them to proteins [78]. This strategy was used to create a GPI microarray chip that simultaneously measures and characterizes antibody responses to GPI structures in the context of malarial infections [79].

Capitalizing on their strategy for synthesis of unsaturated GPI anchors [63, 67], Guo and coworkers have developed a route to alkyne- and azide-functionalized "clickable" GPI structures that can be used for the synthesis of fluorescently labeled GPIs [80, 81]. Guo and coworkers set off to prepare a simple GPI structure carrying an appropriate chemical tag attached to the C2 position of ManI via an ether bond (**116**, Scheme 12.15). A [3+2] glycosylation strategy was envisioned for

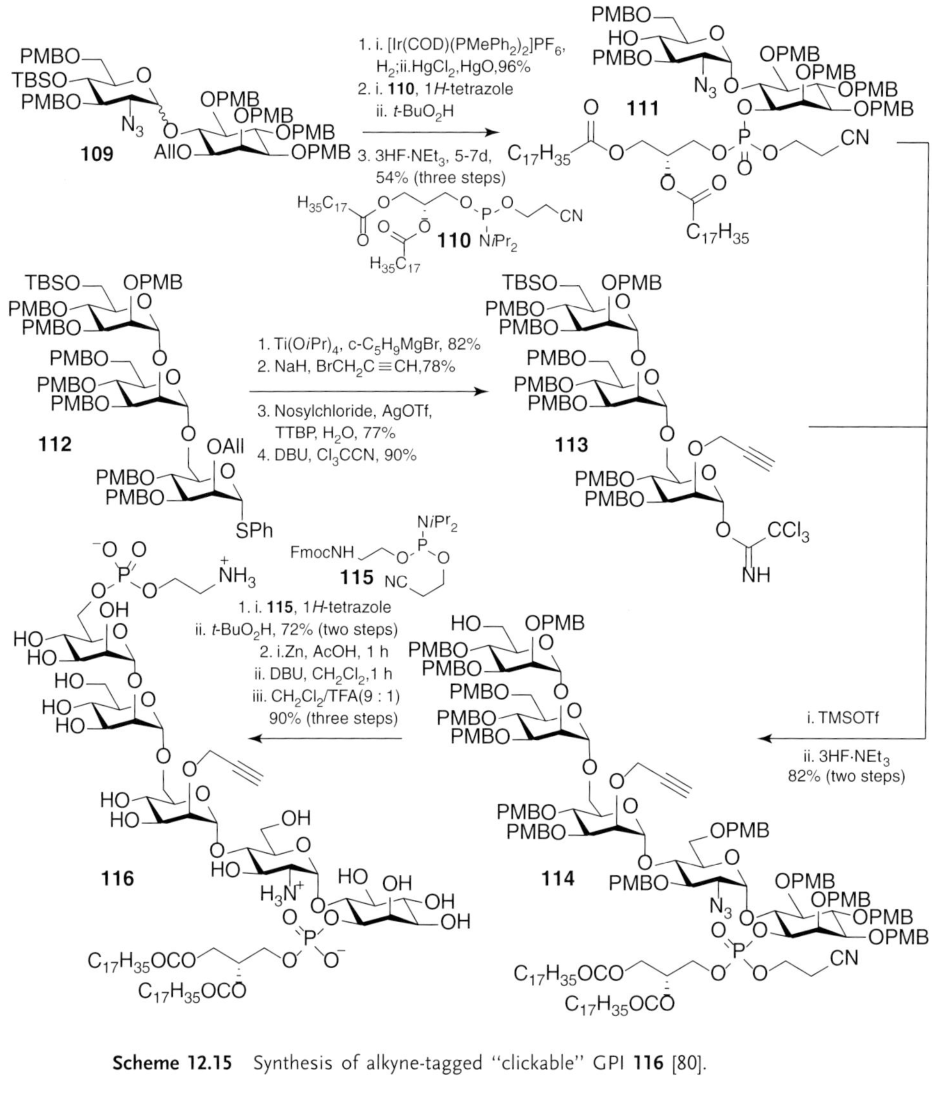

Scheme 12.15 Synthesis of alkyne-tagged "clickable" GPI **116** [80].

construction of the glycan with the appropriate chemical tag already present in the trisaccharide and the lipid phosphodiester present in the pseudodisaccharide fragment prior to coupling.

The requisite PMB-protected trimannoside **112** was prepared via a strategy similar to that developed by Schmidt for benzyl protected analogs. Removal of allyl ether via Ti(IV)-mediated deallylation protocol, followed by alkylation with propargyl bromide installed the alkyne tag. Hydrolysis of the thioglycoside using nosylchloride and AgOTf in the presence of a base followed by formation of trichloroacetimidate afforded the first coupling partner **113**. The pseudodisaccharide **109** was prepared by direct glycosylation of the protected inositol with protected 2-azido-2-deoxy-glucose trichloroacetimidate building block in 80% yield but with poor selectivity (α:$\beta = 2.3:1$). Removal of allyl ether at the C1 position of inositol followed by the phosphitylation/oxidation with phosphoamidite **110** to install the lipid and 3HF·triethylamine treatment to remove the TBS ether provided the second coupling partner **111** in 54% over three steps. Glycosylation of **111** with **113** effected by catalytic amounts of TMSOTf followed by removal of silyl ether cleanly furnished pseudopentasaccharide phosphate **114**. Introduction of the second phosphodiester via the established phosphitylation/oxidation protocol using phosphoamidite **115** afforded the fully protected GPI. Reduction of azide with zinc in acetic acid followed by the global deprotection using TFA in dichloromethane afforded the alkyne-tagged GPI **116**. The tagged GPI was successfully used in a Cu-catalyzed Huisgen cycloaddition with the azide-containing Azide-Fluor 488 fluorophore to produce a fluorescently tagged GPI. An analogous strategy was used to prepare an azide-containing GPI that can engage in copper-free strain-promoted click reactions.

12.7 Synthesis of GPI-Anchored Peptides and Proteins

GPI-APs and glycoproteins can be functionally diverse, ranging from hydrolytic enzymes, adhesion molecules, complementary regulatory proteins, receptors, and protozoan coat proteins to cytokines and prion proteins [82]. While some evidence suggests that the GPI is essential for the structure and activity of such proteins, no information is available on the mechanism through which the GPI influences protein structure and participates in the protein function. Similar to other types of protein glycosylations, some heterogeneity is present in the structure of GPI glycans depending on the enzymatic machinery on the expressing cell. The GPI-APs used in biological studies, aimed at evaluating the effects of the GPI on protein's function, are usually obtained by expression methods. Owing to the heterogeneities in such GPI-AP samples, these studies have been largely limited to the evaluation of the characteristics and functions of the GPI-APs as opposed to the protein without the anchor. The detailed structure–activity relationship (SAR) studies that would link specific GPI structure to a specific function remain elusive. The role of specific modifications in individual GPI glycoforms and their effects on the activity of the attached protein can only be evaluated using pure and well-defined samples of various GPI-APs. Obtaining such samples is burdened by the insurmountable difficulties associated with the isolation of pure GPI-anchored compounds.

With the goal of overcoming the limitations associated with the purity of GPI-APs, the protein semisynthesis has recently emerged as an alternative for the production of well-defined GPI-APs [83]. Besides the problems associated with preparation of suitable GPIs and proteins, semisynthesis of the GPI-APs needs to address the challenge of chemoselective coupling that would link the GPI and the protein. Two different strategies have been used for ligation of GPIs with proteins, both focusing on the creation of the amide bond between the C-terminus of the protein and the amine within the linker at the C6 position at ManIII of the GPI.

12.7.1
Synthesis of the GPI-Anchored Skeleton Structure of Sperm CD52 via Direct Amide Coupling

The first reported synthesis of a GPI-AP was the synthesis of GPI-anchored sperm CD52 skeleton structure **117** accomplished by the Guo group. CD52, a GPI-anchored glycopeptide with roles in the immune and reproductive systems expressed by all human lymphocytes and sperms [84], contains 12 amino acids and an *N*-glycosylation site. In contrast to large proteins, this relatively simple glycopeptide can be accessed using standard synthetic methods (Scheme 12.16a).

The synthesis of the short peptide sequence of CD52 was accomplished by solid-phase peptide synthesis (SPPS). The required *N*-linked glycan was introduced during the peptide synthesis protocol using a presynthesized glycosylated amino acid **118**. The synthesis was performed on polystyrene resin using the acid-labile 2-chloro-trityl linker **119**, which allowed the elongation of the peptide using Fmoc-protected amino acids and a final cleavage from the resin under mild acidic conditions [85]. The glycosylated amino acid **118**, obtained from an activated ester of aspartic acid and the corresponding hexasaccharide aminoglycoside [86], was introduced as a protected building block into the peptide synthesis sequence (Scheme 12.16b). The mild conditions for the release of the glycopeptide from the resin afforded the fully protected peptide with the free carboxylate at the C-terminus. Coupling of the glycopeptide **121** with the primary amine of the phosphoethanolamine in GPI **120** [85] was performed using the active ester of the carboxylate in good yields. After the coupling, the fully protected GPI-anchored glycopeptide was deprotected using a two-step procedure. A Pd-catalyzed hydrogenolysis followed by acidic treatment delivered the CD52 skeleton **121** in good yield after high-performance liquid chromatography (HPLC) purification.

12.7.2
Semisynthesis of GPI-Anchored Cellular Prion Protein via Native Chemical Ligation

The work on CD52 has demonstrated the ability of chemical synthesis to deliver homogeneous GPI-anchored molecules. Limitations of the solid-state peptide

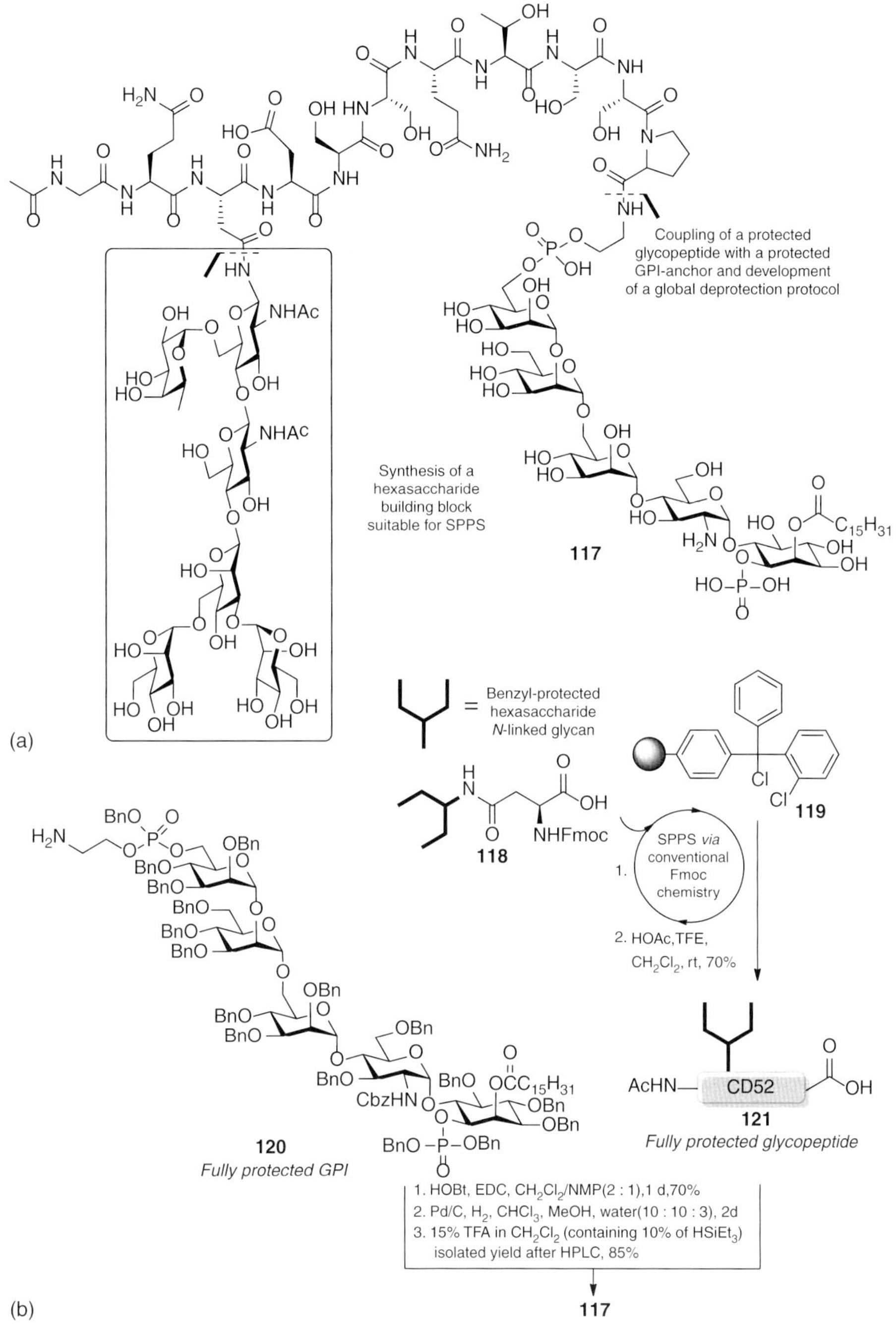

Scheme 12.16 (a) Key disconnections in the synthesis of the sperm CD52 skeleton structure by the Guo group. (b) Total synthesis of GPI-anchored CD52 glycopeptide **117** [85].

synthesis and the requirement for protection of coupling partners prior to the ligation of the GPI and the protein, however, impose limitations to the size and the type of molecules that are accessible in this way. To overcome these limitations, Seeberger and coworkers focused their attention on methods that would circumvent the need for protecting groups in the coupling partners and could accommodate larger proteins. Native chemical ligation (NCL) between a protein's C-terminal thioester and an N-terminal cysteine creates a native amide bond [87] under neutral conditions in the presence of free glycans. NCL does not depend on the size of the coupling partners and the products usually require no additional manipulations. Relying on these advantages, Seeberger and coworkers established a method for the semisynthesis of GPI-APs based on NCL [88] between a synthetic GPI anchor containing a cysteine residue linked to the phosphoethanolamine at the C6 position of ManIII and a recombinant, expressed protein thioester [89].

The synthesis of a GPI-anchored cell prion protein (PrP) was used as a model for the development of the NCL-based protocol to access GPI-APs. Conflicting reports on the contribution of GPI anchoring to the pathogenicity of PrP [90, 91] made this an excellent synthetic target that could help biological research in the field.

After appropriate disconnections, the retrosynthetic analysis of GPI-anchored PrP arrived to the simplified building blocks: GPI fragments **123–125** and the protein thioester **126** (Scheme 12.17a). The cysteine residue in the GPI was incorporated with the H-phosphonate **125** placing it at the C6 position of ManIII.

The two orthogonal protecting groups, allyl and triisopropylsilyl (TIPS), in the protected GPI core were used to install the lipid and the cysteine-containing phosphodiester in the last stage of the synthesis (Scheme 12.17b) [89]. After the final deprotection, *tert*-butylthiol ether used for the protection of the thiol during hydrogenolysis was exchanged for a mixed disulfide in **127**. The NCL of the GPI **127** with the expressed protein 2-sulfanylethanesulfonate thioester **126** in the presence of thiophenol under denaturing conditions (6 M guanidine hydrochloride) afforded the GPI-anchored PrP **122** that could be refolded in 20 mM sodium acetate buffer (pH = 5.5) in 45% yield.

The synthesis of the GPI-anchored PrP by NCL constitutes the first synthesis of a GPI-AP containing more than 100 amino acids. It serves as a proof of principle for the use of NCL in the synthesis of large GPI-APs but it also points out the new challenges in the field such as incorporation of GPI with native lipids and more complex core structures into the GPI-APs and ultimately the development of traceless chemical methods for ligation of GPIs and proteins that will not incorporate an additional cysteine at the protein's C-terminus.

12.8
Conclusions and Outlook

The field of GPI synthesis has seen remarkable progress over the past two decades. With the emergence of novel methodologies and technologies for the synthesis of oligosaccharides and glycolipids, the routes to GPIs have become more succinct and

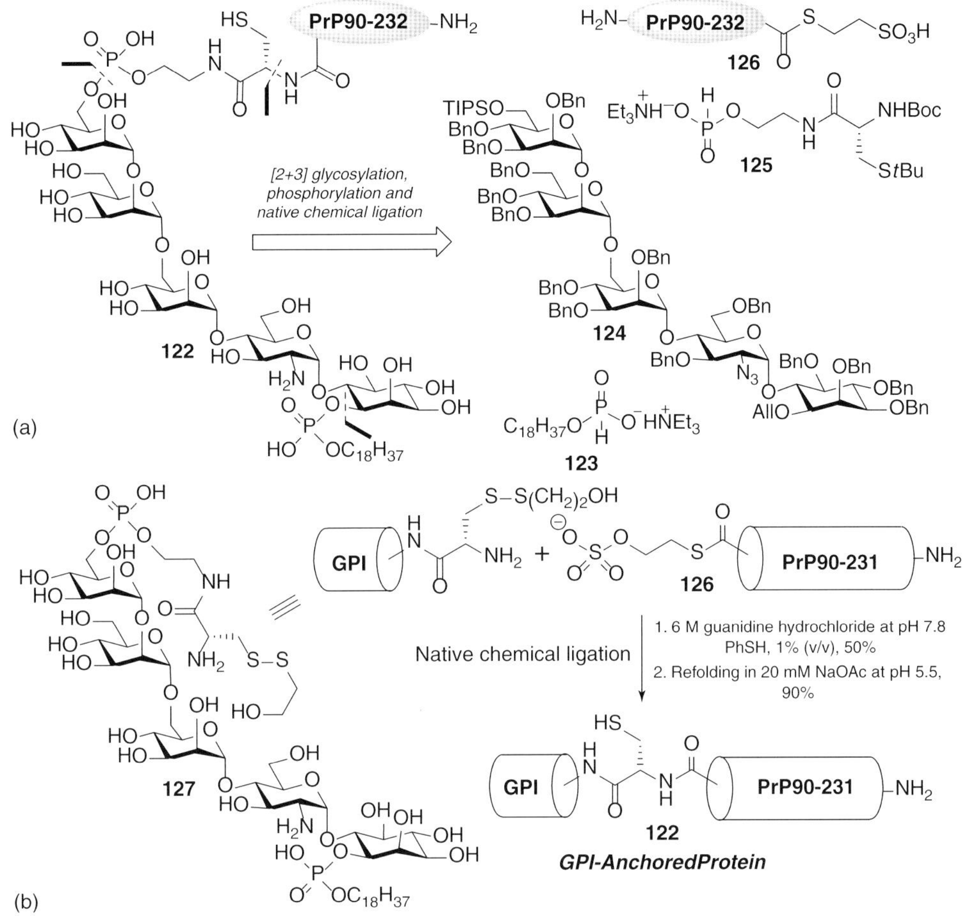

Scheme 12.17 (a) Retrosynthetic plan for the GPI-anchored prion protein. (b) Synthesis of GPI-anchored PrP **122** by the Seeberger group [89].

the complexity of GPI structures accessible via chemical synthesis has increased steadily. The synthesis of these molecules, however, remains laborious and lengthy. Making the synthetic GPIs available to researchers in the biomedical field without access to expert knowledge or instrumentation is still a distant goal. The efforts to develop routes that would allow for late stage diversification of the GPI structures could capitalize on the conserved nature of the GPI core and enable swift access to diverse sets of GPI samples pertinent to SAR studies. With these studies coupled with automated oligosaccharide synthesis technologies and strategies for rapid assembly of oligosaccharides by means of orthogonal glycosylation and/or armed–disarmed multicomponent couplings, this goal could become attainable.

Despite a large body of work on the synthetic GPIs, well-defined synthetic GPI-APs are not yet accessible in their native form via chemical synthesis. Given that the GPIs are typically bound to the protein, deciphering GPI's functional role in many aspects will depend on the ability to produce these constructs. Further developments are necessary to reach this goal and they will be the subject of studies aimed at the traceless ligation methods to produce glycoforms of these proteins in their native form. The ability of synthesis to deliver suitably modified GPIs for traceless ligation to the protein's C-terminus, will likely fuel the progress toward synthetic GPI-APs.

In a similar vein, the demand for functionalized chemically tagged GPIs for biological research, which only recently became available, is expected to grow and present new synthetic challenges. These challenges may be best addressed via unconventional routes or new synthetic strategies that circumvent traditional problems in the synthesis of complex glycolipids. Considering the fundamental importance of GPIs, the products of these and similar efforts are primed to become valuable tools for biological research that will bring about fervor in what appears to be a mature field of research.

Acknowledgments

This work was supported by the Max Planck Society and the RIKEN – Max Planck Joint Research Center. The funding from the European Commission's Seventh Framework Programme FP7/2007–2013 (postdoctoral fellowship to I.V.) is kindly acknowledged.

List of Abbreviations

AAG	1-alkyl-2-acylglycerol
Ac	Acetyl
AEP	Aminoethylphosphonate
AG	1-Alkyl-2-lyso-glycerol
All	Allyl
Bn	Benzyl
Boc	*tert*-Butoxycarbonyl
Bu	*n*-Butyl
Bz	Benzoyl
CAN	Ceric ammonium nitrate
Cbz	Carboxybenzyl
CD52	Cluster of differentiation 52
CER	Ceramide
CSA	Camphorsulfonic acid
ClAc	Chloroacetyl
COD	Cyclooctadiene
Cp	Cyclopentadienyl
DAG	Diacylglycerol

DAST	(Diethylamino)sulfur trifluoride
DBU	1,8-Diazabicyclo[5.4.0]undec-7-ene
DCC	Dicyclohexylcarbodiimide
DMAP	4-Dimethylaminopyridine
DMF	Dimethylformamide
DMSO	Dimethyl sulfoxide
EDC	1-Ethyl-3-(3-dimethylaminopropyl)carbodiimide
Et	Ethyl
Fmoc	9-Fluorenylmethoxycarbonyl
GPI	Glycosylphosphatidylinositol
GPI-AP	Glycosylphosphatidylinositol-anchored protein
HOBt	Hydroxybenzotriazole
HPLC	High-performance liquid chromatography
IL-12	Interleukin 12
Ino	Inositol
*i*Pr	Isopropyl
Lev	Levulinoyl
m-CPBA	*meta*-Chloroperoxybenzoic acid
Me	Methyl
MP	Methoxyphenyl
MS	Molecular sieves
NIS	*N*-Iodosuccinimide
NAP	2-Naphthylmethyl
NCL	Native chemical ligation
NMP	*N*-Methylpyrrolidone
pal.	Palmitoyl
PBS	Phosphate buffer solution
PEtN	Phosphoethanolamine
Ph	Phenyl
Piv	Pivaloyl
Py	Pyridine
PMB	*p*-Methoxybenzyl
PPTS	Pyridinium *p*-toluenesulfonate
PrP	Prion protein
SAR	Structure–activity relationship
SEM	2-Trimethylsilylethoxymethoxy
SPPS	Solid-phase peptide synthesis
TBAB	Tetrabutylammonium bromide
TBAF	Tetrabutylammonium fluoride
TBAI	Tetrabutylammonium iodide
TBDPS	*tert*-Butyldiphenylsilyl
TBS	*tert*-Butyldimethylsilyl
t-Bu	*tert*-Butyl
Tf	Trifluoromethylsulfonyl
TFA	Trifluoroacetic acid

TIPS	Triisopropylsilyl
TMS	Trimethylsilyl
TNF-α	Tumor necrosis factor-α, cachexin
Tr	Trityl
TTBP	2,4,6-Tri-*tert*-butylpyridine
VSG	Variant surface glycoprotein

References

1. Walsh, C.T., Garneau-Tsodikova, S., and Gatto, G.J. (2005) *Angew. Chem. Int. Ed.*, **44**, 7342–7372.
2. Orlean, P. and Menon, A.K. (2007) *J. Lipid Res.*, **48**, 993–1011.
3. Ferguson, M.A. and Williams, A.F. (1988) *Annu. Rev. Biochem.*, **57**, 285–320.
4. Low, M.G. (1989) *Biochim. Biophys. Acta*, **988**, 427–454.
5. Ferguson, M.A. (1999) *J. Cell Sci.*, **112**, 2799–2809.
6. Ferguson, M.A.J., Homans, S.W., Dwek, R.A., and Rademacher, T.W. (1988) *Science*, **239**, 753–759.
7. Paulick, M.G. and Bertozzi, C.R. (2008) *Biochemistry*, **47**, 6991–7000.
8. Fujita, M. and Jigami, Y. (2008) *Biochim. Biophys. Acta*, **1780**, 410–420.
9. Almeida, I.C., Camargo, M.M., Procopio, D.O., Silva, L.S., Mehlert, A., Travassos, L.R., Gazzinelli, R.T., and Ferguson, M.A.J. (2000) *EMBO J.*, **19**, 1476–1485.
10. Tsai, Y.-H., Götze, S., Vilotijevic, I., Grube, M., Silva, D.V., and Seeberger, P.H. (2013) *Chem. Sci.*, **4**, 468–481.
11. Tsai, Y.-H., Liu, X., and Seeberger, P.H. (2012) *Angew. Chem. Int. Ed.*, **51**, 11438–11456.
12. Azzouz, N., Shams-Eldin, H., Niehus, S., Debierre-Grockiego, F., Bieker, U., Schmidt, J., Mercier, C., Delauw, M.F., Dubremetz, J.F., Smith, T.K., and Schwarz, R.T. (2006) *Int. J. Biochem. Cell Biol.*, **38**, 1914–1925.
13. Maeda, Y., Tashima, Y., Houjou, T., Fujita, M., Yoko-o, T., Jigami, Y., Taguchi, R., and Kinoshita, T. (2007) *Mol. Biol. Cell*, **18**, 1497–1506.
14. Stefaniu, C., Vilotijevic, I., Santer, M., Varón Silva, D., Brezesinski, G., and Seeberger, P.H. (2012) *Angew. Chem. Int. Ed.*, **51**, 12874–12878.
15. Rooney, I.A., Atkinson, J.P., Krul, E.S., Schonfeld, G., Polakoski, K., Saffitz, J.E., and Morgan, B.P. (1993) *J. Exp. Med.*, **177**, 1409–1420.
16. Rooney, I.A., Heuser, J.E., and Atkinson, J.P. (1996) *J. Clin. Invest.*, **97**, 1675–1686.
17. Fujita, M., Maeda, Y., Ra, M., Yamaguchi, Y., Taguchi, R., and Kinoshita, T. (2009) *Cell*, **139**, 352–365.
18. Fujita, M., Watanabe, R., Jaensch, N., Romanova-Michaelides, M., Satoh, T., Kato, M., Riezman, H., Yamaguchi, Y., Maeda, Y., and Kinoshita, T. (2011) *J. Cell Biol.*, **194**, 61–75.
19. Ruhela, D., Banerjee, P., and Vishwakarma, R.A. (2012) *Curr. Sci.*, **102**, 194–211.
20. Tsai, Y.-H., Grube, M., Seeberger, P.H., and Varon Silva, D. (2012) *Trends Glycosci. Glycotechnol.*, **24**, 231–243.
21. Homans, S.W., Ferguson, M.A.J., Dwek, R.A., Rademacher, T.W., Anand, R., and Williams, A.F. (1988) *Nature*, **333**, 269–272.
22. Nikolaev, A.V. and Al-Maharik, N. (2011) *Nat. Prod. Rep.*, **28**, 970–1020.
23. Guo, Z. and Bishop, L. (2004) *Eur. J. Org. Chem.*, **2004**, 3585–3596.
24. Ley, S.V., Mio, S., and Meseguer, B. (1996) *Synlett*, 787–788.
25. Jia, Z.J., Olsson, L., and Fraser-Reid, B. (1998) *J. Chem. Soc., Perkin Trans. 1*, 631–632.
26. Baeschlin, D.K., Chaperon, A.R., Charbonneau, V., Green, L.G., Ley, S.V., Lucking, U., and Walther, E. (1998) *Angew. Chem. Int. Ed.*, **37**, 3423–3428.
27. Azzouz, N., Kamena, F., and Seeberger, P.H. (2010) *OMICS*, **14**, 445–454.
28. Weishaupt, M., Eller, S., and Seeberger, P.H. (2010) in *Methods in Enzymology: Glycomics*, Vol. 478 (ed. M. Fukuda),

Academic press, San Diego, CA, pp. 463–484.
29. Hewitt, M.C., Snyder, D.A., and Seeberger, P.H. (2002) *J. Am. Chem. Soc.*, **124**, 13434–13436.
30. Lu, J., Jayaprakash, K.N., and Fraser-Reid, B. (2004) *Tetrahedron Lett.*, **45**, 879–882.
31. Lu, J., Jayaprakash, K.N., Schlueter, U., and Fraser-Reid, B. (2004) *J. Am. Chem. Soc.*, **126**, 7540–7547.
32. Wu, X. and Guo, Z. (2007) *Org. Lett.*, **9**, 4311–4313.
33. Mayer, T.G., Kratzer, B., and Schmidt, R.R. (1994) *Angew. Chem., Int. Ed. Engl.*, **33**, 2177–2181.
34. Schmidt, R.R. and Mayer, T.G. (1999) *Eur. J. Org. Chem.*, **5**, 1153–1165.
35. Kwon, Y.U., Soucy, R.L., Snyder, D.A., and Seeberger, P.H. (2005) *Chem. Eur. J.*, **11**, 2493–2504.
36. Liu, X.Y., Kwon, Y.U., and Seeberger, P.H. (2005) *J. Am. Chem. Soc.*, **127**, 5004–5005.
37. Ali, A. and Vishwakarma, R.A. (2010) *Tetrahedron*, **66**, 4357–4369.
38. Ali, A., Gowda, D.C., and Vishwakarma, R.A. (2005) *Chem. Commun.*, 519–521.
39. Yashunsky, D.V., Borodkin, V.S., Ferguson, M.A.J., and Nikolaev, A.V. (2006) *Angew. Chem. Int. Ed.*, **45**, 468–474.
40. Murakata, C. and Ogawa, T. (1992) *Carbohydr. Res.*, **235**, 95–114.
41. Murakata, C. and Ogawa, T. (1991) *Tetrahedron Lett.*, **32**, 671–674.
42. Murakata, C. and Ogawa, T. (1990) *Tetrahedron Lett.*, **31**, 2439–2442.
43. Baeschlin, D.K., Chaperon, A.R., Green, L.G., Hahn, M.G., Ince, S.J., and Ley, S.V. (2000) *Chem. Eur. J.*, **6**, 172–186.
44. Kwon, Y.U., Liu, X., and Seeberger, P.H. (2005) *Chem. Commun.*, 2280–2282.
45. Tsai, Y.H., Götze, S., Azzouz, N., Hahm, H.S., Seeberger, P.H., and Varon Silva, D. (2011) *Angew. Chem. Int. Ed.*, **50**, 9961–9964.
46. Campbell, A.S. and Fraser-Reid, B. (1995) *J. Am. Chem. Soc.*, **117**, 10387–10388.
47. Pekari, K. and Schmidt, R.R. (2003) *J. Org. Chem.*, **68**, 1295–1308.
48. Gigg, R. and Gigg, J. (1997) in *Glycopeptides and Related Compounds: Synthesis, Analysis, and Applications* (eds D.G. Large and C.D. Warren), Marcel Dekker, New York, pp. 327–392.
49. Mach, M., Schlueter, U., Mathew, F., Fraser-Reid, B., and Hazen, K.C. (2002) *Tetrahedron*, **58**, 7345–7354.
50. Xue, J., Shao, N., and Guo, Z. (2003) *J. Org. Chem.*, **68**, 4020–4029.
51. Grindley, T.B. (1994) *Synthetic Oligosaccharides*, Vol. 560, American Chemical Society, pp. 51–76.
52. Xue, J. and Guo, Z. (2002) *Bioorg. Med. Chem. Lett.*, **12**, 2015–2018.
53. Liu, X., Stocker, B.L., and Seeberger, P.H. (2006) *J. Am. Chem. Soc.*, **128**, 3638–3648.
54. Conrad, R.M., Grogan, M.J., and Bertozzi, C.R. (2002) *Org. Lett.*, **4**, 1359–1361.
55. Boonyarattanakalin, S., Liu, X., Michieletti, M., Lepenies, B., and Seeberger, P.H. (2008) *J. Am. Chem. Soc.*, **130**, 16791–16799.
56. Zhu, X. and Schmidt, R.R. (2009) *Angew. Chem. Int. Ed.*, **48**, 1900–1934.
57. Ruda, K., Lindberg, J., Garegg, P.J., Oscarson, S., and Konradsson, P. (2000) *J. Am. Chem. Soc.*, **122**, 11067–11072.
58. Robert, J.K. and Peng, W. (2012) *Glycobiology and Drug Design*, Vol. 1102, American Chemical Society, pp. 235–263.
59. Garegg, P.J., Konradsson, P., Oscarson, S., and Ruda, K. (1997) *Tetrahedron*, **53**, 17727–17734.
60. Pekari, K., Tailler, D., Weingart, R., and Schmidt, R.R. (2001) *J. Org. Chem.*, **66**, 7432–7442.
61. Ruda, K., Lindberg, J., Garegg, P.J., Oscarson, S., and Konradsson, P. (2000) *Tetrahedron*, **56**, 3969–3975.
62. Yashunsky, D.V., Borodkin, V.S., McGivern, P.G., Ferguson, M.A.J., and Nikolaev, A.V. (2007) in *Frontiers in Modern Carbohydrate Chemistry*, Vol. 960 (ed. A.V. Demchenko), American Chemical Society, pp. 285–306.
63. Swarts, B.M. and Guo, Z. (2010) *J. Am. Chem. Soc.*, **132**, 6648–6650.
64. Fraser-Reid, B., Schlueter, U., and Lu, J. (2003) *Org. Lett.*, **5**, 255–257.
65. Götze, S., Fitzner, R., and Kunz, H. (2009) *Synlett*, 3346–3348.

66. Schmidt, R.R. and Michel, J. (1980) *Angew. Chem., Int. Ed. Engl.*, **19**, 731–732.
67. Burgula, S., Swarts, B.M., and Guo, Z. (2012) *Chem. Eur. J.*, **18**, 1194–1201.
68. Vacca, J.P., de Solms, S.J., Huff, J.R., Billington, D.C., Baker, R., Kulagowski, J.J., and Mawer, I.M. (1989) *Tetrahedron Lett.*, **45**, 5679–5702.
69. Vacca, J.P., de Solms, S.J., Huff, J.R., Billington, D.C., Baker, R., Kulagowski, J.J., and Mawer, I.M. (1991) *Tetrahedron Lett.*, **47**, 907.
70. Ogawa, T. (1994) *Chem. Soc. Rev.*, **23**, 397–407.
71. Cottaz, S., Brimacombe, J.S., and Ferguson, M.A.J. (1993) *J. Chem. Soc., Perkin Trans. 1*, 2945–2951.
72. Sato, S., Mori, M., Ito, Y., and Ogawa, T. (1986) *Carbohydr. Res.*, **155**, C6–C10.
73. Kanie, O., Ito, Y., and Ogawa, T. (1994) *J. Am. Chem. Soc.*, **116**, 12073–12074.
74. Mootoo, D.R., Konradsson, P., Udodong, U., and Fraser-Reid, B. (1988) *J. Am. Chem. Soc.*, **110**, 5583–5584.
75. Grice, P., Ley, S.V., Pietruszka, J., Priepke, H.W.M., and Walther, E.P.E. (1995) *Synlett*, 781–784.
76. Fraser-Reid, B. and Lopez, J.C. (2011) *Top. Curr. Chem.*, **301**, 1–29.
77. Wehle, M., Vilotijevic, I., Lipowsky, R., Seeberger, P.H., Varon Silva, D., and Santer, M. (2012) *J. Am. Chem. Soc.*, **134**, 18964–18972.
78. Seeberger, P.H., Soucy, R.L., Kwon, Y.U., Snyder, D.A., and Kanemitsu, T. (2004) *Chem. Commun.*, 1706–1707.
79. Kamena, F., Tamborrini, M., Liu, X.Y., Kwon, Y.U., Thompson, F., Pluschke, G., and Seeberger, P.H. (2008) *Nat. Chem. Biol.*, **4**, 238–240.
80. Swarts, B.M. and Guo, Z. (2011) *Chem. Sci.*, **2**, 2342–2352.
81. Swarts, B.M. and Guo, Z. (2012) *Adv. Carbohydr. Chem. Biochem.*, **67**, 137–219.
82. Ikezawa, H. (2002) *Biol. Pharm. Bull.*, **25**, 409–417.
83. Durek, T. and Becker, C.F.W. (2005) *Biomol. Eng.*, **22**, 153–172.
84. Kirchhoff, C., Schroter, S., Derr, P., Conrad, H.S., Nimtz, M., and Hale, G. (1999) *J. Biol. Chem.*, **274**, 29862–29873.
85. Shao, N., Xue, J., and Guo, Z. (2003) *J. Org. Chem.*, **68**, 9003–9011.
86. Guo, Z.-W., Nakahara, Y., Nakahara, Y., and Ogawa, T. (1997) *Bioorg. Med. Chem.*, **5**, 1917–1924.
87. Wieland, T., Bokelmann, E., Bauer, L., Lang, H.U., and Lau, H. (1953) *Liebigs Ann. Chem.*, **583**, 129–149.
88. Dawson, P.E., Muir, T.W., Clark-Lewis, I., and Kent, S.B.H. (1994) *Science*, **266**, 776–779.
89. Becker, C.F., Liu, X., Olschewski, D., Castelli, R., Seidel, R., and Seeberger, P.H. (2008) *Angew. Chem. Int. Ed.*, **47**, 8215–8219.
90. Chesebro, B., Trifilo, M., Race, R., Meade-White, K., Teng, C., LaCasse, R., Raymond, L., Favara, C., Baron, G., Priola, S., Caughey, B., Masliah, E., and Oldstone, M. (2005) *Science*, **308**, 1435–1439.
91. Radford, H.E. and Mallucci, G.R. (2010) *Curr. Iss. Mol. Biol.*, **12**, 119–127.

Index

Modern Synthetic Methods in Carbohydrate Chemistry: From Monosaccharides to Complex Glycoconjugates,
First Edition. Edited by Daniel B. Werz and Sébastien Vidal.

h